Endothelial Signaling in Vascular Dysfunction and Disease

and Disease

From Bench to Bedside

Endothelial Signaling in Vascular Dysfunction and Disease
From Bench to Bedside

Edited by

Shampa Chatterjee
University of Pennsylvania Perelman School of Medicine
Philadelphia, PA, United States

ELSEVIER

ACADEMIC PRESS
An imprint of Elsevier

Academic Press is an imprint of Elsevier
125 London Wall, London EC2Y 5AS, United Kingdom
525 B Street, Suite 1650, San Diego, CA 92101, United States
50 Hampshire Street, 5th Floor, Cambridge, MA 02139, United States
The Boulevard, Langford Lane, Kidlington, Oxford OX5 1GB, United Kingdom

Notices
Knowledge and best practice in this field are constantly changing. As new research and experience broaden our understanding, changes in research methods, professional practices, or medical treatment may become necessary.

Practitioners and researchers must always rely on their own experience and knowledge in evaluating and using any information, methods, compounds, or experiments described herein. In using such information or methods they should be mindful of their own safety and the safety of others, including parties for whom they have a professional responsibility.

To the fullest extent of the law, neither the Publisher nor the authors, contributors, or editors, assume any liability for any injury and/or damage to persons or property as a matter of products liability, negligence or otherwise, or from any use or operation of any methods, products, instructions, or ideas contained in the material herein.

British Library Cataloguing-in-Publication Data
A catalogue record for this book is available from the British Library

Library of Congress Cataloging-in-Publication Data
A catalog record for this book is available from the Library of Congress

ISBN: 978-0-12-816196-8

For Information on all Academic Press publications
visit our website at https://www.elsevier.com/books-and-journals

Publisher: Stacy Masucci
Acquisitions Editor: Rafael E. Teixeira
Editorial Project Manager: Samantha Allard
Production Project Manager: Selvaraj Raviraj
Cover Designer: Mark Rogers

Typeset by MPS Limited, Chennai, India

To my Mother

Contents

Section 1
The vascular tree as an Integrative Structure

Section 2
Endothelial signaling in inflammatory diseases

12. Flow-adapted vascular systems: mimicking the vascular network to predict clinical response to radiation 129

Aravindan Natarajan, Mohan Natarajan, Sheeja Aravindan and Sumathy Mohan

13. Effects of space radiation on the endovasculature: implications for future human deep space exploration 147

Melpo Christofidou-Solomidou, Thais Sielecki and Shampa Chatterjee

14. Role of endothelial cells in normal tissue radiation injury 157

Marjan Boerma

Section 4
Infection, vascular signaling and injury

Section 5
Approaches in understanding endothelial function

Section 6
Therapeutic interventions to limit vascular disease

List of contributors

Adakole Sylvanus Adah Department of Veterinary Physiology and Biochemistry, University of Ilorin, Ilorin, Nigeria

Kaitlin Allen Department of Integrative Biology, University of California, Berkeley, Berkeley, CA, United States

Hauwa Motunrayo Ambali Department of Veterinary Medicine, University of Ilorin, Ilorin, Nigeria

Soliu Akanni Ameen Department of Veterinary Medicine, University of Ilorin, Ilorin, Nigeria

Sheeja Aravindan Stephenson Cancer Center, Oklahoma City, OK, United States

Oyebisi Mistura Azeez Department of Veterinary Physiology and Biochemistry, University of Ilorin, Ilorin, Nigeria

Rashidat Bolanle Balogun Veterinary Teaching Hospital, University of Ilorin, Ilorin, Nigeria

André Sales Barreto Department of Physiology, Federal University of Sergipe, São Cristóvão, Brazil; Department of Health Education, Federal University of Sergipe, Lagarto, Brazil

Afisu Basiru Department of Veterinary Physiology and Biochemistry, University of Ilorin, Ilorin, Nigeria

Chhanda Biswas Children's Hospital of Philadelphia, Philadelphia, PA, United States

Marjan Boerma Division of Radiation Health, Department of Pharmaceutical Sciences, University of Arkansas for Medical Sciences, Little Rock, AR, United States

Moshood Bolaji Department of Veterinary Pathology, University of Ilorin, Ilorin, Nigeria

Suzanne C. Cannegieter Department of Clinical Epidemiology, Leiden University Medical Center, Leiden, The Netherlands; Department of Thrombosis and Haemostasis, Leiden University Medical Center, Leiden, The Netherlands

Shampa Chatterjee Department of Physiology, Institute for Environmental Medicine, University of Pennsylvania Perelman School of Medicine, Philadelphia, PA, United States

Melpo Christofidou-Solomidou Pulmonary, Allergy and Critical Care Division, Department of Medicine, University of Pennsylvania Perelman School of Medicine, Philadelphia, PA, United States

Andrea L. DiCarlo Radiation and Nuclear Countermeasures Program (RNCP), Division of Allergy, Immunology and Transplantation (DAIT), National Institute of Allergy and Infectious Diseases (NIAID), National Institutes of Health (NIH), Rockville, MD, United States

Alexander R. Farid Department of Radiology, University of Pennsylvania, Philadelphia, PA, United States

Milene Tavares Fontes Department of Physiology, Federal University of Sergipe, São Cristóvão, Brazil

Kumkum Ganguly Bioscience Division, Los Alamos National Laboratory, Los Alamos, NM, United States

Sirajo Garba Department of Veterinary Medicine, Usmanu Danfodiyo University, Sokoto, Nigeria

Thais Girão-Silva Laboratory of Genetics and Molecular Cardiology, Heart Institute (InCor), University of São Paulo Medical School, São Paulo, Brazil

Denis Glotz Human Immunology, Pathophysiology & Immunotherapy, INSERM UMRS-976, Paris, France; Université Paris Diderot, Paris, France; Department of Transplantation and Nephrology, Saint-Louis Hospital, AP-HP, Paris, France

Xiaohui Guo Department of Physiology, Institute for Environmental Medicine, University of Pennsylvania Perelman School of Medicine, Philadelphia, PA, United States

Madhu Gupta School of Pharmaceutical Sciences, Department of Pharmaceutics, Delhi Pharmaceutical Sciences and Research University, New Delhi, India

Pilar Guzmán-Díaz Biomedical Sciences Department, Calixto García Faculty, University of Medical Sciences of Havana, Havana, Cuba

Daniel C. Kargilis Department of Orthopaedic Surgery, University of Pennsylvania, Philadelphia, PA, United States

Young-Mee Kim Department of Medicine (Cardiology), University of Illinois at Chicago, Chicago, IL, United States

Diana Klein Institute of Cell Biology (Cancer Research), University Hospital Essen, University of Duisburg-Essen, Essen, Germany

Silvia Lacchini Institute of Biomedical Sciences, University of São Paulo, São Paulo, Brazil

Willem M. Lijfering Department of Clinical Epidemiology, Leiden University Medical Center, Leiden, The Netherlands

Fabricio Nunes Macedo Department of Physiology, Federal University of Sergipe, São Cristóvão, Brazil; Department of Physical Education, Estácio University of Sergipe, Aracaju, Brazil

Michael M. Mayer Department of Radiology, University of Pennsylvania, Philadelphia, PA, United States

Sofia M. Miguez Department of Radiology, University of Pennsylvania, Philadelphia, PA, United States

Ela María Céspedes Miranda Biomedical Sciences Department, Calixto García Faculty, University of Medical Sciences of Havana, Havana, Cuba

Ayumi Aurea Miyakawa Laboratory of Genetics and Molecular Cardiology, Heart Institute (InCor), University of São Paulo Medical School, São Paulo, Brazil

Sumathy Mohan Department of Pathology, University of Texas Health Sciences Center at San Antonio, San Antonio, TX, United States

Nuala Mooney Human Immunology, Pathophysiology & Immunotherapy, INSERM UMRS-976, Paris, France; Université Paris Diderot, Paris, France

Aravindan Natarajan Department of Radiation Oncology, University of Oklahoma Health Sciences Center, Oklahoma City, OK, United States; Department of Pathology, University of Oklahoma Health Sciences Center, Oklahoma City, OK, United States

Mohan Natarajan Department of Pathology, University of Texas Health Sciences Center at San Antonio, San Antonio, TX, United States

Folashade Helen Olaifa Department of Veterinary Physiology and Biochemistry, University of Ilorin, Ilorin, Nigeria

Oindrila Paul Department of Physiology, Institute for Environmental Medicine, University of Pennsylvania Perelman School of Medicine, Philadelphia, PA, United States

Nabendu Pore Oncology R&D, AstraZeneca, One MedImmune Way, Gaithersburg, MD, United States

Neha Raina School of Pharmaceutical Sciences, Department of Pharmaceutics, Delhi Pharmaceutical Sciences and Research University, New Delhi, India

Chamith S. Rajapakse Department of Radiology, University of Pennsylvania, Philadelphia, PA, United States; Department of Orthopaedic Surgery, University of Pennsylvania, Philadelphia, PA, United States

Radha Rani School of Pharmaceutical Sciences, Department of Pharmaceutics, Delhi Pharmaceutical Sciences and Research University, New Delhi, India

Roger Rodríguez-Guzmán Biomedical Sciences Department, Calixto García Faculty, University of Medical Sciences of Havana, Havana, Cuba

Frits R. Rosendaal Department of Clinical Epidemiology, Leiden University Medical Center, Leiden, The Netherlands; Department of Thrombosis and Haemostasis, Leiden University Medical Center, Leiden, The Netherlands

Sourav Roy Department of Microbiology and Immunology, Brody School of Medicine, East Carolina University, Greenville, NC, United States

Valter Joviniano Santana-Filho Department of Physiology, Federal University of Sergipe, São Cristóvão, Brazil

Merriline M. Satyamitra Radiation and Nuclear Countermeasures Program (RNCP), Division of Allergy, Immunology and Transplantation (DAIT), National Institute of Allergy and Infectious Diseases (NIAID), National Institutes of Health (NIH), Rockville, MD, United States

F. Sertic Department of Surgery, Hospital of the University of Pennsylvania, Philadelphia, PA, USA

Syed Raza Shah University of Louisville Hospital, Louisville, KY, United States

Thais Sielecki JT-MeSh Diagnostics LLC, Penn Center of Innovation, Philadelphia, PA, United States

Jian Qin Tao Department of Physiology, Institute for Environmental Medicine, University of Pennsylvania Perelman School of Medicine, Philadelphia, PA, United States

Masuko Ushio-Fukai Department of Medicine (Cardiology), Vascular Biology Center, Medical College of Georgia, Augusta University, Augusta, GA, United States

José Pablo Vázquez-Medina Department of Integrative Biology, University of California, Berkeley, Berkeley, CA, United States

Preface

The importance of the vascular system became very obvious early on in human civilization. Hippocrates (460−370 BCE) in ancient Greece was the first to discover that blood supply to the brain was via the carotid artery and that carotid artery disease led to transient ischemic attack. Later, Herophilus of Chalcedon (335−280 BCE) and Erasistratus of Ceos (304−250 BCE) carried out dissections on cadavers to discover arterial and venous blood flow in the brain circulation. During the Roman and Byzantine kingdoms, Oribasius of Pergamum (c. CE 325−405) described the first varicosis surgery and techniques of surgery on aneurysms and Paul of Aegina (CE 700) elaborated on the applications of tourniquets to control blood flow. This continued through the medieval world as scholars such as Haji Abbas (died CE 990) described the carotid arteries and the presence of a capillary network. With the start of Renaissance in Europe, there was growing interest in the architecture of the human body and Leonardo Da Vinci's (CE 1452−1519) dissections of the human corpse revealed the anatomical complexities and networks of the lung vasculature. Later the positions of the major vessels on the human body were mapped by Vesalius (CE 1514−1564). William Harvey (CE 1578−1657), discovered that the blood flow in the vascular network was regulated by valves thus signaling the birth of modern vascular medicine.

In the last few centuries, modern biology and medicine has further unraveled the anatomical and functional complexity of the vascular system, as a conduit and connector of blood flow across organs. It is now well established that the vascular system functions as a network, integrating biochemical and biophysical signals via systemic transport of blood, nutrients, and inflammatory, pathogenic moieties across the body. At the heart of the vascular system is a layer of specialized cells that participate in all major functions of the vascular network. This is the endothelium, a one-cell thick layer of endothelial cells that lines the interior of the entire vascular network and serves as a permeable barrier. The permeability is due to large gaps between endothelial cells (paracellular permeability) as well as due to "openings" on the endothelial cells (transcellular permeability). The permeability is very well controlled, as it is a key step in regulation of tissue−fluid homeostasis, transport of nutrients, and migration of immune cells from the blood to fight pathogens in the tissue. In addition to permeability, endothelial cells express surface and cytosolic receptors and proteins that are part of signaling processes that maintain vessel structure and function and also participate in wound healing, angiogenesis, inflammation, and injury. The naïve or resting endothelium is a tightly regulated layer that has an anticoagulant, antiinflammatory phenotype, and fibrinolytic activity. The endothelial cells synthesize tissue plasminogen activator and plasminogen activator inhibitor type 1, the balance between which determines fibrinolysis. The endothelium also produces nitric oxide (NO) that controls vascular tone.

When the endothelium was first identified a century ago, it was considered as an "inert" layer encasing the tissue and separating it from the blood flow. It is only in the past decades that the crucial role of the endothelium in vascular biology is gradually emerging and it is being recognized as an active interface between the circulating blood and the various underlying tissues; indeed it is now well established to play a key role in maintaining life. In the adult human, the endothelium covers a large surface area of 5000−7000 m^2; thus it can easily be considered as an organ. Endothelial activation by ligands (chemotransduction) or by mechanical forces (mechanotransduction) triggers signaling processes via receptors and sensors that cause increased production of chemokines, cytokines, adhesion molecules, oxidants, and decreased generation of NO. Some of these processes activate transcription factors, while others lead to oxidative injury to the endothelium. Oxidative injury is acknowledged to play a role in the development of endothelial dysfunction. The signaling-injury-dysfunction process often functions as a feed-forward cycle such that endothelial dysfunction can trigger processes that in turn aggravate injury. Thus endothelial dysfunction is a complex process that can (1) facilitate the recruitment and adherence of immune cells (such as neutrophils, monocytes, and T lymphocytes) that lead to inflammation (acute and chronic) that is crucial to fight infection in tissue and (2) impair fibrinolytic activity, thus facilitating platelet activation and aggregation which are factors that precipitate thrombus formation. Endothelial dysfunction would thus aid in inflammation and coagulation processes that in turn affect vascular structure and diameter and alter the dynamics of blood flow. Indeed, as Rudolf Virchow, a 19th-century German scientist and physician (considered as father of modern pathology) surmised from his endeavors in pathology, there is a link between vascular endothelial injury (dysfunction), reduced blood flow, and alterations in the constitution of the blood (Virchow's triad).

During our life span, we are exposed to numerous chemical stimuli such as toxic stimuli, radiation, chemicals, smoke from the environment, mechanical stimuli such as blood flow and alterations in blood flow, and stimuli in the internal environment such as diets (high salt, sugar, or saturated fatty acids intake). All of these strongly impact endothelial cells and their functions. Additionally, pathogens also affect the endothelium and often target the endothelial layer directly (as is being reported with the current SARS-CoV2 virus) leading to endothelial infection or endotheliitis. As long as these stimuli are within physiological limits, the endothelial cells can maintain homeostasis between vasoconstrictors and vasodilators, anti- and pro-inflammatory mediators, anti- and pro-coagulant factors, and anti- and pro-fibrinolytic processes. However, when exposed in a continuous and chronic manner to these stimuli, endothelial dysfunction ensues leading to disease.

Our purpose in this book has been to highlight various aspects of endothelial cell signaling that affect structure and function of the endothelium and form the basis of both vascular health and disease. Our current knowledge on endothelial signaling processes and their complexity means that a lot of ground needs to be covered. We have therefore organized the book thematically into six sections, such that each addresses signaling associated with similar or closely related stimuli and disease manifestations. Thus the chapters of the first section describe the vascular tree in terms of integrative structure and signaling, while Sections 2–5 highlight the role of endothelium in inflammation, infection, injury, and toxicity. Section 6 is focused on therapies in combating vascular diseases.

The vascular tree as an integrative structure

Chapter 1

The vascular system: components, signaling, and regulation

Oindrila Paul, Jian Qin Tao, Xiaohui Guo and Shampa Chatterjee
Department of Physiology, Institute for Environmental Medicine, University of Pennsylvania Perelman School of Medicine, Philadelphia, PA, United States

1.1 Introduction

The vascular system is a network that acts as a conduit for the flow of blood and lymph throughout the body. The blood vessels that comprise the vascular system consist of the arteries that deliver the blood from the heart to the rest of the body, the veins that bring the blood back to the heart from the different parts of the body and the capillaries that are the tiny blood vessels interconnecting the arteries and veins allowing the oxygenated blood to be transported to the body. The vascular tree is thus a closed system formed from a perfect balance of arteries, capillaries, and veins that help the blood and other nutrients to be delivered to every cell in the body.

At the center of the vascular system is the heart, a muscular pumping device, which provides the force necessary to circulate the blood through a closed system of arteries, veins, and capillaries. Blood flows from the heart (right atrium to the right ventricle) and is then pumped to the lungs for oxygenation via gas exchange at the alveolar interface. From the lungs, the blood flows to the left atrium, then to the left ventricle. This oxygen-rich blood leaves the heart through the arteries that further break down into smaller branches that deliver oxygen and other nutrients to every tissue and organs. Eventually, the blood moves through the capillaries bringing oxygen and nutrients into the cells while removing waste from the cells into the capillaries. Substances enter the capillary wall by diffusion, filtration, and osmosis. The closed system ensures that the blood that leaves the capillaries enters the veins that become large in order to carry the blood back to the heart.

In addition to the vascular system, the circulatory system also comprises the lymphatic system that carries the interstitial fluid that leaves the capillaries and accumulates in tissue spaces. Often proteins and other biomolecules may seep out of capillaries and increase the volume of tissue fluid driving edema. This fluid is taken up by the lymph capillaries and returned to the venous circulation. The lymphatic capillaries also absorb fat-soluble vitamins and other fat-soluble substances and transport these to the venous system. Thus the lymphatic system actively regulates tissue fluid homeostasis, absorbs gastrointestinal lipids, and acts as a conduit for lymphocytes (i.e., specific immune cells that are produced by the lymph nodes) to reach the systemic circulation. In contrast to the blood vascular circulation, the lymphatic system is a unidirectional system that connects the extracellular space to the venous system.

A healthy and functioning vascular system is critical for ensuring that sufficient nutrients and oxygen reach the ~ 37 trillion cells in our bodies and the cellular waste products are efficiently removed. To service such a colossal quantity of cells, there is over 100,000 km of vasculature present within our bodies [1], which forms a continuum of vessels from arteries to arterioles, capillaries, venules, and veins [2]. Overall, the vascular network comprising of the heart and interconnected vessels plays an integral role in transport of nutrients, removal of waste, and regulation of immune function; indeed it is crucial in maintenance of life. In this chapter, we will review the various components of this system and their respective roles in molecular signaling that drives vascular function and adaptation that are crucial in health and in triggering the onset of diseases. This chapter will thus set the stage for this volume; subsequent chapters will discuss various aspects of vascular signaling and regulation mechanisms that cover the gamut from homeostasis to its disruption with inflammation, infection, tumorigenesis, and angiogenesis.

Endothelial Signaling in Vascular Dysfunction and Disease. DOI: https://doi.org/10.1016/B978-0-12-816196-8.00023-0

1.2 Components of the vascular tree

1.2.1 Arteries, veins, and lymphatic system

Arteries, veins, and lymphatic vessels are distinguished by structural differences that correspond to their different functions. These defining features are determined during embryogenesis, when blood vessels develop from the mesoderm. At around embryologic day E6.5 (in mice), the mesoderm differentiates into hemangioblasts that give rise to two different cell populations: hematopoietic precursor cells and endothelial precursor cells [3,4].

While the largest artery is the aorta, all arteries sequentially branch and decrease in size into arterioles until they end in a network of smaller vessels called capillaries. The capillary beds are fine networks of endothelial cells (ECs) located in the (interstitial) spaces within tissues. Here capillaries release nutrients, active agents, immune cells and growth factors, and oxygen. The capillaries connect to a series of vessels called venules, which increase in size to form the veins. The veins join together into larger vessels as they transfer blood back to the heart. The largest veins, the superior and inferior vena cava, return the blood to the right atrium.

As mentioned earlier, the lymphatic system is also a network of vessels that run throughout the body. However, these vessels do not form a full circulating system and are not pressurized by the heart. Rather, the lymphatic system is an open system with the fluid moving in one direction from the extremities toward two drainage points into veins just above the heart. Lymphatic fluids move more slowly than blood because they are not pressurized. Small lymph capillaries interact with blood capillaries in the interstitial spaces in tissues. Fluids from the tissues enter the lymph capillaries and are drained away. These fluids, termed lymph, also contain large numbers of white blood cells.

Large arteries such as aorta comprise of several layers of cells (Fig. 1.1): the tunica adventitia or tunica externa is the outermost layer of connective tissue, followed by the tunica media that is a layer of smooth muscle cells. Finally the innermost layer of the blood vessel is the tunica intima that is made up of a single layer of ECs that are in direct contact with blood flow. The endothelial layer is supported by a basement membrane, a subendothelial connective tissue layer that is supported by a layer of elastic tissue (the internal elastic lamina). Smaller arteries or arterioles consist of layers of smooth muscle cells and the internal elastic lamina with the endothelial layer. Capillaries that are smaller vessels consist of only endothelium surrounded by a basement membrane. Capillaries may have continuous endothelium (less permeable as in muscle, lung, and skin tissues), fenestrated endothelium (have gaps between ECs on a continuous basement membrane as in intestinal villi, renal glomeruli, endocrine glands), or discontinuous endothelium with large gaps between cells and a discontinuous basement membrane (as in liver and in lymphoid organs).

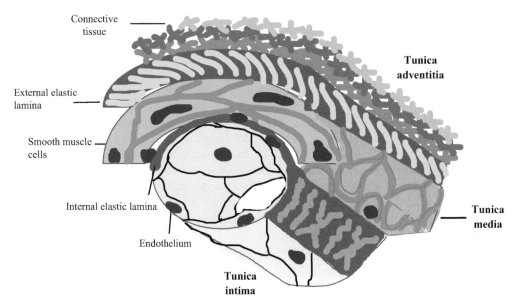

FIGURE 1.1 The structure of blood vessels. The innermost layer, the tunica intima, is simple endothelial layer surrounded by a connective tissue basement membrane with elastic fibers. The middle layer, the tunica media, is primarily smooth muscle and is usually the thickest layer. It provides support for the vessel and facilitates changes to vessel diameter to regulate blood flow and blood pressure. The outermost layer, which attaches the vessel to the surrounding tissue, is the tunica adventitia.

Similar to arteries, veins have multiple layers of tissue; tunica adventitia is thicker but tunica media is thinner than in arteries. The venous endothelial cell–cell junctions can separate or open in the presence of inflammatory or other agents, thus allowing for movement of immune cells and alterations in permeability.

1.2.2   The endothelium

The endothelium is the layer of cells lining the blood vessels in animals. It weighs more than 1 kg in adult humans, and it covers a surface area of 4000–7000 square meters. The endothelium is the cellular interface between the circulating blood and underlying tissue. As the medium between these two sets of tissues, endothelium is part of many normal and diseased processes throughout the body [5]. The endothelium responds to signals from its surrounding environment to help regulate functions like the resistance that blood vessels need to pump blood through the body (vasomotor tone), control the flow of substances trying to enter or exit the blood vessel (blood vessel permeability), and the ability of blood to clot (hemostasis). In addition to diseases like atherosclerosis, endothelium has been indicated as a component in pathologies like cancer, asthma, diabetes, hepatitis, multiple sclerosis, and sepsis. ECs are extremely heterogeneous both structurally and functionally. The different arrangement of cell junctions between ECs and the local organization of the basal membrane generate different type of endothelium with different permeability features and functions. The shape, size, and appearance of ECs, called their phenotypes, vary depending upon which part of the body the cells are from, a property called phenotypic heterogeneity. The endothelium, its properties, and its responses to stimuli are governed largely by the local environment of the cells.

1.2.3   The basement matrix

Basement membranes (BMs) are thin, dense sheets of specialized, extracellular matrix (ECM) that surround tissues. The BM of ECs is organized as a complex network of numerous glycoproteins that include fibronectin (FN), vitronectin (VN), collagens, and von Willebrand factor. ECs assemble the BM matrix that provides critical support for vascular endothelium. The interactions of integrins and other "adhesive" proteins on the EC surface, with the ECM provide a scaffold essential for maintaining the organization of vascular ECs into blood vessels. The thickness of BMs ranges from (100–1000 nm) to nanoscale (1–100 nm) range depending on the location and physical properties of the vessel [6].

1.2.4   Smooth muscle

Vascular smooth muscle cells (VSMCs) an important component of blood vessels are located in the medium part of a blood vessel as tunica media (Fig. 1.1). As the medium part of the blood vessel wall, the tunica media is located between the tunica intima and tunica adventitia, and is separated from these parts by the lamina elastica interna and the lamina elastica externa, respectively. The VSMC layer interacts with the EC layer; indeed the EC-VSMC maintains vessel structure and function in terms of vasodilation and constriction [7]. Aberrant interaction could promote atherogenesis. Endothelial dysfunction, a well-defined pathological state of the endothelium, underlies change in vascular signals and composition of the vascular wall, a common feature of diseases such as atherosclerosis, hypertension, hypercholesterolemia, and diabetes [8]. This will be discussed in detail in the section titled *Regulation of vascular homeostasis*.

1.2.5   Macrovascular and microvascular vessels

Macrovascular endothelium that lines large vessels such as aorta is prone to atherogenic changes arising from risk factors and hemodynamic forces (discussed under Section 1.3.1). The resultant changes in the form of inflammation and remodeling lead to plaque formation and affect the cardiovascular system, the cerebrovascular system, and the peripheral blood supply to the lower limbs. Microvessels, on the other hand, do not face disturbed hemodynamic forces; an important hemodynamic function of the microvascular vessels is the avoidance of large fluctuations in hydrostatic pressure at the capillary level. Thereby the microcirculation determines overall peripheral resistance. Overall microvascular ECs are a key regulator of permeability, vascular tone, and hemostasis [9].

1.3 Endothelial signaling

As the endothelial layer is in direct contact with blood flow it "senses" physical forces associated with the flow and chemical stimuli in the blood. These events termed as mechanotransduction (sensing of mechanical forces) and chemotransduction (sensing of chemotactic, inflammatory, and other agents) play a major role in endothelial structure, function, and disease.

1.3.1 Mechanotransduction

The endothelium has the ability to "sense" physical forces associated with blood flow. This process is known as mechanotransduction; indeed ECs have been reported to possess the "machinery" that senses physical force and converts it into a biochemical signal. The normal endothelium senses "uniform shear stress" associated with laminar flow (Fig. 1.2) and produces nitric oxide (NO), prostacyclin (PGI$_2$), tissue plasminogen activator (t-PA), and reactive oxygen species (ROS) [11−13]. These collectively regulate vasoreactivity by controlling vascular dilatation and constriction as well as the expression of the adherence of platelets, neutrophils, etc. (Fig. 1.2). However, under conditions of altered blood flow, endothelial mechanotransduction or mechanosignaling leads to activation of signaling cascades that facilitate inflammation, injury, or angiogenesis (growth of new vessels) [14,15].

Alterations in shear could be in the form of irregular blood flow in the form of low shear, eddies, recirculation of flow, or by loss of shear (such as would occur with impeded blood flow) (Fig. 1.3) cause the production of ROS that drives an activated endothelial phenotype [16]. The activated endothelium is characterized by increased expression of cellular adhesion molecules (intercellular adhesion molecule, ICAM-1; vascular cell adhesion molecule, VCAM-1) and P and E-selectins that drive the recruitment and adherence of immune cells as well as infiltration of plasma components such as cholesterol, low-density lipoprotein (LDL), and fibrinogen. These changes can promote an atherogenic phenotype [10,17,18]. Overall the signaling associated with blood flow intersects with risk factors such as hypercholesterolemia, inflammatory conditions, and underlying antioxidant, antithrombotic genes that drive the mechanosignaling-remodeling-disease paradigm.

Pivotal to this mechanosignaling paradigm is endothelial dysfunction that arises from the loss of equilibrium between pro- and antioxidant signaling, thereby producing oxidative stress. Oxidative stress or pro-oxidant load of the endothelium occurs via ROS and reactive nitrogen species [such as NO, peroxynitrite (ONOO-), nitrous oxide (NO$_2$), N$_2$O$_3$] that are

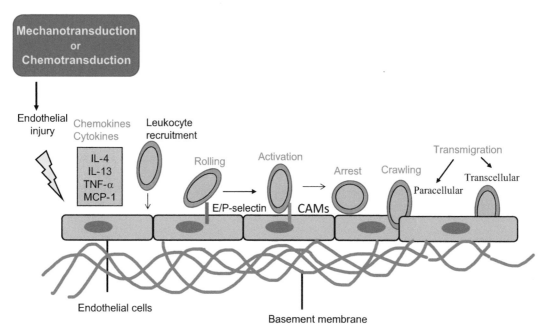

FIGURE 1.2 A simple diagram of the multistep leukocyte adhesion cascade: (1) Endothelial cells are affected by a chemical (chemotransduction) or physical (mechanotransduction) stimulus. This activates inflammatory cytokines such as tumor necrosis factor (TNF-α), interleukins (IL-4 and -13) and monocyte chemoattractant protein (MCP-1). Expression of cellular adhesion molecules and selections also occurs. (2) Leukocytes are recruited from flowing blood by CAMs and selectins. (3) Adhesion is stabilized. Leukocytes adhere to the endothelium. (4) The leukocytes elongate (cytoskeletal remodeling) and migrate through the endothelium (transcellular) or across (paracellular) the junctions of the endothelial cell monolayer into tissue.

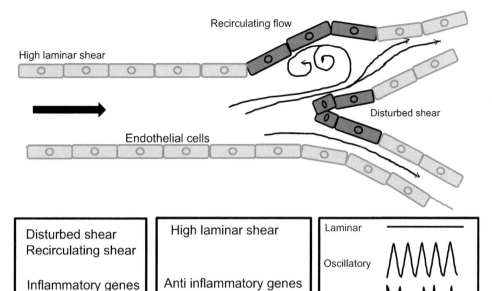

FIGURE 1.3 Schematic diagram to show laminar flow in a major vessel and disturbed flow at a bifurcation. The diagram illustrates how blood flow forms recirculatory patterns and eddies at the curvatures. *Reproduced from Chatterjee S. Endothelial mechanotransduction, redox signaling and the regulation of vascular inflammatory pathways. Front Physiol 2018;9:524 [10].*

generated from alterations in shear stress or from turbulent shear [10,19]. ROS in turn activates pro-inflammatory or pro-angiogenic pathways via transcription factors such as NF-κB, AP-1, and kinases such as p38 MAPK [20].

Work from our group on endothelial mechanosignaling showed that the mechanical stimulus associated with removal of shear associated with stop of blood flow as would occur with a blood clot or surgery, triggered the onset of a signaling cascade, the earliest event of which is the depolarization of the EC membrane [20]. We discovered that depolarization occurred via a K_{ATP} channel (composed of an inwardly rectifying K + channel of 6 family) that closed, leading to activation of a PI3Kinase-Akt pathway that caused assembly of the subunits of the enzyme NADPH oxidase (NOX2) and subsequent production of ROS [21]. We also observed that depolarization of the endothelial membrane altered its potential from -80 to -60 mV to about -30 to -20 mV, at which potential there was activation of voltage-gated calcium channels (VGCCs). VGCC facilitates entry of calcium ions into the endothelium, leading to activation of endothelial nitric oxide synthase (eNOS) and subsequent NO production [5,17].

Studies by other groups showed that turbulent shear occurs at branch points, curvatures of large arteries and the aortic arch, and also with anastomosis/stent insertions (Fig. 1.3) leads to nonuniform or irregular shear along the vessel wall, creating sites of abnormally low and high shear stress. This is sensed by the endothelium and the resultant signaling causes the activation of a cascade that drives an inflammatory phenotype. This is in the form of a procoagulant surface, production of inflammatory cytokines, and cell proliferation. Eventually cell proliferation (of both endothelial and smooth muscle cell) results in remodeling of the vasculature [13,22,23].

1.3.2 Chemotransduction

EC signaling can be initiated by soluble factors that act through either receptor-mediated or receptor-independent mechanisms. This process is termed as chemotransduction, and involves agonists such as angiotensin, thrombin, transforming growth factor-β, platelet-derived growth factor, vascular endothelial growth factor (VEGF), and endotoxins. These activate endothelial signaling either via receptor-mediated or nonreceptor-mediated pathways.

The interaction of angiotensin II (Ang II) with its receptor (angiotensin receptor, ATR1/2) activates phospholipase C, which in turn hydrolyzes phosphatidylinositol-4,5-bisphosphate to inositol 1,4,5-triphosphate (InsP3), and diacylglycerol (DAG) [24]. DAG and Ca^{2+} released by $InsP_3$ activate protein kinase C (PKC) [20,24]. PKC phosphorylates NADPH oxidase 2 leading to the production of ROS. PKC activation through either receptor-mediated or nonreceptor-mediated mechanisms represents a key step in NADPH oxidase 2 activation and ROS [25].

Endotoxins such as lipopolysaccharides (LPS) (a component of the outer membrane in Gram-negative bacteria and is involved in the pathogenesis of sepsis) [26] bind to Toll-like receptor (TLR) family of receptors; we reported that LPS exposure on ECs leads to ROS production via the NADPH oxidase 2 [27]. LPS-TLR4 interaction activates mitogen activated protein kinases that phosphorylate phospholipase A_2: indeed, as we previously demonstrated Prdx6—PLA$_2$ generates lysophosphatidylcholine (LPC), which is converted to lysophosphatidic acid (LPA) by the lysophospholipase D activity of autotaxin. The binding of LPA to its receptor (LPAR) leads to an assembly of NADPH oxidase 2 components on the cell membrane [28]. Thus Prdx6—PLA$_2$ activity facilitates (via LPA production) ROS generation by ECs in response to agonists such as LPS.

VEGF-A (referred to as VEGF) binds to VEGF receptor 1/2 (VEGFR1/2) on the endothelium; VEGFR1 is a high-affinity tyrosine kinase receptor that shows low phosphorylation compared to VEGFR2, which is considered as the main signaling receptor for VEGF [29]. Both VEGFR1 and VEGFR2 are predominantly expressed on ECs. VEGF-A can bind to either VEGFR1 or VEGFR2 or both. VEGFR signal transduction steps include: receptor dimerization; autophosphorylation, a posttranslational modification of carboxy-terminal tyrosines; adapter binding to phospho-tyrosine residues; and adapter phosphorylation. PLCγ and PI3K are activated due to their binding preference with the VEGFR1 phospho-tyrosine sites [30]. VEGFR2 is endocytosed (via a classical clathrin-mediated pathway) into the ECs [31]. Upon entry into cell, VEGFR2 signaling results in decreased phosphorylation of Y1175 site that is critical to activation of phospholipase C gamma-extracellular signal regulated kinase (PLCγ/ERK) signaling. As a result, ERK phosphorylation is reduced [32]. VEGFA/VEGFR-2 signaling appears to mediate cellular responses involved in angiogenesis prominently. This cascade participates in the angiogenic response by increasing microvascular permeability, inducing EC proliferation, migration, survival, and secretion of matrix metalloproteinases [33].

1.3.3   Biochemical signals associated with mechano and chemotransduction

1.3.3.1   Reactive oxygen species and nitric oxide

ROS generation in vivo can occur via various enzymatic pathways and via nonenzymatic processes. The former generate either $O_2^-\bullet$ or H_2O_2 by the transfer of one or two electrons, respectively, whereas the nonenzymatic pathway generates $O_2^-\bullet$ by auto-oxidation of a reduced compound. ROS generated in ECs either exert their physiological (signaling) effects or cause oxidative damage. Signaling involves activation of transcription factors, protein tyrosine phosphatases, protein tyrosine kinases, mitogen-activated protein kinases, and other components of various signaling cascades. Many of these modified proteins transmit their signal through phosphorylation and dephosphorylation of serine, threonine, tyrosine, and histidine. ROS modify these signaling-related proteins through the oxidation of specific residues that result in reversible activation or inactivation of enzymatic activity. Excess ROS produced by these pathways can result in cell injury, and organisms have developed a variety of enzymatic and nonenzymatic defense mechanisms for protection.

The production of NO$\bullet$ involves cleavage of the terminal guanidino nitrogen of L-arginine by nitric oxide synthases (NOS) to form NO$\bullet$ and L-citrulline. Endothelium-derived NO$\bullet$ is produced by the endothelial isoform of NOS (eNOS) in response to various stimuli such as increased shear stress or by agents such as thrombin or acetylcholine. NO$\bullet$ diffuses out of the ECs and interacts with the heme moiety of soluble guanylate cyclase to generate the second messenger, cyclic GMP (cGMP), from guanosine triphosphate (GTP). The soluble cGMP then activates cyclic nucleotide-dependent protein kinase G (PKG or cGKI), to phosphorylate proteins that affect the actin and myosin dynamics, resulting in smooth muscle relaxation and vasodilatation [34].

1.3.3.2   Growth factors

Fibroblast growth factor (FGF) and VEGF are important growth factors for ECs in vitro [35]. Furthermore, basic FGF (bFGF) and VEGF have been shown to increase the development of collateral vessels in ischemic models and to increase endothelial proliferation following arterial injury. FGFs have been shown to bind other cell surface molecules to initiate signaling cascades; in fact FGF directly binds to the integrins (specifically integrin $\alpha v\beta 3$) promoting EC adhesion and spreading and thus facilitating angiogenesis.

The vasculature develops further by angiogenesis, the sprouting of new capillaries from existing vascular structures [36]. In the embryo the first blood vessels are generated from endothelial precursors, which have a common origin with hematopoietic progenitors, a process known as vasculogenesis [37]. Vasculogenesis as well as angiogenesis are primarily regulated through several receptors of the VEGF and bFGF receptors, all of which are expressed in ECs [38].

1.3.3.3   Calcium signaling

Calcium is an essential second messenger in ECs and plays a pivotal role in regulating NO production (via eNOS), cell migration, angiogenesis, barrier function, and inflammation. An increase in intracellular Ca^{2+} concentration can occur either via release of Ca^{2+} from intracellular stores (like endoplasmic reticulum, ER) or from calcium entry through VGCCs.

Quiescent ECs at rest maintain a very low intracellular cytosolic-free calcium concentration, ranging from 30 to 100 nmol/L and there is an approximately 20,000-fold concentration gradient across the plasma membrane [39]. Intracellular calcium concentrations are actively maintained by transmembrane channels expressed in the plasma membrane, such as the plasma membrane calcium ATPase channels, and by transmembrane channels in the ER membrane [40], such as the sarco/ER Ca^{2+}-ATPase channels.

1.4 Regulation of vascular homeostasis: healthy and diseased systems

Endothelial dysfunction implies long-term changes arising from endothelial activation; it represents a switch from a naïve and quiescent phenotype toward one that involves increased expression of chemokines, cytokines, and adhesion molecules. This facilitates an interaction with leukocytes and platelets that may eventually contribute to arterial disease (Fig. 1.3). A detectable long-term change in vascular reactivity is the altered composition of the vascular wall. A crucial change involved in this process is ROS and NO-mediated signaling; ROS and NO can diffuse rapidly throughout the cell and react with cysteine groups in proteins to alter their function. The effects of endothelial dysfunction on VSMCs are reduction of NO bioavailability and/or augmentation of vasoactive constrictors released from the endothelium [41]. Mechanistically, normal and controlled VSMC proliferation is beneficial, while dysregulated VSMC proliferation (often arising from aberrant inflammation) contributes to plaque formation [7]. Outlined below are some of the endothelial signaling processes associated with vascular integrity and function.

1.4.1 Inflammation signaling

Vascular inflammation is accompanied by attraction of circulating leukocytes to the site of injury and their transmigration into the intima of the vessel wall, to clear the tissue from the source of inflammation and dead cells and ultimately resolve the inflammation. These processes play a pivotal role in atherosclerosis, myocardial infarction, pulmonary arterial hypertension, and neurodegeneration.

Atherosclerosis starts with EC dysfunction in response to increased levels of oxidized LDLs in the bloodstream [42]. ECs sense and uptake toxic LDLs, which promote oxidative stress, a reduction in nitric oxide synthesis, and increased apoptosis. Oxidative stress as well as ECs loaded with LDL upregulate adhesion molecules like ICAM-1, VCAM-1, and E- and P-selectins, as well as pro-inflammatory cytokines such as CCL2 or monocyte chemoattractant protein (MCP-1), and interleukin (IL)-1β, and the complement system. These changes in EC phenotype promote attraction, attachment, and slow rolling of circulating immune cells on the vessel wall [43]. Thus the first step is mediated by cellular adhesion molecules (CAMs), selectins, etc.; the selectins on ECs interact with selectins on immune cells to capture the immune cell to the surface of the endothelium. Next, immune cells adhere to the endothelial layer. The immune cells transmigrate into intima of the vessel wall, where monocytes differentiate into macrophages and thus facilitate an inflammatory state and clearance of excessive lipids and dead cells [44] (Fig. 1.2).

1.4.2 Angiogenic signaling

Angiogenesis or the growth of new vessels from existing vasculature is regulated by endothelial signals that drive (1) the arrangement of EC tip cells, (2) the increase in permeability, (3) enzymatic degradation of capillary BM, (4) EC proliferation, and (5) EC tube formation. EC tip cells are leading cells at the tips of vascular sprouts that form filopodia that invade surrounding tissue, leading the path of neo-vessel formation [45]. Filopodia secrete large amounts of proteolytic enzymes, which digest parts of the ECM allowing for a developing sprout. The filopodia of tip cells also express VEGF (VEGF-A) receptor VEGFR2, allowing them to align with the VEGF gradient [46]. VEGF increases permeability thus allowing the blood proteins like fibrin to leak into the microenvironment. The ECs migrate and proliferate as the physical barriers of the vessel are disrupted. These cells become the trunk of the newly formed capillary. When the tip cells of two or more capillary sprouts converge at the source of VEGF-A secretion, the tip cells fuse creating a lumen through which oxygenated blood can flow [47]. The emerging concept is that endothelial redox and ROS initiated signals play a key role during angiogenesis [48].

1.4.3 Endothelial cell polarization

The polarity of ECs is critical in the context of migration (associated with angiogenesis), vascular structure (apical/basolateral polarity), and endocytosis. The directed migration of ECs which is an initial step in angiogenesis requires a front-rear polarized EC. This tip cell leads an angiogenic front. The developing of the front-rear polarity is driven by a growth factor (VEGF) whereby the responsive EC (or selected tip cell) orients the filopodia to the leading front. This is accompanied by significant cytoskeletal changes [49,50]. These cytoskeletal changes involve reorganization of actin-based structures that make up the filopodial extensions such that the EC balanced by forces of EC attachment to the basement matrix and that of focal adhesion proteins at membrane protrusion points [49].

ECs face the blood flow but are also tethered to a basement matrix. Each of these apical and basolateral sides needs to be maintained as two structurally and functionally different entities. The apicobasal polarization enables the transport of small and large solutes from the apical side to the basolateral side either via endocytosis (paracellular) or transcellular transport. Select membrane proteins are localized at the apical and basolateral sides of ECs to facilitate specific cellular processes. Due to differential interaction with blood and the parenchyma, ECs secrete proteins and other factors in a polarized manner either to the apical or the basolateral side. The apical and the basolateral membrane domains are separated by tight junctions. VEGF, histamine, and insulin-like growth factor-binding protein 3 increase the transendothelial permeability when administered to the basolateral, but not to the apical [51]. However, agents such as thrombin and bradykinin are effective on both sides and LPA mostly on the apical side and less on the basolateral side. Endocytosis or transport of molecules into the cells occurs mostly via receptor-mediated (clathrin-mediated) pathways. For instance, VEGFR2 undergoes VEGF-induced endocytosis via clathrin-mediated pathway, whereas FGF2-induced FGFR1 endocytosis is clathrin-independent. Invaginations on the endothelium called caveolae facilitate polarization and participate in β1 integrin endocytosis [52,53]. Caveolin-1 (the main component of caveolae) localizes to the leading edge of ECs in 3D environments suggesting dependence on local contextual cues [54].

1.4.4 Barrier function

The control of the endothelial barrier function is largely mediated by cell-to-cell junctions, which include adherens (AJ) and tight junctions (TJ). The structural and functional integrity of these junctions is a major determinant of paracellular permeability. The paracellular pathway is responsible for the majority of leakage of blood fluid and proteins across the microvascular endothelium under pathophysiological conditions.

Junctional integrity may be compromised through degradation of AJ and TJ, reorganization of the cytoskeleton, or destabilization of attachments to the ECM. All these lead to endothelial hyperpermeability that is a hallmark of several pathologies that are accompanied by edema [55]. Vascular endothelial (VE)−cadherin regulates the function of AJs. Intracellularly, VE−cadherin is connected to the actin cytoskeleton via α-, β-, γ-, and p120-catenins. AJ structure involves VE−cadherin binding to β-catenin and γ-catenin, which in turn are connected to actin via binding to α-catenin. One of the other key mediators of endothelial barrier function are the CAMs that mediate cell−cell and cell−matrix adhesion and transmit signals from the extracellular environment involved in tissue morphogenesis, angiogenesis, and tumor progression [56,57]. The stability of the VE−cadherin−catenin−cytoskeleton complex and CAMs is essential to the maintenance of endothelial barrier function. TJs impart additional barrier function and are therefore expressed in the brain endothelium, preventing the passage of much smaller molecules (<1 kDa), even restricting the flow of small inorganic ions.

1.5 The endothelium in innate and adaptive immunity

The first line of defense against non-self-pathogens is the innate, or nonspecific, immune response. Innate immune cells "recognize" infection via pattern recognition receptors (PRRs). Pathogen-associated molecular patterns (PAMPs) (such as microbial nucleic acids, lipoproteins, and carbohydrates from microbes) or damage-associated molecular patterns released from cells injured by sterile insults (heat, oxidant-induced damage, etc.) bind to PRR; the activated PRRs then oligomerize and assemble large multisubunit complexes that initiate signaling cascades that trigger the release of factors that promote further recruitment and adherence of polymorphonuclear neutrophils and other immune cells. The endothelium is a major player in the innate immune response. PRRs on ECs such as TLRs recognize bacterial and viral PAMPs in the extracellular environment (TLR1, TLR2, TLR4, TLR5, TLR6, and TLR11) and signal the production of pro-inflammatory cytokines.

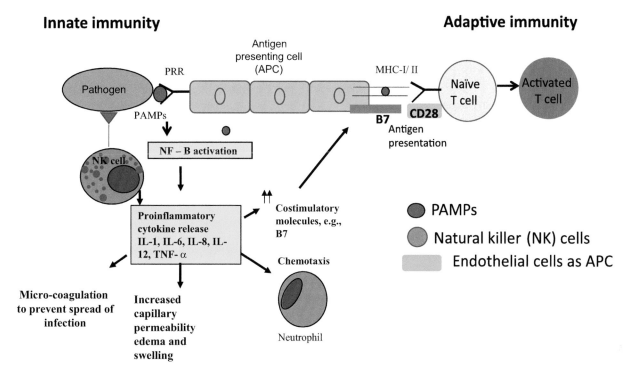

FIGURE 1.4 The endothelium in innate or acquired immunity: The mechanisms mediating innate (left) and adaptive (right) are shown.

The second line of defense against nonself-pathogens is called the adaptive immune response. A key adaptive immune response is antigen presentation. This occurs when a subset of immune cells called dendritic cells (DCs) capture and process antigens emanating from foreign agents and particles. The processed antigens (Ag) are then presented within major and minor histocompatibility complexes (MHC-class I and II) on the cell surface. DCs migrate to the lymphoid organs where they arrive in the T-cell regions. ECs are being increasingly recognized for their ability to act as antigen-presenting cells (APCs) (Fig. 1.4) [58]. Although ECs cannot replace T and B cells, of late it is becoming clear that ECs can express MHC I and II class molecules and process Ag, under some conditions that have the potential of acting as APC. This action has been reported and ECs from different species express various accessory molecules, including CD80, CD86, ICOS-L, programmed death ligand 1, programmed death ligand 2, LFA-3, CD40, and CD134L [59]. Because the total endothelium comprises of a large surface area, its transformation into APC implies that the endothelium can play a major part in the adaptive immune response. However, the mechanism by which the endothelium is transformed into an APC is not very clear.

1.6 Vascular signals in progression of disease

Endothelial signaling arising from changes in the immediate environment of the vascular layer leads to cell growth, differentiation, cell migration, as well as immune cell infiltration along the endothelial wall. This occurs due to intercellular communication between EC and other cell types such as VSMC. Collectively, these processes drive degradation and reorganization of the ECM scaffold in the vessel wall, altered barrier function, and vascular leakage [60]. Adaptive and maladaptive remodeling of the vascular wall represents key processes in vascular diseases and is the major focus of therapeutic interventions [61].

1.7 Conclusion

The endothelium plays a critical role in a variety of human disorders, including peripheral vascular disease, heart disease, diabetes, insulin resistance, chronic kidney failure, tumor growth, venous thrombosis, and viral infectious diseases. This is because endothelial signals that drive dysfunction serve as triggers that progress via communication with other cell types alter vascular structure and function. While endothelial structure is regulated via cytoskeletal elements and barrier signaling, endothelial function is controlled by NO and ROS production. All these elements control the various indices of endothelial function and are therefore key regulators of vascular health. Thus understanding endothelial dysfunction and agents that can lessen it are crucial in effective therapeutic treatments.

References

[1] Pugsley MK, Tabrizchi R. The vascular system. An overview of structure and function. J Pharmacol Toxicol Methods 2000;44(2):333−40.

[2] Aird WC. Spatial and temporal dynamics of the endothelium. J Thromb Haemost 2005;3(7):1392−406.

[3] Davidson AJ, Zon LI. Turning mesoderm into blood: the formation of hematopoietic stem cells during embryogenesis. Curr Top Dev Biol 2000;50:45−60.

[4] Patel-Hett S, D'Amore PA. Signal transduction in vasculogenesis and developmental angiogenesis. Int J Dev Biol 2011;55(4-5):353−63.

[5] Chatterjee S, Fisher AB. Ischemia in the pulmonary microvasculature in endothelium: a comprehensive reference In: William C. Aird (ed.). Cambridge University Press: New York; 2007. p. 1202−13.

[6] Liliensiek SJ, Nealey P, et al. Characterization of endothelial basement membrane nanotopography in rhesus macaque as a guide for vessel tissue engineering. Tissue Eng Part A 2009;15(9):2643−51.

[7] Li M, Qian M, et al. Endothelial-vascular smooth muscle cells interactions in atherosclerosis. Front Cardiovasc Med 2018;5:151.

[8] Hirase T, Node K. Endothelial dysfunction as a cellular mechanism for vascular failure. Am J Physiol Heart Circ Physiol 2012;302(3): H499−505.

[9] Stehouwer CDA. Microvascular dysfunction and hyperglycemia: a vicious cycle with widespread consequences. Diabetes 2018;67(9):1729−41.

[10] Chatterjee S. Endothelial mechanotransduction, redox signaling and the regulation of vascular inflammatory pathways. Front Physiol 2018;9:524.

[11] Bagot CN, Arya R. Virchow and his triad: a question of attribution. Br J Haematol 2008;143(2):180−90.

[12] Barakat AI. Responsiveness of vascular endothelium to shear stress: potential role of ion channels and cellular cytoskeleton (review). Int J Mol Med 1999;4(4):323−32.

[13] Barakat AI, Davies PF. Mechanisms of shear stress transmission and transduction in endothelial cells. Chest 1998;114(1 Suppl.):58S−63S.

[14] Browning E, Wang H, et al. Mechanotransduction drives post ischemic revascularization through K(ATP) channel closure and production of reactive oxygen species. Antioxid Redox Signal 2014;20(6):872−86.

[15] Hwang J, Saha A, et al. Oscillatory shear stress stimulates endothelial production of O_2- from p47phox-dependent NAD(P)H oxidases, leading to monocyte adhesion. J Biol Chem 2003;278(47):47291−8.

[16] Davies PF, Mundel T, et al. A mechanism for heterogeneous endothelial responses to flow in vivo and in vitro. J Biomech 1995;28 (12):1553−60.

[17] Chatterjee S, Chapman KE, et al. Lung ischemia: a model for endothelial mechanotransduction. Cell Biochem Biophys 2008;52(3):125−38.

[18] Davies PF, Civelek M, et al. Endothelial heterogeneity associated with regional athero-susceptibility and adaptation to disturbed blood flow in vivo. Semin Thromb Hemost 2010;36(3):265−75.

[19] Higashi Y, Maruhashi T, et al. Oxidative stress and endothelial dysfunction: clinical evidence and therapeutic implications. Trends Cardiovasc Med 2014;24(4):165−9.

[20] Browning EA, Chatterjee S, et al. Stop the flow: a paradigm for cell signaling mediated by reactive oxygen species in the pulmonary endothelium. Annu Rev Physiol 2012;74:403−24.

[21] Chatterjee S, Browning EA, et al. Membrane depolarization is the trigger for PI3K/Akt activation and leads to the generation of ROS. Am J Physiol Heart Circ Physiol 2012;302(1):H105−14.

[22] Gimbrone Jr. MA, Topper JN, et al. Endothelial dysfunction, hemodynamic forces, and atherogenesis. Ann N Y Acad Sci 2000;902:230−9 discussion 239−240.

[23] Nerem RM, Alexander RW, et al. The study of the influence of flow on vascular endothelial biology. Am J Med Sci 1998;316(3):169−75.

[24] Touyz RM. Intracellular mechanisms involved in vascular remodelling of resistance arteries in hypertension: role of angiotensin II. Exp Physiol 2005;90(4):449−55.

[25] Frey RS, Ushio-Fukai M, et al. NADPH oxidase-dependent signaling in endothelial cells: role in physiology and pathophysiology. Antioxid Redox Signal 2009;11(4):791−810.

[26] Intae Lee CD, Chatterjee S, Feinstein SI, Fisher AB. Protection against LPS-induced acute lung injury by a mechanism based inhibitor of NADPH-oxidase (Type 2). Am J Physiol Lung Cell Mol Physiol 2014;306:L635−44.

[27] Chatterjee S, Feinstein SI, et al. Peroxiredoxin 6 phosphorylation and subsequent phospholipase A2 activity are required for agonist-mediated activation of NADPH oxidase in mouse pulmonary microvascular endothelium and alveolar macrophages. J Biol Chem 2011;286 (13):11696−706.

[28] Vazquez-Medina JP, Dodia C, et al. The phospholipase A2 activity of peroxiredoxin 6 modulates NADPH oxidase 2 activation via lysophosphatidic acid receptor signaling in the pulmonary endothelium and alveolar macrophages. FASEB J 2016;30(8):2885−98.

[29] de Vries C, Escobedo JA, et al. The fms-like tyrosine kinase, a receptor for vascular endothelial growth factor. Science 1992;255 (5047):989−91.

[30] Eichmann A, Simons M. VEGF signaling inside vascular endothelial cells and beyond. Curr Opin Cell Biol 2012;24(2):188−93.

[31] Sorkin A, von Zastrow M. Endocytosis and signalling: intertwining molecular networks. Nat Rev Mol Cell Biol 2009;10(9):609−22.

[32] Apte RS, Chen DS, et al. VEGF in signaling and disease: beyond discovery and development. Cell 2019;176(6):1248−64.

[33] Lohela M, Bry M, et al. VEGFs and receptors involved in angiogenesis versus lymphangiogenesis. Curr Opin Cell Biol 2009;21(2):154−65.

[34] Bian K, Murad F. Nitric oxide (NO)—biogeneration, regulation, and relevance to human diseases. Front Biosci 2003;8:d264−78.

[35] Yang X, Liaw L, et al. Fibroblast growth factor signaling in the vasculature. Curr Atheroscler Rep 2015;17(6):509.

[36] Yancopoulos GD, Davis S, et al. Vascular-specific growth factors and blood vessel formation. Nature 2000;407(6801):242−8.

[37] Coultas L, Chawengsaksophak K, et al. Endothelial cells and VEGF in vascular development. Nature 2005;438(7070):937−45.

[38] Eklund L, Olsen BR. Tie receptors and their angiopoietin ligands are context-dependent regulators of vascular remodeling. Exp Cell Res 2006;312(5):630−41.

[39] Clapham DE. Calcium signaling. Cell 2007;131(6):1047−58.

[40] Tiruppathi C, Minshall RD, et al. Role of Ca^{2+} signaling in the regulation of endothelial permeability. Vascul Pharmacol 2002;39 (4−5):173−85.

[41] Sandoo A, van Zanten JJ, et al. The endothelium and its role in regulating vascular tone. Open Cardiovasc Med J 2010;4:302−12.

[42] Libby P, Buring JE, et al. Atherosclerosis. Nat Rev Dis Prim 2019;5(1):56.

[43] Schnitzler JG, Hoogeveen RM, et al. Atherogenic lipoprotein(a) increases vascular glycolysis, thereby facilitating inflammation and leukocyte extravasation. Circ Res 2020;126(10):1346−59.

[44] Moore KJ, Sheedy FJ, et al. Macrophages in atherosclerosis: a dynamic balance. Nat Rev Immunol 2013;13(10):709−21.

[45] Eilken HM, Adams RH. Dynamics of endothelial cell behavior in sprouting angiogenesis. Curr Opin Cell Biol 2010;22(5):617−25.

[46] Small JV, Stradal T, et al. The lamellipodium: where motility begins. Trends Cell Biol 2002;12(3):112−20.

[47] Senger DR, Davis GE. Angiogenesis. Cold Spring Harb Perspect Biol 2011;3(8):a005090.

[48] Panieri E, Santoro MM. ROS signaling and redox biology in endothelial cells. Cell Mol Life Sci 2015;72(17):3281−303.

[49] Gerhardt H, Golding M, et al. VEGF guides angiogenic sprouting utilizing endothelial tip cell filopodia. J Cell Biol 2003;161(6):1163−77.

[50] Lamalice L, Boeuf F Le, et al. Endothelial cell migration during angiogenesis. Circ Res 2007;100(6):782−94.

[51] Hudson N, Powner MB, et al. Differential apicobasal VEGF signaling at vascular blood-neural barriers. Dev Cell 2014;30(5):541−52.

[52] Beardsley A, Fang K, et al. Loss of caveolin-1 polarity impedes endothelial cell polarization and directional movement. J Biol Chem 2005;280 (5):3541−7.

[53] Shi F, Sottile J. Caveolin-1-dependent beta1 integrin endocytosis is a critical regulator of fibronectin turnover. J Cell Sci 2008;121(Pt 14):2360−71.

[54] Parat MO, Anand-Apte B, et al. Differential caveolin-1 polarization in endothelial cells during migration in two and three dimensions. Mol Biol Cell 2003;14(8):3156−68.

[55] Komarova Y, Malik AB. Regulation of endothelial permeability via paracellular and transcellular transport pathways. Annu Rev Physiol 2010;72:463−93.

[56] Cavallaro U, Christofori G. Cell adhesion and signalling by cadherins and Ig-CAMs in cancer. Nat Rev Cancer 2004;4(2):118−32.

[57] Mehta D, Malik AB. Signaling mechanisms regulating endothelial permeability. Physiol Rev 2006;86(1):279−367.

[58] Pober JS. Immunobiology of human vascular endothelium. Immunol Res 1999;19(2-3):225−32.

[59] Carman CV, Martinelli R. T lymphocyte-endothelial interactions: emerging understanding of trafficking and antigen-specific immunity. Front Immunol 2015;6:603.

[60] Brown IAM, Diederich L, et al. Vascular smooth muscle remodeling in conductive and resistance arteries in hypertension. Arterioscler Thromb Vasc Biol 2018;38(9):1969−85.

[61] Chen PY, Qin L, et al. Endothelial TGF-beta signalling drives vascular inflammation and atherosclerosis. Nat Metab 2019;1(9):912−26.

Endothelial signaling in inflammatory diseases

Chapter 2

Reactive oxygen species-induced reactive oxygen species release in vascular signaling and disease

Young-Mee Kim[1] and Masuko Ushio-Fukai[2]

[1]*Department of Medicine (Cardiology), University of Illinois at Chicago, Chicago, IL, United States,* [2]*Department of Medicine (Cardiology), Vascular Biology Center, Medical College of Georgia, Augusta University, Augusta, GA, United States*

2.1 Introduction

Reactive oxygen species (ROS) such as O_2^- and H_2O_2 function as signaling molecules mediating both physiological and pathophysiological functions. ROS generating systems, antioxidant enzymes, and their subcellular compartments determine the fate of ROS. Production of ROS in the wrong place or time or generation of ROS in excessive amounts results in oxidative stress leading to cellular dysfunction which contributes to various pathophysiology. Accumulating evidence suggest that ROS derived from nicotinamide adenine dinucleotide phosphate (NADPH) oxidase (NOX) and mitochondria play a critical role in vascular signaling and disease including angiogenesis. However, how highly diffusible ROS produced from different sources can coordinate to drive angiogenesis in endothelial cells (ECs) as well as other responses is poorly understood.

Recently, the cross-talk between NOX and mitochondria, termed "ROS-induced ROS release (RIRR)" has been proposed as a mechanism for ROS amplification and localized ROS production in physiological and pathophysiological conditions [1]. It is shown that H_2O_2 activates O_2^- production by phagocytic and nonphagocytic NADPH oxidases; [2] eNOS uncoupling produces O_2^- instead of nitric oxide (NO), resulting in mitochondrial ROS production [3,4]; NADPH oxidase-derived ROS increase mitochondrial ROS [5,6]; and mitochondrial ROS stimulate NADPH oxidase activation [7]. This RIRR mechanism may represent a feed-forward vicious cycle of ROS production, which can be targeted under conditions of oxidative stress or enhanced in physiological condition. In this chapter, we will summarize the recent knowledge regarding the RIRR involved in vascular signaling and disease focusing especially on angiogenesis, a process of new capillary formation from the preexisting vessels.

2.2 Reactive oxygen species

During an aerobic cellular metabolism, ROS are constantly generated from oxygen(O_2) [8], and free radicals, ions, and molecules with a single unpaired electron confering them with high reactivity [9]. ROS are grouped as oxygen-free radicals [hydroxyl radical (OH), superoxide (O_2^-), organic radicals (R·), alkoxyl radicals (RO·), peroxyl radicals (ROO·)], and nonradicals [singlet oxygen ($O_2·$), hydrogen peroxide (H_2O_2), organic hydroperoxides, ozone/trioxygen (O_3), hypochloride] [8]. The superoxide (O_2^-) is dismutated to hydrogen peroxide (H_2O_2) spontaneously or more than a thousand times faster by an enzymatic process involving superoxide dismutase (SOD) [10]. H_2O_2 is scavenged by catalase or glutathione peroxidase or peroxiredoxins (Fig. 2.1).

In this chapter, we will focus on O_2^- and H_2O_2 which are mainly produced from NADPH oxidase and mitochondria and their cross-talk in vascular signaling and disease.

In normal cells, ROS are generated in a highly regulated manner to maintain cellular homeostasis by balancing the steady-state levels between rates of formation by endogenous ROS generating systems and decomposition by endogenous antioxidant enzyme systems and by small molecules that function as antioxidants [10] (Fig. 2.2).

Endothelial Signaling in Vascular Dysfunction and Disease. DOI: https://doi.org/10.1016/B978-0-12-816196-8.00019-9

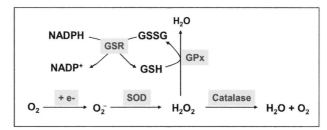

FIGURE 2.1   ROS cycles to maintain cellular homeostasis. Superoxide (O_2^-) forms from oxygen after obtaining an electron from a ROS source. Superoxide is converted to hydrogen peroxide (H_2O_2) by superoxide dismutase (SOD), which is catalyzed by catalase to water (H_2O) and molecular oxygen (O_2). Glutathione peroxidases (GPx) reduce peroxides, such as hydrogen peroxide (H_2O_2), oxidizing glutathione (GSH) to GSSG and glutathione reductase (GSR) consume NADPH to restore GSH.

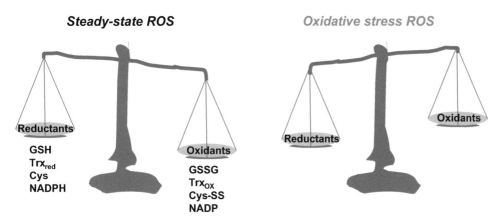

FIGURE 2.2 Biological steady-state balance of ROS and antioxidants.

2.3   Double-edged role of reactive oxygen species

2.3.1   Beneficial reactive oxygen species effects: essential roles in physiology

ROS function as signaling molecules to mediate many biological responses including cell division, immune regulation, autophagy, inflammation, angiogenesis, and stress-related response dependent of concentration [5,8]. ROS at physiological levels act as the mitogens and promote cell proliferation and survival [5], whereas intermediate concentration leads to a transient or permanent cell cycle arrest and induces cell differentiation [11]. However, the signaling dose ranges for H_2O_2 vary from 1 μM to 10 mM in human and animal models [10], and the sensitivity to H_2O_2 is increased with aging in rat coronary arterioles [12].

2.3.2   Harmful reactive oxygen species effects: roles in pathology and disease

At high concentrations, ROS can produce oxidative damages in the DNA (causing mutations which eventually lead to cancer [11]) and other major macromolecules (proteins, lipids etc.) resulting in cellular and systemic dysfunction [13]. Especially, endothelial dysfunction caused by increased ROS has been shown to contribute to the development of atherosclerosis, hypertension, ischemic heart disease, ischemic-reperfusion injury, pulmonary fibrosis, vascular retinopathy, and other vascular diseases [13]. Therefore, many studies have suggested a crucial role of antioxidants in the improvement of pathophysiological conditions [8,13,14]; however several studies till date have failed in the clinical trials [13]. In contrast, the beneficial role of increased ROS in the vascular system depending on the sources and the duration of the subcellular ROS [5] have been well supported by several studies. Although ROS show the double-edged roles (good or bad) in pathophysiology, a complete understanding of the roles of ROS is still lacking [13,15]. We will further discuss the duality of the role of ROS in the vasculature.

2.4 Source of reactive oxygen species

Intracellular ROS originate from different intracellular sources, including NADPH oxidases (NOX), the mitochondrial electron transport chain (ETC), peroxisomes, lysosomes, cytochrome *P*-450 oxygenase, xanthine oxidase (XO), lipoxygenase, cyclooxygenase, and uncoupled nitric oxide (NO) synthase (NOS) [10,13,16] (Table 2.1).

2.4.1 Nicotinamide adenine dinucleotide phosphate oxidase-derived reactive oxygen species

Although there are many different sources of ROS in the vasculature, NOX is a prominent source of ROS having several isoforms localized to different sites within the cell [13,14,17]. NOXes catalyze the transfer of electrons from NADPH to molecular oxygen via their "NOX" catalytic subunit, generating O_2^- and H_2O_2. NOX family consists of Nox1, Nox2, Nox3, Nox4, Nox5, and Duox1/Duox2 [18]. These differ in their expression level, expression control, organ-specific expression, type of ROS release, and in the control of their activity [18]. Nox1, Nox2, Nox3, and Nox4 interact with the small transmembrane protein p22phox, but Nox5 does not interact with p22phox [18]. Duox1 and 2 require Duoxa1 and Duoxa2 as a scaffold for their maturation and proper function [18,19].

Nox2 (gp91phox), the prototype NOX was first discovered as a phagocytic oxidase [14] in phagocytic cells to kill microbes [13,14]. Subsequently, it was detected in several different cell types including vascular cells and cardiomyocytes, suggesting its involvement in physiological processes associated with these cell types. Nox2 complex consists of membrane-bound catalytic subunit gp91phox (Nox2) and regulatory subunit p22phox, and cytosolic components such as p47phox, p67phox, and Rac1 [20−22] (Fig. 2.3). ECs also contain Nox4, Nox1, and Nox5 [14,23−25].

Nox2 activation requires the assembly of cytosolic organizers such as p47phox, p67phox, and small GTPase Rac1 to produce O_2^- that is converted to H_2O_2. By contrast, Nox4, which is the most highly expressed NOX homolog in ECs, requires p22phox but does not require Rac1 or any of the cytosolic organizers needed for Nox2 activation [5,14].

Nox2 and Nox4 are shown to exist in the cell membrane and diverse subcellular compartments, such as the perinuclear membrane and endoplasmic reticulum (ER) [14,26].

Recent studies have shown roles of Nox2 and Nox4 in EC dysfunction or survival depending on their subcellular localization and duration of activation [5,13]. Nox4 activity is determined by mRNA expression levels and ROS elevation induced by Nox4 induction. Of note, Nox4 overexpression increases H_2O_2 [17]. Furthermore, how Nox4 is rapidly activated by vascular endothelial growth factor (VEGF) remains unclear. Recently, Xi et al. reported that growth factor IGF-I stimulation induces rapid Tyr^{491} phosphorylation of Nox4, thereby promoting Nox4 binding to the adaptor protein Grb2, a component of plasma membrane scaffold SHPS-1 complex, which in turn induces localized ROS production and subsequent sustained Src activation [27]. Thus, it is possible that this mechanism may also apply to VEGF-induced rapid activation of Nox4 to increase H_2O_2 that activates Nox2 [28].

Nox2 in the phagocytes and ECs shows several differences; (1) endothelial NOX (EC-NOX) is present in the cell membrane as well as in perinuclear and other intracellular membranes; (2) EC-NOX is preassembled and generates ambient levels of ROS intracellularly, whereas phagocytes-Nox enzyme complex does not have any basal activity and is only formed upon stimulation to generate extracellular (within the phagosome) ROS; and (3) phagocyte-ROS generation is fast (within seconds) and at the mM levels, but nonphagocyte-Nox including ECs is slow to produce ROS at the μM levels. These critical differences should be taken into consideration when we compare the phagocyte-Nox and EC-Nox [14,21].

TABLE 2.1 Sources of ROS and antioxidants.

Sources of ROS	Antioxidants
NADPH oxidase (Nox4, Nox2, Nox1, Nox5)	Cytosolic SOD
Mitochondrial electron transport (ETC)	Cu/ZnSOD, Catalase, glutathione
Nitric oxide synthase	Glutathione peroxidase/reductase
Xanthin oxidase	Peroxiredoxin, thiredoxin reductase,
Lipooxygenase and cyclooxygenase	Heme oxygenase, biliverdin reductase.

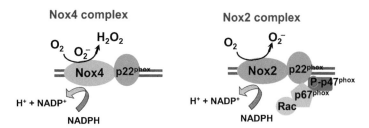

FIGURE 2.3   Assembly and activation of Nox4 and Nox2 oxidase. NADPH oxidase (NOX) is a membrane-bound enzyme complex and Nox4 and Nox2 are major NADPH oxidase expressed in Endothelial cells. Nox2 (gp91phox) activates by binding with cytosolic cofactors (p47phox, p67phox, and Rac1). However, Nox4 does not need to cytosolic cofactors.

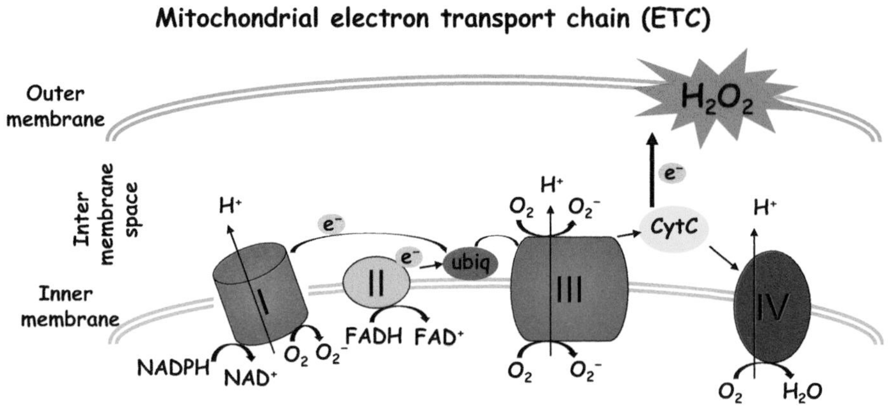

Mitochondrial electron transport chain (ETC)

FIGURE 2.4   Mitochondria electron transport chain. The mitochondrial inner membrane contains five multiprotein complexes with central functions in oxidative phosphorylation. Three of these complexes are redox-driven proton pumps: complex I (NADH-quinone oxidoreductase), complex III (coenzyme Q: cytochrome reductase), and complex IV (cytochrome oxidase). Complex II (succinate dehydrogenase) transfers electrons into the chain from succinate, and complex V (ATP synthase) is a reversible proton pump which uses the electrochemical proton gradient ($\Delta\mu H^+$) to drive ATP synthesis.

2.4.2   Mitochondrial reactive oxygen species

Mitochondria are important sources of ROS production and a common target for the damaging effect of oxidative stress. Mitochondria are the most redox-active compartment of mammalian cells, accounting for more than 90% of electron transfer to O_2 as the terminal electron acceptor [29]. The mitochondrial inner membrane contains five multiprotein complexes with central functions in oxidative phosphorylation. Three of these complexes are redox-driven proton pumps: [30] complex **I** (NADH-quinone oxidoreductase), complex **III** (coenzyme Q: cytochrome reductase), and complex **IV** (cytochrome oxidase). Complex **II** (succinate dehydrogenase) transfers electrons into the chain from succinate, and complex **V** (adenosine triphosphate (ATP) synthase) is a reversible proton pump which uses the electrochemical proton gradient ($\Delta\mu H^+$) to drive ATP synthesis. During the process of cellular respiration, the electrons released from the ETC react with O_2 to produce O_2^- [8]. Complexes **I** and **III** are the main sites of electron transfer to O_2 to produce O_2^- involved in redox signaling [31]. The O_2^- are released into the intermembrane space which comprises 80% of O_2^- generated in the mitochondria and remaining 20% are made by mitochondrial matrix [8] (Fig. 2.4).

In addition, mitochondria functions as a sensor for O_2 and hypoxic signal by releasing ROS to the cytosol [32]. Cytochrome c released from mitochondria by apoptotic signaling also stimulates ROS generation [33]. Hypoxia stimulates H_2O_2 production from mitochondria and the ROS trigger HIF-α stabilization during hypoxia [34]. Interestingly, Zorov et al. demonstrated that inhibition of complex **III** with antimycin at the inward electron transfer site diverted electrons toward the inner-membrane space, leading to O_2^- formation in mitochondria in freshly isolated adult rat cardiac myocytes [1]. After several years later, the same research group reported that in single mitochondria, ROS production occurred in two distinct phases: the initial or "trigger phase," followed by a delayed amplified release of ROS [35]. Zorov et al. described "RIRR" in neighboring mitochondria [35], which suggests that under oxidative stress, simultaneous collapse of the mitochondrial membrane potential and a transient increase in ROS by the electron transfer chain can result in mitochondrial release of ROS to cytosol. Moreover, mitochondrial ROS in ECs are involved in H_2O_2-induced transactivation of VEGFR2 or VEGF-induced cell migration in a Rac1-dependent manner in ECs [5,36,37].

2.5 Reactive oxygen species-induced reactive oxygen species release

The balance between subcellular ROS levels is critical for endothelial and vascular homeostasis. ROS in subcellular compartments such as cytosol, mitochondria, and ER may play different roles within the cell. ROS in the cytosol may affect mitochondrial ROS or vice versa. However, little is known about the communication of ROS derived from distinct compartments and whether endogenous ROS affect cell function differentially depending on their subcellular localization [14]. Two decades ago, Zorov et al. identified that photodynamically produced ROS caused the mitochondrial permeability transition (MPT) induction in cardiac myocytes, which was associated with a burst of mitochondrial ROS generation, termed mitochondrial RIRR [1]. Chung et al. also showed that sodium salicylate-induced ROS are key mediators of mitochondrial membrane potential collapse, which leads to the release of cytochrome c followed by caspase activation [38].

Brady et al. described two modes of RIRR in cardiac cells [39]. The first mode is that increased ROS leads to mitochondrial depolarization via activation of MPT pore, yielding a short-lived burst of ROS originating from the mitochondrial ETC [40]. The second mode is that MPTP-independent increased ROS trigger opening of the inner mitochondrial membrane anion channel, resulting in a brief increase in ETC-derived ROS. Release of this ROS burst to the cytosol could function as a "second messenger" to activate RIRR in neighboring mitochondria. Thus, mitochondria-to-mitochondria RIRR constitutes a positive feedback mechanism for enhanced ROS production, leading to significant mitochondrial and cellular injury [35]. The RIRR phenomenon extends organelle excitability function for electrical and Ca^{2+} signals of mitochondria to include the ability to modulate and amplify ROS signaling [28,41]. Thus, RIRR has been proposed as a mechanism for ROS amplification and localized ROS production [1,5,10].

2.6 Reactive oxygen species-induced reactive oxygen species release in cardiovascular disease

Excess levels of ROS are observed in pathological conditions involving vascular dysfunction [42]. The major causes of morbidity and mortality in the Western world are coronary artery disease and ischemic heart disease, which are critically associated with high ROS levels [43]. H_2O_2 is shown to activate Nox2 to produce O_2^- in fibroblasts and smooth muscle cells with unknown mechanism [2]. Zinkevich et al. recently reported that ROS produced by NOX are an upstream component of the mitochondria-dependent pathway contributing to flow-dependent H_2O_2 generation and dilation in peripheral microvessels from patients with cardiovascular diseases (CAD) [44]. These results indicate that in CAD, both mitochondria and Nox contribute to flow-induced dilation through a redox mechanism in visceral arterioles. In addition, mitochondrial reactive oxygen species (mtROS) trigger the activation of Nox in phagocytes and cardiovascular tissues, which is fundamental for immune cell activation and development of angiotensin II (Ang II)-induced hypertension [7,45,46]. Ang II stimulated Nox1 caused mitochondrial dysfunction by increasing mitochondrial oxidative stress and leads to vascular senescence [47,48]. Moreover, it is shown that mitochondria are not only a target for NOX-derived ROS but also a significant source of ROS, which may stimulate NOXes. This cross-talk between mitochondria and NOX, therefore, may represent a feed-forward vicious cycle of ROS production in pathophysiological process, which can be pharmacologically targeted under conditions of oxidative stress [6].

2.6.1 Dual role of reactive oxygen species in endothelial function

The redox signaling transmit information along a signaling cascade after being oxidized by H_2O_2, and this oxidation may be direct or indirect when the redox sensor is a highly reactive protein that relays the oxidation to a redox switch with low reactivity toward H_2O_2 [49]. Therefore, it may very difficult to titrate the dose of antioxidant to inhibit only the excess ROS present in some diseases without decreasing the "physiological" levels [50]. Recent studies demonstrated that ROS reduction by antioxidants did not improve endothelial and vascular functions [14] but inhibited eNOS activation and NO synthesis in ECs, which resulted in the reduction of endothelial function in cardiovascular events and coronary vasodilatation. These results suggest that ROS in ECs exert a critical positive effect on vascular endothelial function [14]. For example, the nuclear factor erythroid 2-related factor 2 (Nrf2) directly affects the homeostasis of ROS by regulating the antioxidant defense systems [51]. Nezu et al. showed that genetic ablation of an inhibitor of Nrf2 or treatment with an activator of Nrf2 decreased placental angiogenesis, fetal growth, and maternal survival, while deficiency in Nrf2 showed opposite effects [52]. These results suggest that ROS are necessary for angiogenesis that reduces the risk of preeclampsia. Thus, ROS have a concentration-, tissue-, and context-specific effect on endothelial function including angiogenesis [53].

2.6.2 Role of reactive oxygen species in angiogenesis

Multifaceted role of ROS in physiological and pathological angiogenesis is dependent on its concentration [8]. ROS at physiological concentration generally act as signaling molecules to promote cell proliferation, migration and survival, which contribute to therapeutic angiogenesis [39] required for tissue repair and remodeling associated with events such as such as wound healing, skeletal remodeling, and menstrual cycle [54]. For example, in wound healing, ROS produced by tissue hypoxia stimulate VEGF production from macrophages, fibroblasts, ECs, and keratinocytes to induce angiogenesis [55−57].

The high levels of ROS can cause mutations in the DNA which eventually lead to cancer [8]. The oxidative stress is directly related to pathological angiogenesis which proceeds in an uncontrolled and unbalanced manner, resulting in an excessive and abnormal vascular pattern [58]. The most pathological ROS-induced oxidative stress contributes in tumorigenesis, eye diseases, and developing atherosclerotic plaques, hypertension, ischemic heart disease, and vasculopathy [13,44,57]. The retinopathy shows highly elevated ROS production due to high oxygen tension and insufficient ROS scavenging by antioxidants [57,59]. Further, abnormal angiogenesis induced by oxidative stress promotes atherosclerosis by increasing both macrophage infiltration and the thickening of blood vessel wall by oxidized low-density lipoproteins (LDLs), which are generated and accumulated in plasma and the vessel wall of mice and human [13,60−62].

In ECs, VEGF induces angiogenesis by stimulating EC migration and proliferation primary through the VEGF receptor type 2 (VEGFR2) [5,57]. Exogenous ROS stimulate the induction of VEGF expression in various cell types, such as ECs, smooth muscle cells, and macrophages, whereas VEGF induces EC migration and proliferation through an increase in intracellular ROS [57]. Thus, ROS-mediated angiogenesis is associated with VEGF expression [57,63] and then VEGF further stimulates ROS production through the activation of Nox in ECs [60]. VEGF-induced O_2^- production is also regulated by NOX components [57]. In ECs, Noxs, especially Nox2 and Nox4, are important sources of O_2^- and H_2O_2. Several reports showed that Nox2, Nox4, or their regulators play a critical role in ROS-dependent VEGFR2 signaling as well as postnatal angiogenesis using global knock-out mice of Nox4/Nox2; EC-specific Nox4 transgenic mice; dominant negative-Nox4 overexpressing mice; EC-specific catalase overexpressing mice [18,64−66].

p66Shc is an adaptor protein for redox signaling and is phosphorylated at Serine (Ser) 36 residue in N-terminal CH2 domain by oxidative stress. The phosphorylated p66shc leads to mtROS generation by catalyzing electron transfer to O_2 through the oxidation of cytochrome c [5,67,68]. We reported that the nonphosphorylated form of p66Shc is involved in VEGF-induced rapid Rac1 activation and subsequent H_2O_2 production which is required for VEGFR2 phosphorylation at caveolae/lipid rafts and angiogenic responses in ECs [30]. We also demonstrated that VEGF-induced protein kinase C (PKC) and ERK/JNK signaling increase phosphorylation of p66Shc at Ser36 [30], which in turn promotes mitochondrial ROS production (Fig. 2.5) driving angiogenesis [5]. It is also shown that p66Shc knockdown decreased expression of cytosolic NOX organizer p47phox subunit, which in turn decreased ROS generation [13,69].

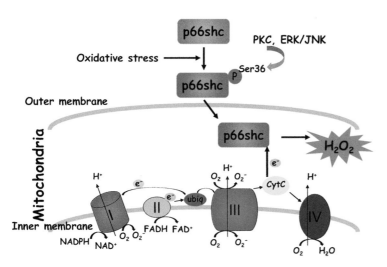

FIGURE 2.5 **Mitochondrial ROS production by pSer36-p66shc.** p66shc is phosphorylated at Serine (Ser) 36 residue by PKC or ERK/JNK pathways under oxidative stress. The phosphorylated p66shc leads to mtROS generation by catalyzing electron transfer to O_2 through the oxidation of cytochrome c.

2.7 Role of sulfenylation (cysteine-hydroxyl radical formation) in angiogenesis

Signaling function of ROS, especially H_2O_2, is also mediated through reversible oxidation of cysteine (Cys) residues of target proteins, called "redox signaling" [70−74]. Cys thiol groups (R-SH) are oxidized to sulfenic acid (R-SOH) by H_2O_2, followed by interaction with other protein thiol groups to create disulfide bonds or further modify to S-glutathionylation [71]. Therefore, cysteine sulfenic acid formation (*Cys-OH*, **sulfenylation**) is a key initial ROS-mediated posttranslational protein modification involved in redox signaling [75] (Fig. 2.6).

Evidence reveal that many proteins are sulfenylated by ROS in mammalian cells [76] and the Cys residues in over 200 human kinase domains are targets for ROS [74]. Recently, we demonstrated that redox regulatory protein, protein disulfide isomerase A1 depletion in ECs induces sulfenylation of Drp1 at Cys644, thereby promoting mitochondrial fission and ROS elevation, which drives EC senescence [77]. It appreciates that many kinases and proteins related to redox signaling are regulated physiologically by sulfenylation. Furthermore, we previously demonstrated that overexpression of extracellular SOD (ecSOD) that dismutases O_2^- to generate H_2O_2 promotes VEGFR2 signaling and angiogenic responses in ECs via sulfenylation-mediated inactivation of protein tyrosine phosphatases (protein tyrosine phosphatase 1B (PTP1B) and density-enhanced protein phosphatase1 (DEP1)) in caveolae/lipid rafts [78]. This represents a compartmentalized VEGFR2 signaling mediated through ecSOD-derived H_2O_2 generated at the specialized plasma membrane signaling domains which facilitates outside in redox signaling promoting angiogenesis.

2.8 Reactive oxygen species-induced reactive oxygen species release via cross-talk between nicotinamide adenine dinucleotide phosphate oxidase and mitochondrial reactive oxygen species involved in angiogenesis

ROS are involved in VEGF-induced angiogenesis in ECs and postischemic revascularization [5,79−81]. However, fundamental questions remain how highly diffusible H_2O_2 derived from different ROS sources such as Noxs and mitochondria can efficiently promote VEGF signaling and angiogenesis [5]. ROS produced by activation of Nox2 and Nox4 activate intracellular signaling pathways that are essential for physiological functions of cardiovascular system including ECs [5,13]. Among several intracellular and extracellular stimuli to mediate NOX-derived ROS generation, VEGF is a major growth factor through binding to VEGF receptor type 2 (VEGFR2) to activate both Nox2 and Nox4. This NOX-derived ROS production is required for VEGF-induced angiogenic responses in ECs [13,82].

To address the temporal and spatial relationship for VEGF-induced cytosolic and mtROS production in ECs, we used cytosol- and mitochondria-targeted redox sensitive green fluorescent proteins biosensor [83] and other redox-sensitive probes [5]. We demonstrated that Nox2 senses Nox4-derived H_2O_2 signal to promote mtROS production, suggesting a critical communication between different Nox enzymes [5,13] driving mitochondrial ROS required for angiogenic responses in ECs. Mechanistically, we found that Nox4-derived H_2O_2 activates Nox2 to promote mtROS production via phosphorylation of p66Shc at Ser36, thereby enhancing VEGFR2 activation, EC migration, and proliferation [5]. This may represent a novel RIRR mechanism orchestrated by the Nox4/Nox2/p-p66Shc/mtROS axis, which drives switching from a quiescent to an angiogenic phenotype [5] (Fig. 2.7). Taken together, the Nox4/Nox2/p-p66Shc/mtROS axis is important mechanism by which VEGF-induced ROS can rapidly and specifically active VEGFR2-mediated redox signaling linked to angiogenesis. In addition, angiopoietin-1 (Ang-1) also triggers NOX2, NOX4, and the mitochondria to release ROS and that ROS derived from these sources play distinct roles in the regulation of the Ang-1/Tie 2 signaling pathway and pro-angiogenic responses [84].

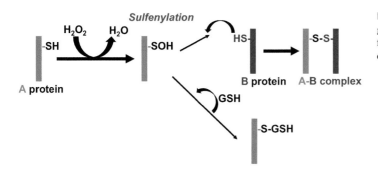

FIGURE 2.6 **Cysteine sulfenic acid formation.** Cys thiol groups (R-SH) are oxidized to sulfenic acid (R-SOH) by H_2O_2, followed by interaction with other protein thiol groups to create disulfide bonds or further modify to S-glutathionylation.

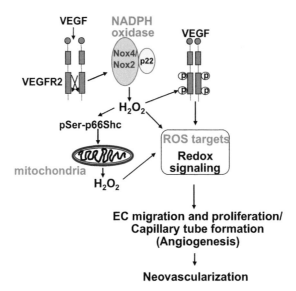

FIGURE 2.7 ROS-induced ROS release orchestrated by Nox4, Nox2 and mitochondria in VEGF signaling and angiogenesis in ECs and postischemic neovascularization.

2.9 Summary, conclusions, and future perspectives

In summary, ROS play an important role in physiological and pathological processes. RIRR representing the functional interaction between NOX- and mitochondria-derived ROS, a feed-forward regulation of the ROS sources, plays crucial roles in determining the outcomes (beneficial vs harmful) in the vascular and other systems [85]. We have addressed two opposing perspectives regarding ROS and their roles in vascular signaling pathways and angiogenesis. It appreciates that ROS at physiological and appropriate level are required to promote physiological angiogenesis and homeostatic maintenance of healthy vasculature. By contrast, ROS at excess level by uncontrolled continuous production or impairment of antioxidant system will ultimately contribute to pathological angiogenesis and tissue damage. Thus a linear query of whether ROS harmful or beneficial is not sufficient to gauge the complex role of ROS in biology [13]. In addition, signals driving angiogenesis etc., are complicated as multiple components of angiogenesis are interlinked with ROS which appears to be carrying out contrasting functions during angiogenic processes. Multifaced pathways and ROS targets such as sulfenylated, S-glutathionylated proteins, or other Cys oxidized proteins involved in ROS-induced physiological or pathological angiogenesis are still poorly understood. Therefore, a systemic approach may be needed whereby there is a focus on temporal and spatial modulation of ROS in vascular angiogenesis. Lack of such information could account for the failure of ROS generating/blocking agents in animal models and clinical trials. Thus, developing the new therapeutic strategy specifically targeting the RIRR mechanisms is important for the treatment of oxidative stress-dependent various vascular and other diseases.

References

[1] Zorov DB, Filburn CR, Klotz LO, Zweier JL, Sollott SJ. Reactive oxygen species (ROS)-induced ROS release: a new phenomenon accompanying induction of the mitochondrial permeability transition in cardiac myocytes. J Exp Med 2000;192:1001−14.

[2] Li WG, Miller Jr. FJ, Zhang HJ, Spitz DR, Oberley LW, Weintraub NL. H_2O_2-induced O_2 production by a non-phagocytic NAD(P)H oxidase causes oxidant injury. J Biol Chem 2001;276:29251−6.

[3] Kuzkaya N, Weissmann N, Harrison DG, Dikalov S. Interactions of peroxynitrite, tetrahydrobiopterin, ascorbic acid, and thiols: implications for uncoupling endothelial nitric-oxide synthase. J Biol Chem 2003;278:22546−54.

[4] Radi R, Cassina A, Hodara R, Quijano C, Castro L. Peroxynitrite reactions and formation in mitochondria. Free Radic Biol Med 2002;33:1451−64.

[5] Kim YM, Kim SJ, Tatsunami R, Yamamura H, Fukai T, Ushio-Fukai M. ROS-induced ROS release orchestrated by Nox4, Nox2, and mitochondria in VEGF signaling and angiogenesis. Am J Physiol Cell Physiol 2017;312:C749−64.

[6] Dikalov S. Cross talk between mitochondria and NADPH oxidases. Free Radic Biol Med 2011;51:1289−301.

[7] Daiber A. Redox signaling (cross-talk) from and to mitochondria involves mitochondrial pores and reactive oxygen species. Biochim Biophys Acta 2010;1797:897−906.

[8] Kumari S, Badana AK, Murali MG, Shailender G, Malla R. Reactive oxygen species: a key constituent in cancer survival. Biomark Insights 2018;13 1177271918755391.

[9] Traverso N, Ricciarelli R, Nitti M, Marengo B, Furfaro AL, Pronzato MA, et al. Role of glutathione in cancer progression and chemoresistance. Oxid Med Cell Longev 2013;2013:972913.

[10] Zinkevich NS, Gutterman DD. ROS-induced ROS release in vascular biology: redox-redox signaling. Am J Physiol Heart Circ Physiol 2011;301:H647−53.

[11] del Rio LA, Sandalio LM, Palma JM, Bueno P, Corpas FJ. Metabolism of oxygen radicals in peroxisomes and cellular implications. Free Radic Biol Med 1992;13:557−80.

[12] Kang LS, Reyes RA, Muller-Delp JM. Aging impairs flow-induced dilation in coronary arterioles: role of NO and H_2O_2. Am J Physiol Heart Circ Physiol 2009;297:H1087−95.

[13] Aldosari S, Awad M, Harrington EO, Sellke FW, Abid MR. Subcellular reactive oxygen species (ROS) in cardiovascular pathophysiology. Antioxidants (Basel), 7. 2018.

[14] Shafique E, Torina A, Reichert K, Colantuono B, Nur N, Zeeshan K, et al. Mitochondrial redox plays a critical role in the paradoxical effects of NAPDH oxidase-derived ROS on coronary endothelium. Cardiovasc Res 2017;113:234−46.

[15] Shafique E, Choy WC, Liu Y, Feng J, Cordeiro B, Lyra A, et al. Oxidative stress improves coronary endothelial function through activation of the pro-survival kinase AMPK. Aging (Albany NY) 2013;5:515−30.

[16] Dikalov S, Griendling KK, Harrison DG. Measurement of reactive oxygen species in cardiovascular studies. Hypertension. 2007;49:717−27.

[17] Serrander L, Cartier L, Bedard K, Banfi B, Lardy B, Plastre O, et al. NOX4 activity is determined by mRNA levels and reveals a unique pattern of ROS generation. Biochem J 2007;406:105−14.

[18] Brandes RP, Weissmann N, Schroder K. Nox family NADPH oxidases: molecular mechanisms of activation. Free Radic Biol Med 2014;76:208−26.

[19] Grasberger H, Refetoff S. Identification of the maturation factor for dual oxidase. Evolution of an eukaryotic operon equivalent. J Biol Chem 2006;281:18269−72.

[20] Abid MR, Tsai JC, Spokes KC, Deshpande SS, Irani K, Aird WC. Vascular endothelial growth factor induces manganese-superoxide dismutase expression in endothelial cells by a Rac1-regulated NADPH oxidase-dependent mechanism. FASEB J 2001;15:2548−50.

[21] Ushio-Fukai M, Tang Y, Fukai T, Dikalov SI, Ma Y, Fujimoto M, et al. Novel role of gp91(phox)-containing NAD(P)H oxidase in vascular endothelial growth factor-induced signaling and angiogenesis. Circ Res 2002;91:1160−7.

[22] Bayraktutan U, Blayney L, Shah AM. Molecular characterization and localization of the NAD(P)H oxidase components gp91-phox and p22-phox in endothelial cells. Arterioscler Thromb Vasc Biol 2000;20:1903−11.

[23] Arbiser JL, Petros J, Klafter R, Govindajaran B, McLaughlin ER, Brown LF, et al. Reactive oxygen generated by Nox1 triggers the angiogenic switch. Proc Natl Acad Sci U S A 2002;99:715−20.

[24] Higashi M, Shimokawa H, Hattori T, Hiroki J, Mukai Y, Morikawa K, et al. Long-term inhibition of Rho-kinase suppresses angiotensin II-induced cardiovascular hypertrophy in rats in vivo: effect on endothelial NAD(P)H oxidase system. Circ Res 2003;93:767−75.

[25] Li JM, Shah AM. ROS generation by nonphagocytic NADPH oxidase: potential relevance in diabetic nephropathy. J Am Soc Nephrol 2003;14: S221−6.

[26] Chen F, Haigh S, Barman S, Fulton DJ. From form to function: the role of Nox4 in the cardiovascular system. Front Physiol 2012;3:412.

[27] Xi G, Shen XC, Wai C, Clemmons DR. Recruitment of Nox4 to a plasma membrane scaffold is required for localized reactive oxygen species generation and sustained Src activation in response to insulin-like growth factor-I. J Biol Chem 2013;288:15641−53.

[28] Aon MA, Cortassa S, O'Rourke B. The fundamental organization of cardiac mitochondria as a network of coupled oscillators. Biophys J 2006;91:4317−27.

[29] Go YM, Jones DP. Redox compartmentalization in eukaryotic cells. Biochim Biophys Acta 2008;1780:1273−90.

[30] Oshikawa J, Kim SJ, Furuta E, Caliceti C, Chen GF, McKinney RD, et al. Novel role of p66Shc in ROS-dependent VEGF signaling and angiogenesis in endothelial cells. Am J Physiol Heart Circ Physiol 2012;302:H724−32.

[31] St-Pierre J, Buckingham JA, Roebuck SJ, Brand MD. Topology of superoxide production from different sites in the mitochondrial electron transport chain. J Biol Chem 2002;277:44784−90.

[32] Chandel NS, McClintock DS, Feliciano CE, Wood TM, Melendez JA, Rodriguez AM, et al. Reactive oxygen species generated at mitochondrial complex III stabilize hypoxia-inducible factor-1alpha during hypoxia: a mechanism of O_2 sensing. J Biol Chem 2000;275:25130−8.

[33] Cai J, Jones DP. Superoxide in apoptosis. Mitochondrial generation triggered by cytochrome closs. J Biol Chem 1998;273:11401−4.

[34] Guzy RD, Hoyos B, Robin E, Chen H, Liu L, Mansfield KD, et al. Mitochondrial complex III is required for hypoxia-induced ROS production and cellular oxygen sensing. Cell Metab 2005;1:401−8.

[35] Zorov DB, Juhaszova M, Sollott SJ. Mitochondrial ROS-induced ROS release: an update and review. Biochim Biophys Acta 2006;1757:509−17.

[36] Chen K, Thomas SR, Albano A, Murphy MP, Keaney Jr. JF. Mitochondrial function is required for hydrogen peroxide-induced growth factor receptor transactivation and downstream signaling. J Biol Chem 2004;279:35079−86.

[37] Wang Y, Zang QS, Liu Z, Wu Q, Maass D, Dulan G, et al. Regulation of VEGF-induced endothelial cell migration by mitochondrial reactive oxygen species. Am J Physiol Cell Physiol 2011;301:C695−704.

[38] Chung YM, Bae YS, Lee SY. Molecular ordering of ROS production, mitochondrial changes, and caspase activation during sodium salicylate-induced apoptosis. Free Radic Biol Med 2003;34:434−42.

[39] Brady NR, Hamacher-Brady A, Westerhoff HV, Gottlieb RA. A wave of reactive oxygen species (ROS)-induced ROS release in a sea of excitable mitochondria. Antioxid Redox Signal 2006;8:1651−65.

[40] Leach JK, Black SM, Schmidt-Ullrich RK, Mikkelsen RB. Activation of constitutive nitric-oxide synthase activity is an early signaling event induced by ionizing radiation. J Biol Chem 2002;277:15400−6.

[41] Aon MA, Cortassa S, O'Rourke B. Mitochondrial oscillations in physiology and pathophysiology. Adv Exp Med Biol 2008;641:98−117.

[42] Luo Z, Teerlink T, Griendling K, Aslam S, Welch WJ, Wilcox CS. Angiotensin II and NADPH oxidase increase ADMA in vascular smooth muscle cells. Hypertension. 2010;56:498−504.

[43] Lassegue B, Griendling KK. NADPH oxidases: functions and pathologies in the vasculature. Arterioscler Thromb Vasc Biol 2010;30:653−61.

[44] Zinkevich NS, Fancher IS, Gutterman DD, Phillips SA. Roles of NADPH oxidase and mitochondria in flow-induced vasodilation of human adipose arterioles: ROS-induced ROS release in coronary artery disease. Microcirculation. 2017;24.

[45] Kroller-Schon S, Steven S, Kossmann S, Scholz A, Daub S, Oelze M, et al. Molecular mechanisms of the crosstalk between mitochondria and NADPH oxidase through reactive oxygen species-studies in white blood cells and in animal models. Antioxid Redox Signal 2014;20:247−66.

[46] Wosniak Jr. J, Santos CX, Kowaltowski AJ, Laurindo FR. Cross-talk between mitochondria and NADPH oxidase: effects of mild mitochondrial dysfunction on angiotensin II-mediated increase in Nox isoform expression and activity in vascular smooth muscle cells. Antioxid Redox Signal 2009;11:1265−78.

[47] Tsai IC, Pan ZC, Cheng HP, Liu CH, Lin BT, Jiang MJ. Reactive oxygen species derived from NADPH oxidase 1 and mitochondria mediate angiotensin II-induced smooth muscle cell senescence. J Mol Cell Cardiol 2016;98:18−27.

[48] Salazar G. NADPH oxidases and mitochondria in vascular senescence. Int J Mol Sci 2018;19:1327.

[49] Antunes F, Brito PM. Quantitative biology of hydrogen peroxide signaling. Redox Biol 2017;13:1−7.

[50] Ghezzi P, Jaquet V, Marcucci F, Schmidt H. The oxidative stress theory of disease: levels of evidence and epistemological aspects. Br J Pharmacol 2017;174:1784−96.

[51] Ma Q. Role of nrf2 in oxidative stress and toxicity. Annu Rev Pharmacol Toxicol 2013;53:401−26.

[52] Nezu M, Souma T, Yu L, Sekine H, Takahashi N, Wei AZ, et al. Nrf2 inactivation enhances placental angiogenesis in a preeclampsia mouse model and improves maternal and fetal outcomes. Sci Signal 2017;10:eaam5711.

[53] Wong W. New connections: the duality of ROS in angiogenesis. Sci Signal 2017;10:eaan64381.

[54] Agarwal A, Gupta S, Sharma RK. Role of oxidative stress in female reproduction. Reprod Biol Endocrinol 2005;3:28.

[55] Schreml S, Szeimies RM, Prantl L, Karrer S, Landthaler M, Babilas P. Oxygen in acute and chronic wound healing. Br J Dermatol 2010;163:257−68.

[56] Knighton DR, Hunt TK, Scheuenstuhl H, Halliday BJ, Werb Z, Banda MJ. Oxygen tension regulates the expression of angiogenesis factor by macrophages. Science. 1983;221:1283−5.

[57] Kim YW, Byzova TV. Oxidative stress in angiogenesis and vascular disease. Blood. 2014;123:625−31.

[58] Chung AS, Ferrara N. Developmental and pathological angiogenesis. Annu Rev Cell Dev Biol 2011;27:563−84.

[59] Cui Y, Xu X, Bi H, Zhu Q, Wu J, Xia X, et al. Expression modification of uncoupling proteins and MnSOD in retinal endothelial cells and pericytes induced by high glucose: the role of reactive oxygen species in diabetic retinopathy. Exp Eye Res 2006;83:807−16.

[60] Ushio-Fukai M, Alexander RW. Reactive oxygen species as mediators of angiogenesis signaling: role of NAD(P)H oxidase. Mol Cell Biochem 2004;264:85−97.

[61] Bochkov VN, Philippova M, Oskolkova O, Kadl A, Furnkranz A, Karabeg E, et al. Oxidized phospholipids stimulate angiogenesis via autocrine mechanisms, implicating a novel role for lipid oxidation in the evolution of atherosclerotic lesions. Circ Res 2006;99:900−8.

[62] Podrez EA, Poliakov E, Shen Z, Zhang R, Deng Y, Sun M, et al. A novel family of atherogenic oxidized phospholipids promotes macrophage foam cell formation via the scavenger receptor CD36 and is enriched in atherosclerotic lesions. J Biol Chem 2002;277:38517−23.

[63] Li J, Wang JJ, Yu Q, Chen K, Mahadev K, Zhang SX. Inhibition of reactive oxygen species by Lovastatin downregulates vascular endothelial growth factor expression and ameliorates blood-retinal barrier breakdown in db/db mice: role of NADPH oxidase 4. Diabetes. 2010;59:1528−38.

[64] Craige SM, Kant S, Keaney Jr. JF. Reactive oxygen species in endothelial function − from disease to adaptation. Circ J 2015;79:1145−55.

[65] Evangelista AM, Thompson MD, Bolotina VM, Tong X, Cohen RA. Nox4- and Nox2-dependent oxidant production is required for VEGF-induced SERCA cysteine-674 S-glutathiolation and endothelial cell migration. Free Radic Biol Med 2012;53:2327−34.

[66] Urao N, Sudhahar V, Kim SJ, Chen GF, McKinney RD, Kojda G, et al. Critical role of endothelial hydrogen peroxide in post-ischemic neovascularization. PLoS One 2013;8:e57618.

[67] Giorgio M, Migliaccio E, Orsini F, Paolucci D, Moroni M, Contursi C, et al. Electron transfer between cytochrome c and p66Shc generates reactive oxygen species that trigger mitochondrial apoptosis. Cell. 2005;122:221−33.

[68] Pinton P, Rimessi A, Marchi S, Orsini F, Migliaccio E, Giorgio M, et al. Protein kinase C beta and prolyl isomerase 1 regulate mitochondrial effects of the life-span determinant p66Shc. Science. 2007;315:659−63.

[69] Bosutti A, Grassi G, Zanetti M, Aleksova A, Zecchin M, Sinagra G, et al. Relation between the plasma levels of LDL-cholesterol and the expression of the early marker of inflammation long pentraxin PTX3 and the stress response gene p66ShcA in pacemaker-implanted patients. Clin Exp Med 2007;7:16−23.

[70] Winterbourn CC, Hampton MB. Redox biology: signaling via a peroxiredoxin sensor. Nat Chem Biol 2015;11:5−6.

[71] Cremers CM, Jakob U. Oxidant sensing by reversible disulfide bond formation. J Biol Chem 2013;288:26489−96.

[72] Svoboda LK, Reddie KG, Zhang L, Vesely ED, Williams ES, Schumacher SM, et al. Redox-sensitive sulfenic acid modification regulates surface expression of the cardiovascular voltage-gated potassium channel Kv1.5. Circ Res 2012;111:842−53.

[73] Corcoran A, Cotter TG. Redox regulation of protein kinases. FEBS J 2013;280:1944−65.

[74] Hourihan JM, Moronetti Mazzeo LE, Fernandez-Cardenas LP, Blackwell TK. Cysteine sulfenylation directs IRE-1 to activate the SKN-1/Nrf2 antioxidant response. Mol Cell 2016;63:553−66.

[75] Shao D, Oka S, Liu T, Zhai P, Ago T, Sciarretta S, et al. A redox-dependent mechanism for regulation of AMPK activation by Thioredoxin1 during energy starvation. Cell Metab 2014;19:232−45.

[76] Yang J, Gupta V, Carroll KS, Liebler DC. Site-specific mapping and quantification of protein S-sulphenylation in cells. Nat Commun 2014;5:4776.

[77] Kim YM, Youn SW, Sudhahar V, Das A, Chandhri R, Cuervo Grajal H, et al. Redox regulation of mitochondrial fission protein Drp1 by protein disulfide isomerase limits endothelial senescence. Cell Rep 2018;23:3565−78.

[78] Oshikawa J, Urao N, Kim HW, Kaplan N, Razvi M, McKinney R, et al. Extracellular SOD-derived H_2O_2 promotes VEGF signaling in caveolae/lipid rafts and post-ischemic angiogenesis in mice. PLoS One 2010;5:e10189.

[79] Ushio-Fukai M. Redox signaling in angiogenesis: role of NADPH oxidase. Cardiovasc Res 2006;71:226−35.

[80] Ushio-Fukai M. VEGF signaling through NADPH oxidase-derived ROS. Antioxid Redox Signal 2007;9:731−9.

[81] Wright GL, Maroulakou IG, Eldridge J, Liby TL, Sridharan V, Tsichlis PN, et al. VEGF stimulation of mitochondrial biogenesis: requirement of AKT3 kinase. FASEB J 2008;22:3264−75.

[82] Abid MR, Kachra Z, Spokes KC, Aird WC. NADPH oxidase activity is required for endothelial cell proliferation and migration. FEBS Lett 2000;486:252−6.

[83] Waypa GB, Marks JD, Guzy R, Mungai PT, Schriewer J, Dokic D, et al. Hypoxia triggers subcellular compartment redox signaling in vascular smooth muscle cells. Circ Res 2010;106:526−35.

[84] Harel S, Mayaki D, Sanchez V, Hussain SNA. NOX2, NOX4, and mitochondrial-derived reactive oxygen species contribute to angiopoietin-1 signaling and angiogenic responses in endothelial cells. Vasc Pharmacol 2017;92:22−32.

[85] Abid MR, Sellke FW. Antioxidant therapy: is it your gateway to improved cardiovascular health? Pharm Anal Acta 2015;6:323.

Chapter 3

Nitric oxide as a vascular modulator to resistance training

André Sales Barreto[1,2], Fabricio Nunes Macedo[1,3], Milene Tavares Fontes[1] and Valter Joviniano Santana-Filho[1]

[1]Department of Physiology, Federal University of Sergipe, São Cristóvão, Brazil, [2]Department of Health Education, Federal University of Sergipe, Lagarto, Brazil, [3]Department of Physical Education, Estácio University of Sergipe, Aracaju, Brazil

3.1 Effects of resistance training on endothelial nitric oxide synthase activity and bioavailability of nitric oxide

The cardiovascular system provides enough blood flow to maintain metabolic needs of organs and tissues. To control the blood flow, vascular beds regulate the vascular resistance, by altering vessel diameter and consequently the blood supply. It is well established that during exercise there is an increase in the muscle requirement for oxygen and other nutrients leading to vasodilatation in the exercised area [1].

Vasodilatory response seen after either aerobic or resistance training is generated by the increase of nitric oxide (NO) bioavailability triggered by the exercise [2,3]. Mechanisms involved in this response encompass a variety of molecular and cellular adjustments including increased endothelium-dependent vasodilation; increase in the expression and/or activity of the enzyme endothelial nitric oxide synthase (eNOS); and appropriate concentrations of cofactors and antioxidant enzymes [4].

Studies in animals demonstrated that resistance exercise also promotes acute vasodilatation that is directly associated with an increase in NO release by the endothelium [5], showing that, even after a single session, resistance exercise is able to promote an increase in NO bioavailability [6] which directly correlated with muscle mass involved on exercise.

Besides the size of trained area, intensity of resistance training also plays an important role in its effects on vascular tone and NO bioavailability. Studies suggest that resistance exercise improves arterial relaxation, in an intensity-dependent way; however, high-intensity resistance exercise, when performed as training, has been associated with deleterious responses on the vascular function promoting structural remodeling [7].

At endothelial level, the production of NO is mainly regulated by the presence of eNOS. This enzyme can be activated by either mechanical or neurohumoral stimulation [8,9]. In the presence of endothelium, a mechanical stimulus on the vessel wall activates the enzyme eNOS that leads to an increase in endothelial NO bioavailability and consequent vasodilatation [8,10]. Due to an increment in cardiac output induced by resistant training, there is an increase in blood flow and cyclic strain that corresponds to an increase in shear stress and transmural pressure, respectively, triggering augmentation in the activity of eNOS [11−13].

3.2 Nitric oxide production during resistance training in nonexercised vascular beds

Despite the action on exercised sites, the effect of resistance exercise can also be seen in nonexercised areas [5,11−16]. NO produced with exercise can freely diffuse and act in other areas. With training that characterizes the summation of acute effects of exercise promoting chronic adaptations, there is an effect of NO on neural circuitry that causes a reduction in sympathetic activity either by action of NO on central areas that controls sympathetic activity or by inhibiting noradrenergic release from postganglionic nerve endings [11].

Resistance training also exerts an effect that improves insulin sensibility in normoglycemic [14] as well as hyperglycemic conditions [15]. Reduction in vascular insulin sensitivity is known as contributor to endothelial dysfunction that

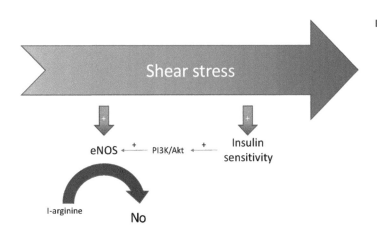

FIGURE 3.1 General mechanism of NO production.

stimulates vasoconstrictor pathways [16,17]. Resistance training is able to improve vascular insulin sensitivity contributing to the improvement of endothelial vasodilatory response of NO during training in nonexercised area. This effect is associated by augment of NO bioavailability generated by either (1) an increase in the activity of phosphatidylinositol 3-kinase (PI3K)/eNOS pathway, promoting an increase in Akt phosphorylation and activity, which consequently promotes an increase in the phosphorylation of serine residues of the eNOS or by (2) an increase of K + -channel activation and Na + /K + -ATPase activity leading to hyperpolarization of smooth muscle cells [14,18] (Fig. 3.1).

3.3 Influence of intensity and volume of resistance training on nitric oxide production

Despite several studies demonstrating the benefits of resistance training [19,20], it is clinically relevant to select and evaluate the correct intensity and volume of resistance exercises, because of deleterious effect on vascular function related to high-intensity training [2].

The main contributor that stimulates NO production during resistance exercises is the shear stress promoted by muscle contractions and redistribution of blood flow to active muscles [21]. In this regard, it is plausible to deduct that higher intensities of exercises promote higher levels of plasmatic NO [22]. While the rational for this deduction is valid [7] long-term high-intensity exercise promotes arterial stiffness and consequent reduction on NO production [23,24].

Thus acutely, high-intensity resistance exercise appears to play an effective role in increasing vascular NO production and bioavailability; however, chronically, this intense effort shows to be deleterious to vascular function. In contrast, low-intensity resistance exercise has a weak shear stress stimulus; however, in long term its cumulative effect also increases eNOS activity and NO bioavailability without any deleterious effect to vascular function [25].

Another exercise parameter that should be controlled to obtain better results on vascular function is volume. Studies have also demonstrated that high blood volume leads to high NO bioavailability. This effect is mediated by metabolites produced during exercise that signalizes the necessity to increase the blood supply to exercised areas [26].

To minimize the deleterious effect of either higher intensities or volume and maximize the effect of resistance training, a strategy that reconciliates low-intensity training with blood flow restriction [27–31] performed at lower volumes, has been suggested. Towards this effort, a cuff pressure is employed that occludes venous return and promotes a turbulent arterial blood flow, resulting in an increase of metabolic stress and activation of fast-twitch muscle fibers [32]. In addition, ischemic reperfusion stimulates shear stress, increasing eNOS activation and NO bioavailability [33,34].

In addition, low-intensity resistance training with blood flow restriction augments vascular endothelial growth factor [35], due to the hypoxic stimulus promoted by the cuff restricting blood flow [36], that also leads to raises in NO bioavailability [37,38]. Thus low-intensity resistance training with blood flow restriction is becoming a new and important strategy to promote muscle hypertrophy, prevent sarcopenia, without promoting the deleterious effects on vascular endothelial function that are observed in traditional resistance training. However, as a new modality of training, additional studies are needed to better understand the effects of resistance training with blood flow restriction.

3.4   Detraining and vascular function

It is important to point out that cessation of resistance training can reverse the functional and morphological benefits achieved by training, a phenomenon known as detraining [39]. It is known that 2 weeks of detraining is enough to significantly decrease the benefits achieved by training [40]. Several factors can influence the velocity of detraining as age, ethnicity, physical fitness, training time and type of exercise [41−43].

Regarding the vascular function, 1 month to 6 weeks of detraining is enough to completely reverse the positive effects on endothelial relaxation promoted by training [44,45]. Cessation or reduction of muscle contraction with detraining decreases pulsatile stretch and shear stress on the arterial wall shifting the endothelium-dependent vasodilatation to the baseline levels [43,46,47] causing a reduction in NO bioavailability. Detraining also decreases NO production generated by metabolic stress and intermittent hypoxia. In addition to reduction in NO bioavailability, there is increase in concentration of vasoconstrictors substances such as endothelin in detraining [48].

Besides the direct effect on vasculature, detraining also acts through indirect pathways mitigating the beneficial actions promoted by training, as seen in the reduction of insulin sensitivity promoted by detraining that stimulates the mitogen-activated protein kinase (MAPK) pathway leading to an increase in concentration of Endothelin 1 (ET-1) contributing to vasoconstriction [49−51].

3.5   Vascular function in pathologic situations

Resistance training promotes benefits to vascular function not only in normal subjects. Its effect seems to be even more pronounced in pathologic conditions such as hypertension, diabetes, and atherosclerosis. Like in healthy individuals, the effect of resistance training in the diseased is related to the increase of eNOS activity generating an increase of NO production or reduction in its degradation.

It is known that hypertension promotes a decrease in eNOS bioavailability due to an increase in oxidative stress especially anion superoxide (O_2^-). This redox unbalance inactivates the enzyme eNOS leading to vasoconstriction [52,53]. On the other hand, when resistant training is performed in hypertensive patients there is an increase in concentration of intracellular calcium that activates the eNOS through the involvement of calcium/calmodulin pathway counteracting the detrimental effect of hypertension [54−57]. Besides, an antioxidant effect of resistance training increasing the activity of superoxide dismutase, catalase, and glutathione peroxidase [53,58−60] cannot be ruled out.

In diabetic patients, vasomotor response is occasioned by insulin resistance that leads to a reduction in eNOS bioavailability. This reduction occurs via the increase in endothelin production that promotes vasoconstriction and increase in oxidative stress [61]. Resistance training augments endothelin receptor sensibility to insulin and enhances eNOS activity through the activation of PI3K/Akt pathway [15,50] and thus augments production of NO.

With regard to atherosclerosis, endothelial dysfunction is related to pro-inflammatory response of the vascular wall. This inflammatory response inactivates eNOS reducing the NO bioavailability. As resistance training reduces pro-inflammatory cytokines and C-reactive protein, it counteracts the deleterious action of these moieties on endothelial function [62−64].

It is important to point out that all the beneficial effects of resistance training in endothelial function in pathological conditions are based on exercises performed in low to moderate intensities and volume. Resistance training protocols with higher intensities have been associated with deleterious effects on vascular functions.

Therefore resistance training emerges as an effective therapeutic modality for mitigating endothelial dysfunction. However, it is important to select the appropriate intensity and volume of exercise. This is not only because inappropriate exercise will be ineffective but also because it can be pathogenic. The knowledge of mechanisms that explain the beneficial effects of resistant training will provide the necessary comprehension to professionals to apply parameters that will extract the maximum benefits with the minimum risk.

3.6   Vascular aging and resistance training

Aging is directly related to decline in several biological processes [65]. Regarding vascular function, aging is associated with deleterious changes in structure and mechanical properties of the vascular wall. This involves reduction in the amount of elastin relative to collagen in artery walls, increase in arterial diameter and wall thickening, with a concomitant increase in the risk of cardiovascular diseases [66].

The aging effect on vascular function has been related to malfunction in regulators of vascular tone promoting an increase in endothelin-1 signaling [67,68] and impairment in endothelial-dependent dilatation [69,70,71].

An important approach to minimize the deleterious effect of aging in vascular function is exercise training, especially resistance training [72,73,74]. Even a single bout of resistance training in elderly patients promotes an increase in blood flow in a volume-dependent way [75] showing an improvement of vascular function that counteracts the deleterious effect produced by aging [76−78]. The mechanisms involved in this beneficial response is mediated by cyclooxygenase products, Ca^{2+} pathways, increased levels of endothelium-derived hyperpolarizing factor, and increased NOS activation [11,79].

3.7 Future perspectives

Historically, resistance training has been associated with increase in bone mineral density, skeletal muscle hypertrophy; however its functional capacity as being "endothelial protective" has received scant attention thus far as the focus has been more on aerobic exercises etc. Indeed, resistance training has not been perceived as a means of improving vascular function because it has been incorrectly associated with impairment of cardiovascular health [80−82]. Such impression arises from studies that applied higher intensities aiming higher performance in athlete populations (i.e., bodybuilders, powerlifters, or Olympic weightlifters) [83].

More recently resistance exercise has emerged as an important approach to improve cardiovascular performance when well prescribed especially with adequate control of its intensity and using large muscle groups with an adequate volume of training [84−89]. It is important to mention that this modality, besides the benefits, is safe and the incidence of cardiovascular events during practice is very low [89,90].

As a result of its greater benefic impact, multiple international guidelines have indicated the addition of resistance training for the most diverse cardiovascular pathological conditions and age groups with both preventive and therapeutic purposes [91−95] aiming short- and long-term benefits to vascular health.

Understanding the mechanisms involved in its action in the vasculature helps to improve the efficacy of training and mitigates the risks involved. Besides, the effective control of this modality of training will allow its safe prescription for special populations such as children, adolescents, and patients with advanced heart disease.

References

[1] Ashor AW, Lara J, Siervo M, Celis-Morales C, Oggioni C, Jakovljevic DG, et al. Exercise modalities and endothelial function: a systematic review and dose-response meta-analysis of randomized controlled trials. Sports Med Auckl NZ 2015;45(2):279−96.

[2] Miyachi M. Effects of resistance training on arterial stiffness: a meta-analysis. Br J Sports Med 2013;47(6):393−6.

[3] Hasegawa N, Fujie S, Horii N, Miyamoto-Mikami E, Tsuji K, Uchida M, et al. Effects of different exercise modes on arterial stiffness and nitric oxide synthesis. Med Sci Sports Exerc 2018;50(6):1177−85.

[4] Newsholme P, Homem De Bittencourt PI, O' Hagan C, De Vito G, Murphy C, Krause MS. Exercise and possible molecular mechanisms of protection from vascular disease and diabetes: the central role of ROS and nitric oxide. Clin Sci 2010;118(5):341−9.

[5] de Oliveira Faria T, Targueta GP, Angeli JK, Almeida EAS, Stefanon I, Vassallo DV, et al. Acute resistance exercise reduces blood pressure and vascular reactivity, and increases endothelium-dependent relaxation in spontaneously hypertensive rats. Eur J Appl Physiol 2010;110(2):359−66.

[6] Mikus CR, Roseguini BT, Uptergrove GM, Matthew Morris E, Scott Rector R, Libla JL, et al. Voluntary wheel running selectively augments insulin-stimulated vasodilation in arterioles from white skeletal muscle of insulin-resistant rats. Microcirculation. 2012;19(8):729−38.

[7] Mota MM, Mesquita TRR, Braga da Silva TLT, Fontes MT, Lauton Santos S, dos Santos Aggum Capettini L, et al. Endothelium adjustments to acute resistance exercise are intensity-dependent in healthy animals. Life Sci 2015;142:86−91.

[8] Gielen S, Schuler G, Adams V. Cardiovascular effects of exercise training: molecular mechanisms. Circulation. 2010;122(12):1221−38.

[9] Quillon A, Fromy B, Debret R. Endothelium microenvironment sensing leading to nitric oxide mediated vasodilation: a review of nervous and biomechanical signals. Nitric Oxide Biol Chem 2015;45:20−6.

[10] Li Y-SJ, Haga JH, Chien S. Molecular basis of the effects of shear stress on vascular endothelial cells. J Biomech 2005;38(10):1949−71.

[11] Harris MB, Slack KN, Prestosa DT, Hryvniak DJ. Resistance training improves femoral artery endothelial dysfunction in aged rats. Eur J Appl Physiol 2010;108(3):533−40.

[12] Olson TP, Dengel DR, Leon AS, Schmitz KH. Moderate resistance training and vascular health in overweight women. Med Sci Sports Exerc 2006;38(9):1558−64.

[13] Collier SR, Diggle MD, Heffernan KS, Kelly EE, Tobin MM, Fernhall B. Changes in arterial distensibility and flow-mediated dilation after acute resistance vs. aerobic exercise. J Strength Cond Res 2010;24(10):2846−52.

[14] Fontes MT, Silva TLBT, Mota MM, Barreto AS, Rossoni LV, Santos MRV. Resistance exercise acutely enhances mesenteric artery insulin-induced relaxation in healthy rats. Life Sci 2014;94(1):24−9.

[15] Mota MM, da Silva TLTB, Fontes MT, Barreto AS, Araújo JE, dos S, et al. Resistance exercise restores endothelial function and reduces blood pressure in type 1 diabetic rats. Arq Bras Cardiol 2014;103(1):25−32.

[16] Arce-Esquivel AA, Bunker AK, Mikus CR, Laughlin MH. Insulin resistance and endothelial dysfunction: macro and microangiopathy. Type 2 Diabetes [Internet] 2013; [cited 2018 Mar 16]. < http://www.intechopen.com/books/type-2-diabetes/insulin-resistance-and-endothelial-dysfunction-macro-and-microangiopathy > .

[17] Manrique C, Lastra G, Sowers JR. New insights into insulin action and resistance in the vasculature. Ann N Y Acad Sci 2014;1311(1):138−50.

[18] Tamaki T, Uchiyama S, Nakano S. A weight-lifting exercise model for inducing hypertrophy in the hindlimb muscles of rats. Med Sci Sports Exerc 1992;24(8):881−6.

[19] Winett RA, Carpinelli RN. Potential health-related benefits of resistance training. Prev Med 2001;33(5):503−13.

[20] Braith RW, Beck DT. Resistance exercise: training adaptations and developing a safe exercise prescription. Heart Fail Rev 2008;13(1):69−79.

[21] Van Citters Robert L, Franklin Dean L. Cardiovascular performance of Alaska Sled dogs during exercise. Circ Res 1969;24(1):33−42.

[22] Güzel NA, Hazar S, Erbas D. Effects of different resistance exercise protocols on nitric oxide, lipid peroxidation and creatine kinase activity in sedentary males. J Sports Sci Med 2007;6(4):417−22.

[23] Miyachi M. Unfavorable effects of resistance training on central arterial compliance: a randomized intervention study. Circulation. 2004;110 (18):2858−63.

[24] Cortez-Cooper MY, DeVan AE, Anton MM, Farrar RP, Beckwith KA, Todd JS, et al. Effects of high intensity resistance training on arterial stiffness and wave reflection in women. Am J Hypertens 2005;18(7):930−4.

[25] Esquivel AAA, Welsch MA. High and low volume resistance training and vascular function. Int J Sports Med 2007;28(3):217−21.

[26] Mota MM, Silva TLTB da, Macedo FN, Mesquita TRR, Quintans Júnior LJ, Santana-Filho VJ de, et al. Effects of a single bout of resistance exercise in different volumes on endothelium adaptations in healthy animals. Arq Bras Cardiol [Internet] 2017; [cited 2017 May 28]. < http://www.gnresearch.org/doi/10.5935/abc.20170060 > .

[27] Takarada Y, Tsuruta T, Ishii N. Cooperative effects of exercise and occlusive stimuli on muscular function in low-intensity resistance exercise with moderate vascular occlusion. Jpn J Physiol 2004;54(6):585−92.

[28] Madarame H, Neya M, Ochi E, Nakazato K, Sato Y, Ishii N. Cross-transfer effects of resistance training with blood flow restriction. Med Sci Sports Exerc 2008;40(2):258−63.

[29] Horiuchi M, Okita K. Blood flow restricted exercise and vascular function. Int J Vasc Med [Internet] 2012; [cited 2019 Apr 30, 2012]. < https://www.ncbi.nlm.nih.gov/pmc/articles/PMC3485988/ > .

[30] Fujita S, Abe T, Drummond MJ, Cadenas JG, Dreyer HC, Sato Y, et al. Blood flow restriction during low-intensity resistance exercise increases S6K1 phosphorylation and muscle protein synthesis. J Appl Physiol 2007;103(3):903−10.

[31] Manini TM, Clark BC. Blood flow restricted exercise and skeletal muscle health. Exerc Sport Sci Rev 2009;37(2):78−85.

[32] Suga T, Okita K, Morita N, Yokota T, Hirabayashi K, Horiuchi M, et al. Intramuscular metabolism during low-intensity resistance exercise with blood flow restriction. J Appl Physiol (1985) 2009;106(4):1119−24.

[33] Patterson SD, Ferguson RA. Increase in calf post-occlusive blood flow and strength following short-term resistance exercise training with blood flow restriction in young women. Eur J Appl Physiol 2010;108(5):1025−33.

[34] Shimizu R, Hotta K, Yamamoto S, Matsumoto T, Kamiya K, Kato M, et al. Low-intensity resistance training with blood flow restriction improves vascular endothelial function and peripheral blood circulation in healthy elderly people. Eur J Appl Physiol 2016;116(4):749−57.

[35] Takano H, Morita T, Iida H, Asada K, Kato M, Uno K, et al. Hemodynamic and hormonal responses to a short-term low-intensity resistance exercise with the reduction of muscle blood flow. Eur J Appl Physiol 2005;95(1):65−73.

[36] Shweiki D, Itin A, Soffer D, Keshet E. Vascular endothelial growth factor induced by hypoxia may mediate hypoxia-initiated angiogenesis. Nature. 1992;359(6398):843−5.

[37] Higashi Y, Yoshizumi M. Exercise and endothelial function: role of endothelium-derived nitric oxide and oxidative stress in healthy subjects and hypertensive patients. Pharmacol Ther 2004;102(1):87−96.

[38] Larkin KA, Macneil RG, Dirain M, Sandesara B, Manini TM, Buford TW. Blood flow restriction enhances post-resistance exercise angiogenic gene expression. Med Sci Sports Exerc 2012;44(11):2077−83.

[39] Kraemer WJ, Ratamess NA. Hormonal responses and adaptations to resistance exercise and training. Sports Med 2005;35(4):339−61.

[40] ACSMs Guidelines for Exercise Testing and Prescription [Internet]. <https://www.acsm.org/read-research/books/acsms-guidelines-for-exercise-testing-and-prescription>; 23 April 2019.

[41] Lemmer JT, Hurlbut DE, Martel GF, Tracy BL, Ivey FM, Metter EJ, et al. Age and gender responses to strength training and detraining. Med Sci Sports Exerc 2000;32(8):1505−12.

[42] Heffernan KS, Jae SY, Vieira VJ, Iwamoto GA, Wilund KR, Woods JA, et al. C-reactive protein and cardiac vagal activity following resistance exercise training in young African-American and white men. Am J Physiol Regul Integr Comp Physiol 2009;296(4):R1098−105.

[43] Alomari MA, Mekary RA, Welsch MA. Rapid vascular modifications to localized rhythmic handgrip training and detraining. Eur J Appl Physiol 2010;109(5):803−9.

[44] Vona M, Codeluppi GM, Iannino T, Ferrari E, Bogousslavsky J, von Segesser LK. Effects of different types of exercise training followed by detraining on endothelium-dependent dilation in patients with recent myocardial infarction. Circulation. 2009;119(12):1601−8.

[45] Hunt JEA, Walton LA, Ferguson RA. Brachial artery modifications to blood flow-restricted handgrip training and detraining. J Appl Physiol (1985) 2012;112(6):956−61.

[46] Okamoto T, Masuhara M, Ikuta K. Effects of low-intensity resistance training with slow lifting and lowering on vascular function. J Hum Hypertens 2008;22(7):509−11.

[47] Tinken TM, Thijssen DHJ, Hopkins N, Dawson EA, Cable NT, Green DJ. Shear stress mediates endothelial adaptations to exercise training in humans. Hypertension 2010;55(2):312−18.

[48]   Maeda S, Miyauchi T, Iemitsu M, Sugawara J, Nagata Y, Goto K. Resistance exercise training reduces plasma endothelin-1 concentration in healthy young humans. J Cardiovasc Pharmacol 2004;44(1):S443−6.

[49]   Kawasaki H, Kuroda S, Mimaki Y. [Vascular effects of insulin]. Nihon Yakurigaku Zasshi Folia Pharmacol Jpn 2000;115(5):287−94.

[50]   Zheng C, Liu Z. Vascular function, insulin action, and exercise: an intricate interplay. Trends Endocrinol Metab TEM 2015;26(6):297−304.

[51]   Andersen JL, Schjerling P, Andersen LL, Dela F. Resistance training and insulin action in humans: effects of de-training. J Physiol 2003;551 (3):1049−58.

[52]   Kojda G. Interactions between NO and reactive oxygen species: pathophysiological importance in atherosclerosis, hypertension, diabetes and heart failure. Cardiovasc Res 1999;43(3):562−71.

[53]   Larsen MK, Matchkov VV. Hypertension and physical exercise: the role of oxidative stress. Medicina (Kaunas) 2016;52(1):19−27.

[54]   Tai YL, Marshall EM, Parks JC, Mayo X, Glasgow A, Kingsley JD. Changes in endothelial function after acute resistance exercise using free weights. J Funct Morphol Kinesiol 2018;3(2):32.

[55]   de Araujo AJS, dos Santos ACV, dos Santos Souza K, Aires MB, Santana-Filho VJ, Fioretto ET, et al. Resistance training controls arterial blood pressure in rats with L-NAME- induced hypertension. Arq Bras Cardiol 2013;100(4):339−46.

[56]   Collier SR, Kanaley JA, Carhart Jr R, Frechette V, Tobin MM, Hall AK, et al. Effect of 4 weeks of aerobic or resistance exercise training on arterial stiffness, blood flow and blood pressure in pre- and stage-1 hypertensives. J Hum Hypertens 2008;22(10):678−86.

[57]   de Oliveira Faria T, Angeli JK, Mello LGM, Pinto GC, Stefanon I, Vassallo DV, et al. A single resistance exercise session improves aortic endothelial function in hypertensive rats. Arq Bras Cardiol 2017;108(3):228−36.

[58]   Dantas FFO, Brasileiro-Santos M, do S, Batista RMF, do Nascimento LS, Castellano LRC, et al. Effect of strength training on oxidative stress and the correlation of the same with forearm vasodilatation and blood pressure of hypertensive elderly women: a randomized clinical trial. PLoS One 2016;11(8):e0161178.

[59]   Johnson P. Antioxidant enzyme expression in health and disease: effects of exercise and hypertension. Comp Biochem Physiol C Toxicol Pharmacol 2002;133(4):493−505.

[60]   Padilla J, Simmons GH, Bender SB, Arce-Esquivel AA, Whyte JJ, Laughlin MH. Vascular effects of exercise: endothelial adaptations beyond active muscle beds. Physiology. 2011;26(3):132−45.

[61]   Caballero AE. Endothelial dysfunction in obesity and insulin resistance: a road to diabetes and heart disease. Obes Res 2003;11(11):1278−89.

[62]   Ribeiro F, Alves AJ, Duarte JA, Oliveira J. Is exercise training an effective therapy targeting endothelial dysfunction and vascular wall inflammation? Int J Cardiol 2010;141(3):214−21.

[63]   Ribeiro AS, Tomeleri CM, Souza MF, Pina FLC, Schoenfeld BJ, Nascimento MA, et al. Effect of resistance training on C-reactive protein, blood glucose and lipid profile in older women with differing levels of RT experience. Age [Internet] 2015;37(6) [cited 2019 Apr 29]. < https://www.ncbi.nlm.nih.gov/pmc/articles/PMC5005848/ > .

[64]   Calle MC, Fernandez ML. Effects of resistance training on the inflammatory response. Nutr Res Pract 2010;4(4):259−69.

[65]   Jani B. Ageing and vascular ageing. Postgrad Med J 2006;82:357−62.

[66]   Hodes RJ, Lakatta EG, McNeil CT. Another modifiable risk factor for cardiovascular disease? Some evidence points to arterial stiffness. J Am Geriatr Soc 1995;43:581−2.

[67]   Thijssen DHJ, Rongen GA, van Dijk A, Smits P, Hopman MTE. Enhanced endothelin-1-mediated leg vascular tone in healthy older subjects. J Appl Physiol (1985) 2007;(103):852−7.

[68]   Thijssen Dick HJ, Hopman Maria TE, Levine Benjamin D. Endothelin and aged blood vessels. Hypertension 2007;50:292−3.

[69]   Stefano Taddei, Agostino Virdis, Lorenzo Ghiadoni, Guido Salvetti, Giampaolo Bernini, Armando Magagna, et al. Age-related reduction of NO availability and oxidative stress in humans. Hypertension 2001;38:274−9.

[70]   van der Loo B, Labugger R, Skepper JN, Bachschmid M, Kilo J, Powell JM, et al. Enhanced peroxynitrite formation is associated with vascular aging. J Exp Med 2000;192:1731−44.

[71]   Donato AJ, Gano LB, Eskurza I, Silver AE, Gates PE, Jablonski K, et al. Vascular endothelial dysfunction with aging: endothelin-1 and endothelial nitric oxide synthase. Am J Physiol Heart Circ Physiol 2009;297:H425−32.

[72]   Antunes BMM, Rossi FE, Cholewa JM, Lira FS. Regular physical activity and vascular aging. Curr Pharm Des 2016;22:3715−29.

[73]   Seals DR, Nagy EE, Moreau KL. Aerobic exercise training and vascular function with aging in healthy men and women. J Physiol 2019;597:4901−14.

[74]   Campbell A, Grace F, Ritchie L, Beaumont A, Sculthorpe N. Long-term aerobic exercise improves vascular function into old age: a systematic review, meta-analysis and meta regression of observational and interventional studies. Front Physiol 2019;10:31.

[75]   de F Brito A, de Oliveira CVC, do Socorro.B Santos M, da C Santos A. High-intensity exercise promotes postexercise hypotension greater than moderate intensity in elderly hypertensive individuals. Clin Physiol Funct Imaging 2014;34:126−32.

[76]   Egaña M, Reilly H, Green S. Effect of elastic-band-based resistance training on leg blood flow in elderly women. Appl Physiol Nutr Metab 2010;35:763−72.

[77]   Rossow LM, Fahs CA, Thiebaud RS, Loenneke JP, Kim D, Mouser JG, et al. Arterial stiffness and blood flow adaptations following eight weeks of resistance exercise training in young and older women. Exp Gerontol 2014;53:48−56.

[78]   Kanegusuku H, Queiroz ACC, Silva VJD, de Mello MT, Ugrinowitsch C, Forjaz CLM. High-intensity progressive resistance training increases strength with no change in cardiovascular function and autonomic neural regulation in older adults. J Aging Phys Act 2015;23:339−45.

[79]   Figard H, Gaume V, Mougin F, Demougeot C, Berthelot A. Beneficial effects of isometric strength training on endothelial dysfunction in rats. Appl Physiol Nutr Metab 2006;31:621−30.

[80] MacDougall JD, Tuxen D, Sale DG, Moroz JR, Sutton JR. Arterial blood pressure response to heavy resistance exercise. J Appl Physiol (1985) 1985;58(3):785–90.

[81] Palatini P, Mos L, Munari L, Valle F, Del Torre M, Rossi A, et al. Blood pressure changes during heavy-resistance exercise. J Hypertens Suppl J Int Soc Hypertens 1989;7(6):S72–3.

[82] Hill DW, Butler SD. Haemodynamic responses to weightlifting exercise. Sports Med Auckl NZ 1991;12(1):1–7.

[83] Haykowsky MJ, Dressendorfer R, Taylor D, Mandic S, Humen D. Resistance training and cardiac hypertrophy: unravelling the training effect. Sports Med Auckl NZ 2002;32(13):837–49.

[84] Yu CC-W, McManus AM, So H-K, Chook P, Au C-T, Li AM, et al. Effects of resistance training on cardiovascular health in non-obese active adolescents. World J Clin Pediatr 2016;5(3):293–300.

[85] Grøntved A, Ried-Larsen M, Møller NC, Kristensen PL, Froberg K, Brage S, et al. Muscle strength in youth and cardiovascular risk in young adulthood (the European Youth Heart Study). Br J Sports Med 2015;49(2):90–4.

[86] Au JS, Oikawa SY, Morton RW, Macdonald MJ, Phillips SM. Arterial stiffness is reduced regardless of resistance training load in young men. Med Sci Sports Exerc 2017;49(2):342–8.

[87] Maeda S, Otsuki T, Iemitsu M, Kamioka M, Sugawara J, Kuno S, et al. Effects of leg resistance training on arterial function in older men. Br J Sports Med 2006;40(10):867–9.

[88] Mandic S, Myers J, Selig SE, Levinger I. Resistance versus aerobic exercise training in chronic heart failure. Curr Heart Fail Rep 2012;9 (1):57–64.

[89] Vincent KR, Vincent HK. Resistance training for individuals with cardiovascular disease. J Cardiopulm Rehabil Prev 2006;26(4):207.

[90] Benton MJ. Safety and efficacy of resistance training in patients with chronic heart failure: research-based evidence. Prog Cardiovasc Nurs 2005;20(1):17–23.

[91] Piepoli MF, Corrà U, Benzer W, Bjarnason-Wehrens B, Dendale P, Gaita D, et al. Secondary prevention through cardiac rehabilitation: from knowledge to implementation. A position paper from the Cardiac Rehabilitation Section of the European Association of Cardiovascular Prevention and Rehabilitation. Eur J Cardiovasc Prev Rehabil 2010;17(1):1–17.

[92] Pollock Michael L, Franklin Barry A, Balady Gary J, Chaitman Bernard L, Fleg Jerome L, Barbara Fletcher, et al. Resistance exercise in individuals with and without cardiovascular disease. Circulation. 2000;101(7):828–33.

[93] Vanhees L, Geladas N, Hansen D, Kouidi E, Niebauer J, Reiner Z, et al. Importance of characteristics and modalities of physical activity and exercise in the management of cardiovascular health in individuals with cardiovascular risk factors: recommendations from the EACPR. Part II. Eur J Prev Cardiol 2012;19(5):1005–33.

[94] Canadian Cardiovascular Society Heart Failure Management Primary Panel, Moe GW, Ezekowitz JA, O'Meara E, Howlett JG, Fremes SE, et al. The 2013 Canadian Cardiovascular Society heart failure management guidelines update: focus on rehabilitation and exercise and surgical coronary revascularization. Can J Cardiol 2014;30(3):249–63.

[95] Selig SE, Levinger I, Williams AD, Smart N, Holland DJ, Maiorana A, et al. Exercise & Sports Science Australia position statement on exercise training and chronic heart failure. J Sci Med Sport 2010;13(3):288–94.

Chapter 4

Reactive oxygen species, redox signaling, and regulation of vascular endothelial signaling

Kaitlin Allen and José Pablo Vázquez-Medina

Department of Integrative Biology, University of California, Berkeley, Berkeley, CA, United States

4.1 Reactive oxygen species generation and scavenging in the healthy endothelium

The endothelium generates low levels of reactive oxygen species (ROS) enzymatically and as a byproduct of mitochondrial metabolism under physiological conditions. Electron leak from the inner mitochondrial membrane partially reduces molecular oxygen to generate (superoxide $O_2\bullet^-$) or hydrogen peroxide (H_2O_2) [1,2]. Basal electron leak at complexes I and II generates $O_2\bullet^-$ in the mitochondrial matrix, while leak at complex III generates $O_2\bullet^-$ in both the matrix and the intermembrane space [1,3]. Decreased mitochondrial activity during periods of low cellular ATP demand increases the nicotinamide adenine dinucleotide (phosphate) (NAD(P)H) pool in the cell, promoting mitochondrial superoxide generation under normal physiological conditions [4]. Increased oxidant generation can result in oxidative damage to DNA, proteins, and lipids through rapid redox reactions [5−8]. Under physiological conditions, however, low basal rates of mitochondrial superoxide formation are offset by the activity of mitochondrial superoxide dismutase (MnSOD; SOD2). MnSOD utilizes manganese as a cofactor in the dismutation of $O_2\bullet^-$ to H_2O_2 and O_2, limiting the potential for the formation of more reactive products such as hydroxyl radical $\bullet OH$ [9]. H_2O_2 is then reduced by catalase and peroxiredoxins, including the mitochondrial enzyme peroxiredoxin 3 (Prdx3) [10]. Mitochondrial H_2O_2 can also be generated at high rates by flavoprotein long-chain acyl-CoA dehydrogenase during β-oxidation of fatty acids [11,12]. Whether or not this particular process is of functional significance in the healthy endothelium remains unknown.

NADPH oxidases (NOX) are widely distributed enzymes that generate ROS as primary products. There are seven known members of the NOX family. NOX enzymes have diverse cellular distributions and differing but homologous protein components [13]. Vascular endothelial cells express NOX1, NOX2, NOX4, and NOX5 [14,15]. NOX1 and NOX2 are quiescent under resting conditions, with the intrinsic membrane (gp91^{phox} and p22^{phox}) and cytosolic (p47^{phox}, p67^{phox}/NoxA1, p40^{phox}/Noxo1, and Rac) subunits confined to their respective cellular compartments. Growth factors, cytokines, high blood pressure, changes in shear stress, and oxidized lipids activate endothelial NOX1 and NOX2, leading to translocation of the cytosolic components to the plasma membrane, assembly of the active oxidase complex, and generation of $O_2\bullet^-$ [15]. NOX4, in contrast, lacks cytosolic subunits; it is constitutively active and primarily generates H_2O_2 rather than $O_2\bullet^-$ [16]. NOX5 is a calcium-sensitive superoxide-generating oxidase that has recently emerged as a key player in cardiovascular disease [17−19].

The endothelium experiences alterations in cyclic stretch and shear forces resulting from physiological changes in blood flow. These flow changes are transduced across the endothelial monolayer to the surrounding smooth muscle as cellular signals regulating vascular tone [20−22]. Alterations in shear stress activate membrane-bound NOX enzymes in the vascular endothelium via the closure of K_{ATP} channels, leading to endothelial cell membrane depolarization [15,20,23,24]. Endothelial nitric oxide synthase (eNOS) is another flow- and stretch-responsive enzyme expressed in endothelial cells. High laminar flow activates dimeric eNOS via phosphorylation and increases calcium sensitivity of the enzyme [1,25]. Generation of NO by eNOS is essential for endothelial function [26−28], but generation of NO alongside $O_2\bullet^-$ can yield peroxynitrite (ONOO$^-$), increasing the risk of endothelial dysfunction [29,30]. Concurrent

Endothelial Signaling in Vascular Dysfunction and Disease. DOI: https://doi.org/10.1016/B978-0-12-816196-8.00011-4

generation of $O_2\bullet^-$ by NOX and NO by eNOS can therefore potentially impair endothelium-dependent vasodilation via the formation of $ONOO^-$.

As discussed earlier, endothelial H_2O_2 is generated directly by NOX4 and indirectly by the dismutation of $O_2\bullet^-$ derived from other enzymes and mitochondria. H_2O_2 is a relatively weak oxidant itself but can contribute to the formation of the highly reactive $\bullet OH$ through Fenton chemistry [31]. Intracellular H_2O_2 levels are actively modulated by peroxiredoxins, which are expressed at high levels in endothelial cells and respond to changes in shear stress and inflammatory signaling mediated by agonists such as angiotensin II, lysophospholipids, and TLR4 ligands [32−39]. Modulation of intracellular H_2O_2 levels in endothelial cells is crucial for sensing and cellular signaling [2,40−42]. Endogenous H_2O_2 regulates stress-responsive mitogen-activated protein (MAP) kinases such as c-Jun N-terminal kinase (JNK), p38, and extracellular signal-regulated kinase (ERK) [43]. Similarly, endogenous H_2O_2 can activate redox-sensitive transcription factors and other intracellular sensors critical to inflammatory and antioxidant responses such as nuclear factor kappa-light-chain-enhancer of activated B cells (NF-kB), nuclear factor erythroid-derived related factor 2 (Nrf2), activator protein-1 (AP-1), and NOD-, LRR- and pyrin domain-containing protein 3 (NLRP3) [44,45].

Xanthine oxidase (XO) is another oxidant-generating enzyme expressed in endothelial cells [46]. Endothelial xanthine oxidase is activated (proteolytically cleaved from xanthine dehydrogenase) in response to hypoxia and inflammatory stimuli including tumor necrosis factor α (TNFα) and N-Formylmethionyl-leucyl-phenylalanine (FMLP) [47−49]. XO generates both $O_2\bullet^-$ and H_2O_2 [50]. Early studies using the XO inhibitor allopurinol implicated XO-generated $O_2\bullet^-$ in endothelial dysfunction induced by cigarette smoking, heart failure, and ischemia/reperfusion [51−56]. Recent evidence, however, suggests that H_2O_2 is the main enzymatic product generated by XO under clinically relevant conditions [57]. Thus, in addition to its established role in endothelial dysfunction, XO may participate in physiological endothelial cell signaling via the generation of H_2O_2.

4.2   Redox signaling in the regulation of angiogenesis, vasomotor tone, and inflammatory responses in the healthy endothelium

Endothelial ROS were originally identified as potent mediators of tissue injury following ischemia/reperfusion events and other inflammatory conditions [49,58−60]. In recent years, however, clinical and laboratory studies have elucidated important roles for ROS in cellular signaling and response systems in the healthy endothelium. The vascular endothelium mediates angiogenesis, the growth of vasculature from existing vessels into tissues [61]. Angiogenesis is critical in supporting local tissue oxygen demands during embryonic development and wound repair. Increased tissue oxygen demands result in local hypoxia/reoxygenation cycles, which stimulate mitochondrial and NOX-derived ROS generation. Consequently, NOX enzymes including NOX2 and NOX4 have long been recognized as important angiogenic mediators [62−64].

The cross talk between NOX2- and NOX4-dependent redox signaling during angiogenesis is still under investigation. Recent evidence suggests that endothelial ROS signaling during angiogenesis is a coordinated feed-forward event in which H_2O_2 derived from NOX4 promotes NOX2-derived $O_2\bullet^-$ generation, which then increases mitochondrial ROS generation, sustaining the activation of the angiogenic signaling program via upregulation of hypoxia-inducible factor-1 (HIF-1α), vascular endothelial growth factor (VEGF), and transforming growth factor β (TGF-β) [65]. HIF-1α is a known transcriptional regulator of VEGF; VEGF expression increases in response to hypoxia, H_2O_2, oxidized phospholipids, and low-density lipoproteins. VEGF induces endothelial cell migration and tube formation and activates NOX enzymes in a positive feedback loop, further increasing oxidant generation and contributing to the pro-angiogenic oxidizing environment of the cell [62,66−68]. Deficiency of either NOX2 or NOX4 impairs the angiogenic potential of flow-adapted microvascular endothelial cells in primary culture and limits postischemic revascularization after hind limb ischemia in vivo [23,69]. Thioredoxin 2 and peroxiredoxin 3 play important roles in angiogenesis through the modulation of mitochondrial oxidant levels [70].

The endothelium transduces flow signals from the circulation to the surrounding smooth muscle cells, regulating vascular tone. Under physiological conditions (laminar flow), $O_2\bullet^-$ generated by NOX enzymes is dismutated to H_2O_2, stimulating eNOS expression and activity [71−74]. Additionally, phosphorylation at key serine, threonine, and tyrosine residues and increased calcium sensitivity of eNOS under high shear stress activates the enzyme, leading to NO generation [25]. NO generated by eNOS diffuses across the basolateral endothelial membrane into vascular smooth muscle cells, modulating intracellular calcium levels and signaling for vasorelaxation via the second messenger cyclic guanosine monophosphate [27,28,75−78]. eNOS can also generate low levels of $O_2\bullet^-$ under physiological conditions, ultimately leading to the production of H_2O_2 that participates in endothelial-dependent vasodilation [79−81]. Concurrent, colocalized generation of $O_2\bullet^-$ and NO yields to the rapid formation of $ONOO^-$, effectively scavenging NO and preventing its participation in vasodilation. However, the extent to which endothelial $ONOO^-$ is formed under physiological conditions remains unclear.

In addition to participating in angiogenic and vasodilatory signaling, redox signaling in the endothelium contributes to host defense and the acute inflammatory response. Recognition of pathogens by circulating leukocytes leads to upregulation of pro-inflammatory mediators such as transcription factor NF-κB and inflammation mediator NLRP3 inflammasome, promoting the transcription, maturation, and secretion of cytokines including Interleukins-6 (IL-6) and IL-1β that mediate neutrophil recruitment. Neutrophils recruited to the site of inflammation generate large amounts of NOX-derived ROS that contribute to sustained endothelial cell activation; the resulting redox signaling activates transcription factor NF-κB, which upregulates expression of endothelial adhesion molecules and increases the permeability of the endothelial monolayer, supporting transmigration of immune cells to injury sites [82−85]. Tight regulation of this process in healthy systems ends with the resolution of inflammation. During endothelial dysfunction, however, chronic inflammatory responses often result in increased oxidative damage and subsequent tissue injury.

Antioxidant enzymes are crucial for the modulation of redox signaling and the maintenance of redox balance in the healthy endothelium. Prominent transcription factors controlling antioxidant gene expression include Nrf2 and activator protein-1 (AP-1). Under basal conditions, Nrf2 is bound to its repressor protein Keap1 in the cytosol and continuously targeted for proteasomal degradation. H_2O_2 and electrophiles such as the lipid peroxidation product 4-hydroxynonenal (4-HNE) oxidize key cysteine residues in Keap1, promoting Nrf2 dissociation, nuclear translocation, and binding to antioxidant response elements (AREs). Nrf2 can also be activated by the noncanonical autophagic degradation of Keap1 and by Keap1-independent phosphorylation [86,87]. Binding of activated Nrf2 to AREs initiates transcription of several antioxidant and anti-inflammatory genes [88−91]. For example, Nrf2 activation increases heme oxygenase-1 (HO-1) expression [92]. HO-1 blocks NF-κB-dependent transcription of vascular adhesion molecules in endothelial cells, limiting immune cell adhesion [93]. Similarly, Nrf2 also limits IL-6 and IL-1β transcription by binding in the proximity of these genes and consequently inhibiting recruitment of RNA polymerase II to these loci [94]. *In vitro*, *in vivo*, and clinical studies of heart and brain—two tissues highly susceptible to endothelial ischemia/reperfusion injury—show that moderate levels of 4-HNE stimulate Nrf2 activity, contributing to neuro- and cardio-protection following ischemic insult [95−98]. Thus redox signaling protects the vascular endothelium from major oxidative injury by activating stress-responsive transcription factors and upregulating expression of endogenous antioxidants. This phenomenon, known in clinical literature as ischemic preconditioning, prepares tissues to cope with future oxidative insults by priming tissue antioxidant capacity in advance of injury.

AP-1 is another oxidant stress-sensitive transcription factor. AP-1 is formed by heterodimers of the Jun, Fos, activating transcription factor (ATF), and transcription factor Maf families of proteins and binds to 12-O-tetradecanoylphorbol-13-acetate (TPA) or cAMP response elements. Circulating cytokines, chemokines, growth factors, hormones, and various stressors − including oxidant stress − can activate AP-1. Activation of the AP-1 complex requires phosphorylation of specific components by stress-responsive MAP kinases such as JNK, p38, and ERK [99]; several studies show that endogenous H_2O_2 in intact cells increases the activity of these kinases [100], activating AP-1 [101]. Recent evidence also suggests a NO-dependent mechanism of AP-1 activation in the vascular endothelium [102]. Activated AP-1 induces the transcription of genes involved in a variety of contrasting cellular responses including the antioxidant defense system. Redox signaling is therefore crucial for the regulation of a variety of processes in the healthy endothelium including angiogenesis, vasomotor tone, and inflammatory responses.

4.3 Reactive oxygen species generation and redox signaling in disease: sources and implications

In addition to the signaling roles described earlier, dysregulated ROS generation can contribute to oxidant stress and tissue injury incurred during a range of conditions associated with endothelial dysfunction such as myocardial infarction, ischemic stroke, organ transplantation, cardiovascular disease, and peripheral vascular diseases. In each of these cases, disruption of physiological blood flow and oxygen tension stimulates chronic or excessive ROS generation that leads to inflammatory conditions and tissue injury. Similarly, increased ROS generation and dysregulated redox signaling is well documented during endothelial dysfunction associated with metabolic disease.

In most tissues, acute or chronic reductions in blood flow (ischemia) constrain tissue oxygen availability, leading to hypoxia. Myocardial infarction, ischemic stroke, and organ transplantation represent extreme acute ischemic events during which blood flow and oxygen delivery to tissues are partially or fully occluded. As discussed earlier, depletion of intracellular ATP stores during tissue hypoxia leads to ATP degradation, yielding the xanthine oxidase substrates xanthine and hypoxanthine [58]. Similarly, dysregulated calcium signaling secondary to ischemia activates calcium-dependent proteases that cleave xanthine dehydrogenase to yield the active form of the enzyme, xanthine oxidase (XO) and increase mitochondrial oxidant generation [103,104]. Reperfusion provides molecular O_2 as the remaining required substrate for XO activity, leading to excessive formation of $O_2\bullet^-$ and H_2O_2 [49,58,105]. Increased intracellular calcium levels during ischemia also activate NOX enzymes and stimulate $O_2\bullet^-$ generation by mitochondria during reperfusion [106,107]. Similarly, mitochondrial accumulation of succinate as a result of ATP degradation during ischemia drives succinate oxidation and $O_2\bullet^-$ generation via reverse electron transport at mitochondrial complex I during reperfusion [108,109].

Induction of chronic turbulent flow by clots and plaques uncouples eNOS, inhibiting endothelial-dependent vasodilation [110–112]. Uncoupled eNOS, the monomeric form of the enzyme, generates $O_2\bullet^-$ rather than NO [81]. As discussed earlier, spatially and temporally colocalized generation of $O_2\bullet^-$ and NO yields $ONOO^-$, a highly reactive species that causes tissue injury via protein nitration and impairs endothelial-mediated vasorelaxation by scavenging NO, sustaining eNOS uncoupling and inducing cell death [23,30,69,113–115]. Oxidized low-density lipoprotein (LDL) also inhibits eNOS activation, further impairing endothelial-mediated vasorelaxation [116,117].

In addition to direct generation of ROS by the endothelium during reperfusion, NOX-mediated redox signaling during ischemia upregulates the expression and activity of inflammatory molecules such as cellular adhesion molecules (CAMs), selectins, NF-κB, and NLRP3 inflammasomes. This redox-driven process represents a "priming" event in the vasculature that enables adhesion of and infiltration by neutrophils upon reperfusion, resulting in significant increases in oxidant generation and propagation of endothelial dysfunction and tissue injury [83]. Deficiency of NOX2, which is activated by loss of shear rather than by hypoxia in endothelial cells during ischemia, suppresses endothelial cell priming and limits ischemic inflammation [118]. Thus, ex vivo perfusion remains an attractive alternative to limit endothelial injury derived from NOX2-dependent ischemic signaling associated with organ transplantation [119].

Chronic inflammatory conditions such as metabolic disease are associated with increased oxidant generation and endothelial dysfunction. Hyperglycemia in type 2 diabetes promotes endothelial oxidative stress by increasing oxidant generation derived from mitochondria and NOX enzymes [120,121]. Similarly, eNOS uncoupling increases peroxynitrite formation, reduces NO synthesis, and contributes to endothelial dysfunction in diabetes and aging [122,123]. Moreover, activation of NOX1, which is widely expressed in vascular smooth muscle and endothelial cells, is associated with the progression of atherosclerosis during diabetes [124]. Although there is still no conclusive evidence showing that a global increase in ROS generation is fundamentally involved in the pathogenesis of cardiovascular disease in humans, vascular oxidative stress is a strong predictor of the risk of cardiovascular events in several pathologies including coronary artery disease [125]. Similarly, clinical confirmation of the effectiveness of antioxidant therapy to treat endothelial dysfunction associated with cardiovascular complications remains pending, but specific inhibitors targeting particular oxidant-generating enzymes without altering complex global redox networks are beginning to help identify individual pathways activated during cardiovascular and metabolic diseases that could be targeted for therapeutic intervention [126].

4.4 Conclusion

The endothelium serves both barrier and communication functions as the primary interface between the body's circulation and tissues. Baseline conditions and physiological fluctuations in blood flow and oxygen availability stimulate endothelial ROS generation. Critically, ROS generation under these conditions is tightly regulated and relatively limited, thereby avoiding extensive oxidative damage to cellular components. Indeed, physiological ROS generation serves redox signaling functions and contributes to protection against oxidative stress and inflammation. Dysregulated endothelial ROS generation is observed in many vascular pathologies including myocardial infarction, ischemic stroke, atherosclerosis, and metabolic, cardiovascular, and peripheral vascular diseases. In each of these conditions, oxidant-derived endothelial dysfunction impairs selectivity of the endothelial barrier and responsiveness to appropriate vasodilatory cues, promoting inflammation, oxidant stress, and tissue injury. Identification of specific ROS sources and the species generated, as well as temporal and spatial (e.g., cellular, subcellular) localization, is crucial to understanding redox signaling and potential complications associated with oxidant-derived endothelial dysfunction in the clinic (Fig. 4.1).

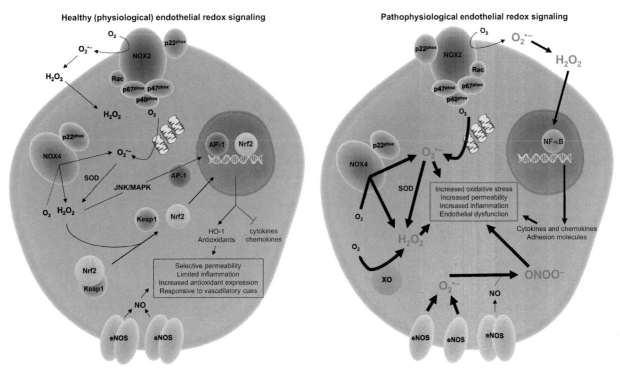

FIGURE 4.1 Physiological and pathophysiological redox signaling in the endothelium. Under physiological conditions, low levels of $O_2\bullet^-$ and H_2O_2 are generated by NADPH oxidases (NOX) and mitochondria. H_2O_2 activates transcription factors including AP-1 and Nrf2, which upregulate expression of antioxidants and decrease expression of inflammatory cytokines and chemokines. eNOS generates NO, which supports endothelial-mediated vasodilation. Under pathophysiological conditions such as inflammation and turbulent or restricted flow, ROS generation by mitochondria and NOX enzymes increases, and NO generation is limited as eNOS becomes uncoupled. $ONOO^-$, the product of NO and $O_2\bullet^-$, as well as other oxidants contribute to activation of NF-κB and upregulated expression of adhesion molecules and inflammatory cytokines and chemokines, leading to endothelial dysfunction via oxidative stress and increased inflammation.

References

[1] Brand MD. Mitchondrial generation of superoxide and hydrogen peroxide as the source of mitochondrial redox signaling. Free Radic Biol Med 2016;100:14−31.

[2] Bretón-Romero R, Lamas S. Hydrogen peroxide signaling in vascular endothelial cells. Redox Biol 2014;2:529−34.

[3] O'Malley Y, Fink BD, Ross NC, Prisinzano TE, Sivitz WI. Reactive oxygen and targeted antioxidant administration in endothelial cell mitochondria. J Biol Chem 2006;281:39766−75.

[4] Murphy MP. How mitochondria produce reactive oxygen species. Biochem J 2009;417:1−13.

[5] Cooke MS, Evans MD, Dizdaroglu M, Lunec J. Oxidative DNA damage: mechanisms, mutation, and disease. FASEB J 2003;17:1195−214.

[6] Davies MJ. Protein oxidation and peroxidation. Biochem J 2016;473:805−25.

[7] Evans MD, Dizdaroglu M, Cooke MS. Oxidative DNA damage and disease: induction, repair and significance. Mutat Res 2004;567:1−61.

[8] Ayala A, Muñoz MF, Argüelles S. Lipid peroxidation: production, metabolism, and signaling mechanisms of malondialdehyde and 4-hydroxy-2-nonenal. Oxid Med Cell Longev 2014;2014.

[9] Candas D, Li JJ. MnSOD in oxidative stress response-potential regulation via mitochondrial protein influx. Antioxid Redox Signal 2014;20:1599−617.

[10] Rhee SG, Chae HZ, Kim K. Peroxiredoxins: a historical overview and speculative preview of novel mechanisms and emerging concepts in cell signaling. Free Radic Biol Med 2005;38:1543−52.

[11] Zhang Y, Bharathi SS, Beck ME, Goetzman ES. The fatty acid oxidation enzyme long-chain acyl-CoA dehydrogenase can be a source of mitochondrial hydrogen peroxide. Redox Biol 2019;26:101253.

[12] Seifert EL, Estey C, Xuan JY, Harper M-E. Electron transport chain-dependent and -independent mechanisms of mitochondrial H_2O_2 emission during long-chain fatty acid oxidation. J Biol Chem 2010;285:5748−58.

[13] Cheng G, Cao Z, Xu X, Van Meir EG, Lambeth JD. Homologs of gp91phox: cloning and tissue expression of Nox3, Nox4, and Nox5. Gene 2001;269:131−40.

[14] Brandes RP, Schröder K. Composition and functions of vascular nicotinamide adenine dinucleotide phosphate oxidases. Trends Cardiovasc Med 2008;18:15−19.

[15] Brandes RP, Weissmann N, Schröder K. Nox family NADPH oxidases: molecular mechanisms of activation. Free Radic Biol Med 2014;76:208−26.

[16] Takac I, Schröder K, Zhang L, Lardy B, Anilkumar N, Lambeth JD, et al. The E-loop is involved in hydrogen peroxide formation by the NADPH oxidase Nox4. J Biol Chem 2011;286:13304−13.

[17] Touyz RM, Anagnostopoulou A, Camargo LL, Rios FJ, Montezano AC. Vascular biology of superoxide-generating NADPH oxidase 5—implications in hypertension and cardiovascular disease. Antioxid Redox Signal 2019;30:1027−40.

[18] do Carmo LS, Berk BC, Harrison DG. NOX5 as a therapeutic target in cerebral ischemic injury. J Clin Invest 2019;130:1530−2.

[19] Jha JC, Watson AMD, Mathew G, de Vos LC, Jandeleit-Dahm K. The emerging role of NADPH oxidase NOX5 in vascular disease. Clin Sci 2017;131:981−90.

[20] Chatterjee S, Fisher AB. Mechanotransduction in the endothelium: role of membrane proteins and reactive oxygen species in sensing, transduction, and transmission of the signal with altered blood flow. Antioxid Redox Signal 2014;20:899−913.

[21] Chien S, Li S, Shyy JY-J. Effects of mechanical forces on signal transduction and gene expression in endothelial cells. Hypertension 1998;31:162−9.

[22] Natarajan M, Aravindan N, Sprague EA, Mohan S. Hemodynamic flow-induced mechanotransduction signaling influences the radiation response of the vascular endothelium. Radiat Res 2016;186:175−88.

[23] Browning E, Wang H, Hong N, Yu K, Buerk DG, DeBolt K, et al. Mechanotransduction drives post ischemic revascularization through K-ATP channel closure and production of reactive oxygen species. Antioxid Redox Signal 2014;20:872−86.

[24] Chatterjee S, Browning EA, Hong N, DeBolt K, Sorokina EM, Liu W, et al. Membrane depolarization is the trigger for PI3K/Akt activation and leads to the generation of ROS. Am J Physiol Heart Circ Physiol 2012;302:H105−14.

[25] Förstermann U, Sessa WC. Nitric oxide synthases: regulation and function. Eur Heart J 2012;33:829−37.

[26] Ignarro LJ, Buga GM, Wood KS, Byrns RE, Chaudhuri G. Endothelium-derived relaxing factor produced and released from artery and vein is nitric oxide. Proc Natl Acad Sci U S A 1987;84:9265−9.

[27] Palmer RMJ, Ferrige AG, Moncada S. Nitric oxide release accounts for the biological activity of endothelium-derived relaxing factor. Nature 1987;327:524−6.

[28] Palmer RMJ, Ashton DS, Moncada S. Vascular endothelial cells synthesize nitric oxide from L-arginine. Nature 1988;333:664−6.

[29] Frey RS, Ushio−Fukai M, Malik AB. NADPH oxidase-dependent signaling in endothelial cells: role in physiology and pathophysiology. Antioxid Redox Signal 2009;11:791−810.

[30] Beckman JS, Beckman TW, Chen J, Marshall PA, Freeman BA. Apparent hydroxyl radical production by peroxynitrite: implications for endothelial injury from nitric oxide and superoxide. Proc Natl Acad Sci U S A 1990;87:1620−4.

[31] Winterbourn CC. Toxicity of iron and hydrogen peroxide: the Fenton reaction. Toxicol Lett 1995;82−83:969−74.

[32] Stacey MM, Vissers MC, Winterbourn CC. Oxidation of 2-Cys peroxiredoxins in human endothelial cells by hydrogen peroxide, hypochlorous acid, and chloramines. Antioxid Redox Signal 2012;17:411−21.

[33] Tao R-R, Wang H, Hong L-J, Huang J-Y, Lu Y-M, Liao M-H, et al. Nitrosative stress induces peroxiredoxin 1 ubiquitination during ischemic insult via E6AP activation in endothelial cells both in vitro and in vivo. Antioxid Redox Signal 2014;21:1−16.

[34] Vázquez-Medina JP, Tao J-Q, Patel P, Bannitz-Fernandes R, Dodia C, Sorokina EM, et al. Genetic inactivation of the phospholipase A2 activity of peroxiredoxin 6 in mice protects against LPS-induced acute lung injury. Am J Physiol Lung Cell Mol Physiol 2019;316:L656−68.

[35] Chatterjee S, Feinstein SI, Dodia C, Sorokina E, Lien Y-C, Nguyen S, et al. Peroxiredoxin 6 phosphorylation and subsequent phospholipase A2 activity are required for agonist-mediated activation of NADPH oxidase in mouse pulmonary microvascular endothelium and alveolar macrophages. J Biol Chem 2011;286:11696−706.

[36] Kang DH, Lee DJ, Lee KW, Park YS, Lee JY, Lee S-H, et al. Peroxiredoxin II is an essential antioxidant enzyme that prevents the oxidative inactivation of VEGF receptor-2 in vascular endothelial cells. Mol Cell 2011;44:545−58.

[37] Mowbray AL, Kang D-H, Rhee SG, Kang SW, Jo H. Laminar shear stress up-regulates peroxiredoxins (PRX) in endothelial cells: PRX1 as a mechanosensitive antioxidant. J Biol Chem 2008;283:1622−7.

[38] Patel P, Chatterjee S. Peroxiredoxin6 in endothelial signaling. Antioxidants 2019;8:63.

[39] Riddell JR, Maier P, Sass SN, Moser MT, Foster BA, Gollnick SO. Peroxiredoxin 1 Stimulates endothelial cell expression of VEGF via TLR4 dependent activation of HIF-1a. PLoS One 2012;7:e50394.

[40] Veal EA, Day AM, Morgan BA. Hydrogen peroxide sensing and signaling. Mol Cell 2007;26:1−14.

[41] Wood ZA, Poole LB, Karplus PA. Peroxiredoxin evolution and the regulation of hydrogen peroxide signaling. Science 2003;300:650−3.

[42] Rhee SG. Redox signaling: hydrogen peroxide as intracellular messenger. Exp Mol Med 1999;31:53−9.

[43] Torres M, Forman HJ. Redox signaling and the MAP kinase pathways. BioFactors 2003;17:287−96.

[44] Suzuki YJ, Forman HJ, Sevanian A. Oxidants as stimulators of signal transduction. Free Radic Biol Med 1997;22:269−85.

[45] Xiang M, Shi X, Li Y, Xu J, Yin L, Xiao G, et al. Hemorrhagic shock activation of NLRP3 inflammasome in lung endothelial cells. J Immunol 2011;187:4809−17.

[46] Jarasch E-D, Bruder G, Heid H. Significance of xanthine oxidase in capillary endothelial cells. Acta Physiol Scand Suppl 1986;548:39−46.

[47] Zhang C, Hein TW, Wang W, Ren Y, Shipley RD, Kuo L. Activation of JNK and xanthine oxidase by TNF-α impairs nitric oxide-mediated dilation of coronary arterioles. J Mol Cell Cardiol 2006;40:247−57.

[48] Friedl HP, Till GO, Ryan US, Ward PA. Mediator-induced activation of xanthine oxidase in endothelial cells. FASEB J 1989;3:2512−18.

[49] Granger DN. Role of xanthine oxidase and granulocytes in ischemia-reperfusion injury. Am J Physiol Heart Circ Physiol 1988;255:H1269−75.

[50] Kellogg EW, Fridovch I. Superoxide, hydrogen peroxide, and singlet oxygen in lipid peroxidation by a xanthine oxidase system. J Biol Chem 1975;250:8812–17.

[51] Doehner W, Schoene N, Rauchhaus M, Leyva-Leon F, Pavitt DV, Reaveley DA, et al. Effects of xanthine oxidase inhibition with allopurinol on endothelial function and peripheral blood flow in hyperuricemic patients with chronic heart failure: results from 2 placebo-controlled studies. Circulation 2002;105:2619–24.

[52] Guthikonda S, Sinkey C, Barenz T, Haynes WG. Xanthine oxidase inhibition reverses endothelial dysfunction in heavy smokers. Circulation 2003;107:416–21.

[53] Beetsch JW, Park TS, Dugan LL, Shah AR, Gidday JM. Xanthine oxidase-derived superoxide causes reoxygenation injury of ischemic cerebral endothelial cells. Brain Res 1998;786:89–95.

[54] Landmesser U, Spiekermann S, Dikalov S, Tatge H, Wilke R, Kohler C, et al. Vascular oxidative stress and endothelial dysfunction in patients with chronic heart failure: role of xanthine-oxidase and extracellular superoxide dismutase. Circulation 2002;106:3073–8.

[55] Meneshian A, Bulkley GB. The physiology of endothelial xanthine oxidase: from urate catabolism to reperfusion injury to inflammatory signal transduction. Microcirculation 2002;9:161–75.

[56] Rieger JM, Shah AR, Gidday JM. Ischemia-reperfusion injury of retinal endothelium by cyclooxygenase- and xanthine oxidase-derived superoxide. Exp Eye Res 2002;74:493–501.

[57] Kelley EE, Khoo NKH, Hundley NJ, Malik UZ, Freeman BA, Tarpey MM. Hydrogen peroxide is the major oxidant product of xanthine oxidase. Free Radic Biol Med 2010;48:493–8.

[58] McCord JM. Oxygen-derived free radicals in postischemic tissue injury. N Engl J Med 1985;312:159–63.

[59] McCord JM. Superoxide radical: a likely link between reperfusion injury and inflammation. Adv Free Rad Biol Med 1986;2:325–45.

[60] McCord JM, Roy RS. The pathophysiology of superoxide: roles in inflammation and ischemia. Can J Physiol Pharmacol 1982;60:1346–52.

[61] Muñoz-Chápuli R, Quesada AR, Medina MÁ. Angiogenesis and signal transduction in endothelial cells. Cell Mol Life Sci 2004;61:2224–43.

[62] Ushio-Fukai M. Redox signaling in angiogenesis: role of NADPH oxidase. Cardiovasc Res 2006;71:226–35.

[63] Ushio-Fukai M, Alexander RW. Reactive oxygen species as mediators of angiogenesis signaling: role of NAD(P)H oxidase. Mol Cell Biochem 2004;264:85–97.

[64] Schröder K, Zhang M, Benkhoff S, Mieth A, Pliquett R, Kosowski J, et al. Nox4 is a protective reactive oxygen species generating vascular NADPH oxidase. Circ Res 2012;110:1217–25.

[65] Kim Y-M, Kim S-J, Tatsunami R, Yamamura H, Fukai T, Ushio-Fukai M. ROS-induced ROS release orchestrated by Nox4, Nox2, and mitochondria in VEGF signaling and angiogenesis. Am J Physiol Cell Physiol 2017;312:C749–64.

[66] Wang Y, Zang QS, Liu Z, Wu Q, Maass D, Dulan G, et al. Regulation of VEGF-induced endothelial cell migration by mitochondrial reactive oxygen species. Am J Physiol Cell Physiol 2011;301:C695–704.

[67] Chua CC, Hamdy RC, Chua BHL. Upregulation of vascular endothelial growth factor by H_2O_2 in rat heart endothelial cells. Free Radic Biol Med 1998;25:891–7.

[68] Kim Y-W, Byzova TV. Oxidative stress in angiogenesis and vascular disease. Blood 2014;123:625–31.

[69] Noel J, Wang H, Hong N, Tao J-Q, Yu K, Sorokina EM, et al. PECAM-1 and caveolae form the mechanosensing complex necessary for NOX2 activation and angiogenic signaling with stopped flow in pulmonary endothelium. Am J Physiol Lung Cell Mol Physiol 2013;305:L805–18.

[70] Chen C, Li L, Zhou HJ, Min W. The role of NOX4 and TRX2 in angiogenesis and their potential cross-talk. Antioxidants 2017;6:42.

[71] Thomas SR, Chen K, Keaney JF. Hydrogen peroxide activates endothelial nitric-oxide synthase through coordinated phosphorylation and dephosphorylation via a phosphoinositide 3-kinase-dependent signaling pathway. J Biol Chem 2002;277:6017–24.

[72] Bretón-Romero R, González de Orduña C, Romero N, Sánchez-Gómez FJ, de Álvaro C, Porras A, et al. Critical role of hydrogen peroxide signaling in the sequential activation of p38 MAPK and eNOS in laminar shear stress. Free Radic Biol Med 2012;52:1093–100.

[73] Cai H, Li Z, Davis ME, Kanner W, Harrison DG, Dudley SC. Akt-dependent phosphorylation of serine 1179 and mitogen-activated protein kinase kinase/extracellular signal-regulated kinase 1/2 cooperatively mediate activation of the endothelial nitric-oxide synthase by hydrogen peroxide. Mol Pharmacol 2003;63:325–31.

[74] Drummond GR, Cai H, Davis ME, Ramasamy S, Harrison DG. Transcriptional and posttranscriptional regulation of endothelial nitric oxide synthase expression by hydrogen peroxide. Circ Res 2000;86:347–54.

[75] Furchgott RF, Cherry PD, Zawadzki JV, Jothianandan D. Endothelial cells as mediators of vasodilation of arteries. J Cardiovasc Pharmacol 1984;6:S336–43.

[76] Knowles RG, Palacios M, Palmer RMJ, Moncada S. Formation of nitric oxide from L-arginine in the central nervous system: a transduction mechanism for stimulation of the soluble guanylate cyclase. Proc Natl Acad Sci U S A 1989;86:5159–62.

[77] Rapoport RM, Draznin MB, Murad F. Endothelium-dependent relaxation in rat aorta may be mediated through cyclic GMP-dependent protein phosphorylation. Nature 1983;306:174–6.

[78] Sandoo A, Veldhuijzen van Zanten JJCS, Metsios GS, Carroll D, Kitas GD. The endothelium and its role in regulating vascular tone. TOCMJ 2010;4:302–12.

[79] Yokoyama M, Hirata K. Endothelial nitric oxide synthase uncoupling: is it a physiological mechanism of endothelium-dependent relaxation in cerebral artery? Cardiovasc Res 2007;73:8–9.

[80] Drouin A, Thorin-Trescases N, Hamel E, Falck J, Thorin E. Endothelial nitric oxide synthase activation leads to dilatory H_2O_2 production in mouse cerebral arteries. Cardiovasc Res 2007;73:73–81.

[81] Luo S, Lei H, Qin H, Xia Y. Molecular mechanisms of endothelial NO synthase uncoupling. CPD 2014;20:3548–53.

[82] Chatterjee S. Endothelial mechanotransduction, redox signaling and the regulation of vascular inflammatory pathways. Front Physiol 2018;9:524.

[83] Eltzschig HK, Eckle T. Ischemia and reperfusion—from mechanism to translation. Nat Med 2011;17:1391—401.

[84] Iadecola C, Anrather J. The immunology of stroke: from mechanisms to translation. Nat Med 2011;17:796—808.

[85] Mittal M, Siddiqui MR, Tran K, Reddy SP, Malik AB. Reactive oxygen species in inflammation and tissue injury. Antioxid Redox Signal 2014;20:1126—67.

[86] Komatsu M, Kurokawa H, Waguri S, Taguchi K, Kobayashi A, Ichimura Y, et al. The selective autophagy substrate p62 activates the stress responsive transcription factor Nrf2 through inactivation of Keap1. Nat Cell Biol 2010;12:213—23.

[87] Rada P, Rojo AI, Chowdhry S, McMahon M, Hayes JD, Cuadrado A. SCF/B-TrCP promotes glycogen synthase kinase 3-dependent degradation of the Nrf2 transcription factor in a Keap1-independent manner. Mol Cell Biol 2011;31:1121—33.

[88] Ahmed SMU, Luo L, Namani A, Wang XJ, Tang X. Nrf2 signaling pathway: pivotal roles in inflammation. Biochim Biophys Acta Mol Basis Dis 2017;1863:585—97.

[89] Itoh K, Chiba T, Takahashi S, Ishii T, Igarashi K, Katoh Y, et al. An Nrf2/small Maf heterodimer mediates the induction of phase II detoxifying enzyme genes through antioxidant response elements. Biochem Biophys Res Commun 1997;236:313—22.

[90] Itoh K, Wakabayashi N, Katoh Y, Ishii T, Igarashi K, Engel JD, et al. Keap1 represses nuclear activation of antioxidant responsive elements by Nrf2 through binding to the amino-terminal Neh2 domain. Genes Dev 1999;13:76—86.

[91] Raghunath A, Sundarraj K, Nagarajan R, Arfuso F, Bian J, Kumar AP, et al. Antioxidant response elements: discovery, classes, regulation and potential applications. Redox Biol 2018;17:297—314.

[92] Alam J, Stewart D, Touchard C, Boinapally S, Choi AMK, Cook JL. Nrf2, a Cap'n'Collar transcription factor, regulates induction of the heme oxygenase-1 gene. J Biol Chem 1999;274:26071—8.

[93] Soares MP, Seldon MP, Gregoire IP, Vassilevskaia T, Berberat PO, Yu J, et al. Heme oxygenase-1 modulates the expression of adhesion molecules associated with endothelial cell activation. J Immunol 2004;172:3553—63.

[94] Kobayashi EH, Suzuki T, Funayama R, Nagashima T, Hayashi M, Sekine H, et al. Nrf2 suppresses macrophage inflammatory response by blocking proinflammatory cytokine transcription. Nat Commun 2016;7:11624.

[95] Thielmann M, Kottenberg E, Kleinbongard P, Wendt D, Gedik N, Pasa S, et al. Cardioprotective and prognostic effects of remote ischaemic preconditioning in patients undergoing coronary artery bypass surgery: a single-centre randomised, double-blind, controlled trial. Lancet 2013;382:597—604.

[96] Zhang Y, Sano M, Shinmura K, Tamaki K, Katsumata Y, Matsuhashi T, et al. 4-Hydroxy-2-nonenal protects against cardiac ischemia—reperfusion injury via the Nrf2-dependent pathway. J Mol Cell Cardiol 2010;49:576—86.

[97] Calvert JW, Jha S, Gundewar S, Elrod JW, Ramachandran A, Pattillo CB, et al. Hydrogen sulfide mediates cardioprotection through Nrf2 signaling. Circ Res 2009;105:365—74.

[98] Bell KF, Al-Mubarak B, Fowler JH, Baxter PS, Gupta K, Tsujita T, et al. Mild oxidative stress activates Nrf2 in astrocytes, which contributes to neuroprotective ischemic preconditioning. Proc Natl Acad Sci U S A 2011;108:E1—2.

[99] Karin M, Takahashi T, Kapahi P, Delhase M, Chen Y, Makris C, et al. Oxidative stress and gene expression: the AP-1 and NF-κB connections. BioFactors 2001;15:87—9.

[100] Whitmarsh AJ, Davis RJ. Transcription factor AP-1 regulation by mitogen-activated protein kinase signal transduction pathways. J Mol Med 1996;74:589—607.

[101] Iles KE, Dickinson DA, Watanabe N, Iwamoto T, Forman HJ. AP-1 activation through endogenous H_2O_2 generation by alveolar macrophages. Free Radic Biol Med 2002;32:1304—13.

[102] Srivastava M, Saqib U, Naim A, Roy A, Liu D, Bhatnagar D, et al. The TLR4—NOS1—AP1 signaling axis regulates macrophage polarization. Inflamm Res 2017;66:323—34.

[103] Arnould T, Michiels C, Alexandre I, Remacle J. Effect of hypoxia upon intracellular calcium concentration of human endothelial cells. J Cell Physiol 1992;152:215—21.

[104] Berna N, Arnould T, Remacle J, Michiels C. Hypoxia-induced increase in intracellular calcium concentration in endothelial cells: role of the Na + -glucose cotransporter. J Cell Biochem 2002;84:115—31.

[105] Chambers D, Parks D, Patterson G, Roy R, Mccord J, Yoshida S, et al. Xanthine oxidase as a source of free radical damage in myocardial ischemia. J Mol Cell Cardiol 1985;17:145—52.

[106] Brookes PS, Yoon Y, Robotham JL, Anders MW, Sheu S-S. Calcium, ATP, and ROS: a mitochondrial love-hate triangle. Am J Physiol Cell Physiol 2004;287:C817—33.

[107] Granger DN, Kvietys PR. Reperfusion injury and reactive oxygen species: the evolution of a concept. Redox Biol 2015;6:524—51.

[108] Chouchani ET, Pell VR, Gaude E, Aksentijević D, Sundier SY, Robb EL, et al. Ischaemic accumulation of succinate controls reperfusion injury through mitochondrial ROS. Nature 2014;515:431—5.

[109] Chouchani ET, Pell VR, James AM, Work LM, Saeb-Parsy K, Frezza C, et al. A unifying mechanism for mitochondrial superoxide production during ischemia-reperfusion injury. Cell Metab 2016;23:254—63.

[110] Warboys C, Narges A, de Luca A, Evans PC. The role of blood flow in determining the sites of atherosclerotic plaques. F1000 Med Rep 2011;3:5.

[111] Gimbrone MA, García-Cardeña G. Endothelial cell dysfunction and the pathobiology of atherosclerosis. Circ Res 2016;118:620—36.

[112] Karbach S, Wenzel P, Waisman A, Munzel T, Daiber A. eNOS uncoupling in cardiovascular diseases — the role of oxidative stress and inflammation. CPD 2014;20:3579−94.

[113] Cassuto J, Dou H, Czikora I, Szabo A, Patel VS, Kamath V, et al. Peroxynitrite disrupts endothelial caveolae leading to eNOS uncoupling and diminished flow-mediated dilation in coronary arterioles of diabetic patients. Diabetes 2014;63:1381−93.

[114] Al-Mehdi AB, Zhao G, Dodia C, Tozawa K, Costa K, Muzykantov V, et al. Endothelial NADPH oxidase as the source of oxidants in lungs exposed to ischemia or high K + . Circ Res 1998;83:730−7.

[115] Fisher AB, Chien S, Barakat AI, Nerem RM. Endothelial cellular response to altered shear stress. Am J Physiol Lung Cell Mol Physiol 2001;281:L529−33.

[116] Hirata K, Akita H, Yokoyama M. Oxidized low density lipoprotein inhibits bradykinin-induced phosphoinositide hydrolysis in cultured bovine aortic endothelial cells. FEBS 1991;287:181−4.

[117] Inoue N, Hirata K, Yamada M, Hamamori Y, Matsuda Y, Akita H, et al. Lysophosphatidylcholine inhibits bradykinin-induced phosphoinositide hydrolysis and calcium transients in cultured bovine aortic endothelial cells. Circ Res 1992;71:1410−21.

[118] Tao J-Q, Sorokina EM, Vázquez-Medina JP, Mishra MK, Yamada Y, Satalin J, et al. Onset of inflammation with ischemia: implications for donor lung preservation and transplant survival. Am J Transpl 2016;16:2598−611.

[119] Zhu B, Suzuki Y, DiSanto T, Rubin S, Penfil Z, Pietrofesa RA, et al. Applications of out of body lung perfusion. Acad Radiol 2019;26:404−11.

[120] De Vriese AS, Verbeuren TJ, Van de Voorde J, Lameire NH, Vanhoutte PM. Endothelial dysfunction in diabetes. Br J Pharmacol 2000;130:963−74.

[121] Rask-Madsen C, King GL. Mechanisms of disease: endothelial dysfunction in insulin resistance and diabetes. Nat Clin Pract Endocrinol Metab 2007;3:46.

[122] Yang Y-M, Huang A, Kaley G, Sun D. eNOS uncoupling and endothelial dysfunction in aged vessels. Am J Physiol Heart Circ Physiol 2009;297:H1829−36.

[123] Zou M-H, Cohen RA, Ullrich V. Peroxynitrite and vascular endothelial dysfunction in diabetes mellitus. Endothelium 2004;11:89−97.

[124] Gray SP, Di Marco E, Okabe J, Szyndralewiez C, Heitz F, Montezano AC, et al. NADPH oxidase 1 plays a key role in diabetes mellitus−accelerated atherosclerosis. Circulation 2013;127:1888−902.

[125] Griendling KK, Touyz RM, Zweier JL, Dikalov S, Chilian W, Chen Y-R, et al. Measurement of reactive oxygen species, reactive nitrogen species, and redox-dependent signaling in the cardiovascular system: a scientific statement from the American Heart Association. Circ Res 2016;119:e39−75.

[126] Brandes RP, Rezende F, Schröder K. Redox regulation beyond ROS: why ROS should not be measured as often. Circ Res 2018;123:326−8.

Chapter 5

Cross talk between the endothelium and bone: vascular endothelial cells in bone development

Michael M. Mayer[1], Daniel C. Kargilis[2], Alexander R. Farid[1], Sofia M. Miguez[1] and Chamith S. Rajapakse[1,2]

[1]Department of Radiology, University of Pennsylvania, Philadelphia, PA, United States, [2]Department of Orthopaedic Surgery, University of Pennsylvania, Philadelphia, PA, United States

5.1 Introduction

The skeletal system forms a cage-like infrastructure for the human body. Composed of bones that form a cohesive network, it enables movement and serves crucial roles in mineral storage, hematopoiesis, and energy storage. Bone is a complex tissue consisting of several cell types—osteoblasts, osteoclasts, and osteocytes, among others. Osteoblasts are bone-forming cells, osteoclasts break down bone, and osteocytes are mature bone cells. Bone tissue is a highly vascularized structure that receives around 10%−15% of resting cardiac output [1].

It is important to note that there are two types of bone tissue: compact and spongy. Compact bone consists of closely packed osteons, also known as Haversian systems. Osteons are cylindrical structures that contain a mineral matrix and living osteocytes connected by canaliculi, which transport blood. Each osteon consists of lamellae, which are layers of compact matrix that surround a central canal called the Haversian canal. Meanwhile, its counterpart, spongy bone, consists of plates of bone, called trabeculae, around irregular spaces that contain red bone marrow. Regardless of tissue type, bone is a highly vascularized tissue, characterized by an intense turnover of neoformation and resorption. This vasculature serves several functions—skeletal development and growth, bone modeling and remodeling, and healing processes. Further, the endothelium is an integral part of bone tissue and has a role in the interaction with bone cells in each of the aforementioned processes. The development of new blood vessels from preexisting vessels is integral to proper bone development; termed as angiogenesis, this process is crucial for bone repair through the secretion of regulatory signals from endothelial cells. The blood vessels of bone are lined with endothelial cells that regulate angiogenesis through increasingly understood molecular and functional properties. For example, the release of chemotactic factors such as vascular endothelial growth factor (VEGF) induces both invasion of bone vessels and migration of precursor cells into developing bone. Here in this chapter, we discuss the dynamic, yet incompletely clear, interplay between vascular endothelial cells and bone.

5.2 Angiogenesis in bone development

Bone development and remodeling continuously take place throughout our lifetimes. Every bone in the human body is formed through a process known as ossification, which exists in two forms depending on the type of bone being formed, whether intramembranous or endochondral [2,3]. Intramembranous ossification results in the formation of the compact and spongy bone that compose the flat bones of the face, most of the cranial bones, and the clavicles. The process begins with precursor mesenchymal cells in the embryonic skeleton differentiating into specialized cells known as osteoblasts, which aggregate in an ossification center. Osteoblasts secrete an unmineralized matrix known as osteoid that promotes bone formation. Osteoid forms deposits around blood vessels, resulting in a trabecular matrix that eventually forms marrow [4,5]. Endochondral ossification forms all other bones in the body, including long bones. During embryonic development, mesenchymal cells differentiate into chondrocytes that form an avascular cartilage template.

Endothelial Signaling in Vascular Dysfunction and Disease. DOI: https://doi.org/10.1016/B978-0-12-816196-8.00013-8

Subsequently, in a complex process, the avascular cartilage is replaced by highly vascularized bone as chondrocytes hypertrophy, produce a calcified extracellular matrix, and induce blood vessel invasion. A primary ossification center develops similar to intramembranous ossification, which replaces cartilage with bone. Simultaneously, bone increases in length as chondrocytes grow outward, releasing signals that promote vessel growth and ossification along the longitudinal axis. Interestingly, distinguishable metaphyseal and diaphyseal capillary networks are formed around this time [6]. After birth, secondary ossification centers form in the epiphysis of bone through a similar process, as vessels invade the epiphyseal chondrocytes. At this point, cartilage is only found at the growth plate and joint surfaces [4,5].

The organization of bone vasculature consists of an afferent arteriole feeding into a capillary network that drains into a venous system in the diaphysis of bone, with the main shaft of the bone containing a marrow of hematopoietic stem cells. Endothelial cells of bone vasculature are permeable to growth factors that allow specific differentiation of bone precursor cells. A 2014 study by Kusumbe et al. [6] characterized two subtypes of bone capillaries, H and L, based on marker expression and functional characteristics. In brief, type H capillaries express higher levels of hypoxia-inducible factor 1-α (HIF-1α) and localize to the bone metaphysis, which contains the avascular growth plate. In contrast, type L endothelial cells localize to sinusoidal vessels in the diaphysis. Experiments conducted by Kusumbe et al. demonstrated that while the number of L endothelial cells remains relatively constant, the number of H endothelial cells is age-dependent. Furthermore, precursor cells involved in bone regeneration interact mostly with the type H endothelial cells, and age-related decrease in bone precursor cell corresponds to decreased H endothelial cells, suggesting a critical role in bone functioning [6,7].

Ultimately, the skeletal system is complex and ever-changing. The role of endothelial cells, regulatory cells, and growth factors in the disease, repair, and development processes of bone will be explored in depth below.

5.3 Interplay between endothelial cells, regulatory factors, and bone development

There is a relationship between blood vessels and bone formation, frequently referred to as "angiogenic-osteogenic coupling." This cross talk between the endothelial wall of skeletal blood vessels and bone cells involves interplay between several cell types and signaling molecules with key roles in bone development.

5.3.1 Vascular endothelial growth factor, a critical regulator in bone development

Vascular endothelial growth factor-A, commonly referred to as simply VEGF, is a critical regulator of angiogenesis [8]. VEGF exists in multiple isoforms and is secreted by cell types within or surrounding blood vessels, including osteoblasts. VEGF binds to receptors of two receptor tyrosine kinases, VEGF receptor 1 (VEGFR1) and VEGF receptor 2 (VEGFR2), with differing signaling properties and downstream physiological effects. The expression of both VEGFRs has been detected in primary human osteoblasts, and the chemotactic and proliferative effects of VEGF on osteoblastic cells have been demonstrated through recombinant stimulation [9,10]. VEGF is critical in endochondral ossification, playing key roles in both early angiogenic invasive events and the differentiation of bone precursor cells. Early in endochondral ossification, VEGF is first expressed by cells in mesenchymal condensations and perichondrial cells of avascular cartilage models of the future bone in HIF-1α [11]. In subsequent steps of endochondral ossification, chondrocytes express high levels of VEGF during their process of maturation and hypertrophy. The expression of VEGF induces an invasion of osteoblast precursors, osteoclasts, blood vessels, and hematopoietic cells from the perichondrium into the primary ossification center [12]. Impaired angiogenesis interferes with the resorption of cartilage required for development of the primary ossification center, resulting in a thickened growth plate [13].

Animal models have allowed investigation into the embryological consequences of VEGF deletion. Inactivation of the VEGF120 isoform during early embryological development demonstrates a delay in osteoclast and blood vessel invasion into the primary ossification center, cartilage removal, and reduced bone mineralization [12]. Inactivation of VEGF in osteoblastic precursors positive for osterix, a transcription factor essential for osteoblast differentiation that upregulates VEGF expression, results in reduced blood vessels in the perichondrium and impaired osteoblastic differentiation in long bones [14]. Additionally, VEGF expression in primary ossification centers is required for chondrocyte survival. Mice with selective expression of the VEGF188 isoform have sufficient angiogenesis in the primary ossification center but reported chondrocyte necrosis within the epiphyseal region of long bones [15]. The conditional deletion of VEGF in chondrocytes results in delayed blood vessel invasion into chondrocytes, impaired chondrocyte resorption, and massive chondrocyte death in developing joint and epiphyseal region [16].

Additionally, VEGF has been shown to play a regulatory role in the bone homeostasis in development, specifically the relationship between osteoblast and adipocyte differentiation in bone marrow mesenchymal stem cells [9,10].

In conditional VEGF knockout mice, there is a demonstrated reduction in bone mass and an increase in bone marrow fat in osteoblastic progenitor cells [17]. While the role of VEGF in blood vessel invasion and bone precursor cell migration appears to function through an intracellular mechanism, VEGF appears to influence osteoblast development through a paracrine mechanism. In support of this, the addition of recombinant VEGF failed to affect osteoblast or adipocyte differentiation in a VEGF-deficient mesenchymal stem cell culture, but the viral-mediated expression of VEGF normalized osteoblast and adipocyte expression [17]. Finally, numerous studies have demonstrated the role of osteoblast-derived VEGF exerts on osteoclast differentiation [18−20], and a recent study has demonstrated a paracrine effect of VEGF on osteoclast formation [17].

Hypoxia is a major regulator of the expression of VEGF. A hypoxic environment triggers the release of hypoxia-inducible factor-1 (HIF-1α), a transcriptional regulator that increases the expression of VEGF. VEGF can then bind to endothelial cells and induce angiogenesis, leading to increased levels of oxygen and other nutrients required for osteogenesis [21]. Conditional deletion of HIF-1α resulted in notable chondrocyte apoptosis in central epiphyseal regions of developing cartilage [21]. Chondrocyte apoptosis can be partially rescued in mice lacking HIF-1α, consistent with a model of VEGF as a downstream effector of HIF-1 [22].

5.3.2 Matrix metalloproteinases

Matrix metalloproteinases (MMPs) are a class of degrading enzymes that are secreted by osteoclasts and vascular cells. Generally, MMPs exhibit proteolytic degradation of the extracellular matrix, leading to matrix remodeling that is essential for angiogenesis and bone remodeling [23,24]. While there are a reported 23 versions of MMPs in humans, MMP-2 (gelatinase A), MMP-9 (gelatinase B), MMP-13 (collagenase 3), MMP-14, and MMP-16 specifically have been found to be important for the two aforementioned functions [24−26].

Various studies have demonstrated these effects. MMP-2 knockout mice show osteopenia in their long bones after relatively normal early development [27]. In 1998, Vu et al. [24] reported that MMP-9-deficient mice showed delayed vascularization and ossification of the skeletal growth plate; administration of MMP-9 in these mice rescued these deficits. MMP-13 mutant mice, meanwhile, show increased growth plates and tabular bone, but fail to demonstrate any evident change in the vasculature [26]. Double-mutant MMP-13/MMP-9 mice are found to have reduced endochondral angiogenesis, decreased extracellular matrix remodeling, and complications regarding bone formation [26]. Lastly, MMP-14 knockout mice demonstrate significantly decreased bone mass, particularly via cartilage loss, due to greater bone resorption and diminished bone development [28,29]; MMP-14 is also a known activator of MMP-2, meaning that defects caused by an MMP-14 knockout may come as a result, at least in part, of deficient MMP-2 activity [30,31].

It is important to note the interplay between MMP and VEGF in angiogenesis and bone remodeling. VEGF, as previously mentioned, exerts various effects on bone angiogenesis via binding of the extracellular matrix [32], while matrix-bound VEGF contributes to extended activation of receptor VEGFR2 [33]; these findings illuminate an avenue by which the matrix-targeting MMPs can interact with VEGF. Administration of VEGF in MMP-9-deficient mice rescues the defects such mice had previously demonstrated [23]. This, among other similar studies, suggests that MMP-9 is likely involved in the release of VEGF via the extracellular matrix [13]. Overall, evidence indicates that MMPs likely impact aspects of angiogenesis and bone remodeling via the degradation and remodeling of the extracellular matrix (ECM).

5.3.3 Fibroblast growth factor

Fibroblast growth factor (FGF) signaling refers to a complex family of ligands and receptors involved in cell survival, proliferation, migration, and differentiation [34]. There are 23 different ligands, with 18 total FGF receptors (FGFR) [34]. FGF signaling is relatively versatile, occurring through paracrine, autocrine, and even endocrine modes of communication depending on function. Due to the wide array of effects related to these ligands, their interactions are highly complex and can often impact cellular functions in an interrelated manner [34]. During early skeletal development, FGF signaling plays a regulatory role in limb bud development and mesenchymal condensation [35]. Beyond development, FGF continues to play a critical role in the continuous processes of chondrogenesis, osteogenesis, angiogenesis, and homeostasis of bone [35].

FGFs are primarily secreted from chondrocytes and osteogenic cells for functions related to osteogenesis [36]. Two receptors, FGFR1 and FGFR2, are expressed in bone vasculature. Inactivation of the respective genes encoding for these receptors causes changes in vascular permeability that suggest a role for FGF in maintaining arterial function and integrity [37]. FGF signaling has previously been shown to trigger VEGF expression and expansion of arterial bone vasculature [38]. Additionally, mutant mouse lines deficient in FGF2 display reduced trabecular bone volume and rate of bone formation [39]. Furthermore, FGFs are known to play a critical role in bone repair. Expression of FGFs and FGFRs are increased during fracture healing processes, and both FGF2 and FGF9 have been shown to increase angiogenesis and osteogenesis in various models [40,41]. Furthermore, mutant mouse lines deficient in FGF9 display impaired bone healing and expression of VEGF near lesion sites [42]. These lines of evidence point to the significant role of FGF signaling in bone and vasculature development and repair.

5.4 Endothelial cells in bone repair

Angiogenesis and the wide array of signaling processes associated with endothelial cells are important components of bone fracture healing. In this section, we review the various involvements of endothelial cell types and their signaling mechanisms, including endothelial progenitor differentiation and inflammation in bone repair.

5.4.1 Endothelial progenitor cells in fracture healing

In addition to roles in the continuous processes of osteogenesis and angiogenesis, endothelial cells have recently been shown to activate in response to injury or bone fracture. These observations have led to a recent surge in clinical applications based on endothelial progenitor cells (EPCs), or cell populations that can differentiate into adult endothelial cells [43]. In 1997, EPCs isolated from human peripheral blood were shown to incorporate into active sites of angiogenesis, suggesting potential for use in treatments for endothelial repair [44]. Additionally, EPCs have been shown to mobilize in response to ischemia of vasculature or exposure to exogenous cytokine stimulation [45]. In light of these observations, the potential for EPCs in neo-angiogenic based therapies has been explored for patients with ischemic damage [43,46].

Recently, with the growing linkage between endothelial cells and osteogenesis, investigations have focused on the potential for EPCs in bone repair following sustained fracture. Fracture healing is a highly complex process and involves many similar mechanisms to embryogenesis, including the circulation of soluble signaling factors and incompletely differentiated cells [47]. Vascular function is essential for the delivery of these components, and the number of circulating peripheral EPCs is elevated significantly in subjects with fractures compared to age-matched controls [48]. Traumatic fractures are therefore suggested to activate EPC circulation and contribution in fracture healing processes. Recent investigations have specifically focused on CD34 + peripheral blood cells, which contain EPCs and hematopoietic stem cells [49]. CD34 + cells have been shown to encourage fracture healing by secreting VEGF, which plays key roles in both angiogenesis and osteogenesis as previously discussed [49,52]. Interestingly, EPCs from CD34 + cells are also capable of differentiating into osteoblasts in addition to endothelial cells, suggesting an alternate mechanism by which CD34 + cells can facilitate fracture healing [49].

Previous interventions in fracture healing have utilized VEGF, platelet-derived growth factors (PDGF) , and FGFs to encourage angiogenesis, but potential drawbacks around cost and toxicity compared to direct cellular implantation have arisen [51]. In addition to EPCs, mesenchymal stem cells (MSCs) have been utilized in research promoting bone regeneration [51,52]. Notably, interventions with cotransplantation of both EPCs and MSCs have resulted in greater bone regeneration than either cell type alone [51]. This effect may be due to beneficial interactions between endothelial cells and MSCs that encourage bone development. Furthermore, endothelial colony-forming cells have been shown to secrete bone morphogenetic protein and transforming growth factor-β1, which play critical roles in the regulation of tissue and bone formation [17,53]. Taken together, these findings demonstrate the critical role of EPCs in fracture healing processes. Future cell-based therapies implementing EPCs show therapeutic potential due to the overlapping interactions between EPCs and various osteogenic cell types.

5.4.2 Role of vascular endothelial growth factor in bone repair

Sufficient blood supply is essential for bone repair following injury. Considering this fact, appropriate vascularization is critical for ensuring sufficient blood supply. As previously discussed, VEGF promotes angiogenesis, and via such function, has a notable role in bone repair. Fracture healing can be described in three phases (Fig. 5.1). In phase 1, inflammatory processes are triggered. In phase 2, endochondral ossification leads to the formation of new cartilage. Finally, in phase 3, bone is repaired through intramembranous ossification. VEGF, as an angiogenic factor that leads to formation of new blood vessels, is involved throughout the repair and healing process. It was reported that treating fracture areas with VEGF in rabbits led to an improved fracture healing process via increased vascularization of the region [54].

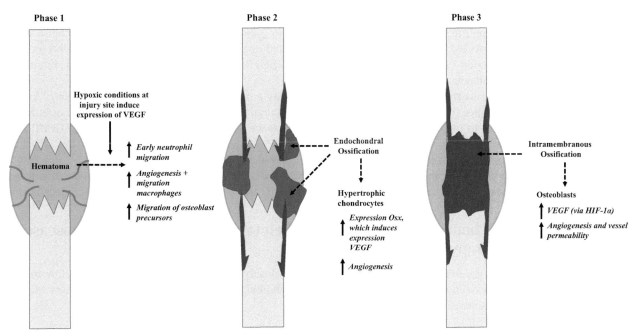

FIGURE 5.1 The three phases of fracture healing. Phase 1: Bone injury results in a hematoma. Local hypoxia induces nearby bone and inflammatory cells to express VEGF, which plays a prominent role in early neutrophil migration, angiogenesis, and subsequent migration of macrophages. Phase 2: New cartilage is formed through endochondral ossification. Hypertrophic chondrocytes express Osx, an inducer of VEGF. Phase 3: Bone repair occurs via intramembranous ossification. Mesenchymal progenitor cells differentiate into osteoblasts, which secrete VEGF. Angiogenesis and increases in vessel permeability allow for delivery of progenitor cells, nutrition, and oxygen to promote healing of injured bone.

In phase 1, during which local vessels have ruptured to create a hematoma, nearby bone and inflammatory cells are induced to express VEGF in the area by hypoxia [55,56]. VEGF has also been found in such hematomas. Neutrophils migrate to the area of injury, removing bone debris and small pathogens while osteoblast precursors and hematopoietic cells remain near the bone endothelial layer; macrophages move to the region via newly forming blood vessels. VEGF plays a role in many of these processes. More specifically, VEGF likely induces neutrophil migration and increases permeability of the nearby endothelium for osteoblast precursors and hematopoietic cells to act on the injury site, promoting the vessel formation necessary to allow the movement of macrophages [57,58]. It has been proposed that macrophages are recruited by the formation of new blood vessels, which in turn encourages further angiogenesis by recruiting VEGF via a positive feedback mechanism. In a study by Hu and Olsen [59], they reported that *Vegfa* knockout mice demonstrated lower levels of neutrophil release, delaying the early-stage inflammatory response. They also found that *Vegfa* knockout mice generally showed lower VEGF levels at injury sites, leading to diminished angiogenesis and a subsequent lower recruitment of macrophages, further delaying the repair process [59].

During phase 2, new cartilage is formed through endochondral ossification. In comparison with intramembranous ossification, which forms bone tissue, endochondral ossification is more prevalent form when a fracture is unstable and is likely facilitated by low blood supply [60]. It was found that inhibition of VEGF in skeletal progenitor cells, led to cartilage formation rather than bone formation [61], suggesting a role for low blood supply via diminished angiogenesis and low VEGF activity in cartilage formation. Hypertrophic chondrocytes, which are found in cartilage and direct cartilage matrix formation, express Osx, an inducer of VEGF [62,63]. By stimulating VEGF, angiogenesis increases in these areas of cartilage, allowing osteoclast and chondroclast movement to occur. In mice with *Vegfa* knockdown in hypertrophic chondrocytes, new vessel formation and osteoclast movement were delayed following bone injury [17,59]. By inhibiting VEGF in mice by blocking VEGFR1, a delay in cartilage turnover and impaired healing were both reported, resulting in an inhibition of fracture healing [64]. Similarly, mice with a deleted gene for placental growth factor, a member of the VEGF family, were found to have reduced inflammation and callus remodeling, contributing to worsened fracture healing processes [65].

Lastly, in phase 3, bone repair occurs via intramembranous ossification. During this process, mesenchymal progenitor cells differentiate to osteoblasts. VEGF is synthesized and secreted by osteoblasts via induction by HIF-1α [66]; this leads to increased proliferation and migration of endothelial cells, as well as increased vessel permeability [56]. Vessels are also continually being formed via induction by VEGF; increased permeability and angiogenesis together thus allow a heightened opportunity for osteoblast progenitor cells, nutrition, oxygen, and osteogenic factors to reach and repair the injured bone [67]. To confirm that osteoblasts release VEGF to induce angiogenesis, it was reported that a *Vegfa* knockout in mouse osteoblasts led to a decoupling of osteogenic and angiogenic processes, as well as a delay in bone repair [59]. Blocking the action of VEGF via neutralization of VEGFRs in mice similarly demonstrated diminished angiogenesis and intramembranous bone formation [68].

There are other mechanisms suggested by which VEGF may play a role in the injury repair process. *Vegfr2* knockout in Osx-positive cells in mice resulted in increased osteoblast maturation during repair [59]. Relatedly, in osteoblasts, low VEGFR2 activation led to increased mineralization in the presence of bone morphogenetic protein 2 (BMP-2), a growth factor; a potential explanation is that VEGFR2 inhibits BMP pathways, which potentially elucidates why high VEGFR activation has been found to inhibit BMP-2- and 4-induced bone repair [59,69].

It is important to note that there are certainly many other factors and pathways that control and regulate angiogenesis and subsequent behavior of bone-healing cells. Nonetheless, VEGF appears integral via several distinct mechanisms.

5.5 Dysfunction of bone endothelial cells in skeletal and systemic disease: disease presentation and therapy

The angiogenesis and bone development and repair pathways are closely intertwined. This is discussed below in both skeletal diseases (namely avascular necrosis, osteoporosis, and mandibular hypoplasia) and systemic diseases (namely diabetes mellitus and atherosclerosis).

5.5.1 Skeletal and systemic pathology

Osteonecrosis, also called avascular necrosis, is characterized by reduced blood flow and bone cell death, and can potentially lead to bone collapse. Osteonecrosis can affect any bone, and commonly affected areas include the hips and knees and less commonly the shoulders, hands, feet, and jaw. The severity of degenerative changes that follow osteonecrosis depends on the extent of the interrupted blood supply and the ability to develop collateral circulation. The etiology of osteonecrosis can be traumatic or atraumatic. In patients with atraumatic osteonecrosis, chronic steroid is a common culprit, although other causes are less understood. It has been proposed that the blood flow interruption in osteonecrosis results from a damaged endothelial cell membrane in the setting of an individual with increased propensity to blood clot formation [70]. Additionally, studies of patients with atraumatic avascular necrosis have demonstrated the importance of bone marrow-derived EPCs in the body's response to osteonecrosis. EPCs have been proposed to function in the processes of neovascularization, and approximately 25% of endothelial cells in animal models consist of EPCs [71]. Patients with avascular necrosis of the femoral head compared to control subjects had a reduced number of EPCs with decreased migratory capacities and increased senescence [72]. Relatedly, patients with bisphosphonates-associated osteonecrosis of the jaw have reduced circulating EPCs [73].

Another skeletal disease associated with bone vasculature is osteoporosis, which is a disease of improper bone homeostasis where the density and quality of bone are reduced. As a result, the risk of bone fracture is significantly increased. Postmenopausal osteoporosis, specifically, is associated with a reduced number of sinusoidal and arterial capillaries in the bone marrow and decreased bone perfusion [74]. It has been found that activation of the HIF-1α signaling pathway via disruption of the HIF degrading protein von Hippel—Lindau resulted in increased trabecular and cortical bone volume in ovariectomized mice compared to control mice that did not receive increased activation of HIF-1α [74]. That is, activation of the HIF signaling pathways in osteoblasts prevented estrogen-deficiency-induced bone loss and a decrease in blood vessels in the bone marrow. A similar observation was reported by Liu et al. when the HIF signaling pathway was activated via the inhibition of prolyl hydroxylases (PHD) in mice. This suggests that estrogen-deficiency—induced bone loss can be prevented through the activation of the HIF/VEGF pathway and that bone formation and angiogenesis are coupled through the HIF signaling in osteoblasts [74].

Additionally, studying craniofacial defects such as cleft palate and mandibular hypoplasia provides insight into the interplay between angiogenesis and intramembranous ossification. Cleft palate is one of the most common craniofacial congenital anomalies, and mandibular hypoplasia, which can be congenital, developmental, or acquired, is a condition where the jaw is undersized or underdeveloped. As discussed earlier, intramembranous ossification is the process by which the flat bones of the face, most of the cranial bones, and the clavicles are formed. In craniofacial bones, VEGF is expressed by the neural crest cell -derived progenitor cells that directly differentiate into osteoblasts. Numerous studies have demonstrated that impairment in CNC-derived VEGF secretion impairs proliferation, vascularization, and ossification during craniofacial development. Conditional deletion of VEGF164 in mice resulted in craniofacial defects such as cleft lip and shorter jaws [75]. In another study, conditional deletion of VEGF in CNC-derived cells resulted in mice with reduced proliferation of cells within the palatal shelves with abnormal palatal shelf formation and inability to undergo fusion. Furthermore, both vascular development and intramembranous ossification of maxillary and palatal mesenchyme were impaired [76].

It is hypothesized that systemic diseases can also negatively affect bone vasculature. Diabetes mellitus is a disease in which the body's ability to produce or respond to insulin is impaired, which results in the abnormal regulation of blood glucose levels. As a result, bone health and cardiovascular function are compromised in individuals with diabetes mellitus and these individuals often develop diabetes-induced osteoporosis [77]. The details of this interaction are still being investigated, though research has found that diabetes mellitus induces bone marrow microangiopathy in mice with type 1 diabetes mellitus (T1DM) [78]. Mice with T1DM were found to have reduced blood flow and microvascular density in the bone marrow, and higher levels of oxidative stress in endothelial cells [78]. Additionally, through the study of mice models, it has been found that the progression of type 2 diabetes mellitus causes vasoconstriction and impairs the blood flow within long bones [77]. Research suggests that this interaction may be the cause of diabetes-induced-osteopenia [77]. T1DM has also been associated with the impairment of angiogenesis and researchers believe that this is the main cause of decreased bone formation associated with diabetes [79]. Additionally, recent studies have uncovered that men presenting with osteoporosis are at an increased risk for atherosclerotic disease [80], suggesting a link between osteoporosis and atherosclerosis. Prasad et al. concluded that postmenopausal women with coronary microvascular endothelial dysfunction were twice as likely to develop osteoporosis as women without endothelial dysfunction [81]. That is, in postmenopausal women older than 50, coronary atherosclerosis with endothelial dysfunction is correlated with increased risk of developing osteoporosis [81].

5.5.2   A brief overview of therapeutic approaches to engineering vascularized bone

Our increased understanding of the molecular biology of bone has allowed tremendous advancement in the field of bone tissue engineering. Bone tissue engineering involves the successful implantation of progenitor cells, growth factors, and scaffolds. It also involves adopting strategies to enhance or accelerate vascularization. Prominent strategies in bone tissue engineering are discussed in this section.

One cell-based therapy for enhancing vascularity in bone involves the mobilization of EPCs via the harvesting of CD34 + cells in adult blood [82]. EPCs express both hematopoietic progenitor markers (CD34 and CD133) and an endothelial marker (VEGF receptor 2), and are involved in vascular repair, fracture healing, and bone regeneration [51,83]. CD34 + cells are more abundant in adult blood than EPCs, and therefore easier to harvest. CD34 + cells have demonstrated both angiogenic and osteogenic characteristics similar to EPCs in vitro and in vivo [82]. A study evaluating the efficacy of EPC therapy on bone regeneration concluded that local EPC therapy significantly enhanced bone regeneration in rats with femur fractures [83]. This provides encouraging evidence that EPCs can be harvested with CD34 + cells and effectively enhances bone healing.

Early animal models demonstrated the efficacy of applying recombinant proteins such as VEGF in improving bone regeneration. However, injection of recombinant proteins is unideal as therapeutic agents, due to their short biological half-life, rapid degradation in vivo, and costs [84]. Gene therapy has thus emerged as a highly efficacious approach to bone tissue engineering, with unique advantages over recombinant proteins, such as the release of regulatory factors at a selected site of injury over a set period. In other words, gene therapy allows for spatially and temporally regulated gene expression. Ex vivo gene therapy approaches for bone engineering demonstrate healing of bone injury in mice with skull defects through coimplantation of bone morphogenic protein 4 (BMP-4) transduced muscle-derived stem cells and VEGF transduced muscle-derived stem cells. The effects were synergistic, with enhanced angiogenesis, improved cartilage resorption, and decreased mineralization in comparison to mice with only BMP-induced bone regeneration [69,85]. In vivo gene therapy approaches involving the direct injection into a bone lesion have demonstrated repair in a rabbit model of avascular necrosis of the femoral head. Coexpression of BMP-7 and VEGF through an adenovirus vector improved healing capacity by promoting the metabolism of the necrotic region by inducing angiogenesis [89]. Furthermore, clinical trials in gene therapy for bone repair are underway. Recent trials are taking advantage of gene plasmid-based therapy to treat alveolar bone loss, utilizing a scaffold loaded with VEGF plasmids [90].

Another approach is cell-free scaffolds that can incorporate bioactive factors to promote osteoconduction, osteoinduction, and/or angiogenesis as well as to provide mechanical stability for implanted material [82,88]. These scaffolds are designed to be either gradually degraded by the body and replaced by bone as it grows or to become integrated within the new bone as it grows [89,90].

5.5.3 Concluding remarks

Vasculature is a crucial component of bone structure and plays a pivotal role in both bone development and repair. Vascular signaling is closely associated with bone growth. In particular, the release of chemotactic factors such as VEGF, is critical for angiogenesis. In turn, the invasion of blood vessels allows the migration of precursor cells into developing bone. This angiogenic-osteogenic coupling is critical in cell repair as well, with impairments leading to skeletal and systemic disease. Furthering our understanding of this cross talk between bone and vasculature will allow us to continue advancing the field of bone tissue engineering. Overall, the answer to accelerating bone repair and growth also lies in understanding the cascade of vascular signals.

References

[1] Tomlinson RE, Silva MJ. Skeletal blood flow in bone repair and maintenance. Bone Res 2013;1(4):311−22.

[2] Hayashi S, Kim JH, Hwang SE, Shibata S, Fujimiya M, Murakami G, et al. Interface between intramembranous and endochondral ossification in human foetuses. Folia Morphol (Warsz) 2014;73(2):199−205.

[3] Berendsen AD, Olsen BR. Bone development. Bone 2015;80:14−18.

[4] Filipowska J, Tomaszewski KA, Niedźwiedzki Ł, Walocha JA, Niedźwiedzki T. The role of vasculature in bone development, regeneration and proper systemic functioning. Angiogenesis 2017;20(3):291−302.

[5] Sivaraj KK, Adams RH. Blood vessel formation and function in bone. Development 2016;143(15):2706−15.

[6] Kusumbe AP, Ramasamy SK, Adams RH. Coupling of angiogenesis and osteogenesis by a specific vessel subtype in bone. Nature 2014;507:323−8.

[7] Ramasamy SK, Kusumbe AP, Wang L, Adams RH. Endothelial Notch activity promotes angiogenesis and osteogenesis in bone. Nature 2014;507(7492):376−80.

[8] Ferrara N, Gerber HP, LeCouter J. The biology of VEGF and its receptors. Nat Med 2003;9(6):669−76.

[9] Deckers MM, Karperien M, van der Bent C, Yamashita T, Papapoulos SE, Löwik CW. Expression of vascular endothelial growth factors and their receptors during osteoblast differentiation. Endocrinology 2000;141(5):1667−74.

[10] Mayr-Wohlfart U, Waltenberger J, Hausser H, et al. Vascular endothelial growth factor stimulates chemotactic migration of primary human osteoblasts. Bone 2002;30:472−7.

[11] Amarilio R, Viukov SV, Sharir A, Eshkar-Oren I, Johnson RS, Zelzer E. HIF1α regulation of Sox9 is necessary to maintain differentiation of hypoxic prechondrogenic cells during early skeletogenesis. Development 2007;134(21):3917−28.

[12] Zelzer E, McLean W, Ng YS, Fukai N, Reginato AM, Lovejoy S, et al. Skeletal defects in VEGF(120/120) mice reveal multiple roles for VEGF in skeletogenesis. Development 2002;129:1893−904.

[13] Gerber HP, Vu TH, Ryan AM, Kowalski J, Werb Z, Ferrara N. VEGF couples hypertrophic cartilage remodeling, ossification and angiogenesis during endochondral bone formation. Nat Med 1999;5(6):623−8.

[14] Duan X, Murata Y, Liu Y, Nicolae C, Olsen BR, Berendsen AD. Vegfa regulates perichondrial vascularity and osteoblast differentiation in bone development. Development 2015;142(11):1984−91.

[15] Maes C, Stockmans I, Moermans K, Van Looveren R, Smets N, Carmeliet P, et al. Soluble VEGF isoforms are essential for establishing epiphyseal vascularization and regulating chondrocyte development and survival. J Clin Invest 2004;113(2):188−99.

[16] Zelzer E, Glotzer DJ, Hartmann C, Thomas D, Fukai N, Soker S, et al. Tissue specific regulation of VEGF expression during bone development requires Cbfa1/Runx2. Mech Dev 2001;106:97−106.

[17] Liu Y, Berendsen AD, Jia S, Lotinun S, Baron R, Ferrara N, et al. Intracellular VEGF regulates the balance between osteoblast and adipocyte differentiation. J Clin Invest 2012;122:3101−13.

[18] Nakagawa M, Kaneda T, Arakawa T, et al. Vascular endothelial growth factor (VEGF) directly enhances osteoclastic bone resorption and survival of mature osteoclasts. FEBS Lett 2000;473:161−4.

[19] Niida S, Kaku M, Amano H, et al. Vascular endothelial growth factor can substitute for macrophage colony-stimulating factor in the support of osteoclastic bone resorption. J Exp Med 1999;190:293−8.

[20] Yang Q, McHugh KP, Patntirapong S, et al. VEGF enhancement of osteoclast survival and bone resorption involves VEGF receptor-2 signaling and beta3-integrin. Matrix Biol 2008;27:589−99.

[21] Schipani E, Maes C, Carmeliet G, Semenza GL. Regulation of osteogenesis-angiogenesis coupling by HIFs and VEGF. J Bone Miner Res 2009;24:1347−53.

[22] Maes C, Araldi E, Haigh K, Khatri R, Van Looveren R, Giaccia AJ, et al. VEGF-independent cell-autonomous functions of HIF-1α regulating oxygen consumption in fetal cartilage are critical for chondrocyte survival. J Bone Miner Res 2012;27(3):596−609.

[23] Ortega N, Wang K, Ferrara N, Werb Z, Vu TH. Complementary interplay between matrix metalloproteinase-9, vascular endothelial growth factor and osteoclast. 2010.

[24] Vu TH, Shipley JM, Bergers G, et al. MMP-9/gelatinase B is a key regulator of growth plate angiogenesis and apoptosis of hypertrophic chondrocytes. Cell. 1998;93(3):411−22.

[25] Liang HPH, Xu J, Xue M, Jackson CJ. Matrix metalloproteinases in bone development and pathology: current knowledge and potential clinical utility. Metalloproteinases Med 2016;3:93−102.

[26] Stickens D, Behonick DJ, Ortega N, Heyer B, Hartenstein B, Yu Y, et al. Altered endochondral bone development in matrix metalloproteinase 13-deficient mice. Development 2004;131:5883−95.

[27] Inoue K, Mikuni-Takagaki Y, Oikawa K, et al. A crucial role for matrix metalloproteinase 2 in osteocytic canalicular formation and bone metabolism. J Biol Chem 2006;281(44):33814−24.

[28] Holmbeck K, Bianco P, Pidoux I, et al. The metalloproteinase MT1-MMP is required for normal development and maintenance of osteocyte processes in bone. J Cell Sci 2005;118(Pt 1):147−56.

[29] Zhao W, Byrne MH, Wang Y, Krane SM. Osteocyte and osteoblast apoptosis and excessive bone deposition accompany failure of collagenase cleavage of collagen. J Clin Invest 2000;106(8):941−9.

[30] Oblander SA, Zhou Z, Galvez BG, et al. Distinctive functions of membrane type 1 matrix-metalloprotease (MT1-MMP or MMP-14) in lung and submandibular gland development are independent of its role in pro-MMP-2 activation. Dev Biol 2005;277(1):255−69.

[31] Strongin AY, Collier I, Bannikov G, Marmer BL, Grant GA, Goldberg GI. Mechanism of cell surface activation of 72-kDa type IV collagenase. Isolation of the activated form of the membrane metalloprotease. J Biol Chem 1995;270(10):5331−8.

[32] Allerstorfer D, Longato S, Schwarzer C, Fischer-Colbrie R, Hayman AR, Blumer MJ. VEGF and its role in the early development of the long bone epiphysis. J Anat 2010;216:611−24.

[33] Avraamides CJ, Garmy-Susini B, Varner JA. Integrins in angiogenesis and lymphangiogenesis. Nat Rev Cancer 2008;8:604−17.

[34] Turner N, Grose R. Fibroblast growth factor signalling: from development to cancer. Nat Rev Cancer 2010;10:116−29.

[35] Ornitz DM, Marie PJ. Fibroblast growth factor signaling in skeletal development and disease. Genes Dev 2015;29:1463−86.

[36] Kozhemyakina E, Lassar AB, Zelzer E. A pathway to bone: signaling molecules and transcription factors involved in chondrocyte development and maturation. Development 2015;142:817−31.

[37] Itkin T, Gur-Cohen S, Spencer JA, Schajnovitz A, Ramasamy SK, Kusumbe AP, et al. Distinct bone marrow blood vessels differentially regulate haematopoiesis. Nature 2016;532:323−8.

[38] Seghezzi G, Patel S, Ren CJ, Gualandris A, Pintucci G, Robbins ES, et al. Fibroblast growth factor-2 (FGF-2) induces vascular endothelial growth factor (VEGF) expression in the endothelial cells of forming capillaries: an autocrine mechanism contributing to angiogenesis. J Cell Biol 1998;141:1659−73.

[39] Montero A, Okada Y, Tomita M, Ito M, Tsurukami H, Nakamura T, et al. Disruption of the fibroblast growth factor-2 gene results in decreased bone mass and bone formation. J Clin Invest 2000;105:1085−93.

[40] Dirckx N, Van Hul M, Maes C. Osteoblast recruitment to sites of bone formation in skeletal development, homeostasis, and regeneration. Birth Defects Res C Embryo Today 2013;99:170−91.

[41] Wallner C, Schira J, Wagner JM, Schulte M, Fischer S, Hirsch T, et al. Application of VEGFA and FGF-9 enhances angiogenesis, osteogenesis and bone remodeling in type 2 diabetic long bone regeneration. PLoS One 2015;10:e0118823.

[42] Behr B, Leucht P, Longaker MT, Quarto N. Fgf-9 is required for angiogenesis and osteogenesis in long bone repair. Proc Natl Acad Sci U S A 2010;107:11853−8.

[43] Chong MS, Ng WK, Chan JK. Concise review: endothelial progenitor cells in regenerative medicine: applications and challenges. Stem Cells Transl Med 2016;5(4):530−8.

[44] Asahara T, Murohara T, Sullivan A, Silver M, van der Zee R, Li T, et al. Isolation of putative progenitor endothelial cells for angiogenesis. Science 1997;275(5302):964−6.

[45] Takahashi T, Kalka C, Masuda H, Chen D, Silver M, Kearney M, et al. Ischemia-and cytokine-induced mobilization of bone marrow-derived endothelial progenitor cells for neovascularization. Nat Med 1999;5(4):434−8.

[46] Kocher AA, Schuster MD, Szabolcs MJ, Takuma S, Burkhoff D, Wang J, et al. Neovascularization of ischemic myocardium by human bone-marrow−derived angioblasts prevents cardiomyocyte apoptosis, reduces remodeling and improves cardiac function. Nat Med 2001; 7(4):430−6.

[47] Gerstenfeld LC, Cullinane DM, Barnes GL, Graves DT, Einhorn TA. Fracture healing as a post-natal developmental process: molecular, spatial, and temporal aspects of its regulation. J Cell Biochem 2003;88(5):873−84.

[48] Ma XL, Sun XL, Wan CY, Ma JX, Tian P. Significance of circulating endothelial progenitor cells in patients with fracture healing process. J Orthop Res 2012;30(11):1860−6.

[49] Matsumoto T, Kuroda R, Mifune Y, Kawamoto A, Shoji T, Miwa M, et al. Circulating endothelial/skeletal progenitor cells for bone regeneration and healing. Bone 2008;43(3):434−9.

[50] Matsumoto T, Kawamoto A, Kuroda R, Ishikawa M, Mifune Y, Iwasaki H, et al. Therapeutic potential of vasculogenesis and osteogenesis promoted by peripheral blood CD34-positive cells for functional bone healing. Am J Pathol 2006;169(4):1440−57.

[51] Keramaris NC, Kaptanis S, Lucy Moss H, Loppini M, Pneumaticos S, Maffulli N. Endothelial progenitor cells (EPCs) and mesenchymal stem cells (MSCs) in bone healing. Curr Stem Cell Res Ther 2012;7(4):293−301.

[52] Furuta T, Miyaki S, Ishitobi H, Ogura T, Kato Y, Kamei N, et al. Mesenchymal stem cell-derived exosomes promote fracture healing in a mouse model. Stem Cells Transl Med 2016;5(12):1620−30.

[53] Sandell LJ. Bone morphogenetic protein and transforming growth factor-β interactions with extracellular matrix in health and disease. Curr Opin Orthop 2006;17(5):412−17.

[54] Kleinheinz J, Stratmann U, Joos U, Wiesmann H-P. VEGFactivated angiogenesis during bone regeneration. J Oral Maxillofac Surg 2005;63:1310−16.

[55] Krock BL, Skuli N, Simon MC. Hypoxia-induced angiogenesis: good and evil. Genes Cancer 2011;2:1117−33.

[56] Wang Y, Wan C, Deng L, Liu X, Cao X, Gilbert SR, et al. The hypoxia-inducible factor alpha pathway couples angiogenesis to osteogenesis during skeletal development. J Clin Invest 2007;117:1616−26.

[57] Ancelin M, Chollet-Martin S, Herve MA, Legrand C, El Benna J, Perrot-Applanat M. Vascular endothelial growth factor VEGF189 induces human neutrophil chemotaxis in extravascular tissue via an autocrine amplification mechanism. Lab Invest 2004;84:502−12.

[58] Lim S, Zhang Y, Zhang D, Chen F, Hosaka K, Feng N, et al. VEGFR2-mediated vascular dilation as a mechanism of VEGF-induced anemia and bone marrow cell mobilization. Cell Rep 2014;9:569−80.

[59] Hu K, Olsen BR. Vascular endothelial growth factor control mechanisms in skeletal growth and repair. Dev Dyn 2016;246(4):227−34. Available from: https://doi.org/10.1002/dvdy.24463.

[60] Dimitriou R, Tsiridis E, Giannoudis PV. Current concepts of molecular aspects of bone healing. Injury 2005;36:1392−404.

[61] Chan CK, Seo EY, Chen JY, Lo D, Mcardle A, Sinha R, et al. Identification and specification of the mouse skeletal stem cell. Cell 2015;160:285−98.

[62] Carlevaro MF, Cermelli S, Cancedda R, Descalzi Cancedda F. Vascular endothelial growth factor (VEGF) in cartilage neovascularization and chondrocyte differentiation: auto-paracrine role during endochondral bone formation. J Cell Sci 2000;113(Pt 1):59−69.

[63] Zelzer E, Glotzer DJ, Hartmann C, Thomas D, Fukai N, Soker S, et al. Tissue specific regulation of VEGF expression during bone development requires Cbfa1/Runx2. Mech Dev 2001;106:97−106.

[64] Street J, Bao M, Deguzman L, Bunting S, Peale Jr FV, Ferrara N, et al. Vascular endothelial growth factor stimulates bone repair by promoting angiogenesis and bone turnover. Proc Natl Acad Sci U S A 2002;99:9656−61.

[65] Maes C, Coenegrachts L, Stockmans I, Daci E, Luttun A, Petryk A, et al. Placental growth factor mediates mesenchymal cell development, cartilage turnover, and bone remodeling during fracture repair. J Clin Invest 2006;116:1230−42.

[66] Wan C, Gilbert SR, Wang Y, Cao X, Shen X, Ramaswamy G, et al. Activation of the hypoxia-inducible factor-1alpha pathway accelerates bone regeneration. Proc Natl Acad Sci U S A 2008;105:686−91.

[67] Matsubara H, Hogan DE, Morgan EF, Mortlock DP, Einhorn TA, Gerstenfeld LC. Vascular tissues are a primary source of BMP2 expression during bone formation induced by distraction osteogenesis. Bone 2012;51:168−80.

[68] Carvalho RS, Einhorn TA, Lehmann W, Edgar C, Al-Yamani A, Apazidis A, et al. The role of angiogenesis in a murine tibial model of distraction osteogenesis. Bone 2004;34:849−61.

[69] Peng H, Wright V, Usas A, Gearhart B, Shen HC, Cummins J, et al. Synergistic enhancement of bone formation and healing by stem cell-expressed VEGF and bone morphogenetic protein-4. J Clin Invest 2002;110:751−9.

[70] Pouya F, Kerachian MA. Avascular necrosis of the femoral head: are any genes involved? Arch Bone Joint Surg 2015;3(3):149.

[71] Kunz GA, Liang G, Cuculi F, Gregg D, Vata KC, Shaw LK, et al. Circulating endothelial progenitor cells predict coronary artery disease severity. Am Heart J 2006;152(1):190−5.

[72] Feng Y, Yang SH, Xiao BJ, Xu WH, Ye SN, Xia T, et al. Decreased in the number and function of circulation endothelial progenitor cells in patients with avascular necrosis of the femoral head. Bone 2010;46(1):32−40.

[73] Allegra A, Oteri G, Nastro E, Alonci A, Bellomo G, Del Fabro V, et al. Patients with bisphosphonates-associated osteonecrosis of the jaw have reduced circulating endothelial cells. Hematol Oncol 2007;25(4):164−9.

[74] Zhao Q, Shen X, Zhang W, Zhu G, Qi J, Deng L. Mice with increased angiogenesis and osteogenesis due to conditional activation of HIF pathway in osteoblasts are protected from ovariectomy induced bone loss. Bone 2012;50(3):763−70.

[75] Stalmans I, Lambrechts D, De Smet F, Jansen S, Wang J, Maity S, et al. VEGF: a modifier of the del22q11 (DiGeorge) syndrome? Nat Med 2003;9:173−82.

[76] Hill C, Jacobs B, Kennedy L, Rohde S, Zhou B, Baldwin S, et al. Cranial neural crest deletion of VEGFa causes cleft palate with aberrant vascular and bone development. Cell Tissue Res 2015;361:711−22.

[77] Stabley JN, Prisby RD, Behnke BJ, Delp MD. Type 2 diabetes alters bone and marrow blood flow and vascular control mechanisms in the ZDF rat. J Endocrinol 2015;225(1):47−58 68.

[78] Oikawa A, Siragusa M, Quaini F, Katare RG, Caporali A, Buul JD, et al. Diabetes mellitus induces bone marrow microangiopathy. Arterioscler Thromb Vasc Biol 2013;30(3):498−508 69.

[79] Peng J, Hui K, Hao C, Peng Z, Gao QX, Jin Q, et al. Low bone turnover and reduced angiogenesis in streptozotocin-induced osteoporotic mice. Connect Tissue Res 2016;57(4):277−89 67.

[80] Lange V, Dörr M, Schminke U, Völzke H, Nauck M, Wallaschofski H, et al. The association between bone quality and atherosclerosis: results from two large population-based studies. Int J Endocrinol 2017;2017:3946569.

[81] Prasad M, Reriani M, Khosla S, Gössl M, Lennon R, Gulati R, et al. Coronary microvascular endothelial dysfunction is an independent predictor of development of osteoporosis in postmenopausal women. Vasc Health Risk Manag 2014;10:533−8.

[82] Collignon AM, Lesieur J, Vacher C, Chaussain C, Rochefort GY. Strategies developed to induce, direct, and potentiate bone healing. Front Physiol 2017;8:927.

[83] Atesok K, Li R, Stewart DJ, Schemitsch EH. Endothelial progenitor cells promote fracture healing in a segmental bone defect model. J Orthop Res 2010;28(8):1007−14.

[84] Talwar R, Di Silvio L, Hughes FJ, King GN. Effects of carrier release kinetics on bone morphogenetic protein-2-induced periodontal regeneration in vivo. J Clin Periodontol 2001;28:340−7.

[85] Peng H, Usas A, Olshanski A, et al. VEGF improves, whereas sFlt1 inhibits, BMP2-induced bone formation and bone healing through modulation of angiogenesis. J Bone Miner Res 2005;20:2017−27.

[86]   Zhang C, Ma J, Li M, Li XH, Dang XQ, Wang KZ. Repair effect of coexpression of the hVEGF and hBMP genes via an adeno-associated virus vector in a rabbit model of early steroid-induced avascular necrosis of the femoral head. Transl Res 2015;166:269−80.

[87]   Betz VM, Kochanek S, Rammelt S, Müller PE, Betz OB, Messmer C. Recent advances in gene-enhanced bone tissue engineering. J Gene Med 2018;20(6):e3018.

[88]   Bueno EM, Glowacki J. Cell-free and cell-based approaches for bone regeneration. Nat Rev Rheumatol 2009;5(12):685−97.

[89]   Panseri S, Russo A, Sartori M, Giavaresi G, Sandri M, Fini M, et al. Modifying bone scaffold architecture in vivo with permanent magnets to facilitate fixation of magnetic scaffolds. Bone 2013;56(2):432−9.

[90]   Ros-Tárraga P, Mazón P, Rodríguez MA, Meseguer-Olmo L, De Aza PN. Novel resorbable and osteoconductive calcium silicophosphate scaffold induced bone formation. Materials 2016;9(9):785.

Chapter 6

Endothelial signaling in coronary artery disease

F. Sertic

Department of Surgery, Hospital of the University of Pennsylvania, Philadelphia, PA, USA

6.1 Introduction

Coronary artery disease (CAD) represents one of the leading causes of death worldwide and the most common cause of morbidity and mortality in the developed countries of western society [1]. Despite an overall decrease, over the last four decades, in incidence and lethality following major improvements in prevention and early intervention [2], CAD still accounts for one third of all deaths in individuals over 35 years of age [3]. It is estimated that over 15 million adults in the United States of America (USA) suffer from CAD and, every year, around 900,000 people suffer acute myocardial infarction or die of CAD [4,5].

CAD is a complex multifactorial pathophysiological process and it revolves around the development of the atherosclerotic plaque. Historically, CAD was considered a cholesterol storage disorder leading to progressive narrowing of the coronary arteries until complete occlusion. Over the last three decades, however, it has become clear that this is not only a simple mechanical obstructive process but the result of complex interaction between several factors. These include genetic, metabolic, endocrine, and epigenetic factors (Fig. 6.1).

The endothelium lines every single blood vessel in the body. Rather than being an inactive barrier which separates the blood from the vessel wall, it plays an active and essential role in regulating vascular tone, cellular adhesion, thrombogenesis, smooth muscle cell proliferation, vessel wall inflammation, oxygen supply, and organ perfusion [6–9]. With a growing body of supporting evidence [7,10–18], endothelial dysfunction is now thought to be initial event triggering

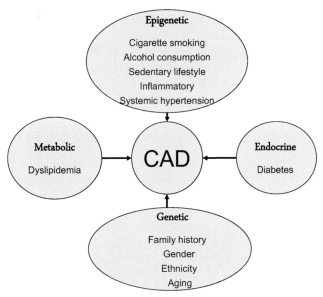

FIGURE 6.1 Multifactorial nature of coronary artery disease (CAD).

Endothelial Signaling in Vascular Dysfunction and Disease. DOI: https://doi.org/10.1016/B978-0-12-816196-8.00022-9

vascular inflammation, injury, and oxidative stress leading to cardiovascular events, including CAD. Endothelial dysfunction can be described as a syndrome exhibiting systemic manifestations associated with morbidity and mortality [14]. The understanding of the linkage between CAD and endothelial dysfunction has created new opportunities for the development of novel techniques for early diagnosis and new potential therapeutic targets for primary prevention.

In this chapter we will focus on the potential mechanisms through which endothelial injury and dysfunction contribute to CAD. We will highlight the close relationship between vascular inflammation, progression of the disease, and its clinical manifestations. Finally, we will discuss the future implications for new diagnostics and early therapeutic interventions.

6.2 Coronary artery disease: specifics and characterization

CAD includes a wide range of diseases which are the result of atheromatous plaque formation and progression in the lumen of the coronary arteries. The clinical spectrum can be acute or chronic and can range from completely asymptomatic individuals unaware of the disease to patients who suffer acute myocardial infarction manifesting as *angina pectoris* (chest pain) resulting in sudden cardiac death, in the most severe cases. This is mostly dependent on the type of atheromatous plaque which can be stable (lipid-poor with thick fibrous cap) or unstable (lipid-rich with a thin fibrous plaque).

The atherosclerotic plaque has an amorphous central core made of lipid and necrotic cells (foam cells; lipid-engorged macrophages and smooth muscle cells) surrounded by a capsule. Plaques can also be classified as fibrosclerotic, fibro-cellular, or calcified depending on the cellularity, collagen density, and presence of calcifications [19].

Stable plaques typically manifest as chest pain during exertion only (*stable angina*) meaning that during physical activity the demand of blood exceeds the supply (*demand ischemia*). An unstable plaque usually manifests as chest pain during exertion and at rest (*unstable angina*). In this case, the blood supply to the heart does not meet the demand even at rest (*supply ischemia*). When a plaque ruptures an inflammatory and thrombogenic cascade is activated, leading to platelets activation and aggregation within the lumen of the coronary artery resulting in acute myocardial infarction (*acute coronary syndrome*).

6.3 Risk factors and vascular endothelium

Risk factors for development and progression of CAD can be classified in modifiable and nonmodifiable. The modifiable risk factors include systemic arterial hypertension, elevated low-density lipoprotein (LDL), decreased high-density lipoprotein, increased triglycerides, diabetes mellitus, cigarette smoking, excessive alcohol consumption, and sedentary lifestyle. The nonmodifiable include factors include aging, familiar history of CAD, ethnicity, and male gender [1]. Despite their variety and heterogeneity, all the modifiable factors share the same mechanism in initiating the atherosclerotic process. The central event of this process is endothelial injury and the subsequent inflammatory response. Endothelial injury stimulates the expression of endothelial adhesion molecules, proinflammatory cytokines, chemotactic proteins, thrombogenic factors, and vasoconstrictors. As a result, the endothelium starts recruiting and binding monocytes and T lymphocytes which then transmigrate and accumulate in the subendothelial space [19−21]. This creates a self-sustained local proinflammatory environment which manifests itself clinically years or even decades after the initial injury. For this reason, in recent years, major primary prevention efforts and campaigns, targeting modifiable risk factors, have been implemented in an attempt to decrease the incidence and the prevalence of the disease.

6.4 Endothelial signaling and dysfunction in coronary artery disease

The recent concept is that CAD is not a simple cholesterol storage disorder but is should be considered as an inflammatory process at the endothelial level [22,23]. Local and systemic inflammation have an essential role during the development and maintenance of the atherosclerotic disease: from endothelial injury and foam cell infiltration to fibrous plaque organization, until plaque rupture, intraluminal thrombosis, and coronary occlusion [12]. The vicious circle of endothelial injury and uncontrolled inflammation represents the central event of the subintimal fibroproliferative process, which leads to CAD and acute coronary syndrome [24,25]. This sequence of events is supported by large clinical studies, including clinical trials, which showed that higher levels of serum C-reactive protein (CRP, a marker of systemic inflammation) were associated with higher risk of cardiovascular disease, irrespective of the cholesterol blood levels [26−28]. These novel insights created the foundation for the modern knowledge of atherosclerosis as a multifactorial disease that revolves around endothelial injury, local and systemic inflammation, and immune cell activation.

In the presence of risk factors for CAD, there is a high expression and activation of lectin-like oxidized low-density lipoprotein receptor-1 (LOX-1) in the endothelium [18,29]. LOX-1 is a type 2 membrane protein which acts as the main receptor for oxidized LDL; it also binds several ligands including products of apoptosis and activated platelets [30]. Further, it is expressed by macrophages and smooth muscle cells [31]. When ligands bind to receptors in cells, they can be endocytosed into the cell. In the presence of increased levels of LOX-1, a greater amount of LDL can be endocytosed inside the endothelial cells triggering the atherosclerotic process. The more LDL are phagocytosed, the more LOX-1 is overexpressed generating a self-sustained microenvironment at the subendothelial level [18]. Moreover, LOX-1 stimulates endothelial cell apoptosis and injury by activation of nuclear factor κB(NFκB) [32] and induces the expression of intercellular adhesion molecule-1 (ICAM-1), vascular cell adhesion molecule (VCAM) and P-selectin. This initiates the process of leukocyte adhesion, rolling, and transmigration in the subendothelial layer. As a result of risk factors for CAD with subsequent overexpression of LOX-1, we now have a perpetual inflammatory environment within the vessel wall which maintain itself through continuous LOX-1 upregulation, cytokine and chemokines production (tumor necrosis factor or TNF-a, Interleukins or IL-1, IL-6, and others), expression of leukocyte adhesion molecules, monocyte recruitment, and subsequent transformation into the so-called foam cells after LDL endocytosis [18,33]. This will eventually lead to the thrombotic events characteristic of CAD and its clinical manifestations (Fig. 6.2).

Vasoconstriction and vasodilation play an important role in maintaining organ perfusion and meeting blood and oxygen demands of peripheral tissues. The endothelial layer is a key player in maintaining a balanced vascular tone through the production of several important mediators including nitric oxide (NO), prostacyclin, endothelium-derived hyperpolarizing factors (EDHFs), natriuretic peptides, endothelin-1 (ET-1), and thromboxane. With endothelial dysfunction, this delicate balance can be disrupted resulting in inflammation, cellular proliferation, and thrombosis leading to atherosclerosis and CAD [7].

NO is a potent vasodilator created from l-arginine through endothelial NO synthase (eNOS). NO mediates its action on vascular smooth muscle cells by increasing the intracellular concentration of cyclic guanosine monophosphate which induces relaxation of these cells resulting in vasodilation. One of the most important eNOS activators is vessel wall shear stress. The increase in shear stress leads to an increase in the production of NO to help maintain appropriate perfusion in response to changes in cardiac output [34]. It is known that laminar flow is the ideal condition and that turbulent flow is thrombogenic but it is also thought to be atherogenic [35]. Loss of laminar flow in the coronary arteries may decrease the production of NO and increase elastase and collagenase resulting in vessel remodeling and expansion toward the lumen [36,37]. Further, NO, through the process of protein nitrosylation, promotes inhibition of inflammation, cellular proliferation, and thrombosis [38]. Other vasodilators, produced by the endothelium, which work synergically with NO are prostacyclin (a metabolite of the arachidonic acid), EDHFs, and C-type natriuretic peptide which also has antiproliferative effects on smooth muscle cells [39]. Endothelial vasoconstrictors are ET-1, which maintains the basal vascular tone, and thromboxane A-2, another metabolic of the arachidonic acid, which is released by activated

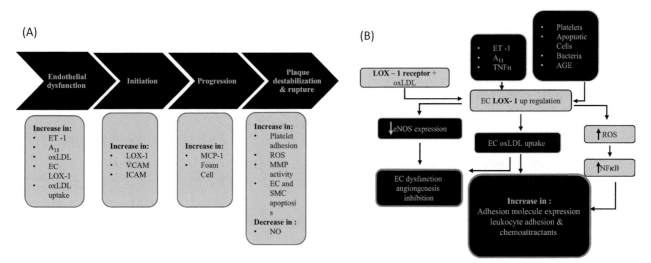

FIGURE 6.2 Oxidized low-density lipoprotein receptor-1 (LOX-1) pathway in endothelial dysfunction. (A) LOX in the initiation and progression of atherosclerotic lesions. (B) LOX-1 initiated signaling leads to induction of inflammation moieties via oxidants that causes activation of transcription factor NFκB which in turn drives increased expression of adhesion molecules.

platelets. The harmonious interaction between all the aforementioned factors can be disrupted by the presence of risk factors for cardiovascular disease. The chronic inflammation that characterizes the atherosclerotic process is associated with an upregulation of nicotinamide adenine dinucleotide phosphate-oxidases which leads to an increased production of reactive oxygen species (ROS) [40]. The disruption of the balance between ROS production and mitochondrial superoxide dismutase during oxidative phosphorylation leading to the production of peroxynitrite ($\cdot ONOO^-$) is considered to be the initial step of NO metabolism dysregulation which characterizes endothelial dysfunction [7,41,42]. Increased oxidative stress contributes to the proinflammatory state by upregulation of the expression of leukocyte adhesion molecules (ICAM-1 and VCAM-1) and chemokines.

Cardiovascular diseases are characterized by the presence of an imbalance between the production of ROS and antioxidant agents resulting in an accumulation of hydrogen peroxide, superoxide, peroxynitrite, and other molecules [17,43]. As mentioned above, this imbalance causes a disruption in the NO pathway at the endothelial level leading to decreased NO bioavailability. This process has been demonstrated in most of the risk factors for CAD including aging, hypercholesterolemia, hypertension, and cigarette smoking [44−49]. ROS mediates signaling pathways responsible for smooth muscle cells growth, differentiation, and apoptosis [41]. The dysregulation of the NO pathway and lipid peroxidation and protein nitration by peroxynitrite are considered as the the progenitors initiating the atherosclerotic process. In addition, ROS upregulation has been associated with activation, recruitment, and tissue infiltration of immune cells including monocytes and T-lymphocytes with an increase in phagocytic nicotinamide adenine dinucleotide phosphate (NADPH) oxidase activity [50−52]. Despite a large body of evidence confirming the association between ROS imbalance and cardiovascular disease all major clinical trials have failed to demonstrate a convincing effect of antioxidant agents, such as ascorbic acid, on cardiovascular morbidity and mortality [53,54]. Only one prospective cohort study demonstrated lower all-cause and cardiovascular mortality in men and women with higher plasma concentrations of ascorbic acid [55].

Other factors that are now considered to contribute to the onset and progression of cardiovascular disease include psychosocial factors such as emotional stress, social inequities, racial disparities, depression, and others [12,56−60]. The hypothalamic pituitary adrenal (HPA) axis and the autonomic nervous system (ANS) are the main pathways through which psychosocial factors affect the cardiovascular system (Fig. 6.3). As an example, the dysregulation of the HPA axis seen in metabolic syndrome and obesity, can result in increased cortisol level, increased plasma norepinephrine, and IL-6 concentrations which lead to endothelial activation, hypercoagulable state, and thrombosis [61,62]. Alterations to the ANS by psychosocial stressors can lead to an imbalance in favor of sympathetic activity with increased adrenaline and noradrenaline release. Chronic sympathetic overstimulation can then lead to systemic hypertension [63], one of the major risk factors for CAD. It is also known that ischemic damage of the neuro-cardiac axis, at any level, is associated with arrhythmias and myocardial injury. This is mainly mediated by the sympathetic neuropeptide Y, another regulator of vascular tone, and endothelial activity also involved in angiogenesis following ischemic insult [64,65]. Finally, disruption of the circadian rhythms and cycles can contribute to cardiovascular morbidity and mortality. Even though the underlying mechanisms are still poorly understood, it has been demonstrated that night-shift workers and poor sleep quality are associated with an increased incidence of hypertension, CAD, obesity, and diabetes [66]. It is thought that sleep deprivation increases sympathetic activity and alters the neuroendocrine balance leading to oxidative stress and endothelial activation [67].

The concept of CAD has evolved from the idea of an accumulating cholesterol plaque toward a multifactorial pathophysiological process involving a large number of physiological, immunological, endocrine, biochemical, metabolic, and psychological pathways. It is now clear how the endothelial layer plays an essential role in the mediation and interaction of these pathways. This recent advances in a deeper understanding of the key mechanism have opened new frontiers for early diagnosis, treatment, and more importantly prevention of CAD.

6.5 Coronary artery stenting: endothelial cells dysfunction in restenosis

Percutaneous coronary intervention (PCI) represents one of the most effective treatment options for acute coronary syndrome (ACS) and early treatment has been associated with improved outcomes, especially in ST segment elevation MI (STEMI) [68,69]. One of the most feared and potentially lethal complications following PCI is in-stent thrombosis and restenosis. It is thought that endothelial function plays an important role in this process as well. Patients who undergo PCI and develop restenosis have a significant impairment in endothelium-dependent vasodilation, before and after the lesion, compared to those who do not develop restenosis [70]. Following PCI significant changes are observed at the endothelial level. These include intimal hyperplasia, usually transient, injury to the arterial wall, more pronounced after stenting than balloon angioplasty, and prolonged impairment of endothelial-dependent vasodilation [71−75]. Further,

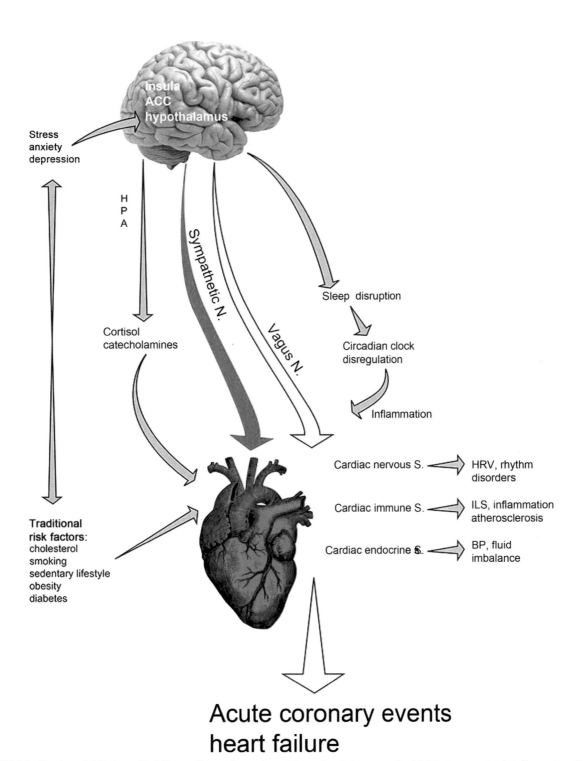

FIGURE 6.3 **Psychosocial factors affect the cardiovascular system.** Atherosclerosis is now well established as a systemic inflammatory disease. Cardiac tissue contains resident immune cells and is able to synthesize and release cytokines and hormones after acute myocardial infarction in the ischemic area, influencing the healing phase. It is also becoming clear that chronic depression has ranked among the most important cardiovascular risk factors for poor prognosis in patients with myocardial infarction. *Reproduced from Fioranelli M, Bottaccioli AG, Bottaccioli F, Bianchi M, Rovesti M, Roccia MG. Stress and Inflammation in coronary artery disease: a review psychoneuroendocrineimmunology-based. Front Immunol 2018;9:2031.*

drug-eluting stents (DES) which are often the preferred choice during PCI, have been associated with adverse effect at the local endothelial level and vasomotor abnormalities [76]. A better understanding of the long-term effects of stent deployment in the coronary arteries on endothelial function may provide useful insights for the identification of therapies that aim to restore endothelial function following PCI and prevent in-stent thrombosis. Promising results have been shown with the use of gene therapy with intracoronary vascular endothelial growth factor (VEGF) transfer [77] and 17-beta-estradiol eluting stents [78] which will require further investigation.

6.6 Clinical events in coronary artery disease: how relevant is mitigation of endothelial dysfunction

The clinical correlation between endothelial dysfunction and cardiovascular disease, including CAD, is now well established and described [79,80]. In a large analysis, including almost 2500 patients, Lerman and Zeiher, demonstrated a strong and independent association between endothelial dysfunction and cardiac death, acute myocardial infarction (MI), and need for revascularization [14]. Early detection of endothelial dysfunction could play an essential role in the prediction of CAD and other cardiovascular events. Unfortunately, current methodology for assessment of endothelial function require invasive procedures and highly trained personnel. The introduction of a noninvasive method for early detection of endothelial dysfunction could represent a major step forward in the diagnosis, treatment, and prevention of CAD.

Most antihypertensive drugs, except beta-blockers and diuretics, and lipid-lowering drugs, used in the treatment of cardiovascular risk factors, have a positive effect at the endothelial level [81]. This is due to an antioxidant effect which increases the bioavailability of NO. For example, angiotensin II is known to increase the production of oxidative products and vasoconstrictors including superoxide, prostanoid, and endothelin-1 [82]. Therefore angiotensin converting enzyme (ACE) inhibitors and angiotensin receptor blocker, regardless of their antihypertensive effect, induce endothelium-dependent vasodilation, mostly in medium and large size vessels, by blocking this pathway [82,83]. Statins, which are widely used in primary and secondary prevention of cardiovascular disease, have a positive effect on endothelial function, regardless of their lipid-lowering effect. This beneficial effect may be due to increased eNOS expression and NO release and decreased endothelin-1 expression [84]. Similarly, calcium channel blockers, glitazones, and nebivolol improve endothelial function [81].

Considering how relevant and crucial endothelial function is in the development and progression of CAD and cardiovascular diseases, it seems to represent the perfect diagnostic and therapeutic target. Prevention and correction of endothelial dysfunction should theoretically improve outcomes of patients, especially of those with cardiovascular risk factors. Data in this sense, however, are still scarce and not convincing enough to make these claims. Specific therapeutic and diagnostic targets still need to be identified and it will also be important to standardize the definition of endothelial dysfunction. As we have shown, the endothelial function involves a large number of pathways, cells, and molecules that interact with each other. It is still unclear when to call the endothelial layer "dysfunctional." First, it will be important to understand the most important beneficial effect of antihypertensive and lipid-lowering drugs. Is it the lower blood pressure and the lower LDL cholesterol or is it the beneficial effect on the endothelium that make these drugs effective? Further, current large-scale studies typically follow-up patients for 5 or 10 years and are not particularly suitable for understanding CAD that appears to develop over decades. Therefore these "short-term" studies depict a limited picture of the disease. Another crucial improvement would be the development of novel noninvasive diagnostic tools quantitatively assess endothelial function, risk stratify patients, and predict development of CAD. A simple blood test measuring circulating markers of endothelial activation and dysfunction and evaluating its correlation to noninvasivel markers of vasodilation etc. may represent the ideal solution. Finally, the development of a new drug targeting the endothelial layer to restore its function or even prevent further damage could represent the most effective solution.

It is evident that the understanding of atherosclerosis, endothelial function, and cardiovascular diseases have evolved and greatly improved over the last few decades; however, future research is needed to understand how to translate this improved knowledge to the clinical world in order to improve diagnostics, treatments, prevention, and ultimately patients' outcomes.

6.7 Conclusions

In conclusion, CAD remains one of the most common causes of morbidity and mortality worldwide. Over the last four decades, the available knowledge on atherosclerosis and cardiovascular diseases has shifted from the concept of a cholesterol storage disorder toward a multifactorial disease that revolves around endothelial function and inflammation.

Onset of endothelial dysfunction may be the initial trigger of a self-sustained proinflammatory local environment which leads to the development of the atherosclerotic plaque. The endothelium could represent a novel target for early diagnosis, treatment, and prevention of CAD.

ET-1, endothelin-1; *A II*, angiotensin II; *VCAM-1*, vascular cell adhesion molecule; *ICAM-1*, intercellular adhesion molecule; *MCP-1*, monocyte chemoattractant protein-1; *ROS*, reactive oxygen species; *NO*, nitric oxide; *SMC*, smooth muscle cells; *TNF-α*, tumor necrosis factor; *MMP*, matrix metalloproteinase; *AGE*, advanced glycation end products.

References

[1] Ashley EA, Niebauer J. Coronary artery disease [Internet]. Cardiology Explained. Remedica; 2004 [cited 2020 Apr 27]. <https://www-ncbi-nlm-nih-gov.proxy.library.upenn.edu/books/NBK2216/>.

[2] Tunstall-Pedoe H, Kuulasmaa K, Mähönen M, Tolonen H, Ruokokoski E, Amouyel P. Contribution of trends in survival and coronary-event rates to changes in coronary heart disease mortality: 10-year results from 37 WHO MONICA project populations. Monitoring trends and determinants in cardiovascular disease. Lancet. 1999;353(9164):1547−57.

[3] Sanchis-Gomar F, Perez-Quilis C, Leischik R, Lucia A. Epidemiology of coronary heart disease and acute coronary syndrome Ann Transl Med 2016;4(13):256[cited 2020 Apr 28]. Available from: https://www.ncbi.nlm.nih.gov/pmc/articles/PMC4958723/>.

[4] Writing Group Members, Mozaffarian D, Benjamin EJ, Go AS, Arnett DK, Blaha MJ, et al. Heart disease and stroke statistics-2016 update: a report from the American Heart Association. Circulation. 2016;133(4):e38−e360.

[5] Writing Group Members, Mozaffarian D, Benjamin EJ, Go AS, Arnett DK, Blaha MJ, et al. Executive summary: heart disease and stroke statistics − 2016 update: a report from the American Heart Association. Circulation. 2016;133(4):447−54.

[6] Schechter AN, Gladwin MT. Hemoglobin and the paracrine and endocrine functions of nitric oxide. N Engl J Med 2003;348(15):1483−5.

[7] Veerasamy M, Bagnall A, Neely D, Allen J, Sinclair H, Kunadian V. Endothelial dysfunction and coronary artery disease: a state of the art review. Cardiol Rev 2015;23(3):119−29.

[8] Kinlay S, Libby P, Ganz P. Endothelial function and coronary artery disease. Curr Opin Lipidol 2001;12(4):383−9.

[9] Lüscher TF, Barton M. Biology of the endothelium. Clin Cardiol 1997;20(11 Suppl 2):II-3−II-10.

[10] Barton M. Prevention and endothelial therapy of coronary artery disease. Curr Opin Pharmacol 2013;13(2):226−41.

[11] Daiber A, Steven S, Weber A, Shuvaev VV, Muzykantov VR, Laher I, et al. Targeting vascular (endothelial) dysfunction. Br J Pharmacol 2017;174(12):1591−619.

[12] Fioranelli M, Bottaccioli AG, Bottaccioli F, Bianchi M, Rovesti M, Roccia MG. Stress and inflammation in coronary artery disease: a review psychoneuroendocrineimmunology-based. Front Immunol 2018;9:2031.

[13] Ganz P, Hsue PY. Endothelial dysfunction in coronary heart disease is more than a systemic process. Eur Heart J 2013;34(27):2025−7.

[14] Lerman A, Zeiher AM. Endothelial function: cardiac events. Circulation. 2005;111(3):363−8.

[15] Matsuzawa Y, Lerman A. Endothelial dysfunction and coronary artery disease: assessment, prognosis, and treatment. Coron Artery Dis 2014;25 (8):713−24.

[16] Shaw J, Anderson T. Coronary endothelial dysfunction in non-obstructive coronary artery disease: risk, pathogenesis, diagnosis and therapy. Vasc Med 2016;21(2):146−55.

[17] Steven S, Frenis K, Oelze M, Kalinovic S, Kuntic M, Bayo Jimenez MT, et al. Vascular inflammation and oxidative stress: major triggers for cardiovascular disease. Oxid Med Cell Longev 2019;2019:7092151.

[18] Szmitko PE, Wang C-H, Weisel RD, Jeffries GA, Anderson TJ, Verma S. Biomarkers of vascular disease linking inflammation to endothelial activation: part II. Circulation. 2003;108(17):2041−8.

[19] Rognoni A, Cavallino C, Veia A, Bacchini S, Rosso R, Facchini M, et al. Pathophysiology of atherosclerotic plaque development. Cardiovasc Hematol Agents Med Chem 2015;13(1):10−13.

[20] Miller MJ, Kuntz RE, Friedrich SP, Leidig GA, Fishman RF, Schnitt SJ, et al. Frequency and consequences of intimal hyperplasia in specimens retrieved by directional atherectomy of native primary coronary artery stenoses and subsequent restenoses. Am J Cardiol 1993;71(8):652−8.

[21] Witztum JL, Steinberg D. Role of oxidized low density lipoprotein in atherogenesis. J Clin Invest 1991;88(6):1785−92.

[22] Alexander RW. Inflammation and coronary artery disease. N Engl J Med 1994;331(7):468−9.

[23] Ross R. Atherosclerosis − an inflammatory disease. N Engl J Med 1999;340(2):115−26.

[24] Li J-J. Inflammation in coronary artery diseases. Chin Med J 2011;124(21):3568−75.

[25] Libby P. Inflammation and cardiovascular disease mechanisms. Am J Clin Nutr 2006;83(2):456S−60S.

[26] Ridker PM, Rifai N, Rose L, Buring JE, Cook NR. Comparison of C-reactive protein and low-density lipoprotein cholesterol levels in the prediction of first cardiovascular events. N Engl J Med 2002;347(20):1557−65.

[27] Ridker PM, Danielson E, Fonseca FAH, Genest J, Gotto AM, Kastelein JJP, et al. Rosuvastatin to prevent vascular events in men and women with elevated C-reactive protein. N Engl J Med 2008;359(21):2195−207.

[28] Lindahl B, Toss H, Siegbahn A, Venge P, Wallentin L. Markers of myocardial damage and inflammation in relation to long-term mortality in unstable coronary artery disease. FRISC Study Group. Fragmin during instability in coronary artery disease. N Engl J Med 2000;343 (16):1139−47.

[29] Mehta JL, Li D. Identification, regulation and function of a novel lectin-like oxidized low-density lipoprotein receptor. J Am Coll Cardiol 2002;39(9):1429−35.

[30] Cominacini L, Fratta Pasini A, Garbin U, Pastorino A, Rigoni A, Nava C, et al. The platelet-endothelium interaction mediated by lectin-like oxidized low-density lipoprotein receptor-1 reduces the intracellular concentration of nitric oxide in endothelial cells. J Am Coll Cardiol 2003;41(3):499−507.

[31] Aoyama T, Chen M, Fujiwara H, Masaki T, Sawamura T. LOX-1 mediates lysophosphatidylcholine-induced oxidized LDL uptake in smooth muscle cells. FEBS Lett 2000;467(2−3):217−20.

[32] Li D, Mehta JL. Upregulation of endothelial receptor for oxidized LDL (LOX-1) by oxidized LDL and implications in apoptosis of human coronary artery endothelial cells: evidence from use of antisense LOX-1 mRNA and chemical inhibitors. Arterioscler Thromb Vasc Biol 2000;20 (4):1116−22.

[33] Li D, Liu L, Chen H, Sawamura T, Mehta JL. LOX-1, an oxidized LDL endothelial receptor, induces CD40/CD40L signaling in human coronary artery endothelial cells. Arterioscler Thromb Vasc Biol 2003;23(5):816−21.

[34] Corson MA, James NL, Latta SE, Nerem RM, Berk BC, Harrison DG. Phosphorylation of endothelial nitric oxide synthase in response to fluid shear stress. Circ Res 1996;79(5):984−91.

[35] Asakura T, Karino T. Flow patterns and spatial distribution of atherosclerotic lesions in human coronary arteries. Circ Res 1990;66(4):1045−66.

[36] Chatzizisis YS, Coskun AU, Jonas M, Edelman ER, Feldman CL, Stone PH. Role of endothelial shear stress in the natural history of coronary atherosclerosis and vascular remodeling: molecular, cellular, and vascular behavior. J Am Coll Cardiol 2007;49(25):2379−93.

[37] Chatzizisis YS, Baker AB, Sukhova GK, Koskinas KC, Papafaklis MI, Beigel R, et al. Augmented expression and activity of extracellular matrix-degrading enzymes in regions of low endothelial shear stress colocalize with coronary atheromata with thin fibrous caps in pigs. Circulation. 2011;123(6):621−30.

[38] Stamler JS, Lamas S, Fang FC. Nitrosylation. the prototypic redox-based signaling mechanism. Cell. 2001;106(6):675−83.

[39] Porter JG, Catalano R, McEnroe G, Lewicki JA, Protter AA. C-type natriuretic peptide inhibits growth factor-dependent DNA synthesis in smooth muscle cells. Am J Physiol 1992;263(5 Pt 1):C1001−6.

[40] Griendling KK, Sorescu D, Ushio-Fukai M. NAD(P)H oxidase: role in cardiovascular biology and disease. Circ Res 2000;86(5):494−501.

[41] Griendling KK, FitzGerald GA. Oxidative stress and cardiovascular injury: part I: basic mechanisms and in vivo monitoring of ROS. Circulation. 2003;108(16):1912−16.

[42] Koppenol WH, Moreno JJ, Pryor WA, Ischiropoulos H, Beckman JS. Peroxynitrite, a cloaked oxidant formed by nitric oxide and superoxide. Chem Res Toxicol 1992;5(6):834−42.

[43] Sies H. Oxidative stress: a concept in redox biology and medicine. Redox Biol 2015;4:180−3.

[44] Ohara Y, Peterson TE, Harrison DG. Hypercholesterolemia increases endothelial superoxide anion production. J Clin Invest 1993;91(6):2546−51.

[45] Harrison DG, Ohara Y. Physiologic consequences of increased vascular oxidant stresses in hypercholesterolemia and atherosclerosis: implications for impaired vasomotion. Am J Cardiol 1995;75(6):75B−81B.

[46] Dikalova AE, Itani HA, Nazarewicz RR, McMaster WG, Flynn CR, Uzhachenko R, et al. Sirt3 impairment and SOD2 hyperacetylation in vascular oxidative stress and hypertension. Circ Res 2017;121(5):564−74.

[47] Dikalov SI, Dikalova AE. Crosstalk between mitochondrial hyperacetylation and oxidative stress in vascular dysfunction and hypertension. Antioxid Redox Signal 2019;31(10):710−21.

[48] Dikalov S, Itani H, Richmond B, Vergeade A, Rahman SMJ, Boutaud O, et al. Tobacco smoking induces cardiovascular mitochondrial oxidative stress, promotes endothelial dysfunction, and enhances hypertension. Am J Physiol Heart Circ Physiol 2019;316(3):H639−46.

[49] Harman D. Aging: a theory based on free radical and radiation chemistry. J Gerontol 1956;11(3):298−300.

[50] Dikalov SI, Li W, Doughan AK, Blanco RR, Zafari AM. Mitochondrial reactive oxygen species and calcium uptake regulate activation of phagocytic NADPH oxidase. Am J Physiol Regul Integr Comp Physiol 2012;302(10):R1134−42.

[51] Kossmann S, Schwenk M, Hausding M, Karbach SH, Schmidgen MI, Brandt M, et al. Angiotensin II-induced vascular dysfunction depends on interferon-γ-driven immune cell recruitment and mutual activation of monocytes and NK-cells. Arterioscler Thromb Vasc Biol 2013;33 (6):1313−19.

[52] Guzik TJ, Hoch NE, Brown KA, McCann LA, Rahman A, Dikalov S, et al. Role of the T cell in the genesis of angiotensin II induced hypertension and vascular dysfunction. J Exp Med 2007;204(10):2449−60.

[53] Münzel T, Gori T, Bruno RM, Taddei S. Is oxidative stress a therapeutic target in cardiovascular disease? Eur Heart J 2010;31(22):2741−8.

[54] Schmidt HHHW, Stocker R, Vollbracht C, Paulsen G, Riley D, et al. Antioxidants in translational medicine. Antioxid Redox Signal 2015;23 (14):1130−43.

[55] Khaw KT, Bingham S, Welch A, Luben R, Wareham N, Oakes S, et al. Relation between plasma ascorbic acid and mortality in men and women in EPIC-Norfolk prospective study: a prospective population study. European prospective investigation into cancer and nutrition. Lancet. 2001;357(9257):657−63.

[56] Yusuf S, Hawken S, Ounpuu S, Dans T, Avezum A, Lanas F, et al. Effect of potentially modifiable risk factors associated with myocardial infarction in 52 countries (the INTERHEART study): case-control study. Lancet. 2004;364(9438):937−52.

[57] Marmot MG, Rose G, Shipley M, Hamilton PJ. Employment grade and coronary heart disease in British civil servants. J Epidemiol Community Health (1978). 1978;32(4):244−9.

[58] Brunner EJ. Social factors and cardiovascular morbidity. Neurosci Biobehav Rev 2017;74(Pt B):260−8.

[59] Sims M, Redmond N, Khodneva Y, Durant RW, Halanych J, Safford MM. Depressive symptoms are associated with incident coronary heart disease or revascularization among blacks but not among whites in the Reasons for Geographical and Racial Differences in Stroke study. Ann Epidemiol 2015;25(6):426−32.

[60] O'Brien EC, Greiner MA, Sims M, Hardy NC, Wang W, Shahar E, et al. Depressive symptoms and risk of cardiovascular events in blacks: findings from the Jackson Heart Study. Circ Cardiovasc Qual Outcomes 2015;8(6):552−9.

[61] Le-Ha C, Herbison CE, Beilin LJ, Burrows S, Henley DE, Lye SJ, et al. Hypothalamic-pituitary-adrenal axis activity under resting conditions and cardiovascular risk factors in adolescents. Psychoneuroendocrinology. 2016;66:118−24.

[62] Brunner EJ, Hemingway H, Walker BR, Page M, Clarke P, Juneja M, et al. Adrenocortical, autonomic, and inflammatory causes of the metabolic syndrome: nested case-control study. Circulation. 2002;106(21):2659−65.

[63] Esler M, Eikelis N, Schlaich M, Lambert G, Alvarenga M, Kaye D, et al. Human sympathetic nerve biology: parallel influences of stress and epigenetics in essential hypertension and panic disorder. Ann N Y Acad Sci 2008;1148:338−48.

[64] Esler M. Mental stress and human cardiovascular disease. Neurosci Biobehav Rev 2017;74(Pt B):269−76.

[65] Zukowska-Grojec Z, Karwatowska-Prokopczuk E, Rose W, Rone J, Movafagh S, Ji H, et al. Neuropeptide Y: a novel angiogenic factor from the sympathetic nerves and endothelium. Circ Res 1998;83(2):187−95.

[66] Tobaldini E, Costantino G, Solbiati M, Cogliati C, Kara T, Nobili L, et al. Sleep, sleep deprivation, autonomic nervous system and cardiovascular diseases. Neurosci Biobehav Rev 2017;74(Pt B):321−9.

[67] Dumaine JE, Ashley NT. Acute sleep fragmentation induces tissue-specific changes in cytokine gene expression and increases serum corticosterone concentration. Am J Physiol Regul Integr Comp Physiol 2015;308(12):R1062−9.

[68] Hochman JS, Sleeper LA, Webb JG, Sanborn TA, White HD, Talley JD, et al. Early revascularization in acute myocardial infarction complicated by cardiogenic shock. SHOCK Investigators. Should we emergently revascularize occluded coronaries for cardiogenic shock. N Engl J Med 1999;341(9):625−34.

[69] Armstrong PW, Gershlick AH, Goldstein P, Wilcox R, Danays T, Lambert Y, et al. Fibrinolysis or primary PCI in ST-segment elevation myocardial infarction. N Engl J Med 2013;368(15):1379−87.

[70] Thanyasiri P, Kathir K, Celermajer DS, Adams MR. Endothelial dysfunction and restenosis following percutaneous coronary intervention. Int J Cardiol 2007;119(3):362−7.

[71] Badimon L, Badimon JJ, Penny W, Webster MW, Chesebro JH, Fuster V. Endothelium and atherosclerosis. J Hypertens Suppl 1992;10(2): S43−50.

[72] Waller BF. Pathology of transluminal balloon angioplasty used in the treatment of coronary heart disease. Cardiol Clin 1989;7(4):749−70.

[73] Shimokawa H, Flavahan NA, Shepherd JT, Vanhoutte PM. Endothelium-dependent inhibition of ergonovine-induced contraction is impaired in porcine coronary arteries with regenerated endothelium. Circulation. 1989;80(3):643−50.

[74] van Beusekom HM, Whelan DM, Hofma SH, Krabbendam SC, van Hinsbergh VW, Verdouw PD, et al. Long-term endothelial dysfunction is more pronounced after stenting than after balloon angioplasty in porcine coronary arteries. J Am Coll Cardiol 1998;32(4):1109−17.

[75] Celik T, Iyisoy A, Kursaklioglu H, Celik M. The forgotten player of in-stent restenosis: endothelial dysfunction. Int J Cardiol 2008;126 (3):443−4.

[76] Togni M, Windecker S, Cocchia R, Wenaweser P, Cook S, Billinger M, et al. Sirolimus-eluting stents associated with paradoxic coronary vasoconstriction. J Am Coll Cardiol 2005;46(2):231−6.

[77] Hedman M, Hartikainen J, Syvänne M, Stjernvall J, Hedman A, Kivelä A, et al. Safety and feasibility of catheter-based local intracoronary vascular endothelial growth factor gene transfer in the prevention of postangioplasty and in-stent restenosis and in the treatment of chronic myocardial ischemia: phase II results of the Kuopio Angiogenesis Trial (KAT). Circulation. 2003;107(21):2677−83.

[78] Abizaid A, Albertal M, Costa MA, Abizaid AS, Staico R, Feres F, et al. First human experience with the 17-beta-estradiol-eluting stent: the estrogen and stents to eliminate restenosis (EASTER) trial. J Am Coll Cardiol 2004;43(6):1118−21.

[79] Widlansky ME, Gokce N, Keaney JF, Vita JA. The clinical implications of endothelial dysfunction. J Am Coll Cardiol 2003;42(7):1149−60.

[80] Ganz P, Vita JA. Testing endothelial vasomotor function: nitric oxide, a multipotent molecule. Circulation. 2003;108(17):2049−53.

[81] Versari D, Daghini E, Virdis A, Ghiadoni L, Taddei S. Endothelial dysfunction as a target for prevention of cardiovascular disease. Diabetes Care 2009;32(Suppl 2):S314−321.

[82] Böhm M. Angiotensin receptor blockers versus angiotensin-converting enzyme inhibitors: where do we stand now? Am J Cardiol 2007;100 (3A):38J−44J.

[83] Taddei S, Virdis A, Ghiadoni L, Sudano I, Salvetti A. Effects of antihypertensive drugs on endothelial dysfunction: clinical implications. Drugs. 2002;62(2):265−84.

[84] Beckman JA, Creager MA. The nonlipid effects of statins on endothelial function. Trends Cardiovasc Med 2006;16(5):156−62.

Chapter 7

An activated endothelium after organ transplantation: the pathogenesis of rejection

Nuala Mooney[1,2] and Denis Glotz[1,2,3]

[1]Human Immunology, Pathophysiology & Immunotherapy, INSERM UMRS-976, Paris, France, [2]Université Paris Diderot, Paris, France, [3]Department of Transplantation and Nephrology, Saint-Louis Hospital, AP-HP, Paris, France

Organ transplantation is a widely used therapy, which while applied to different organs including heart, lung, and liver, has been most thoroughly investigated in the setting of kidney transplantation.

Kidney transplantation is a treatment of choice for patients with end-stage renal disease although the incidence of allograft dysfunction due to chronic antibody-mediated rejection (AMR) remains a major cause of graft dysfunction. It is estimated that AMR is responsible for the loss of up to 50% of allografts within the 10 years following transplantation [1]. The hallmark of AMR is the presence of microvascular inflammatory lesions, ultimately leading to chronic lesions of the allograft endothelium. Chronic AMR may therefore be considered as an inflammatory disease of the allograft endothelium.

AMR is by definition related to the presence of donor-specific antibodies (DSAs) and the vast majority of DSAs are specifically directed against the protein products of the highly polymorphic human leukocyte antigen (HLA) gene locus. HLA molecules are classified as either class I or class II according to their sequences and structures. Classical HLA class I (HLA-I) molecules (e.g., human HLA-A, B, C) are glycoproteins associated with β_2-microglobulin (β_2m) and expressed on the cell surface of nucleated cells. The extracellular region of HLA-I heavy chains is composed of three domains (α1, α2, and α3) and the membrane-distal α1-α2 domains form two α-helices bordering an antiparallel β-sheet platform [2].

Cell surface HLA class II (HLA-II) molecules are expressed in the form of noncovalently associated heterodimers composed of two glycoproteins, a 35-kDa-α chain and a 28-kDa-β chain. Each chain is formed by two extracellular domains, a hydrophobic transmembrane sequence and a short intracytoplasmic tail [3]. In humans HLA class II molecules are expressed as three different isotypes termed HLA-DR, HLA-DP, and HLA-DQ and are encoded by closely linked genes within the HLA gene locus. The outcome of the extremely high level of polymorphism of the HLA-I and HLA-II genes is that more than 21,000 alleles of the HLA molecules are currently known and these alleles are the major underlying reason for the development of DSAs.

Antibodies directed against the HLA molecules of the donor can be present either pretransplant or can be developed de novo by the allograft recipient posttransplant. These anti-HLA DSA are principally, but not exclusively, responsible for AMR. Patients who developed de novo DSA had more endothelial lesions than those with existing DSA pretransplant [4]. Five years after the appearance of de novo DSAs, up to forty percent of patients lose their renal allografts, compared with survival of more than eighty percent of patients without DSAs [5]. The resting or steady-state kidney endothelium expresses HLA-I and a low level of HLA-II antigens. However, when a renal allograft is exposed to an inflammatory environment, probably resulting from ischemia and reperfusion in addition to the surgical stress of transplantation, there is a major increase in expression of the HLA-II molecules. Although this is the case for HLA-DR, -DP, and -DQ, the degree of expression between the three isotypes, as well as their time-course of expression, is according to the hierarchy of HLA-DR, followed by -DP and finally -DQ. The allograft endothelium thereby becomes a key target of activation mediated by the binding of anti-HLA antibodies and certain studies have successfully eluted anti-HLA antibodies bound to the allograft [6,7].

Endothelial Signaling in Vascular Dysfunction and Disease. DOI: https://doi.org/10.1016/B978-0-12-816196-8.00021-7

While it has been most extensively documented following renal transplantation, AMR is increasingly reported to be a major cause of allograft dysfunction following transplantation of other organs including cardiac, pulmonary, and liver, and strongly associates with the detection of circulating DSA in all of the above [4,8,9] and reviewed in ref. [10].

This review will discuss evidence for activation of the allograft endothelium following organ transplantation and the underlying mechanisms by which such activation may play a role in the pathogenesis of rejection.

7.1 Evidence for endothelial activation after organ transplantation

Although high levels of immunosuppressive agents are administered following organ transplantation, and are continued throughout the lifetime of the transplant recipient, the allograft endothelium is activated as revealed by different approaches including histological, proteomic, and transcriptomic studies.

Initially, histological studies revealed steady-state expression of HLA-DR by the allograft endothelium as well as the Class II transcription factor that is largely responsible for coordinated expression of all HLA-II molecules. A marked increase in HLA-II expression was particularly evident in the inflammatory environment of allograft rejection [11]. Both HLA-DR and -DQ expressions were increased in the rejecting kidney. The observation of increased HLA-II proteins is important because of the surge in reports describing the prevalence of DSA directed against the HLA-II antigens and their association with allograft dysfunction. These observations lead to the question of how DSA interacts with the HLA-II expressing endothelium and the outcome of such interactions for the endothelium.

Antibody-mediated activation of the complement cascade following DSA binding was the first potential mechanism of antibody-mediated damage to be described and indeed the deposition of the complement component C4d was initially considered as a defining marker of AMR [12]. Recent studies have continued to underline the importance of complement activating DSA in mediating allograft damage [13,14]. Full activation of the classical complement cascade was considered to lead to endothelial cell lysis but recent data have revealed that endothelial cell activation rather than lysis can ensue from formation of complexes of complement on the endothelial cell [15]. The expression of complement regulatory proteins (e.g., CD55, CD59, CD46) by endothelial cells may afford sufficient protection to prevent full lytic activity.

7.2 Complement-independent activation of allograft endothelium

In addition to complement activation, there is a large body of data demonstrating the ability of antibodies directed against HLA class I to initiate complement-independent intracellular signaling in the endothelial cell (reviewed in refs. [16,17]). The clinical relevance of such signaling was underlined by the ex vivo detection of S6 ribosomal protein (S6RP) phosphorylation in endomyocardial biopsies from patients with cardiac allografts who had developed AMR. An association between phosphorylated S6RP staining of capillary endothelial cells and allograft outcome further supported the relevance of this signaling pathway [18]. These observations led to the concept of complement-independent rejection, now an integral part of the Banff classification [19].

Overall, fewer studies of signal transduction following the binding of HLA II-specific antibodies have been reported. In macrovascular endothelial cells, binding of a HLA-DR-specific monoclonal antibody resulted in the phosphorylation of S6RP, thereby indicating activation of the mammalian target of rapamycin (mTOR) pathway [18].

Activation of macrovascular endothelial cells by binding of either HLA-DR-specific monoclonal or allele-specific polyclonal antibodies demonstrated activation of the serine/threonine kinase protein kinase C and of Akt [20]. Signaling following HLA-DR-specific monoclonal antibody binding leading to phosphorylation of Akt, Erk, and MEK was substantiated in microvascular endothelial cells [21].

An extensive study of the HLA class II signaling pathway activated by mouse monoclonal antibody binding to macrovascular endothelial cells was recently reported in which a wide range of signaling targets along the mTOR1 or mTOR2 pathways were identified. The majority of the targets of phosphorylation were downstream of the activation of focal adhesion kinase (FAK) and Src [22].

7.3 Functional outcome of human leukocyte antigen antibody-mediated signaling

The functional outcome of endothelial signaling activated by anti-HLA antibodies and the implications in AMR has been more difficult to define than the signaling itself.

The exocytosis and release of Von Willebrand factor as well as externalization of P-selectin by endothelial cells [23] revealed the potentially pro-thrombotic role of HLA-I signaling. The interaction of complement-dependent and independent pathways was also suggested because addition of the C5a complement component further increased exocytosis and externalization.

A proliferative response to the binding of HLA class I-specific antibody implicating activation of the mTOR pathway was also reported [24].

Regarding HLA-II antigens, activation of macrovascular endothelial cells by HLA-DR-specific monoclonal antibody also resulted in proliferation as well as migration of endothelial cells (demonstrated by either wound healing or Transwell experiments).

In an attempt to map HLA-DR-activated intracellular pathways of signaling involved in proliferation and wound healing, an extensive study using pharmacological inhibitors and siRNAs was carried out. This approach allowed the identification of multiple signaling targets including Src, FAK, PI3/Akt, and MEK/Erk. The mTOR pathway was therefore highly implicated. Focal adhesion kinase and Src are both nonreceptor protein tyrosine kinases that functionally interact. Impairment of activation of FAK led to lower Src activation indicating that activation of FAK may be one of the very first steps in the HLA-II-mediated pathway of endothelial activation [22].

Paradoxically, using the same monoclonal HLA-DR-specific antibody as for the proliferation studies described earlier, and in models of either microvascular or macrovascular endothelial cells; a necrotic pathway of cell death has recently been unmasked. Cell death of HLA-II expressing cells has been widely reported in different cell types [25−28] but this is the first report in endothelial cells.

This may be explained by the observation that endothelial cell death was only apparent after a short period of antibody binding to the endothelial cells (3 hours) and was not observed after 48 h: indeed when investigated after 48 hours of incubation with the HLA-DR-specific antibody, cell viability was comparable to that of nonactivated endothelial cells [29]. Necrotic cell death is characterized by loss of plasma membrane integrity and subsequent release of proinflammatory damage-associated molecular patterns (DAMPs) [30]. Following activation by HLA-DR-specific antibody binding, the death pathway involved lysosomal membrane disruption, mitochondrial membrane depolarization, and modification of the actin cytoskeleton.

The absence of detectable endothelial cell death following a longer period of activation by HLA-DR-specific antibody binding is in agreement with results of previous studies, in which no significant apoptosis (cell death) by HLA-DR-specific antibody (as examined after either 24 or 48 hours of antibody incubation with macrovascular endothelial cells) was observed. Nevertheless, signal transduction was activated by antibody binding under the latter conditions leading to the modulation of tyrosine phosphorylation and protein kinase (PK)C-alpha/beta and PKB/Akt activation [20].

How HLA-DR ligation by the same antibody can lead to the distinct outcomes of programmed cell death and proliferation in the endothelial cell remains to be explored and the point in the endothelial cell signaling pathway at which the decision to proliferate or to undergo necrotic death has not been identified. Because only a subset of cells are sensitive to cell death, the surviving cells continue to proliferate; it is conceivable that in the setting of AMR, HLA-II-specific alloantibody binding could lead to selection of a HLA-DR-mediated cell death-resistant subpopulation of endothelial cells in comparison with the starting population composed of a mixture of cell death sensitive and resistant cells. The potential clinical relevance of the cell death pathway was highlighted by the observation of endothelial cell death after incubation with sera from allosensitized patients containing relevant allele-specific antibodies [29]. This study also revealed a distinction between the outcome of HLA-I versus HLA-II-specific antibody binding to the endothelial cell as there was no indication of necrotic cell death following binding of a monomorphic HLA-I-specific antibody (W6.32). This distinction between HLA-II and HLA-I has been reported regarding cell death of human dendritic cells [25]. While evidence for the clinical importance of this distinction in endothelial cells is not available, it is possible to speculate that the necrotic cell death pathway and release of DAMPs may contribute to the deleterious outcome of high levels of HLA-II-specific DSA. HLA-I antibody binding may lead to cell proliferation and migration, while HLA-II antibody binding may activate the former in addition to cell death in a fraction of cells leading to release of DAMPs, further activating both innate and adaptive responses and therefore amplification of the inflammatory environment of the allograft.

7.4 Endothelial activation in an allogeneic environment

Following organ transplantation, DSA binding to the allograft endothelium takes place in an allogeneic environment and HLA-specific antibody-mediated activation of the endothelial cell in the presence of non-HLA-matched leukocytes has therefore been analyzed. The importance of the interaction between endothelial cells of the allograft, alloantibodies directed against HLA molecules expressed by endothelium, and circulating leukocytes of the recipient in the context of AMR can thereby be modeled.

DSAs are a robust and independent risk factor for cardiac allograft vasculopathy and Jane-wit et al. examined the outcome of alloantibody binding to the endothelial cell on its allogenicity [18]. The use of panel reactive antibodies enabled antibody binding to a spectrum of HLA molecules in primary aortic endothelial cells. This study revealed that antibody binding resulted in activation of the complement cascade and formation of membrane attack complexes on the endothelial cell. While the membrane attack complexes did not have lytic activity, their assembly on the endothelial cell led to the activation of a noncanonical pathway of NF-κB signaling leading to higher expression of inflammatory genes as well as increased recruitment and activation of allogeneic interferon-γ (IFNγ)-producing CD4$^+$-T lymphocytes (Th1).

The clinical significance was indicated from the detection of noncanonical NF-κB signaling in endothelial cells of allograft biopsies of renal transplant patients with AMR. This study was important firstly, because it revealed a new signaling pathway activated by HLA-specific alloantibodies, secondly, because it demonstrated both complement-dependent and independent signaling, and finally because it showed how complement activation did not induce cell lysis but did alter allogenicity [15]. This study also underlined the importance of the interaction between vascular endothelial cells and inflammatory leukocytes.

Lion et al. probed the vascular-circulating leukocyte interface in a study of HLA-DR-specific antibody-mediated signaling in microvascular endothelial cells, which were later exposed to non-HLA-matched peripheral blood mononuclear cells [31]. Endothelial cell production of the pro-inflammatory cytokine IL-6 was activated by contact with allogeneic peripheral blood mononuclear cells and this was considerably increased by preactivation of the endothelial cell by antibody or by an F(ab′)$_2$ antibody fragment specific for HLA-DR. Although HLA-DR signaling activated Erk, MEK, and Akt, studies with highly selective pharmacological inhibitors indicated that an Akt-dependent pathway activated IL-6 production while Erk and MEK were not implicated. The clinical pertinence of the model was supported by the increased IL-6 observed after endothelial cell binding of patient serum enriched in DSA directed against endothelial cell−expressed HLA-DR. We had previously reported that under inflammatory conditions, endothelial cell expression of HLA−DR-activated CD4$^+$-T proliferation and differentiation into pro-inflammatory Th17 or pro-tolerance FoxP3^{hi} regulatory T (Treg) [32]. Anti-HLA-DR antibody-mediated increases in endothelial cell IL-6 production altered its immunogenicity, resulting in an imbalance in the differentiation of CD4$^+$-T in favor of pro-inflammatory Th17 cells [21].

HLA class II-specific antibody signaling in the endothelial cell therefore not only provides an activation signal for the endothelial cell but also acts indirectly to modify the pro- or antiinflammatory CD4$^+$-T cell population in the allogeneic environment [15,21].

Alloantibodies that arise only after organ transplantation have been termed "de novo" and de novo DSA specific for HLA-DQ are particularly highly represented within this population and strongly associate with allograft damage (reviewed in ref. [33]). The outcome of HLA-DQ antibody binding to endothelial cells was recently examined. Even under conditions of prolonged exposure of microvascular endothelial cells to a mixture of IFNγ and tumor necrosis factor-α, HLA-DQ did not achieve expression comparable with that of HLA-DR; this concords with previous studies in macrovascular endothelial cells [20]. Nonetheless, HLA-DQ-specific antibody or HLA-DQ-specific DSA binding still activated signaling along the mTOR pathway. In an allogeneic environment anti-HLA-DQ antibody-mediated signaling in endothelial cells, prior to coculture with peripheral blood mononuclear cells, led to increased production of the pro-inflammatory cytokine IL-6 as well as the chemoattractant regulated on activation, normal T cell−expressed and secreted (RANTES). Finally, prior activation of endothelial cells with HLA-DQ-specific antibody abrogated endothelial cell−mediated amplification of antiinflammatory Treg while maintaining differentiation of pro-inflammatory Th17 and Th1. These data provide new insights, the first demonstration of HLA-DQ-specific signaling and modification of endothelial cell allogenenicity in favor of a pro-inflammatory response [34].

In contrast to either HLA-DR or HLA-DQ, while alloantibodies directed against HLA-DP have been detected, their specific role in endothelial cell activation has not been reported. Both in vivo and in vitro studies suggest that while the expression of HLA-DP is lower than that of HLA-DR, the time-course of expression resembles that of HLA-DR in vitro and in vivo (reviewed in ref. [35]).

Non-HLA-specific antibodies have also been implicated in AMR. Elevated levels of antibodies directed against angiotensin Type 1 receptor were initially identified in AMR because of their presence in transplanted patients lacking detectable DSAs. Their ability to cause vascular damage was indicated by increased tissue factor expression and the signaling pathways initiated by antibody binding to AT1R implicated Erk, AP-1, and NF-κB in endothelial cells [36]. The overlap between the signaling pathways activated by HLA and non-HLA antibody targets leads to a higher level of allograft dysfunction than by either pathway alone after either renal or cardiac transplantation (reviewed in ref. [37]). Other non-HLA targets include laminin like globular domains of perlecan [38] and collagen [39] but the signaling pathways mediated by antibody binding to these molecules have not yet been mapped.

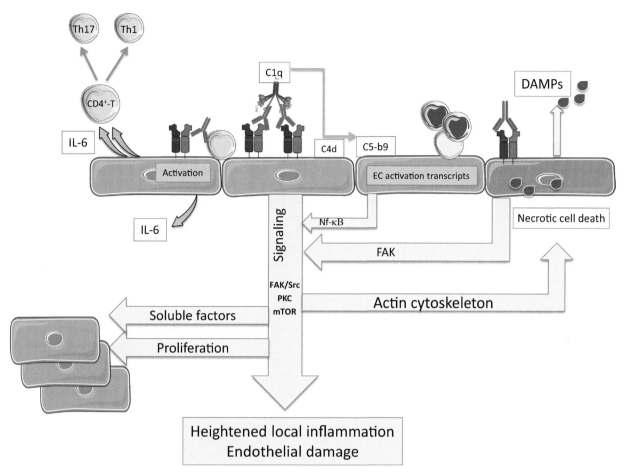

FIGURE 7.1   Binding of donor-specific antibodies directed against HLA-II antigens on the endothelial cell may lead to activation of the complement cascade and of complement-dependent or independent signaling pathways. Different outcomes may ensue including actin cytoskeleton rearrangement preceding either cell death or proliferation and soluble factor production. The fraction of endothelial cells that enter the cell death pathway produce the cell death pathway produces DAMPs (potentially increasing inflammation), complement activation can result in NF-κB signaling following deposition of C5b-9 on the endothelial cell (leading to increased Th1), HLA-DR-mediated signaling activates the endothelial cell including an Akt-dependent pathway of IL-6 secretion (leading to increased Th17). Other soluble factors are produced in an allogeneic setting (e.g., monocyte chemoattractant protein-1, RANTES). Part of the endothelial cell population proliferates in response to activation of the mTOR pathway.

The Notch signaling pathway is highly conserved in mammals and has been particularly implicated in cellular differentiation. Activation of this pathway has also been studied in endothelial cells in the context of AMR. In the inflammatory conditions following cardiac transplantation, a decrease in Notch 4 and an upregulation of Delta-like ligand 4 (DLL4) expression in allograft endothelial cells and in monocytes was noted in graft biopsies. In studies in which the interface between monocytes and endothelial cells was modeled, expression of DLL4 by endothelial cells promoted polarization of monocytes toward a pro-inflammatory macrophage phenotype and increased production of IL-6 [40] (Fig. 7.1).

7.5   Transcriptomic evidence for endothelial cell activation in antibody-mediated rejection

In addition to proteomic studies, large-scale transcriptomic studies comparing cardiac transplant patients with or without DSA confirmed endothelial activation associated with DSA. Differential expression of 132 transcripts was initially identified, which were associated with HLA-II DSA (but not HLA-I DSA). The majority of differentially expressed transcripts in AMR were associated with endothelial cells; others associated with IFNγ effects, macrophages, natural killer cells, endothelial cells, inflammation, and immunoglobulins and reflected the inflammatory state of the allograft endothelium. Twenty-three transcripts were identified by their selective expression in DSA-positive patients during rejection;

almost half of these transcripts were expressed by the endothelium. These data underline the association between the presence of DSA and the activation of endothelial cells in the course of rejection [36] and have led to including these transcripts in the diagnosis of AMR [41].

In an in vitro experimental model of AMR, the transcriptome of endothelial cells, which had been activated by panel reactive antibodies, was examined and revealed increased inflammatory gene expression mediated by the noncanonical NF-κB pathway [15].

7.6 New therapeutics in antibody-mediated rejection

Studies dissecting signaling pathways activated following HLA- or non-HLA-specific antibody binding and the consequences for the endothelial cell will help in the design of novel therapies for use in AMR. Further potential targets include signaling proteins, inhibitors of programmed necrosis, soluble factors, and inhibitors of the complement cascade. Some new therapeutic agents based upon the latter are already under clinical investigation.

Because the mTOR pathway was activated after binding of HLA-I-specific antibodies, effort has gone into studying the outcome of mTOR inhibitors in clinical inhibitors. This group of therapeutic agents is important because it is considered to cause less renal-damage than calcineurin inhibitors. The outcome of inhibiting anti-HLA-I antibody-mediated signaling and proliferation was examined using everolimus. Signaling mediated by both the mTOR complex 1 (mTORC1) and the mTOR complex 2 (mTORC2) was disrupted, as well as endothelial cell proliferation [42].

A humanized monoclonal antibody directed against the complement component C1s (TNT003) has been examined in the context of endothelial activation. Thomas et al. described how the anti-C1 antibody acted to reduce deposition of the complement components C3a and C5a in endothelial cells activated by allosera from sensitized patients or by allele-specific antibodies [43]. Other inhibitors of the complement pathway currently under study include an anti-C5 monoclonal antibody (Eculizumab) and a C1 esterase inhibitor (Berinert) but neither has as yet been tested in the context of endothelial cell activation by alloantibodies.

Intragraft Th17 cells have been associated with shorter graft survival in a study of detransplanted renal allografts. This may be mediated by their ability to secrete IL-21 that was associated with the expression level of a transcription factor (activation-induced cytidine deaminase) implicated in B cell differentiation and thus the humoral alloresponse [44]. A further role of interleukin-17 is in the promotion of intragraft neutrophils allowing the activation of pericytes [45].

The recently published data indicating a HLA class II-mediated pathway of programmed necrotic cell death will lead to interest in the development of inhibitors of this pathway in the context of AMR.

7.7 Immunomodulation of the allograft endothelium

Intravenous immunoglobulins (IVIg) have been administered to allosensitized renal transplant patients and parallel in vitro studies have unveiled multiple effects on circulating immune cells [46]. We have recently determined whether the endothelium is also a potential target of IVIg. Endothelial cells were incubated with concentrations of IVIg (resembling those used in clinical practice) in an inflammatory environment. In contrast with conventional immunosuppressors (cyclosporine A or mycophenolic acid), HLA-DR expression on endothelial cells was increased, and the immunogenicity of the endothelial cells was modified, resulting in amplified expansion of allogeneic Treg cells [31].

The cytokines implicated in Th17 differentiation may be considered as potential therapeutic targets. We and others have demonstrated the role of IL-6 in the differentiation of alloTh17 [32,47] as well as the increase in endothelial cell production of IL-6 in the presence of relevant DSA [21]. A recent study examined the possibility of rescue therapy with an anti-IL-6 receptor humanized monoclonal antibody (Tocilizumab) and reported significant lower circulating DSAs and stabilization of renal function after 2 years [48]. The potential for either anti-IL-6 or anti-IL-6 receptor antibodies to favorably alter the course of disease in patients with AMR is currently under examination.

Finally, because the pathogenesis of AMR is so strongly linked with DSA targeting HLA-II molecules in the microvessels of the allograft [49], the recent demonstration that the deletion of endothelial HLA-II resulted in abrogation of CD4$^+$-T cell helped and prevented rejection in a humanized mouse arterial graft model of acute rejection may point to the possibility of specifically extinguishing HLA expression in the endothelium of the allograft as a long-term objective for improving graft function and survival [50].

Acknowledgments

We are very grateful to Dr. Julien Lion for his critical review of this chapter. We appreciate the support of the Agence pour la Biomedicine, INSERM, and Vaincre La Mucoviscidose.

References

[1]   Sellares J, de Freitas DG, Mengel M, Reeve J, Einecke G, Sis B, et al. Understanding the causes of kidney transplant failure: the dominant role of antibody-mediated rejection and nonadherence. Am J Transplant 2012;12:388−99.

[2]   Bjorkman PJ, Saper MA, Samraoui B, Bennett WS, Strominger JL, Wiley DC. Structure of the human class I histocompatibility antigen, HLA-A2. Nature 1987;329:506−12.

[3]   Cresswell P. Assembly, transport, and function of MHC class II molecules. Annu Rev Immunol 1994;12:259−93.

[4]   Aubert O, Loupy A, Hidalgo L, Duong van Huyen JP, Higgins S, Viglietti D, et al. Antibody-mediated rejection due to preexisting versus *De Novo* donor-specific antibodies in kidney allograft recipients. J Am Soc Nephrol 2017;28:1912−23.

[5]   Wiebe C, Gareau AJ, Pochinco D, Gibson IW, Ho J, Birk PE, et al. Evaluation of C1q status and titer of *De Novo* donor-specific antibodies as predictors of allograft survival. Am J Transplant 2017;17:703−11.

[6]   Visentin J, Chartier A, Massara L, Linares G, Guidicelli G, Blanchard E, et al. Lung intragraft donor-specific antibodies as a risk factor for graft loss. J Heart Lung Transplant 2016;35:1418−26.

[7]   Neau-Cransac M, Le Bail B, Guidicelli G, Visentin J, Moreau K, Quinart A, et al. Evolution of serum and intra-graft donor-specific anti-HLA antibodies in a patient with two consecutive liver transplantations. Transpl Immunol 2015;33:58−62.

[8]   Mangiola M, Marrari M, Feingold B, Zeevi A. Significance of anti-HLA antibodies on adult and pediatric heart allograft outcomes. Front Immunol 2017;8:4.

[9]   Kim PT, Demetris AJ, O'Leary JG. Prevention and treatment of liver allograft antibody-mediated rejection and the role of the 'two-hit hypothesis'. Curr Oporgan Transplant 2016;21:209−18.

[10] Lefaucheur C, Loupy A. Antibody-mediated rejection of solid-organ allografts. N Engl J Med 2018;379:2580−2.

[11] Muczynski KA, Cotner T, Anderson SK. Unusual expression of human lymphocyte antigen class II in normal renal microvascular endothelium. Kidney Int 2001;59:488−97.

[12] Feucht HE, Lederer SR, Kluth B. Humoral alloreactivity in recipients of renal allografts as a risk factor for the development of delayed graft function. Transplantation 1998;65:757−8.

[13] Loupy A, Lefaucheur C, Vernerey D, Prugger C, Duong van Huyen JP, Mooney N, et al. Complement-binding anti-HLA antibodies and kidney-allograft survival. N Engl J Med 2013;369:1215−26.

[14] Sicard A, Ducreux S, Rabeyrin M, Couzi L, McGregor B, Badet L, et al. Detection of C3d-binding donor-specific anti-HLA antibodies at diagnosis of humoral rejection predicts renal graft loss. J Am Soc Nephrol 2015;26:457−67.

[15] Jane-Wit D, Manes TD, Yi T, Qin L, Clark P, Kirkiles-Smith NC, et al. Alloantibody and complement promote T cell-mediated cardiac allograft vasculopathy through noncanonical NF-κB signaling in endothelial cells. Circulation 2013;128:2504−16.

[16] Zhang X, Reed EF. Effect of antibodies on endothelium. Am J Transplant 2009;9:2459−65.

[17] Valenzuela NM, Reed EF. Antibody-mediated rejection across solid organ transplants: manifestations, mechanisms, and therapies. J Clin Invest 2017;127:2492−504.

[18] Lepin EJ, Zhang Q, Zhang X, Jindra PT, Hong LS, Ayele P, et al. Phosphorylated S6 ribosomal protein: a novel biomarker of antibody-mediated rejection in heart allografts. Am J Transplant 2006;6:1560−71.

[19] Haas M, Sis B, Racusen LC, Solez K, Glotz D, Colvin RB, et al. Banff 2013 meeting report: inclusion of c4d-negative antibody-mediated rejection and antibody-associated arterial lesions. Am J Transplant 2014;14:272−83.

[20] Le Bas-Bernardet S, Coupel S, Chauveau A, Soulillou JP, Charreau B. Vascular endothelial cells evade apoptosis triggered by human leukocyte antigen-DR ligation mediated by allospecific antibodies. Transplantation 2004;78:1729−39.

[21] Lion J, Taflin C, Cross AR, Robledo-Sarmiento M, Mariotto E, Savenay A, et al. HLA class II antibody activation of endothelial cells promotes Th17 and disrupts regulatory T lymphocyte expansion. Am J Transplant 2016;16:1408−20.

[22] Jin YP, Valenzuela NM, Zhang X, Rozengurt E, Reed EF. HLA class II-triggered signaling cascades cause endothelial cell proliferation and migration: relevance to antibody-mediated transplant rejection. J Immunol 2018;200:2372−90.

[23] Yamakuchi M, Kirkiles-Smith NC, Ferlito M, Cameron SJ, Bao C, Fox-Talbot K, et al. Antibody to human leukocyte antigen triggers endothelial exocytosis. Proc Natl Acad Sci U S A 2007;104:1301−6.

[24] Jindra PT, Jin YP, Rozengurt E, Reed EF. HLA class I antibody-mediated endothelial cell proliferation via the mTOR pathway. J Immunol 2008;180:2357−66.

[25] Bertho N, Drenou B, Laupeze B, Berre CL, Amiot L, Grosset JM, et al. HLA-DR-mediated apoptosis susceptibility discriminates differentiation stages of dendritic/monocytic APC. J Immunol 2000;164:2379−85.

[26] Drenou B, Blancheteau V, Burgess DH, Fauchet R, Charron DJ, Mooney NA. A caspase-independent pathway of MHC class II antigen-mediated apoptosis of human B lymphocytes. J Immunol 1999;163:4115−24.

[27] Carmagnat M, Drenou B, Chahal H, Lord JM, Charron D, Estaquier J, et al. Dissociation of caspase-mediated events and programmed cell death induced via HLA-DR in follicular lymphoma. Oncogene 2006;25:1914−21.

[28] Al-Daccak R, Mooney N, Charron D. MHC class II signaling in antigen-presenting cells. Curr Opin Immunol 2004;16:108–13.

[29] Aljabri A, Vijayan V, Stankov M, Nikolin C, Figueiredo C, Blasczyk R, et al. HLA class II antibodies induce necrotic cell death in human endothelial cells via a lysosomal membrane permeabilization-mediated pathway. Cell Death Dis 2019;10:235.

[30] Land WG, Agostinis P, Gasser S, Garg AD, Linkermann A. DAMP-induced allograft and tumor rejection: the circle is closing. Am J Transplant 2016;16:3322–37.

[31] Lion J, Burbach M, Cross A, Poussin K, Taflin C, Kaveri S, et al. Endothelial cell amplification of regulatory T cells is differentially modified by immunosuppressors and intravenous immunoglobulin. Front Immunol 2017;8:1761.

[32] Taflin C, Favier B, Baudhuin J, Savenay A, Hemon P, Bensussan A, et al. Human endothelial cells generate Th17 and regulatory T cells under inflammatory conditions. Proc Natl Acad Sci U S A 2011;108:2891–6.

[33] Cross AR, Lion J, Loiseau P, Charron D, Taupin JL, Glotz D, et al. Donor specific antibodies are not only directed against HLA-DR: minding your Ps and Qs. Hum Immunol 2016;77:1092–100.

[34] Cross AR, Lion J, Poussin K, Assayag M, Taupin JL, Glotz D, et al. HLA-DQ alloantibodies directly activate the endothelium and compromise differentiation of FoxP3 high regulatory T lymphocytes. Kidney Int. 2019;96(3):689–98.

[35] Cross AR, Glotz D, Mooney N. The role of the endothelium during antibody-mediated rejection: from victim to accomplice. Front Immunol 2018;9:106.

[36] Sis B, Einecke G, Chang J, Hidalgo LG, Mengel M, Kaplan B, et al. Cluster analysis of lesions in nonselected kidney transplant biopsies: microcirculation changes, tubulointerstitial inflammation and scarring. Am J Transplant 2010;10:421–30.

[37] Dragun D, Catar R, Philippe A. Non-HLA antibodies against endothelial targets bridging allo- and autoimmunity. Kidney Int 2016;90:280–8.

[38] Cardinal H, Dieude M, Brassard N, Qi S, Patey N, Soulez M, et al. Antiperlecan antibodies are novel accelerators of immune-mediated vascular injury. Am J Transplant 2013;13:861–74.

[39] Angaswamy N, Klein C, Tiriveedhi V, Gaut J, Anwar S, Rossi A, et al. Immune responses to collagen-IV and fibronectin in renal transplant recipients with transplant glomerulopathy. Am J Transplant 2014;14:685–93.

[40] Pabois A, Devalliere J, Quillard T, Coulon F, Gerard N, Laboisse C, et al. The disintegrin and metalloproteinase ADAM10 mediates a canonical Notch-dependent regulation of IL-6 through Dll4 in human endothelial cells. Biochem Pharmacol 2014;91:510–21.

[41] Haas M, Loupy A, Lefaucheur C, Roufosse C, Glotz D, Seron D, et al. The Banff 2017 Kidney Meeting Report: revised diagnostic criteria for chronic active T cell-mediated rejection, antibody-mediated rejection, and prospects for integrative endpoints for next-generation clinical trials. Am J Transplant 2018;18:293–307.

[42] Jin YP, Valenzuela NM, Ziegler ME, Rozengurt E, Reed EF. Everolimus inhibits anti-HLA I antibody-mediated endothelial cell signaling, migration and proliferation more potently than sirolimus. Am J Transplant 2014;14:806–19.

[43] Thomas KA, Valenzuela NM, Gjertson D, Mulder A, Fishbein MC, Parry GC, et al. An anti-C1s monoclonal, TNT003, inhibits complement activation induced by antibodies against HLA. Am J Transplant 2015;15:2037–49.

[44] Deteix C, Attuil-Audenis V, Duthey A, Patey N, McGregor B, Dubois V, et al. Intragraft Th17 infiltrate promotes lymphoid neogenesis and hastens clinical chronic rejection. J Immunol 2010;184:5344–51.

[45] Liu R, Lauridsen HM, Amezquita RA, Pierce RW, Jane-Wit D, Fang C, et al. IL-17 promotes neutrophil-mediated immunity by activating microvascular pericytes and not endothelium. J Immunol 2016;197:2400–8.

[46] Maddur MS, Kaveri SV, Bayry J. Circulating normal IgG as stimulator of regulatory T cells: lessons from intravenous immunoglobulin. Trends Immunol 2017;38:789–92.

[47] Fogal B, Yi T, Wang C, Rao DA, Lebastchi A, Kulkarni S, et al. Neutralizing IL-6 reduces human arterial allograft rejection by allowing emergence of CD161+ CD4+ regulatory T cells. J Immunol 2011;187:6268–80.

[48] Choi J, Aubert O, Vo A, Loupy A, Haas M, Puliyanda D, et al. Assessment of tocilizumab (anti-interleukin-6 receptor monoclonal) as a potential treatment for chronic antibody-mediated rejection and transplant glomerulopathy in HLA-sensitized renal allograft recipients. Am J Transplant 2017;17:2381–9.

[49] Stegall MD, Chedid MF, Cornell LD. The role of complement in antibody-mediated rejection in kidney transplantation. Nat Rev Nephrol 2012;8:670–8.

[50] Abrahimi P, Qin L, Chang WG, Bothwell AL, Tellides G, Saltzman WM, et al. Blocking MHC class II on human endothelium mitigates acute rejection. JCI Insight 2016;1:e85293.

Chapter 8

Angiogenesis: Aspects in wound healing

Neha Raina, Radha Rani and Madhu Gupta

School of Pharmaceutical Sciences, Department of Pharmaceutics, Delhi Pharmaceutical Sciences and Research University, New Delhi, India

8.1 Introduction

Angiogenesis or neovascularization is the process in which sprouting of newly formed blood vessels occurs from the preexisting ones. Skin is a major external organ acting as barrier for protecting internal structures of body against hazards of environment. The exceptional property of regeneration is possessed by skin in addition to healing injuries via an eminently coordinated cascade of physiological events. Yet, in certain circumstances, regeneration is lessened and wounds do not heal in a timely fashion, placing the patients at a significant health risk. Usually, wounds that do not heal in 90 days are referred to as chronic wounds [1]. The treatment of chronic wounds and large burns is expensive and laborious as they are liable to infection and often require surgical treatment. Furthermore the burden of chronic wounds is growing due to the increasing incidence of obesity and diabetes [2,3]. Chronic, poorly healing wounds represent an ever-increasing problem worldwide, which mainly occurs in older patients. In the process of wound healing, a number of individual processes take place at the same time; they overlap and have multiple interactions with each other [4]. Angiogenesis, a crucial event in the healing of wounds, is essential in view of the fact that granulation tissue is formed by new capillaries [5–8]. In intact tissues, the state of homeostasis is maintained in microvasculature where nutrients and oxygen are transported in adequate amount to the tissues in balance with elimination of waste products along with carbon dioxide. Upon injury, there is disruption in microvasculature, which leads to accumulation of fluid, the development of hypoxia as well as inflammation. Endothelial cells are activated through inflammatory cytokines and hypoxia, causing further the conscription of immune cells. One extensively believed view is that healing of wound necessitates a strong and active angiogenic response [9,10]. Deficient angiogenesis has been incriminated in chronic and poorly healing wounds that appear in persons having venous stasis disease, in individuals having diabetes, and also in the old age group [11–14]. These nonhealing wounds show evidence of an angiogenic response that is inadequate or nonfunctional. In comparison, the amplified angiogenesis that is seen typically in healing tissue may be excessive. Several studies have revealed that under particular conditions that lessen (but do not eradicate) the angiogenic response, wounds consisting of maximum thickness in normal animals heal entirely. Strategies used in capillary growth blockade are (1) use of antibodies to vascular endothelial growth factor (VEGF) an antiangiogenic agent and (2) use of blockade of integrin signaling [15–20]. The concept that a lowered but efficient angiogenic response may perhaps fulfill the needs of the healing wound is well proven via studies of wounds that heal exceptionally well. Oral mucosal wounds, a site proven to heal very quickly, display a less robust angiogenic response in comparison to wounds of skin [21]. Further studies have revealed that nonscar-forming wounds formed on the second-trimester fetus show evidence of a reduced angiogenic response [22]. A number of studies now indicate that various capillaries that are produced in early healing wounds are not very much functional. A cautiously executed and well-designed experiment by Bluff et al. [23] reveals that it is primarily the newly formed and histologically visible vasculature of the healing wound that is not efficiently perfused. Furthermore, the vasculature that is created under conditions of elevated proangiogenic pressure, for instance, that originated in solid malignancies, has been demonstrated to be tortuous, leaky, and is every so often extremely ineffectual in delivering adequate blood flow [24,25]. Nonetheless, a few reports propose that most capillaries produced in the early healing wound though undeveloped are extremely permeable [26]. Altogether, these investigations indicate the typical pattern of wound angiogenesis that includes an initial compact bed of capillaries that are not very functional. Therapeutic agents that are used for partially blocking this capillary growth may possibly generate an abridged yet functional vasculature that is capable of providing tremendous perfusion. Furthermore, a decrease in the initial burst of capillary growth would reduce both edema in addition lessens the demand for vascular regression.

Endothelial Signaling in Vascular Dysfunction and Disease. DOI: https://doi.org/10.1016/B978-0-12-816196-8.00010-2

8.2   Physiological control of angiogenesis

Angiogenesis plays a key role in the healing of wounds. This occurs via the development of capillary sprouts that digest endothelial cells and invade the extracellular matrix (ECM) stroma after penetration through underlying vascular basement membrane (VBM) and finally form tube-like structures that continue to extend, branch and form vascular networks. Endothelial cell proliferation leads to capillary advancement in extracellular matrix (ECM) while the direction of growth is guided by target region chemotaxis. The interactions amid endothelial cells, angiogenesis factors, and surrounding ECM proteins are harmonized temporally and spatially [27]. Angiogenic processes are guided by a balance between the pro- and anti-angiogenic factors; proangiogenic factors comprise of thrombin, fibrinogen fragments, thymosin-β4, and other growth factors. These are stored in platelets and inflammatory cells that are in the systemic circulation, and are sequestered within the ECM. Genes expressed in reaction to hypoxia and inflammation, such as hypoxia-inducible variables (HIF) and cyclooxygenase-2 (COX-2) regulate the development of these variables [28–30]. Antiangiogenic factors inhibiting angiogenesis suppress the development of the blood vessels [31,32] circulate at low physiological levels in the bloodstream, whereas others are stored in the ECM enclosing blood vessels. Vascular development is suppressed when physiological balance exists between angiogenesis stimulators and inhibitors [33]. However, angiogenic stimuli are released into the wound bed immediately after injury and a shift occurs in regulators supporting vascular growth (Fig. 8.1).

8.3   The intersection of inflammation and angiogenesis

The two processes that appear to be intricately interlinked in injuries are the inflammatory response to injury and subsequent capillary development (Fig. 8.2). Injury in tissue results in rapid acute inflammatory response intended for clearing microbes, dying cells, and debris. This inflammatory response leads to elevated levels of proangiogenic factors (such as vascular endothelial growth factor or VEGF) that are produced by stimulated macrophages and keratinocytes [34,35]. In the skin, several advanced depletion studies have shown that macrophages are a significant source of general proangiogenic stimulus [36–38]. Thus, inflammation (in addition to the eventual angiogenic reaction) seems to be associated with the healing wound. In support of this notion, the relatively lower inflammation in both fetal and oral mucosal wounds, correlates with a lessened angiogenic response at these locations. Hence, both decreased inflammation and angiogenesis are characteristics of optimum healing and reduced scar formation. Because inflammation is decreased in fetal and oral mucosal wounds, the role of inflammation itself in dictating scar formation in wounds has received a great deal of experimental attention [39]. It is revealed from many experimental studies that a decrease in inflammation can enhance skin wound healing outcomes and

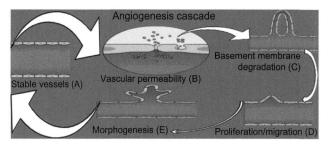

FIGURE 8.1   The cellular events in the angiogenesis cascade. Permeability in vessels (B) leads to influx of inflammatory cells and proangiogenic agents that degrade the extracellular matrix (C). This allows endothelial cell sprouting through the vascular layer followed by proliferation (D) to form new vessels (E).

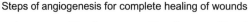

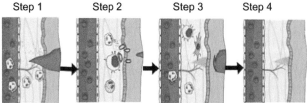

FIGURE 8.2   Steps of angiogenesis for complete healing of wounds. Steps 1 and 2: A wound and the inflammatory response to it in the form of recruitment of immune cells. Steps 3 and 4: wound closure.

reduction in formation of scars. It has been shown that modulation of particular inflammatory mediators or inflammatory cells, together with macrophage-derived mediators, epithelial mediators and mast cells, results in decreased scar formation and enhanced healing [40−46]. The association between inflammation, angiogenesis, and scar formation clearly indicates that both inflammation and angiogenesis directly influence the final scar outcome. While considering the process of angiogenesis of the wound, it is essential to distinguish between circumstances where angiogenesis can be excessive, for instance in normal healing of skin, and situations where angiogenesis is completely deficient, for instance in diabetic wound healing [47]. Diabetes is marked with delayed healing of wounds, and inadequate process of angiogenesis has been shown to play a role in the pathobiology. It is therefore not surprising that treatment with proangiogenic factors or cells to boost angiogenesis has been shown to enhance the recovery or healing of diabetic wounds, at least in animal models [48].

8.3.1   Angiogenesis cascade

Angiogenesis proceeds as an orderly cascade of events, namely molecular and cellular events in the wound bed. These are as follows:

1. The surface of endothelial cells has receptors that are used for binding of angiogenic growth factors in preexisting venules (parent vessels).
2. Signaling pathways within endothelial cells are triggered by growth factor receptor binding.
3. Proteolytic enzymes liberated by activated endothelial cells dissolve the adjacent parent vessels' basement membrane.
4. Basement membrane is used for proliferation and sprouting outward of endothelial cells.
5. Endothelial cells move into the wound bed with the help of integrins ($\alpha v \beta 3$, $\alpha v \beta 5$, and $\alpha v \beta 1$) that are molecules of cell surface adhesion.
6. Surrounding tissue matrix is dissolved by matrix metalloproteinases (MMPs) in the path of sprouting vessels.
7. Vascular loops are formed from connection of tubular channels that are formed by vascular sprouts.
8. Vascular loops differentiate into afferent (arterial) and efferent (venous) limbs.
9. By mural cells recruitment newly formed blood vessels mature (smooth muscle cells and pericytes) for stabilizing the vascular architecture.
10. Blood flow starts in the mature stable vessel (Fig. 8.2).

These complex growth factor receptor, cell−cell, and cell−matrix interactions characterize the angiogenesis process, regardless of the stimuli or its location in the body.

8.4   Growth factors involved in process of angiogenesis

Angiogenesis is an event that is complex and controlled through diverse soluble factors acting in a successive, intensive, or synergistic manner. Extensive research using growth factors aimed both at inducing angiogenesis and augmenting closure of wounds in chronic wounds has shown that several of these factors are crucial in driving angiogenesis:.

8.4.1   Platelet-derived growth factor

Platelet-derived growth factors (PDGFs) are vital chemoattractants in the process of healing of wounds. Currently, it is prescribed as becaplermin and is the only drug that was sanctioned by FDA for treating growth factor driven healing of wounds [49,50]. In some clinical trials, becaplermin has been shown to boost healing of wounds in venous leg ulcers [51,52]. A clinical trial on becaplermin aimed at treating diabetic foot ulcers, 48% of patients (29/61) had proper healing of ulcers versus 25% (14/57) in the placebo group. Clinical trials conducted in the year 1998 and 1999 displayed statistically considerable efficacy in wound healing upon use of becaplermin [53,54]. Decreased wound area besides faster healing rates were displayed in the groups treated once a day with becaplermin. Indeed, this was the case for significant healing of diabetic foot ulcers over the placebo controls. These positive results can be ascribed to the capability of PDGF-BB to augment conscription of cells to the site of wound thereby stimulating angiogenic response. However, these results were inconclusive as other clinical trials showed no improvement in healing of wounds with becaplermin treatment [55]. This discrepancy could be attributed to therapeutic resistance in averting the efficacy of PDGF-BB therapies, consequently requiring higher concentrations, also increasing the risk of side effects of the growth factor therapy. In addition, the black box warning was issued by FDA for becaplermin because of a fivefold increase in death rate in patients that used three tubes of becaplermin and those with preexisting malignancy [56]. The concurrent application of multiple growth factors in addition to proteins possibly will help in providing the additional advantage required to see therapies functional in a clinical environment. The concurrent application of using PDGF-BB besides epidermal growth factor (EGF) has been illustrated to direct pericyte conscription to support in stabilizing vessels of blood [57,58]. The other studies conducted on multiple growth factors have displayed that the concurrent application of PDGF-BB and fibroblast growth factor (FGF)-2 enhanced vascular stability in hind limb ischemia models in rats and rabbits [59]. The twofold delivery of PDGF-BB and VEGF-A from nanofiber scaffolds that are prepared by electrospinning technique for quicker healing of wound in rats also promoted the process of angiogenesis at the site of wound [60]. Similar outcomes were attained through the concurrent delivery of VEGF-A and PDGF-BB bound via laminin heparin connecting domains from matrices of fibrin [61]. In one study, researchers formed nanofiber scaffold comprising of collagen and hyaluronic acid (HA) that was prepared using an electrospinning technique and released FGF-2, EGF, VEGF-A, and PDGF-BB in which release kinetics varied. This construct accelerated closure of wounds and the development of blood vessels in the wound beds of rats with streptozotocin-induced diabetes [62]. Finally, combining PDGF-BB with a syndecan-4 proteoliposome significantly increased its role in wound closure, angiogenesis and increased the proportion of M2 macrophages in a wound healing model of the diabetic mouse [63].

8.4.2 Epidermal growth factor

An eminent mediator of angiogenesis process and other capable growth factors for wound healing therapies is Epidermal Growth Factor (EGF). Clinical trials displayed reduced healing times and re-epithelialization in wounds that were treated by means of recombinant EGF [64]: still, concerns over the function of EGF in cancer have generated considerable interest in these treatments [65]. Numerous forms of recombinant human EGF are commercially available outside of the United States, comprising of Easyef (Nepidermin), Regen-D 150, and Heberprot-P. Easyef is the formulation available in the form of a spray or in the form of ointment and has demonstrated advantages in diabetic foot ulcers in an open-label trial, prospective, crossover study with 89 patients [66]. A gel-based formulation Regen-D 150 containing EGF has demonstrated improvements by reducing the time of wound healing and reducing the risk of amputation in patients suffering from chronic diabetic foot ulcers [67]. A lyophilized formulation of EGF, Heberprot-P, has also been reported to reduce wound healing time and risk of amputation in chronic diabetic foot ulcers in a 100,000 patient study [68]. Lately, methods of new delivery have been discovered to aid in localizing EGF besides increasing its time of circulation [69]. Silk biomaterials for the delivery of EGF were developed to promote angiogenesis and wound closure in cutaneous excisional wounds in mice [70,71]. The experimental study conducted by Gil et al. [70] examined three material structures including silk biomaterials that were different along with two different methods of drug incorporation [72]. A silk biomaterial was used in formulating a porous film, solid film, or electrospun nanofibers. It was revealed from the study that the use of silk biomaterials lead to higher percentages of wound closure as compared to control wounds in a dermal wound model of full thickness. Reduced scar tissue formation and increase in epithelization over a Tegaderm control were caused due to delivery of EGF from biomaterials of silk.

8.4.3 Fibroblast growth factors

Fibroblast Growth Factors (FGFs) come under the category of growth factors that display immense potential in the process of dermal wound healing as various physical processes are influenced by them [73]. In the course of healing process various different FGF family members expressed are FGF-1, FGF-2, FGF-7, FGF-10, and FGF-22 [74,75]. In general, binding of FGF ligands to their receptors takes place in the presence of heparin or heparan sulfate and in endothelial cells proliferation/migration and angiogenesis can be promoted by these signals [76]. FGF-2 is the most widely studied member of the FGF family employed in applications of wound healing [77], due to its strong association with wound angiogenesis. The role of other FGFs is allied with the augmentation of re-epithelialization [78−80], the utmost prominent being FGF-7. FGF-7 has been demonstrated to be integral in the keratinocytes migration to the area of the wound [81] and cutaneous wound healing is delayed in nonexistence of FGF-7 in mice. Additionally, it is shown that skin wound healing is enhanced with the dual delivery of FGF-7 and FGF-2, displaying vascular networks that are more developed than the scaffold control [82]. Novel methods for the delivery of FGF-2 delivery methods have been discovered for enhancing the aforementioned activity of wound healing. FGF-2 delivery from sheets of gelatin having crosslinked with the crosslinker glutaraldehyde enhanced re-epithelialization, granulation tissue formation, and angiogenesis process in contrast to control wounds in a mouse model [83].

8.4.4 Vascular endothelial growth factor

The family of VEGF covers individual growth factors that are five in number namely, VEGF-A, -B, -C, -D, and placental growth factor (PlGF) [84]. In both embryonic and adult development, lymphangiogenesis as well as angiogenesis are regulated chiefly via these growth factors [85]. Moreover a momentous role is played by VEGF family in the vasodilation, blood vessels permeability, and these have been demonstrated to stabilize the development of newly formed blood vessels in the course of wound healing. The function of VEGF treatments in clinical settings has been partially successful, by reason of the vasopermeability brought by means of the treatment [86].

8.5 Phases of angiogenesis process in wound healing

Healing of wounds takes place majorly in coinciding stages that are four in number namely: (1) hemostatic, (2) inflammatory phase, (3) proliferative period, and (4) remodeling stage. Even though granulation is allocated to the stage of proliferation, the process of angiogenesis is instigated instantly after injury to the tissue and is interceded during the course of wound healing process.

8.5.1 Phase 1: Angiogenesis commencement or initiation

Basic fibroblast growth factor (bFGF) is stored inside intact cells and release of ECM takes place from tissue that is damaged [87]. Angiogenesis is also initiated by bleeding and hemostasis in a wound. In the wound, thrombin upregulates cellular receptors meant for VEGF [88]. The thrombin-exposed endothelial cells also release gelatinase A matrix-metalloproteinase 2 (MMP-2), which promotes local dissolution of the basement membrane, a necessary early stage of angiogenesis [89]. Multiple growth factors are released by platelets, comprising VEGF, PDGF, transforming growth factor (TGF-α, TGF-β), bFGF, platelet-derived endothelial cell growth factor, and angiopoietin-1 (Ang-1) and these mentioned factors are responsible for stimulating endothelial proliferation, migration, and tube formation [90−93].

8.5.2 Phase 2: Angiogenesis amplification

In this step a number of angiogenic factors are released by macrophages and monocytes into the wound bed comprising VEGF, PDGF, Ang-1, TGF-α, bFGF, and interleukin-8 (IL-8) besides tumor necrosis factor- α in the course of the inflammatory phase augmenting angiogenesis further [94,95]. Various growth factors (PDGF, VEGF, and bFGF) synergize in their ability to vascularize tissues [96]. Damaged tissue matrix is broken down by proteases additionally releases matrix-bound angiogenic stimulators. Enzymatic cleavage of fibrin fragment E is generated by enzymatic cleavage of fibrin, which directly promotes angiogenesis in addition to enhancing the effects of VEGF and bFGF [97]. Expression of the inducible COX-2 enzyme for the period of the inflammatory stage of healing furthermore leads to production of VEGF and other promoters of the process of angiogenesis [98].

8.5.3 Phase 3: Vascular proliferation

The vital driving force for wound angiogenesis is hypoxia. Expression of gene HIF-1α, by reason of hypoxic gradient between damaged and healthy tissue, prompts the production of VEGF [99,100]. The wound tissue and exudate contain considerable amount of VEGF [101,102]. VEGF is additionally designated as vascular permeability factor in view of the fact that it enhances permeability of capillaries [103]. Hypoxia directs endothelial cell production of nitric oxide (NO). Vasodilation along with angiogenesis is promoted by NO for improving blood flow locally [104].

8.5.4 Phase 4: Vascular stabilization

Vascular stabilization is the process that is facilitated through Ang-1, tyrosine kinase in conjunction with immunoglobulin-like and EGF-like domains 2 (Tie-2), and smooth muscle cells in addition to pericytes. The binding of Ang-1 to its receptor Tie-2 on endothelial cells that are activated [105−107] is a signal for smooth muscle cells and pericytes to be conscripted into the forming vasculature. An inadequacy of PDGF leads to the formation of immature blood vessels [108].

8.5.5 Phase 5: Angiogenesis suppression

Suppression of angiogenic process occurs at the end phase of healing [109]. As hypoxia and inflammation in tissue subsides, the expression level of growth factors subsequently decreases in the area of wound. Vascular proliferation is impeded by the inhibitory form of activated TGF-β secreted by endothelial cells, which are stabilized through pericytes [110–112]. A cleavage product of collagen XVIII is named as endostatin that exists in the surrounding of the VBM that hinders wound vascularity [113,114].

8.6 Wound angiogenic stimulators and inhibitors (regulators)

Numerous angiogenic stimulators have been known to play a vital role in wound repair. The stimulators that are present in fluid of wound are growth factors that are well known for increasing migration of endothelial cells in addition to proliferation *in vitro* [115]. The FGF encompasses homologous structures that are 23 in number, small polypeptides along with a central core possessing 140 amino acids. Acidic FGF and bFGF (FGF-1 and FGF-2, respectively) [26] have been implicated in the course of action of angiogenesis [116,117]. These compounds with a weight of approximately 18 kDa are single-chained and nonglycosylated polypeptides. The intense interactions of FGF-1 and FGF-2 with glycosaminoglycans, for instance, heparin sulfate existing in the ECM [118] make the FGFs stable to the thermal, proteolytic denaturation and confine its diffusibility. ECM thus performs the function of reservoir for proangiogenic factors. Members of the FGF family mostly perform action as broad-spectrum mitogens that help in promoting the proliferation of mesenchymal cells originating from mesoderm, besides ectodermal and endodermal cells.

Synthesis of FGF-1 and FGF-2 occurs by a variety of cell types comprising inflammatory cells, dermal fibroblasts, etc. that are entailed in the process of angiogenesis and wound healing. FGF acts on the endothelial cells either in a paracrine manner, or after release from endothelial cell in an autocrine manner, furthering endothelial cell proliferation along with differentiation. In the course of granulation tissue formation, FGF-2 stimulates migration of cells by means of surface receptors for integrins, which mediate the binding of endothelial cells to ECM [119].

The function of VEGF is to increase vascular permeability by augmenting the fenestration and hydraulic conductivity. This permits leakage of fibronectin and fibrinogen, which are vital for provisional ECM formation [58,120,121]. In the course of wound healing, the ECM is generated by means of the epidermis in large quantities [122]. During tissue hypoxia that occurs in wounds, low-oxygen tension is a chief inducer of VEGF [123] and its receptors [124]. Therefore cell disruption along with hypoxia seems to be powerful initial inducers of potent angiogenesis factors at the site of wound. VEGF family, which plays a major role in angiogenesis comprises of VEGF-A, VEGF-B, VEGF-C, VEGF-D, VEGF-E, and PlGF [125]. VEGF-A is a glycoprotein, a structural homodimer in which subunits are connected via two disulfide bonds and synthesis of VEGF-A occurs from internal rearrangements ("alternative splicing") of mRNA. Thus seven isoforms are produced having amino acids from 121 to 206 [126–128]. Amongst these most important isoforms are VEGF165, VEGF121, VEGF189 and VEGF206 [129]. Analogous biological activity is demonstrated by these isoforms; however, their binding properties vary in case of heparin and ECM [130].

VEGF is an effective vascular endothelial cell–specific mitogen that promotes proliferation of endothelial cells, microvascular permeability, and regulation of numerous endothelial integrin receptors in the course of sprouting of new blood vessels [131]. Additionally, for endothelial cells, VEGF also functions as survival factor by way of inducing the expression of an antiapoptotic protein B-cell lymphoma 2 [132].

Another important player in angiogenesis is transforming growth factor-β (TGF-β) that promotes granulation tissue formation by acting as a chemoattractant for macrophages, neutrophils, and fibroblasts. For this reason, TGF-β is a crucial modulator of the angiogenesis throughout the process of wound healing by regulation of cell proliferation, migration, and formation of capillary tube formation along with ECM deposition [133,134].

Other factors in regulating angiogenesis are other members of the VEGF family comprising angiopoietins, which are chiefly specialized for vascular endothelium. Agonist, Ang-1, and antagonist, Ang-2, act by binding to the Tie-2 receptor on endothelial cells. Ang-3 in mice and Ang-4 in humans have also been identified although their purpose in the process of angiogenesis is not known [135].

Storage of mast cell tryptase, in activated mast cells granules, it is an additional factor of angiogenesis that is capable of directly degrading the components of ECM or emancipate growth factors that are bound to matrix through its proteolytic activity [136,137]. Tryptase is another angiogenc factor that induces angiogenesis by release of stored angiogenic factors bound to the extracellular matrix. The adding tryptase to microvascular endothelial cells cultured on a basement membrane matrix (Matrigel) produced a discernable enhancement in the growth of capillaries. Moreover, tryptase has capability of inducing endothelial cell proliferation in a dose-dependent manner; however, certain tryptase inhibitors repress the growth of capillaries. Table 8.1 summarizes the list of proangiogenic agents that are delivered via nanocarriers to accelerate the process of angiogenesis.

TABLE 8.1 List of nanocarriers used in the process of angiogenesis.

Carrier system	Composition	Advantages	References
Nanoparticle composite scaffold	Mesoporous silica nanoparticle loaded with vascular endothelial growth factor (VEGF) incorporated into type I collagen sponge	Enhanced cell proliferation due to release of VEGF, caused increase in number of blood vessel complexes as compared to non-VEGF scaffolds, good compatibility, and potential ability to stimulate angiogenesis and tissue repair	[138]
Hydrogels	Chitosan, deionized water, acetic acid, gelatin, genipin, DFO, and HYSA	Coadministration of DFO and HYSA upregulate secretion of HIF-1α and promote the process of angiogenesis as compared to the administration of single drug. Potentially enhanced process of tissue regeneration identical to normal skin	[139]
Nanoparticles (electrospun bilayer nanofibers)	Nanoparticles were formed through condensation of siFKBPL with RALA (novel, cationic 30 mer amphipathic peptide)	Potential enhancement in cell migration and process of endothelial tubule formation in vitro	[140]
Hydrogels	Formation of hydrogels glass composite was carried out with a mixture of polyethylene glycol, glucose, tetramethyl orthosilicate, sodium nitrate, and chitosan in 0.5 mol/L sodium phosphate buffer of pH 7	Nitric-oxide releasing nanoparticles considerably speed up the process of angiogenesis in wound healing	[141]
Hydrogel-glass composite	Formation of hydrogels glass composite was carried out with a mixture of polyethylene glycol, glucose, tetramethyl orthosilicate, sodium nitrate, and chitosan in 0.5 mol/L sodium phosphate buffer of pH 7, and blank hydrogels also prepared with similar method	Sustained release of nitric oxide from nitric oxide nanoparticles, enhanced angiogenesis process of wound healing in both infected and uninfected murine wound models	[142]
Gold nanoparticles	AuNPs, epigallocatechin gallate (EGCG), α-lipoic acid (ALA)	AuEA application by topical route accelerates the process of wound healing and showed significant increase of angiopoietin-1 protein expression and no change of angiopoietin-2 protein expression	[143]
Liposomes	Encapsulated glycyl-L-histidyl-L-lysine (GHK)-Cu	Promoted HUVECs proliferation, increase cell number in G1 stage and decrease cell number in G2 stage, enhanced expression of fibroblast growth factor-2, and VEGF, better angiogenesis in burned skin in mice treated with GHK-Cu encapsulated in liposomes as compared to free GHK-Cu	[144]
Liposomes	Encapsulated epidermal growth factor (EGF) in liposomes, chitosan gel	Considerable increase in the cell proliferation after applying the EGF-containing liposomes in chitosan gel on the site of burn wounds in rats	[145]
Liposomes	Encapsulated Danggui Buxue Extract in liposomes	Potentially enhanced dermal wound healing in rats by stimulating collagen synthesis and angiogenesis	[146]
Liposomes	Insulin like growth factor I (IGF-I), platelet-derived growth factor isoform BB (PFGF-BB), and mixed PDGF-BB/IGF-I encapsulated in liposomes and phosphate-buffered saline	Considerable higher number of blood vessels and high percentage of bone trabeculae were seen in IL, PDL, and PDIL groups. GFs carried by liposomes enhanced healing process of wound in tooth sockets of rats by enhanced VEGF expressions in both forms	[147]
Multifunctional nanofibers	Platelet-derived growth factor-BB (PDGF-BB) and VEGF were encapsulated in poly (lactic-co-glycolic acid) (PLGA) nanofibers	Quick release of VEGF and slow-expelling of PDGF-BB helped angiogenesis in early stages of wound healing process results in accelerated wound healing process	[59]

(*Continued*)

TABLE 8.1 (Continued)

Carrier system	Composition	Advantages	References
Novel nanoscale bioactive glasses (NBG)	Extracts of 58S-NBG and 80S-NBG	Ability to stimulate angiogenesis in vitro, cell proliferation, and wound healing	[148]
Nanoflowers	Zinc oxide, hydrogen peroxide	Promotes angiogenesis and infection control in angiogenesis	[149]
Dendrimer	VEGF, arginine-grafted cationic dendrimer, PAM-RG4	Enhanced angiogenesis in diabetic murine wounds	[150]
Hydrogels scaffolds	Dextran, allyl isocyanate, polyethylene glycol-diacrylate, DMSO, dibutyltin dilaurate, 2-bromoethylamine hydrobromide, triethylamine, acryloyl chloride, PEG, photoinitiator 2-hydroxy-1-[4-(hydroxyethoxy)phenyl]-2-methyl-1-propanone	Induced increased flow of blood to the area of burn wound and significantly promoted the process of angiogenesis in wound healing	[151]
Chitosan scaffolds	Plasmid DNA encoding the perlecan domain I and VEGF189 loaded in chitosan scaffolds	Increased number of subepithelial connective tissue matrix components and blood vessels in wound area showing potential ability to promote angiogenesis in wound healing	[152]

PEG, polyethylene glycol; *GFs*, growth factors; *DMSO*, dimethylsulfoxide.

8.7   Impaired angiogenesis in chronic wounds

The process of angiogenesis in all chronic wounds is impaired with certain pathologies. This leads to further tissue damage arising due to hypoxia and impaired delivery of micronutrients. Specialized (angiogenesis) defects have been detected in ischemic ulcers, diabetic ulcers, and venous insufficiency ulcers.

8.7.1   Diabetic ulcers

Patients with diseases such as diabetes demonstrate atypical angiogenesis in several organs. Vasculopathies allied with diabetes comprise abnormal blood vessel formation (e.g., retinopathy, nephropathy), atherosclerosis, cerebrovascular disease, coronary artery disease, and peripheral vascular disease [153]. In diabetics, the process of angiogenesis is reduced [154] causing poor formation of newly formed blood vessels and reduced entry of inflammatory cells and growth factors. These growth factors such as FGF-2 and PDGF are indispensable for healing of wounds, and have been found to be lowered in experimental diabetic wounds models [155−158]. Moreover, in rat models, administering high glucose topically in wounds has shown to impede the normal angiogenic processes [159] clearly signifying a direct role for elevated glucose levels in declined angiogenesis.

8.7.2   Venous insufficiency ulcers

Venous insufficiency ulcers arise from incompetent valves present in lower extremity veins, further causing venous stasis and hypertension that makes the skin liable to ulceration. Pathological findings show that venous stasis ulcers consist of fibrin "cuffing," microangiopathy, and entrapment of leukocytes in the interior of microvasculature [160,161]. All these lead to compromised angiogenesis.

8.7.3   Ischemic ulcers

Peripheral arterial disease may give rise to ischemia of extreme severity [162]. Reduced tissue perfusion due to ischemia results in progressive tissue hypoxia, ischemia, necrosis, and skin breakdown that causes haphazard or insufficient angiogenesis.

8.8   Conclusion

A physiological process that is fundamental for healing of normal wounds is the process of angiogenesis. Numerous factors control wound angiogenesis, such as inflammation, hypoxia, and growth factors. The cellular and molecular events in angiogenesis have been explicated, but gaps in knowledge exist in our understanding of angiogenesis in chronic wounds. On the basis of this knowledge, there is emergence of innovative strategies for wound healing intended for delivering growth factors to the wound bed. This knowledge can be used by wound-care specialists for identifying defects and selecting interventions that can potentially stimulate wound granulation and healing.

References

[1] Saghazadeh S, Rinoldi C, Schot M, Kashaf SS, Sharifi F, Jalilian E, et al. Drug delivery systems and materials for wound healing applications. Adv Drug Deliv Rev 2018;127:138−66.

[2] Sen CK, Gordillo GM, Roy S, Kirsner R, Lambert L, Hunt TK, et al. Human skin wounds: a major and snowballing threat to public health and the economy. Wound Repair Regen 2009;17(6):763−71.

[3] Fife CE, Carter MJ, Walker D, Thomson B. Wound care outcomes and associated cost among patients treated in US outpatient wound centers: data from the US wound registry. Wounds 2012;24(1):10.

[4] Vidmar J, Chingwaru C, Chingwaru W. Mammalian cell models to advance our understanding of wound healing: a review. J Surg Res 2017;210:269−80.

[5] Li WW, Li VW, Tsakayannis D. Angiogenesis therapies. Concepts, clinical trials, and considerations for new drug development. N Angiother 2002;547−71.

[6] Folkman J. Clinical applications of research on angiogenesis. N Engl J Med 1995;333(26):1757−63.

[7] Valko M, Leibfritz D, Moncol J, Cronin MT, Mazur M, Telser J. Free radicals and antioxidants in normal physiological functions and human disease. Int J Biochem Cell Biol 2007;39(1):44−84.

[8] Rees M, Hague S, Oehler MK, Bicknell R. Regulation of endometrial angiogenesis. Climacteric 1999;2(1):52−8.

[9] Mg T, Feng X, Clark RA. Angiogenesis in wound healing. J Invest Dermatol Symp Proc 2000;5(1):40−6.

[10] Eming SA, Brachvogel B, Odorisio T, Koch M. Regulation of angiogenesis: wound healing as a model. Prog Histochem Cytochem 2007;42 (3):115−70.

[11] Brem HA, Tomic-Canic M, Tarnovskaya A, Ehrlich HP, Baskin-Bey E, Gill K, et al. Healing of elderly patients with diabetic foot ulcers, venous stasis ulcers, and pressure ulcers. Surg Technol Int 2001;11:161−7.

[12] Reed MJ, Edelberg JM. Impaired angiogenesis in the aged. Sci Aging Knowl Environ 2004;2004(7):pe7.

[13] Falanga V. Wound healing and its impairment in the diabetic foot. Lancet 2005;366(9498):1736−43.

[14] Bennett NT, Schultz GS. Growth factors and wound healing: part II. Role in normal and chronic wound healing. Am J Surg 1993;166 (1):74−81.

[15] Lange-Asschenfeldt B, Velasco P, Streit M, Hawighorst T, Detmar M, Pike SE, et al. The angiogenesis inhibitor vasostatin does not impair wound healing at tumor-inhibiting doses. J Invest Dermatol 2001;117(5):1036−41.

[16] Berger AC, Feldman AL, Gnant MF, Kruger EA, Sim BK, Hewitt S, et al. The angiogenesis inhibitor, endostatin, does not affect murine cutaneous wound healing. J Surg Res 2000;91(1):26−31.

[17] Roman CD, Choy H, Nanney L, Riordan C, Parman K, Johnson D, et al. Vascular endothelial growth factor-mediated angiogenesis inhibition and postoperative wound healing in rats. J Surg Res 2002;105(1):43−7.

[18] Klein SA, Bond SJ, Gupta SC, Yacoub OA, Anderson GL. Angiogenesis inhibitor TNP-470 inhibits murine cutaneous wound healing. J Surg Res 1999;82(2):268−74.

[19] Bloch W, Huggel K, Sasaki T, Grose R, Bugnon P, Addicks K, et al. The angiogenesis inhibitor endostatin impairs blood vessel maturation during wound healing. FASEB J 2000;14(15):2373−6.

[20] Jang YC, Arumugam S, Gibran NS, Isik FF. Role of αv integrins and angiogenesis during wound repair. Wound Repair Regen 1999;7 (5):375−80.

[21] Szpaderska AM, Walsh CG, Steinberg MJ, DiPietro LA. Distinct patterns of angiogenesis in oral and skin wounds. J Dental Res 2005;84 (4):309−14.

[22] Wilgus TA, Ferreira AM, Oberyszyn TM, Bergdall VK, DiPietro LA. Regulation of scar formation by vascular endothelial growth factor. Lab Invest 2008;88(6):579.

[23] Bluff JE, O'Ceallaigh S, O'Kane S, Ferguson MW, Ireland G. The microcirculation in acute murine cutaneous incisional wounds shows a spatial and temporal variation in the functionality of vessels. Wound Repair Regen 2006;14(4):434−42.

[24] Brown LF, Berse B, Jackman RW, Tognazzi K, Manseau EJ, Dvorak HF, et al. Increased expression of vascular permeability factor (vascular endothelial growth factor) and its receptors in kidney and bladder carcinomas. Am J Pathol 1993;143(5):1255.

[25] Nagy JA, Brown LF, Senger DR, Lanir N, Van De Water L, Dvorak AM, et al. Pathogenesis of tumor stroma generation: a critical role for leaky blood vessels and fibrin deposition. Biochim Biophys Acta Rev Cancer 1989;948(3):305−26.

[26] Nagy JA, Benjamin L, Zeng H, Dvorak AM, Dvorak HF. Vascular permeability, vascular hyperpermeability and angiogenesis. Angiogenesis 2008;11(2):109−19.

[27] Clark RA. Wound repair. Overview and general considerations. In: Clark RAF, editor. The molecular and cellular biology of wound repair. New York, NY: Plenum; 1996. p. 3–50.

[28] Morgan MR, Humphries MJ, Bass MD. Synergistic control of cell adhesion by integrins and syndecans. Nat Rev Mol Cell Biol 2007;8(12):957.

[29] Semenza GL. Signal transduction to hypoxia-inducible factor 1. Biochem Pharmacol 2002;64(5–6):993–8.

[30] Majima M, Hayashi I, Muramatsu M, Katada J, Yamashina S, Katori M. Cyclo-oxygenase-2 enhances basic fibroblast growth factor-induced angiogenesis through induction of vascular endothelial growth factor in rat sponge implants. Br J Pharmacol 2000;130(3):641–9.

[31] Pugh CW, Ratcliffe PJ. Regulation of angiogenesis by hypoxia: role of the HIF system. Nat Med 2003;9(6):677.

[32] Miles KA. Perfusion CT for the assessment of tumour vascularity: which protocol? Br J Radiol 2003;76(suppl_1):S36–42.

[33] Hanahan D, Weinberg RA. Hallmarks of cancer: the next generation. Cell 2011;144(5):646–74.

[34] Brown LF, Yeo KT, Berse B, Yeo TK, Senger DR, Dvorak HF, et al. Expression of vascular permeability factor(vascular endothelial growth factor) by epidermal keratinocytes during wound healing. J Exp Med 1992;176:1375–9.

[35] Thakral KK, Goodson WH, Hunt TK. Stimulation of wound blood vessel growth by wound macrophages. J Surg Res 1979;26(4):430–6.

[36] Spiller KL, Anfang RR, Spiller KJ, Ng J, Nakazawa KR, Daulton JW, et al. The role of macrophage phenotype in vascularization of tissue engineering scaffolds. Biomaterials 2014;35(15):4477–88.

[37] Goren I, Allmann N, Yogev N, Schürmann C, Linke A, Holdener M, et al. A transgenic mouse model of inducible macrophage depletion: effects of diphtheria toxin-driven lysozyme M-specific cell lineage ablation on wound inflammatory, angiogenic, and contractive processes. Am J Pathol 2009;175(1):132–47.

[38] Koh TJ, DiPietro LA. Inflammation and wound healing: the role of the macrophage. Expert Rev Mol Med 2011;13.

[39] Lucas T, Waisman A, Ranjan R, Roes J, Krieg T, Müller W, et al. Differential roles of macrophages in diverse phases of skin repair. J Immunol 2010;184(7):3964–77.

[40] Stramer BM, Mori R, Martin P. The inflammation–fibrosis link? A Jekyll and Hyde role for blood cells during wound repair. J Invest Dermatol 2007;127(5):1009–17.

[41] Peranteau WH, Zhang L, Muvarak N, Badillo AT, Radu A, Zoltick PW, et al. IL-10 overexpression decreases inflammatory mediators and promotes regenerative healing in an adult model of scar formation. J Invest Dermatol 2008;128(7):1852–60.

[42] Mori R, Shaw TJ, Martin P. Molecular mechanisms linking wound inflammation and fibrosis: knockdown of osteopontin leads to rapid repair and reduced scarring. J Exp Med 2008;205(1):43–51.

[43] Zgheib C, Xu J, Liechty KW. Targeting inflammatory cytokines and extracellular matrix composition to promote wound regeneration. Adv Wound Care 2014;3(4):344–455.

[44] Gallant-Behm CL, Du P, Lin SM, Marucha PT, DiPietro LA, Mustoe TA. Epithelial regulation of mesenchymal tissue behavior. J Invest Dermatol 2011;131(4):892–9.

[45] Gallant-Behm CL, Hildebrand KA, Hart DA. The mast cell stabilizer ketotifen prevents development of excessive skin wound contraction and fibrosis in red Duroc pigs. Wound Repair Regen 2008;16(2):226–33.

[46] Chen L, Schrementi ME, Ranzer MJ, Wilgus TA, DiPietro LA. Blockade of mast cell activation reduces cutaneous scar formation. PLoS One 2014;9(1):e85226.

[47] Costa PZ, Soares R. Neovascularization in diabetes and its complications. Unraveling the angiogenic paradox. Life Sci 2013;92(22):1037–45.

[48] Demidova-Rice TN, Durham JT, Herman IM. Wound healing angiogenesis: innovations and challenges in acute and chronic wound healing. Adv Wound Care 2012;1(1):17–22.

[49] Papanas N, Maltezos E. Becaplermin gel in the treatment of diabetic neuropathic foot ulcers. Clin Interv Aging 2008;3(2):233.

[50] Woodley DT, Wysong A, DeClerck B, Chen M, Li W. Keratinocyte migration and a hypothetical new role for extracellular heat shock protein 90 alpha in orchestrating skin wound healing. Adv Wound Care 2015;4(4):203–12.

[51] Wieman TJ. Clinical efficacy of becaplermin (rhPDGF-BB) gel. Am J Surg 1998;176(2):74S–79SS.

[52] Fang RC, Galiano RD. A review of becaplermin gel in the treatment of diabetic neuropathic foot ulcers. Biol Targets Ther 2008;2(1):1.

[53] Smiell JM, Wieman TJ, Steed DL, Perry BH, Sampson AR, Schwab BH. Efficacy and safety of becaplermin (recombinant human platelet-derived growth factor-BB) in patients with nonhealing, lower extremity diabetic ulcers: a combined analysis of four randomized studies. Wound Repair Regen 1999;7(5):335–46.

[54] Veith AP, Henderson K, Spencer A, Sligar AD, Baker AB. Therapeutic strategies for enhancing angiogenesis in wound healing. Adv Drug Deliv Rev 2018;146:97–125.

[55] Gilligan AM, Waycaster CR, Milne CT. Cost effectiveness of becaplermin gel on wound closure for the treatment of pressure injuries. Wounds 2018;30(6):197–204.

[56] Bergers G, Song S. The role of pericytes in blood-vessel formation and maintenance. Neuro Oncol 2005;7(4):452–64.

[57] Stratman AN, Schwindt AE, Malotte KM, Davis GE. Endothelial-derived PDGF-BB and HB-EGF coordinately regulate pericyte recruitment during vasculogenic tube assembly and stabilization. Blood 2010;116(22):4720–30.

[58] Cao R, Bråkenhielm E, Pawliuk R, Wariaro D, Post MJ, Wahlberg E, et al. Angiogenic synergism, vascular stability and improvement of hindlimb ischemia by a combination of PDGF-BB and FGF-2. Nat Med 2003;9(5):604.

[59] Xie Z, Paras CB, Weng H, Punnakitikashem P, Su LC, Vu K, et al. Dual growth factor releasing multi-functional nanofibers for wound healing. Acta Biomater 2013;9(12):9351–9.

[60] Ishihara J, Ishihara A, Fukunaga K, Sasaki K, White MJ, Briquez PS, et al. Laminin heparin-binding peptides bind to several growth factors and enhance diabetic wound healing. Nat Commun 2018;9(1):2163.

[61] Lai HJ, Kuan CH, Wu HC, Tsai JC, Chen TM, Hsieh DJ, et al. Tailored design of electrospun composite nanofibers with staged release of multiple angiogenic growth factors for chronic wound healing. Acta Biomater 2014;10(10):4156−66.

[62] Das S, Majid M, Baker AB. Syndecan-4 enhances PDGF-BB activity in diabetic wound healing. Acta Biomater 2016;42:56−65.

[63] Brown GL, Nanney LB, Griffen J, Cramer AB, Yancey JM, Curtsinger III LJ, et al. Enhancement of wound healing by topical treatment with epidermal growth factor. N Engl J Med 1989;321(2):76−9.

[64] Falanga V, Eaglstein WH, Bucalo B, Katz MH, Harris B, Carson P. Topical use of human recombinant epidermal growth factor (h-EGF) in venous ulcers. J Dermatol Surg Oncol 1992;18(7):604−6.

[65] Bodnar RJ. Epidermal growth factor and epidermal growth factor receptor: the Yin and Yang in the treatment of cutaneous wounds and cancer. Adv Wound Care 2013;2(1):24−9.

[66] Hong JP, Jung HD, Kim YW. Recombinant human epidermal growth factor (EGF) to enhance healing for diabetic foot ulcers. Ann Plast Surg 2006;56:394−8 discussion 399−400.

[67] Mohan VK. Recombinant human epidermal growth factor (REGEN-D™ 150): effect on healing of diabetic foot ulcers. Diabetes Res Clin Pract 2007;78(3):405−11.

[68] Berlanga J, Fernández JI, López E, López PA, Río AD, Valenzuela C, et al. Heberprot-P: a novel product for treating advanced diabetic foot ulcer. MEDICC Rev 2013;15:11−15.

[69] Park JW, Hwang SR, Yoon IS. Advanced growth factor delivery systems in wound management and skin regeneration. Molecules 2017;22 (8):1259.

[70] Gil ES, Panilaitis B, Bellas E, Kaplan DL. Functionalized silk biomaterials for wound healing. Adv Healthc Mater 2013;2(1):206−17.

[71] Schneider A, Wang XY, Kaplan DL, Garlick JA, Egles C. Biofunctionalized electrospun silk mats as a topical bioactive dressing for accelerated wound healing. Acta Biomater 2009;5(7):2570−8.

[72] Imamura T. Physiological functions and underlying mechanisms of fibroblast growth factor (FGF) family members: recent findings and implications for their pharmacological application. Biol Pharm Bull 2014;37(7):1081−9.

[73] Komi-Kuramochi A, Kawano M, Oda Y, Asada M, Suzuki M, Oki J, et al. Expression of fibroblast growth factors and their receptors during full-thickness skin wound healing in young and aged mice. J Endocrinol 2005;186(2):273−89.

[74] Werner S, Peters KG, Longaker MT, Fuller-Pace F, Banda MJ, Williams LT. Large induction of keratinocyte growth factor expression in the dermis during wound healing. Proc Natl Acad Sci U S A 1992;89(15):6896−900.

[75] Ronca R, Giacomini A, Rusnati M, Presta M. The potential of fibroblast growth factor/fibroblast growth factor receptor signaling as a therapeutic target in tumor angiogenesis. Expert Opin Ther Targets 2015;19(10):1361−77.

[76] Barrientos S, Brem H, Stojadinovic O, Tomic-Canic M. Clinical application of growth factors and cytokines in wound healing. Wound Repair Regen 2014;22(5):569−78.

[77] Werner S, Smola H, Liao X, Longaker MT, Krieg T, Hofschneider PH, et al. The function of KGF in morphogenesis of epithelium and reepithelialization of wounds. Science 1994;266(5186):819−22.

[78] Schreiber AB, Winkler ME, Derynck R. Transforming growth factor-alpha: a more potent angiogenic mediator than epidermal growth factor. Science 1986;232(4755):1250−3.

[79] Tuyet HL, Quynh NTT, Vo Hoang Minh H, Thi Bich DN, Do Dinh T, Le Tan D, et al. The efficacy and safety of epidermal growth factor in treatment of diabetic foot ulcers: the preliminary results. Int Wound J 2009;6:159−66.

[80] Blaber SI, Diaz J, Blaber M. Accelerated healing in NONcNZO10/LtJ type 2 diabetic mice by FGF-1. Wound Repair Regen 2015;23 (4):538−49.

[81] Seeger MA, Paller AS. The roles of growth factors in keratinocyte migration. Adv Wound Care 2015;4(4):213−24.

[82] Peng C, Chen B, Kao HK, Murphy G, Orgill DP, Guo L. Lack of FGF-7 further delays cutaneous wound healing in diabetic mice. Plast Reconstr Surg 2011;128(6):673e−684ee.

[83] Qu Y, Cao C, Wu Q, Huang A, Song Y, Li H, et al. The dual delivery of KGF and b FGF by collagen membrane to promote skin wound healing. J Tissue Eng Regen Med 2018;12(6):1508−18.

[84] Sakamoto M, Morimoto N, Ogino S, Jinno C, Taira T, Suzuki S. Efficacy of gelatin gel sheets in sustaining the release of basic fibroblast growth factor for murine skin defects. J Surg Res 2016;201(2):378−87.

[85] Ferrara N, Gerber HP, LeCouter J. The biology of VEGF and its receptors. Nat Med 2003;9(6):669−76.

[86] Simons M, Ware JA. Therapeutic angiogenesis in cardiovascular disease. Nat Rev Drug Discov 2003;2(11):863.

[87] Tsopanoglou NE, Maragoudakis ME. On the mechanism of thrombin-induced angiogenesis potentiation of vascular endothelial growth factor activity on endothelial cells by up-regulation of its receptors. J Biol Chem 1999;274(34):23969−76.

[88] Nguyen M, Arkell J, Jackson CJ. Human endothelial gelatinases and angiogenesis. Int J Biochem Cell Biol 2001;33(10):960−70.

[89] Hellberg C, Ostman A, Heldin CH. PDGF and vessel maturation. Recent Results Cancer Res 2010;180:103−14.

[90] Pintucci G, Froum S, Pinnell J, Mignatti P, Rafii S, Green D. Trophic effects of platelets on cultured endothelial cells are mediated by platelet-associated fibroblast growth factor-2 (FGF-2) and vascular endothelial growth factor (VEGF). Thromb Haemost 2002;88(11):834−42.

[91] Li JJ, Huang YQ, Basch R, Karpatkin S. Thrombin induces the release of angiopoietin-1 from platelets. Thromb Haemost 2001;85(02):204−6.

[92] Nath SG, Raveendran R. An insight into the possibilities of fibroblast growth factor in periodontal regeneration. J Indian Soc Periodontol 2014;18:289−92.

[93] Yoshida S, Yoshida A, Matsui H, Takada Y, Ishibashi T. Involvement of macrophage chemotactic protein−1 and interleukin−1beta during inflammatory but not basic fibroblast growth factor−dependent neovascularization in the mouse cornea. Lab Invest 2003;83:927−38.

[94] Grimm D, Bauer J, Schoenberger J. Blockade of neoangiogenesis, a new and promising technique to control the growth of malignant tumors and their metastases. Curr Vasc Pharmacol 2009;7:347−57.

[95] Lutolf MP, Hubbell JA. Synthetic biomaterials as instructive extracellular microenvironments for morphogenesis in tissue engineering. Nat Biotechnol 2005;23:47−55.

[96] Bootle−Wilbraham CA, Tazzyman S, Thompson WD, Stirk CM, Lewis CE. Fibrin fragment E stimulates the proliferation, migration and differentiation of human microvascular endothelial cells in vitro. Angiogenesis 2001;4:269−75.

[97] Ji K, Tsirka SE. Inflammation modulates expression of laminin in the central nervous system following ischemic injury. J Neuroinflammation 2012;9:159.

[98] Acker T, Plate KH. Role of hypoxia in tumor angiogenesis−molecular and cellular angiogenic crosstalk. Cell Tissue Res 2003;314:145−55.

[99] Howdieshell TR, Webb WL, Sathyanarayana, McNeil PL. Inhibition of inducible nitric oxide synthase results in reductions in wound vascular endothelial growth factor expression, granulation tissue formation, and local perfusion. Surgery 2003;133:528−37.

[100] Leonardi R, Caltabiano M, Pagano M, Pezzuto V, Loreto C, Palestro G. Detection of vascular endothelial growth factor/vascular permeability factor in periapical lesions. J Endod 2003;29:180−3.

[101] Smith Jr RS, Gao L, Bledsoe G, Chao L, Chao J. Intermedin is a new angiogenic growth factor. Am J Physiol Heart Circ Physiol 2009;297:H1040−7.

[102] Inoki I, Shiomi T, Hashimoto G, Enomoto H, Nakamura H, Makino K, et al. Connective tissue growth factor binds vascular endothelial growth factor (VEGF) and inhibits VEGF−induced angiogenesis. FASEB J 2002;16:219−21.

[103] Ma J, Wang Q, Fei T, Han JD, Chen YG. MCP−1 mediates TGF−beta−induced angiogenesis by stimulating vascular smooth muscle cell migration. Blood 2007;109:987−94.

[104] Korff T, Kimmina S, Martiny−Baron G, Augustin HG. Blood vessel maturation in a 3−dimensional spheroidal coculture model: direct contact with smooth muscle cells regulates endothelial cell quiescence and abrogates VEGF responsiveness. FASEB J 2001;15:447−57.

[105] Onimaru M, Yonemitsu Y, Fujii T, Tanii M, Nakano T, Nakagawa K, et al. VEGF−C regulates lymphangiogenesis and capillary stability by regulation of PDGF−B. Am J Physiol Heart Circ Physiol 2009;297:H1685−96.

[106] Kumar I, Staton CA, Cross SS, Reed MW, Brown NJ. Angiogenesis, vascular endothelial growth factor and its receptors in human surgical wounds. Br J Surg 2009;96:1484−91.

[107] Darland DC, D'Amore PA. TGF beta is required for the formation of capillary−like structures in three−dimensional cocultures of 10T1/2 and endothelial cells. Angiogenesis 2001;4:11−20.

[108] McCarty MF, Bielenberg DR, Nilsson MB, Gershenwald JE, Barnhill RL, Ahearne P, et al. Epidermal hyperplasia overlying human melanoma correlates with tumour depth and angiogenesis. Melanoma Res 2003;13:379−87.

[109] Michaels 5th J, Dobryansky M, Galiano RD, Bhatt KA, Ashinoff R, Ceradini DJ, et al. Topical vascular endothelial growth factor reverses delayed wound healing secondary to angiogenesis inhibitor administration. Wound Repair Regen 2005;13:506−12.

[110] Lange−Asschenfeldt B, Velasco P, Streit M, Hawighorst T, Pike SE, Tosato G, et al. The angiogenesis inhibitor vasostatin does not impair wound healing at tumor−inhibiting doses. J Invest Dermatol 2001;117:1036−41.

[111] Van der Bilt JD, Borel Rinkes IH. Surgery and angiogenesis. Biochim Biophys Acta 2004;1654:95−104.

[112] Hiromatsu Y, Toda S. Mast cells and angiogenesis. Microsc Res Tech 2003;60:64−9.

[113] Ornitz DM, Itoh N. Fibroblast growth factors. Genome Biol 2001;2 REVIEWS3005.

[114] Barrientos S, Stojadinovic O, Golinko MS, Brem H, Tomic−Canic M. Growth factors and cytokines in wound healing. Wound Repair Regen 2008;16:585−601.

[115] Plum SM, Vu HA, Mercer B, Fogler WE, Fortier AH. Generation of a specific immunological response to FGF−2 does not affect wound healing or reproduction. Immunopharmacol Immunotoxicol 2004;26:29−41.

[116] Breier G, Blum S, Peli J, Groot M, Wild C, Risau W, et al. Transforming growth factor−beta and Ras regulate the VEGF/VEGF−receptor system during tumor angiogenesis. Int J Cancer 2002;97:142−8.

[117] Bates DO, Heald RI, Curry FE, Williams B. Vascular endothelial growth factor increases Rana vascular permeability and compliance by different signalling pathways. J Physiol 2001;533:263−72.

[118] Failla CM, Odorisio T, Cianfarani F, Schietroma C, Puddu P, Zambruno G. Placenta growth factor is induced in human keratinocytes during wound healing. J Invest Dermatol 2000;115:388−95.

[119] Hemmerlein B, Kugler A, Ozisik R, Ringert RH, Radzun HJ, Thelen P. Vascular endothelial growth factor expression, angiogenesis, and necrosis in renal cell carcinomas. Virchows Arch 2001;439:645−52.

[120] Zachary I, Gliki G. Signaling transduction mechanisms mediating biological actions of the vascular endothelial growth factor family. Cardiovasc Res 2001;49:568−81.

[121] Efron PA, Moldawer LL. Cytokines and wound healing: the role of cytokine and anticytokine therapy in the repair response. J Burn Care Rehabil 2004;25:149−60.

[122] Bates DO, Harper SJ. Regulation of vascular permeability by vascular endothelial growth factors. Vasc Pharmacol 2002;39:225−37.

[123] Ferrara N. Vascular endothelial growth factor: basic science and clinical progress. Endocr Rev 2004;25:581−611.

[124] Kessler T, Fehrmann F, Bieker R, Berdel WE, Mesters RM. Vascular endothelial growth factor and its receptor as drug targets in hematological malignancies. Curr Drug Targets 2007;8:257−68.

[125] Roth D, Piekarek M, Paulsson M, Christ H, Bloch W, Krieg T, et al. Plasmin modulates vascular endothelial growth factor−A−mediated angiogenesis during wound repair. Am J Pathol 2006;168:670−84.

[126] Primo L, Seano G, Roca C, Maione F, Gagliardi PA, Sessa R, et al. Increased expression of alpha6 integrin in endothelial cells unveils a proangiogenic role for basement membrane. Cancer Res 2010;70:5759−69.

[127] Rao X, Zhong J, Zhang S, Zhang Y, Yu Q, Yang P, et al. Loss of methyl−CpG−binding domain protein 2 enhances endothelial angiogenesis and protects mice against hind−limb ischemic injury. Circulation 2011;123:2964−74.

[128] Brunner G, Blakytny R. Extracellular regulation of TGF−beta activity in wound repair: growth factor latency as a sensor mechanism for injury. Thromb Haemost 2004;92:253−61.

[129] Verrecchia F, Mauviel A. Transforming growth factor−beta and fibrosis. World J Gastroenterol 2007;13:3056−62.

[130] Tsigkos S, Koutsilieris M, Papapetropoulos A. Angiopoietins in angiogenesis and beyond. Expert Opin Invest Drugs 2003;12:933−41.

[131] Solovyan VT, Keski-Oja J. Apoptosis of human endothelial cells is accompanied by proteolytic processing of latent TGF-beta binding proteins and activation of TGF-beta. Cell Death Differ 2005;12:815−26.

[132] Iddamalgoda A, Le QT, Ito K, Tanaka K, Kojima H, Kido H. Mast cell tryptase and photoaging: possible involvement in the degradation of extra cellular matrix and basement membrane proteins. Arch Dermatol Res 2008;300(Suppl 1):S69−76.

[133] Martin A, Komada MR, Sane DC. Abnormal angiogenesis in diabetes mellitus. Med Res Rev 2003;23:117−45.

[134] Brem H, Jacobs T, Vileikyte L, Weinberger S, Gibber M, Gill K, et al. Wound healing protocols for diabetic foot and pressure ulcers. Surg Technol Int 2003;11:85−92.

[135] Keswani SG, Katz AB, Lim FY, Zoltick P, Radu A, Alaee D, et al. Adenoviral mediated gene transfer of PDGF-B enhances wound healing in type I and type II diabetic wounds. Wound Repair Regen 2004;12:497−504.

[136] Altavilla D, Saitta A, Cucinotta D, Galeano M, Deodato B, Colonna M, et al. Inhibition of lipid peroxidation restores impaired vascular endothelial growth factor expression and stimulates wound healing and angiogenesis in the genetically diabetic mouse. Diabetes 2001;50:667−74.

[137] Wicke C, Halliday B, Allen D, Roche NS, Scheuenstuhl H, Spencer MM, et al. Effects of steroids and retinoids on wound healing. Arch Surg 2000;135:1265−70.

[138] Kim JH, Kim TH, Kang MS, Kim HW. Angiogenic effects of collagen/mesoporous nanoparticle composite scaffold delivering VEGF165. BioMed Res Int 2016;2016.

[139] Gao SQ, Chang C, Li JJ, Li Y, Niu XQ, Zhang DP, et al. Co-delivery of deferoxamine and hydroxysafflor yellow A to accelerate diabetic wound healing via enhanced angiogenesis. Drug Deliv 2018;25(1):1779−89.

[140] Mulholland Eoghan J, Ahlam A, Tracy R, Dunne Nicholas J, McCarthy Helen O. Delivery of RALA/siFKBPL nanoparticles via electrospun bilayer nanofibres: an innovative angiogenic therapy for wound repair. J Control Release 2019;316:53−65.

[141] Han G, Nguyen LN, Macherla C, Chi Y, Friedman JM, Nosanchuk JD, et al. Nitric oxide−releasing nanoparticles accelerate wound healing by promoting fibroblast migration and collagen deposition. Am J Pathol 2012;180(4):1465−73.

[142] Blecher K, Martinez LR, Tuckman-Vernon C, Nacharaju P, Schairer D, Chouake J, et al. Nitric oxide-releasing nanoparticles accelerate wound healing in NOD-SCID mice. Nanomedicine 2012;8(8):1364−71.

[143] Leu JG, Chen SA, Chen HM, Wu WM, Hung CF, Yao YD, et al. The effects of gold nanoparticles in wound healing with antioxidant epigallocatechin gallate and α-lipoic acid. Nanomedicine 2012;8(5):767−75.

[144] Wang X, Liu B, Xu Q, Sun H, Shi M, Wang D, et al. GHK-Cu-liposomes accelerate scald wound healing in mice by promoting cell proliferation and angiogenesis. Wound Repair Regen 2017;25(2):270−8.

[145] Değim Z, Çelebi N, Alemdaroğlu C, Deveci M, Öztürk S, Özoğul C. Evaluation of chitosan gel containing liposome-loaded epidermal growth factor on burn wound healing. Int Wound J 2011;8(4):343−54.

[146] Cui MD, Pan ZH, Pan LQ. Danggui Buxue extract-loaded liposomes in thermosensitive gel enhance in vivo dermal wound healing via activation of the VEGF/PI3K/Akt and TGF-β/Smads signaling pathway. Evid Based Compl Alt Med 2017;2017.

[147] Abreu FA, Ferreira CL, Silva GA, Paulo CD, Miziara MN, Silveira FF, et al. Effect of PDGF-BB, IGF-I growth factors and their combination carried by liposomes in tooth socket healing. Braz Dental J 2013;24(4):299−307.

[148] Mao C, Chen X, Miao G, Lin C. Angiogenesis stimulated by novel nanoscale bioactive glasses. Biomed Mater 2015;10(2):025005.

[149] Barui AK, Veeriah V, Mukherjee S, Manna J, Patel AK, Patra S, et al. Zinc oxide nanoflowers make new blood vessels. Nanoscale 2012;4(24):7861−9.

[150] Kwon MJ, An S, Choi S, Nam K, Jung HS, Yoon CS, et al. Effective healing of diabetic skin wounds by using nonviral gene therapy based on minicircle vascular endothelial growth factor DNA and a cationic dendrimer. J Gene Med 2012;14(4):272−8.

[151] Sun G, Zhang X, Shen YI, Sebastian R, Dickinson LE, Fox-Talbot K, et al. Dextran hydrogel scaffolds enhance angiogenic responses and promote complete skin regeneration during burn wound healing. Proc Natl Acad Sci U S A 2011;108(52):20976−81.

[152] Lord MS, Ellis AL, Farrugia BL, Whitelock JM, Grenett H, Li C, et al. Perlecan and vascular endothelial growth factor-encoding DNA-loaded chitosan scaffolds promote angiogenesis and wound healing. J Control Release 2017;250:48−61.

[153] Peplow PV, Baxter GD. Gene expression and release of growth factors during delayed wound healing: a review of studies in diabetic animals and possible combined laser phototherapy and growth factor treatment to enhance healing. Photomed Laser Surg 2012;30:617−36.

[154] Stavrou D. Neovascularisation in wound healing. J Wound Care 2008;17:298−300 2.

[155] Johnson KE, Wilgus TA. Vascular endothelial growth factor and angiogenesis in the regulation of cutaneous wound repair. Adv Wound Care (N Rochelle) 2014;3:647−61.

[156] Hoffman M, Monroe DM. Wound healing in haemophilia-breaking the vicious cycle. Haemophilia 2010;16(Suppl 3)):13−18.

[157] Goova MT, Li J, Kislinger T, Qu W, Lu Y, Bucciarelli LG, et al. Blockade of receptor for advanced glycation end−products restores effective wound healing in diabetic mice. Am J Pathol 2001;159:513−25.

[158]   Corral CJ, Siddiqui A, Wu L, Farrell CL, Lyons D, Mustoe TA. Vascular endothelial growth factor is more important than basic fibroblastic growth factor during ischemic wound healing. Arch Surg 1999;134:200−5.

[159]   Howdieshell TR, Callaway D, Webb WL, Gaines MD, Procter Jr CD, Sathyanarayana PJS, et al. Antibody neutralization of vascular endothelial growth factor inhibits wound granulation tissue formation. J Surg Res 2001;96:173−82.

[160]   Lerman OZ, Galiano RD, Armour M, Levine JP, Gurtner GC. Cellular dysfunction in the diabetic fibroblast: impairment in migration, vascular endothelial growth factor production, and response to hypoxia. Am J Pathol 2003;162:303−12.

[161]   Franzeck UK, Haselbach P, Speiser D, Bollinger A. Microangiopathy of cutaneous blood and lymphatic capillaries in chronic venous insufficiency (CVI). Yale J Biol Med 1993;66:37−46.

[162]   Jünger M, Steins A, Hahn M, Häfner HM. Microcirculatory dysfunction in chronic venous insufficiency (CVI). Microcirculation 2000;7:S3−12.

Chapter 9

Normalization of the tumor vasculature

Diana Klein

Institute of Cell Biology (Cancer Research), University Hospital Essen, University of Duisburg-Essen, Essen, Germany

9.1 The vascular normalization concept

The cellular demands of the rapidly growing neoplastic cells require the formation of new blood vessels within the tumor mass in order to ensure the needs of oxygen and nutrients as well as to remove waste products [1,2]. Tumor neovascularization by angiogenesis and postnatal vasculogenesis is initiated from an imbalance of pro- and antiangiogenic growth factors— a process known as the angiogenic switch [3,4]. Herein, tumor and stromal cells secrete various amounts of signaling molecules, including vascular endothelial growth factor (VEGF), platelet-derived growth factor (PDGF), transforming growth factor (TGFβ) families, angiopoietins, and thrombospondins, which, among others, regulate the formation of an abnormal vascular network in tumors. Compared to adult, usually quiescent blood vessels, tumor blood vessels were characterized to be functionally abnormal because of their immature phenotype (Fig. 9.1). An irregular endothelial lining, a defective or discontinuous basement membrane, and the lack of vessel stabilizing pericytes and smooth muscle cells (SMCs) cause the dilated, tortuous, and disorganized vascular network of solid tumors [7–9]. The vascular immaturity of tumor vessels functionally results in tumor vessel instability and increased vessel permeability that, together with the chaotically vessel organization, foster an irregular blood, an increase in the intratumor fluid pressure, vessel collapse, hypoxia, and increased risks of metastatic dissemination [7,10–12]. Concomitantly, the efficacy of cancer therapy is reduced because of the inefficient distribution of oxygen and resultant hypoxia, which limit the response to radiotherapy, and because of the inefficient distribution of intravenously applied cancer therapeutics that can potentially limit the efficiency of chemotherapy and/or immunotherapy.

Apart from that, the fact that tumors dependent on angiogenesis has lead to the discovery of angiogenesis inhibitors, and thus to the establishment of alternative therapy methods that aimed to starve tumors by blocking neovascularization [13,14]. The first antiangiogenic agent approved for clinical use was bevacizumab, a humanized monoclonal antibody that neutralizes all isoforms of the main proangiogenic growth factor VEGF, thereby preventing the interactions of these ligands with the corresponding VEGF receptors [15,16]. Upon treatment, a dramatic tumor regression was observed. However, the clinical efficacy of antiangiogenesis (mono-) therapy was not as successfully as initially hoped. In fact improved clinical outcomes have been reported only when antiangiogenesis therapy was combined with systemic chemotherapy treatment [17–21]. This could be explained by the "tumor vascular normalization" concept which was provided by Jain in 2001 [22]. Vascular normalization in terms of blood vessel maturation results from blood vessel remodeling when pericytes and SMCs were recruited to envelop the newly formed tumor endothelium resulting in blood vessel stabilization [9]. The hitherto heterogeneous and abnormal tumor vascular bed that is characterized by hyperpermeability, limited perfusion, increased interstitial fluid pressure, and/or severe hypoxia thereby normalizes upon vascular remodeling (Fig. 9.2). This process is accelerated when antiangiogenic agents were used [3,12,23–25]. Histologically, antiangiogenic treatment-induced tumor regression was accompanied by a loss of less mature and highly proliferative small-caliber vessels and thus a reduction in tumor vascularity [5,26–28]. The remaining stabilized vessels showed an increase in vessel diameter by the association with and integration of pericytes and SMC (Figs. 9.1 and 9.2).

Although antiangiogenic therapy triggers vascular stabilization and vessel normalization, accompanied endothelial quiescence could impair tumor vessel regression in response to antiangiogenic treatment. In contrast, the restored endothelial quiescence reverses the unresponsiveness of the angiogenic tumor endothelium to inflammatory signals (endothelial anergy) a phenomenon that causes lymphocyte tolerance [29–32]. Antiangiogenic therapy-mediated neutralizing of angiogenic growth factors that also were shown to suppress adhesion molecule expressions involved in leukocyte binding in tumor endothelial cells, would then result in restored leukocyte–vessel wall interactions and infiltration of tumor leukocytes that exert antitumor effects necessary to eradicate the tumor [7,33–35].

Endothelial Signaling in Vascular Dysfunction and Disease. DOI: https://doi.org/10.1016/B978-0-12-816196-8.00015-1

Vessel normalization

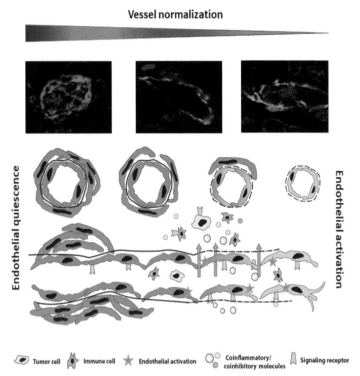

FIGURE 9.1 Structural and functional abnormalities of tumor blood vessels. Angiogenic endothelial cells (left side; shown as green cells) were characterized by an immature and activated phenotype with an increased migration and proliferation potential. Irregular endothelial cell arrangements, a missing or defective basement membrane and the lack of vascular mural cells (smooth muscle cells, SMC, and pericytes; shown as red cells) then cause increased leakage for blood stream components as well as structural instability of the respective tumor endothelium. This complicates the efficient distribution of nutrition and oxygen as well as the effective administration of intravenously applied cancer therapeutics. Tumor endothelial cells further contribute to an immune-privileged tumor microenvironment by the lack of response to inflammatory activation and an altered expression of costimulatory and coinhibitory molecules can promote immune tolerance. Upon tumor progression, the vascular network of chaotically organized, immature, and unstable vessels can mature by recruitment, association, and integration of vascular mural cells (vascular normalization) that stabilize these immature vessels; and this process is accelerated in cancer therapy when antiangiogenic agents are used. As a result, the stabilized and thereby normalized blood vessels are more perfused, which in turn improves tumor oxygenation as well as the efficient distribution of applied drugs. Angiogenic tumor endothelium and normalized vessel structures were visualized in human PC-3 prostate carcinoma xenografts (upper fluorescence photographs). Endothelial cells were stained in green (CD31/PECAM-1) and pericytes/SMC were stained in red (SMA/ACTA2). For details see [5,6]. *Adapted from Klein D. The tumor vascular endothelium as decision maker in cancer therapy. Front Oncol 2018;8:367.*

Thus, vascular normalization of the tumor vascular bed with a restored and more regular blood perfusion is supposed to improve the delivery of oxygen and intravenously applied drugs, and consequently the efficiency of the previously established therapeutic approaches, such as chemo-, radio-, and immunotherapy would be improved [7,36−39].

9.2 Vascular normalization and chemotherapy

In contrast to the original rationale for antiangiogenic therapy to starve tumors by cutting of the tumor blood supply, antiangiogenic treatment-induced vascular normalization turned out to improve the outcome of chemotherapy [40,41] by improving tumor perfusion, reducing vascular permeability, and consequently reducing hypoxia [42,43]. Functional imaging data of glioblastoma, non small-cell lung cancer, and breast cancer patients already suggested that improved vascular function and the resulting increase in tumor oxygenation are associated with response to antiangiogenic treatment [44−46]. In addition, antiangiogenic therapy with anti-VEGF (bevacizumab) therapy combined with chemotherapy improved survival in previously untreated metastatic colorectal cancer patients [47]. Mechanistically, at least four molecular pathways are involved in regulating the process of vessel maturation and stabilization by recruiting mural cells: PDGFB−PDGF receptor (PDGFRβ), sphingosine-1-phosphate-1 (SPP)−endothelial differentiation gene (Edg) family of G protein−coupled receptors (sphingosine 1-phosphate receptor-1, S1PR1), a TEK tyrosine kinase acting as angiopoietin receptor (ANG1/2-TIE2), and TGFβ [9]. Antiangiogenic drugs that mainly target these signaling pathways have been administered in combination with chemotherapeutic drugs in many types of cancers.

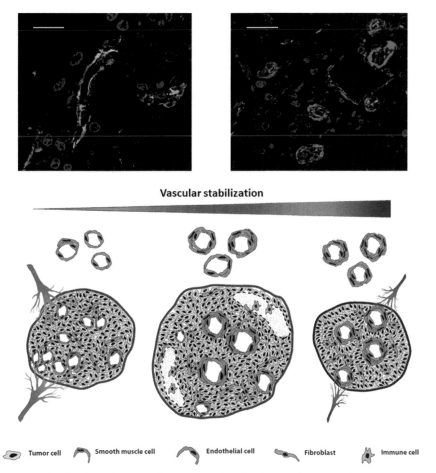

FIGURE 9.2 Tumor vessel normalization in solid tumors. Upon angiogenic activation, small a-vascularized tumors become endowed with newly formed blood vessels. These immature vessels are organized chaotically, structurally instable, and are characterized by increased permeability as well as endothelial anergy. Thus, the oxygen supply is inefficient which in turn might lower the response, for example, of radiotherapy and the distribution of intravenously applied drugs in cancer therapy might be limited. Inadequate tumor blood perfusion might foster hypoxia, which stimulates the metastatic potential of tumor cells to escape and metastasize in distant organs. In addition, due to the reduced expression levels of leukocyte adhesion molecules, these angiogeneic blood vessels contribute to the tumor immune escape. Upon tumor progression vascular remodeling angiogenic blood vessels become stabilized upon association with and integration of pericytes and smooth muscle cells. This vessel normalization could improve respective therapeutic approaches by a more homogenous drug and oxygen distribution, and/or by overcoming endothelial anergy. A normalization of the tumor vasculature is fostered upon antiangiogeneic therapy. For example, upon antiangiogenic treatment using the humanized monoclonal anti-VEGF bevacizumab, the major part of tumor vessels showed increased size and were characterized by a structural stabilization. Vascular normalization was visualized in animal model and xenograft transplantation of human PC-3 prostate carcinoma cells in combination with anti-VEGF therapy (bevacizumab treatment; upper fluorescence photographs). Vessels were stained for endothelial cells (CD31/PECAM-1; green) and pericytes/SMC (SMA/ACTA2; red). For details see [5,6]. *Adapted from Klein D. The tumor vascular endothelium as decision maker in cancer therapy. Front Oncol 2018;8:367.*

Some years ago studies from Goel et al. revealed that through TIE2 activation, tumor growth can be delayed, distant metastasis can be reduced and the response to concomitant cytotoxic treatments can be enhanced [48]. In this study, a potent and selective inhibitor of vascular endothelial protein tyrosine phosphatase (VE-PTP), termed AKB-9778 was used to treat mammary tumor growth. By inhibiting VE-PTP and thus inhibiting its ability to selectively dephosphorylate and inactivate TIE2, AKB-9778 treatment normalized the structure and function of tumor vessels through TIE2 activation. The promoted maturation of tumor vessels reduced tumor growth and significantly improved the efficiency of concomitant doxorubicin treatment [48,49]. Park et al. used then an angiopoietin 2 (ANG2)-binding and TIE2-activating antibody and an ANG2-blocking antibody for the treatment of glioma, Lewis lung carcinoma (LLC), and breast cancer mouse models [50]. TIE2 is usually expressed in endothelial cells in order to stabilize these vessels by tightening endothelial cell junctions. TIE2 and can be activated or inhibited by angiopoietins in a context-dependent manner: ANG1 usually acts as stabilizer for blood vessels through TIE2 activation, whereas ANG2 acts as destabilizer

[51]. The simultaneous TIE2 activation and ANG2 inhibition-induced tumor vascular normalization that in turn resulted in enhanced blood perfusion and chemotherapeutic drug delivery (temozolomide delivery in glioma and cisplatin delivery in LLC and MMTV-PyMT mouse models) thus reduced tumor growth and metastasis [50].

Another tumor entity which is characterized by a high microvascular density, impaired microvessel integrity, and poorly perfused vessels is pancreatic cancer [52]. Thus, the pancreatic cancer vasculature is prone for therapies with the purpose to induce normalization of tumor blood vessels, resulting in improved tumor perfusion, reduced hypoxia, and improved drug delivery [53,54]. Awasthi et al. investigated the antitumor activity of nintedanib alone or in combination with the cytotoxic agent gemcitabine in experimental pancreatic ductal adenocarcinoma (PDAC) [55]. Nintedanib is a triple angiokinase inhibitor that targets VEGFR1/2/3, FGFR1/2/3, and PDGFRα/β signaling. These receptors and their respective ligands VEGF, PDGF, FGF are expressed at high levels and correlate with poor prognosis in human PDAC [52,55]. Although the exact mechanisms for the enhancement in antitumor activity of gemcitabine by nintedanib addition remains unclear, normalization of tumor microvessels and increased gemcitabine delivery into the tumor microenvironment due to reduced interstitial pressure were the most likely mechanisms by which gemcitabine in the combined treatment exerts the antitumor effects [55,56]. These preclinical studies further highlighted the benefits of combining polymechanistic, multitargeting antiangiogenic agents, as the single-target anti-VEGF agent bevacizumab did not exert any significant clinical activity in pancreatic cancer in combination with gemcitabine [57].

The multitargeting antiangiogenic agent nintedanib was also shown to interfere with phosphoinositide 3-kinase (PI3K)/mitogen-activated protein kinase signaling, and by blocking the respective activity to induce apoptosis in pancreatic tumor cells in vitro and in vivo [55,58]. PI3K signaling can further affect tumor angiogenesis both directly and indirectly because PI3K activation could upregulate VEGF and promote vessel formation [59]. Thus, targeting PI3K signaling was shown foster antiangiogenic activities in preclinical tumor models [60,61]. In line with the findings that inhibition of the PI3K pathway would suppress tumor cells and affect the tumor the vasculature, Kim et al. investigated whether PI3K inhibitors (HS-173 and BEZ235) are able to improve the vascular structure and function to achieve effective vessel normalization [60]. Using pancreatic and melanoma (PaCa-2 and B16) tumor xenograft models, it was shown that PI3K inhibitor treatment improved stability of vessel structure, and normalized tumor vessels by increasing vascular maturity, pericyte coverage, basement membrane thickness, and tight junctions. Restrained tumor growth and metastasis was further improved upon concomitant doxorubicin treatment, caused by an enhanced drug delivery into the tumor upon vascular normalization [60]. It was already shown that treatment of mice bearing tumors with PI3K inhibitors alters the vascular structure in a manner analogous to vascular normalization resulting in increased tumor perfusion [62]. Transgenic mice with spontaneous breast tumors were treated here with the class I PI3K inhibitor GDC-0941 that mediated increased perfusion levels, substantially reduced hypoxia and vascular normalization. Although GDC-0941 alone had no effect on tumor growth, significantly increased amounts of coadministered doxorubicin were delivered to the tumors correlating with synergistic tumor growth delay [62].

Glioblastomas (GBM), the most common brain tumors, are characterized by high levels of VEGFs that correlate on the one hand to the aggressiveness and prognosis of GBM and at the same time with dense and highly disorganized vessel structures [63]. This leads to the speculation that the inhibition of VEGF might slow down tumor growth and enhance the effects of radiotherapy and chemotherapy [64]. In line with these findings, analysis of four clinical trials (including 607 patients) for the efficacy and safety of bevacizumab when combined with chemotherapy revealed that bevacizumab, when combined to chemotherapy, improved progression-free survival of patients [65]. Mathivet et al. used longitudinal intravital imaging in a mouse glioma model (with tumors raised from CT2A and GL261 cells) to follow up the dynamic sprouting and functional morphogenesis of a highly branched vessel network, which is characterized by leakage and loss of branching complexity in glioblastoma at advanced tumor stages [66]. At these later stages, macrophage in situ repolarization to VEGF-producing M2-like macrophages that relocate to perivascular areas was observed. Targeting macrophages using anti-CSF1 or decreasing macrophage VEGF production similarly accelerated tumor growth but restored a functional and perfused blood vessel network. The restored normal blood vessel patterning and function improved the delivery of temozolomide [66]. These results suggest that macrophage-derived VEGF plays a key role in glioma neovascularization and thus in glioma progression, and that targeting macrophages might present an additional option to improve the efficacy of chemotherapeutic agents through vascular normalization.

Endothelial cells promote vascular mural cell survival and vice versa. Endothelial cells secrete PDGFB, presumably in response to VEGF, that in turn facilitates the recruitment of pericytes and/or SMC which express PDGFRβ, the receptor being responsible for mural cell proliferation and migration during vascular maturation [9,67]. Thus, pericyte-targeted therapies aim to reach a balance between proangiogenic and antiangiogenic function for tumor vascular normalization [68]. Accordingly, in an orthotopic tumor mice model of ovarian cancer, it was shown that dual targeting of endothelial cells and mural cells via VEGFR inhibition (with the VEGFR inhibitor AEE788) and STI571 (a PDGFRβ inhibitor) treatment is more effective than the use of AEE788 alone [69]. Microvessel density was significantly reduced in treated tumors. The use of both agents then improved the effectiveness of paclitaxel, resulting in a nearly complete inhibition of tumor growth [69].

Fan et al. produced a fusion protein [tumor necrosis factor α (Z-TNFα)], by fusing the PDGFRβ-antagonistic affibody ZPDGFRβ to TNFα that specifically binds to PDGFRβ-expressing pericytes [70]. Z-TNFα treatment improved the delivery of doxorubicin and enhanced the antitumor effect of doxorubicin in B16F1 melanoma and S180 sarcoma mouse models via a Z-TNFα induced normalization of the tumor vessels [70]. Simultaneously, Z-TNFα treatment triggered the interaction between perivascular macrophages and pericytes by an elevated expression of a cellular adhesion molecule (CAM), specifically intercellular cell adhesion molecule-1 (ICAM-1) in pericytes. These results strongly suggest that also mural cells could be considered as novel target cells for vessel normalization and that Z-TNFα might be developed as a potential tool for antitumor combination therapy [70,71].

The analysis of a phase II study of neoadjuvant bevacizumab (single dose) treatment followed by combined bevacizumab and adriamycin/cyclophosphamide/paclitaxel chemotherapy in HER2-negative breast cancer patients further suggested that the clinical response to bevacizumab may occur through increased vascular normalization primarily in patients with a high baseline tumor microvessel density [45]. Thus, the structural features of tumor vessels might be used to predict the response to antiangiogenic therapy and to identify patients who would benefit from that combinatorial cancer therapy.

In line with these findings, Yapp et al. used antiangiogenic agent DC101, the murine analog of ramucirumab, a human monoclonal antibody targeting the VEGFR2, to compare the efficacy of metronomic chemotherapy versus combinatorial chemotherapy in a mouse model of pancreatic cancer [72]. Upon treatment, vessel density increased, tumor perfusion transiently improved, and hypoxia decreased. Of note, metronomic gemcitabine-treated tumors had even higher perfusion rates and a more uniformly distributed blood flow within tumors, as compared to the perfusion rates in DC101-treated tumors [72]. In two syngeneic mouse models of metastatic colorectal cancer (SL4 and CT26 cells orthotopically implanted in C57BL/6 and BALB/c mice), it was further shown that combined anti-VEGF therapy and chemotherapy with fluorouracil resulted in improved tumor perfusion and treatment efficacy and prolonged survival of mice [73,74].

Similarly, the epidermal growth factor receptor (EGFR)-specific tyrosine kinase inhibitor erlotinib was shown to improve delivery of cisplatin and synergistically inhibit tumor growth in non small-cell lung cancer by decreasing tumor microvascular densities and increasing vessel perfusion [75]. Mechanistically the combination treatment targeted angiogenesis through downregulation of the c-MYC/hypoxia inducible factor 1-alpha (HIF-1α) pathway [75,76]. In general, HIFs are the key transcriptional factors that mediate the cellular adaptive responses to hypoxia. Expression levels of HIFs are indirectly correlated to the local oxygen concentration that determines the activity of the HIF regulating protein prolyl hydroxylase domain (PHD). Local upregulation of HIF1 $\propto$ then promotes direct or indirect angiogenesis and vasculogenesis [77]. Inhibition of PHD by 2-oxoglutarate dimethyloxalyl glycine (DMOG) results in the stabilization of HIFs and the promotion of neovascularization. The blood vessel's endothelial cells harbor oxygen sensors which allow the vessels to readjust their shape to optimize blood flow in response to HIF signaling pathways [78,79]. Koyama et al. applied DMOG to LLC model mice and observed that a transient PHD inhibition led to the normalization of tumor blood vessels [78]. Similar to DMOG, a transient PHD inhibition by FG4592 mediates vascular normalization as revealed by promoted pericyte coverage, tight junction formation, vessel maturation, and barrier tightening. After the administration of the PHD inhibitor, an enhanced efficiency of chemotherapy (cisplatin) without accelerating the primary tumor growth was observed [78].

HIF activity under hypoxic conditions can also be modulated by the sphingosine kinase 1/sphingosine-1-phosphate (SphK1/S1P) pathway, which in turn decreases proteasome degradation of the HIF-1α subunit mediated by the Akt/GSK3β pathway in various cancer cell models [80]. Neutralizing S1P was shown to inhibit intratumoral hypoxia by inducing vascular remodeling in terms of vessel normalization and thereby sensitizing prostate cancer to chemotherapy [81]. Here, a monoclonal antibody neutralizing extracellular S1P (sphingomab) was used to inhibit S1P extracellular signaling, which in turn blocked HIF-1α accumulation in an orthotopic xenograft model of prostate cancer. These findings strongly suggest that a pharmacological SphK1/S1P pathway inhibition can serve as potential vascular normalization strategy that could contribute to successful sensitization of hypoxic tumors to chemotherapy [81]. HIF signaling inhibition can also be mediated by FTY720 (fingolimod), an analog of sphingosine that has been approved by the FDA for the treatment of relapsing-remitting multiple sclerosis [82]. As the SphK1/S1P-HIF axis is also decisive for the progression clear cell renal cell carcinoma, Gstalder et al. could show that FTY720 transiently decreases intratumoral HIF levels and modifies the tumor vessel architecture in terms of vascular normalization using a heterotopic clear cell renal cell carcinoma xenograft model [83]. Interestingly, a significant effect on tumor size without toxicity of FTY720 was observed when the treatment was combined with a gemcitabine-based chemotherapy, which alone only displayed a limited effect [83]. In line with these findings, the sphingosine 1-phosphate receptor-1 (S1PR1) was identified as key receptor for S1P binding in the tumor vasculature among the family of G-protein-coupled sphingosine 1-phosphate receptors 1 to 5 by using genetic mouse models in which S1PRs have been modified either alone or in combination [84]. Using subcutaneous LLC xenografts, it could be shown that S1PR1 fostered tumor vascular normalization and enhanced the chemotherapeutic efficiency of doxorubicin, suggesting that stimulation of this pathway enhances chemotherapeutic efficacy [84].

In an alternative approach, Cantelmo et al. used the novel paradigm to heal the abnormal tumor vasculature by tumor vessel normalization via targeting glycolysis in vascular endothelial cells [85]. The authors showed that a genetic or pharmacologic inhibition of the glycolytic enzyme 6-phosphofructo-2-kinase/fructose-2,6-bisphosphatase 3 (PFKFB3) in hyperglycolytic tumor endothelial cells resulted in improved perfusion and lower tumor hypoxia of respective B16-F10 melanomas, which was due to a tightened endothelial cell barrier and an increased pericyte coverage of the neovasculature, features associated with vessel normalization [86]. Concurrently, an increased delivery of cisplatin to the tumor cells, a correspondingly decreased tumor growth and nearly completely prevented metastasis were observed in melanoma-bearing mice [86,87]. Using two metastatic melanoma models, Maes et al. were able to show that the lysosomo-tropic antimalarial agent chloroquine exerts its anticancer activity not only through blockage of prosurvival autophagy in the cancer cells but also affects the tumor vessels [88]. Here chloroquine modified the tumor milieu by improving tumor perfusion and oxygenation, which finally resulted in reduced tumor hypoxia, cancer cell invasion and spreading, and improved the delivery and efficacy of chemotherapeutics (cisplatin) [88].

Together, antiangiogenic therapy-induced vascular normalization provides the basis for an enhanced delivery of chemotherapeutic agents into tumors (Table 9.1).

9.3 Vascular normalization and radiotherapy

In line with the increased therapeutic efficiencies of combined antiangiogenic treatment with chemotherapy, it was reported that antiangiogenesis therapy ameliorated the efficacy of radiotherapy [89]. Thus, tumor vessel normalization gained attraction also as a novel paradigm for combination strategies to overcome the resistance of hypoperfused tumors to conventional radiotherapy [39,90]. Herein, the antiangiogenic therapy reduced vascular density but increased vessel stability that in turn improved vessel perfusion and mitigated hypoxia finally results in improved oxygen delivery in tumors [22,25,39].

Myers et al. tested this hypothesis using the anti-VEGF antibody bevacizumab in rhabdomyosarcoma (RMS) xenografts [91]. After bevacizumab treatment, a significant decrease in tumor microvessel density and a significant increase in vessel maturity could be observed, finally resulting in decreased vessel permeability and a significantly increased intratumoral oxygen tension. Of note, the greatest antitumor activity was observed in the preclinical cohort who had the highest intratumoral oxygen tension prior to radiation treatment [91]. Similar results were reported from the same group using the macrolide antibiotic rapamycin, a mTOR inhibitor that has strong antiangiogenic properties related to the suppression of VEGF [92]. Combination therapy with rapamycin given prior to irradiation to mice bearing orthotopic Rh30 alveolar rhabdomyosarcomas was shown to efficiently normalize the tumor vasculature and thereby improving tumor oxygenation that increased the sensitivity of RMS xenografts to adjuvant irradiation [92].

Electron paramagnetic resonance with in vivo-compatible paramagnetic tracers capable of mapping tissue pO2 was used to study tumor oxygenation for correlation with microvessel densities in a squamous cell carcinoma mouse model upon combined antiangiogenic/radiation treatment [93]. This allowed the authors to determine noninvasively the "normalization window" for increased tumor oxygenation in response to antiangiogenic treatment with sunitinib. Radiation treatment during this narrow time period resulted then in a synergistic tumor growth delay [93]. Sunitinib (Sutent, SU11248) prevents the activation of a wide spectrum of receptors including VEGFR, PDGFR, FLT3 (fms-like tyrosine kinase receptor 3), and c-KIT (α and β, stem cell growth factor), and is one of the few FDA approved antiangiogenic agents for the treatment of renal cell carcinoma and imatinib-resistant gastrointestinal stromal tumors [94]. MDA-MB-231 breast cancer xenografts that were treated with radiotherapy alone or in combination sunitinib as pharmacological modulator of the vasculature further suspected that this small molecule tyrosine-kinase inhibitor leads to an enhanced oxygenation and an enhanced overall tumor response via vascular normalization [95,96].

An improved vessel perfusion and thus tumor oxygenation achieved by vascular normalization using antiangiogenic therapy was further reported by Koo et al. using heterotopic human lung cancer xenografts in combination with radiotherapy and the second-generation multitargeted receptor tyrosine kinase inhibitor sunitinib malate, which inhibits PDGF and VEGF [97]. This treatment combination resulted in reduced tumor growth, as vessel normalization achieved an efficient tumor perfusion, that in turn significantly improved tumor oxygenation that is prerequisite for the tumoricidal effects of radiotherapy [97].

GBM are also characterized by structurally and functionally abnormal blood vessels, but high microvascular densities that result from high levels of proangiogenic factors, in particular VEGF and ANG2 [98–100]. Despite an intensive treatment, which is safe resection, followed by radiotherapy and concomitant and adjuvant temozolomide chemotherapy, the overall survival remains limited [101]. Of note, improved intratumoral oxygenation through vascular normalization achieved

TABLE 9.1 Angiogenesis inhibitors used in preclinical studies achieving vessel normalization and thus an improved therapeutically response when combined to chemo-, radio-, or immunotherapy.

Antiangiogenic drug	Brand	Target/mechanism	Final effect (vascular normalization)
Bevacizumab	Avastin	VEGF isoforms/monoclonal anti-VEGF antibody	• Decreased vessel density • Decreased vessel permeability • Improved tumor perfusion • Decreased hypoxia • Overcame endothelial anergy
VEGF-Trap/aflibercept	Zaltrap	VEGF-A, VEGF-B, PlGF/recombinant fusion VEGF protein (Ig domain of VEGFR-IgG Fc fragment)	• Decreased vessel density • Improved tumor perfusion
Ramucirumab	Cyramza	VEGFR/Monoclonal anti-VEGFR antibody	• Increased vessel density • Improved tumor perfusion • Decreased hypoxia
Nintedanib	Vargatef	VEGFRs, FGFRs, PDGFR/TKI	• Decreased vessel density • Decreased vessel permeability • Improved tumor perfusion
Sorafenib	Nexavar	VEGFR-1−3, PDGFR-β, RAF, KIT, FLT3, RET/TKI	• Decreased vessel density • Decreased vessel permeability • Improved tumor perfusion
Sunitinib	Sutent	VEGFR-1−3, PDGFR-α/β, KIT, CSF-1R, FLT3, RET/TKI	• Decreased vessel density • Decreased vessel permeability • Improved tumor perfusion
AKB-9778/Razuprotafib		VE-PTP inhibitor/TIE2 activator	• Decreased vessel density • Decreased vessel permeability • Improved tumor perfusion
Erlotinib	Tarceva	EGFR/TKI	• Decreased vessel density • Improved tumor perfusion
AEE788		HER-1/2, VEGFR-1/2/TKI	• Decreased vessel density • Improved tumor perfusion
Imantinib/STI571	Glivec	ABL, BCR-ABL, KIT, PDGFR/TKI	• Decreased vessel density • Improved tumor perfusion
Endostatin	Endostar	C-terminal fragment of collagen XVIII (recombinant protein)/Inhibits angiogenesis by binding α5β1 integrin	• Decreased vessel density • Decreased vessel permeability • Improved tumor perfusion • Overcame endothelial anergy
Angiostatin		internal fragment of plasminogen/inhibits endothelial cell proliferation and migration	• Decreased vessel density • Decreased vessel permeability • Improved tumor perfusion • Overcame endothelial anergy
Fingolimod/FTY720	Gilenya	chemical modification of the fungal metabolite myriocin/S1P analog	• Decreased vessel permeability • Improved tumor perfusion

ABL, Abelson tyrosine kinase; *BCR-ABL*, chimeric fusion protein; *EGFR*, epidermal growth factor receptor; *FLT3*, FMS-like tyrosine kinase 3; *HER*, human epidermal receptor; *VEGFR*, vascular endothelial growth factor receptor; *KIT*, 1/2 receptor stem cell factor receptor; *PlGF*, placental growth factor; *PDGFR*, platelet-derived growth factor receptors; *RAF*, rapidly accelerated fibrosarcoma; *RET*, rearranged during transfection; *S1P*, sphingosine 1-phosphate; *TKI*, tyrosine kinase inhibitor; *VEGF*, vascular endothelial growth factor; *VEGFR2*, VEGF receptor 2; *VE-PTP*, vascular endothelial protein tyrosine phosphatase.

by VEGF neutralization was shown to increase glioma sensitivity to ionizing radiation [102]. Here orthotopic U87 xenografts treated with either continuous interferon-beta (IFN-β) or bevacizumab, alone, or in combination with cranial irradiation, resulted in normalization of the dysfunctional tumor vascular bed, increased intratumoral oxygenation, and an improved anti-tumor activity of radiation treatment [102]. Solecki et al. investigated different antiangiogenic strategies in combinations with radio- and chemotherapy in glioblastoma and found that anti-VEGF antibody treatment was the optimal combination partner for radiotherapy, while a bispecific antibody inhibiting both ANG2 and VEGF was the best for chemotherapy in a glioblastoma xenograft mouse model using human U-87MG glioblastoma cells [99]. Tumor progression and therapy resistance correlated here with morphological and functional vascular normalization, further supporting that the concept of vascular normalization therapeutic relevance for primary brain tumors.

The mutational profile often evidences a gain of function or hyperactivity of PI3Ks in tumors, and this frequently occurring activation of the PI3K/AKT pathway in tumors correlated with radioresistance [103]. Thus, targeting the PI3K/AKT pathway is supposed to be an effective strategy for improving radiotherapy. The PI3K inhibitor, HS-173, was already shown to exert therapeutic effects against pancreatic cancer cells and xenograft models [104]. Park et al. investigated the radiosensitizing effects of HS-173 in human pancreatic tumor xenograft models and revealed that HS-173 significantly increased the sensitivity of pancreatic cancer cells to radiation, an effect that was associated with G2/M cell cycle arrest and significantly attenuated DNA damage repair [105]. Although the author did not investigate vessel morphology and functionality in this study, it was already shown by the same group that PI3K inhibition using HS-173 improved the structure and function of tumor blood vessels via vessel normalization using the same pancreatic cancer mouse model [60]. Thus, restrained tumor growth and metastasis upon concomitant PI3K inhibition and radiation treatment might further result from improved tumor oxygenation upon vascular normalization.

Conclusively, oxygen is a potent radiosensitizer and combining antiangiogenic treatment to achieve a better tumor oxygenation with radiation therapy increases the therapeutic response (Table 9.1). This may further result in the use of lower radiation doses, as thus minimizing treatment-related normal tissue toxicity [106].

9.4 Vascular normalization and immunotherapy

A proper functional vascular bed of tumors is not only prerequisite for the efficient distribution of nutrients and oxygen but also for the recruitment of circulating immune cells prior tissues extravasation [7,9,107,108]. The interaction of leukocytes with endothelial cells of the functional and structural abnormal tumor blood vessels is limited because of the reduced expression of adhesion molecules involved in leukocyte binding (e.g., ICAM-1/2, VCAM-1, E-selectin, and CD34). This phenomenon, also called endothelial cell anergy, limits the recruitment of immune effector cells, either induced or adoptively transferred, to the tumors that subsequently fail to contribute to tumor eradication [34,35]. Beside the lack of response to inflammatory activation, tumor endothelial cells can further contribute to an immune-privileged tumor microenvironment by an altered expression of costimulatory and coinhibitory molecules that can promote immune tolerance [7,37,109,110]. Thus, the angiogenic tumor endothelium contributes to the tumor's immune escape and limits the effectiveness of cancer immunotherapies [7,108,111]. Antiangiogenic therapy-induced vascular normalization in terms of vessel maturation would then restore endothelial anergy and thus the unresponsiveness of the tumor endothelium to inflammatory stimuli caused by proangiogenic factors, and promote then immune cell infiltration.

A tumor vessel normalization caused influx of immune effector T cells was further reported for the genetically induced vascular normalization mouse model, the regulator of G protein signaling 5-deficient (RGS5) mouse [112]. Herein, matured and stabilized blood vessels were characterized by normalized expression levels of adhesion molecules on the luminal surface of the normalized tumor endothelium and a more uniform distribution of those adhesion molecules. In addition, neutrophils from RGS5-deficient mice were mobilized much more efficiently to the site of tissue inflammation, which is due to increased chemotaxis and endothelial leukocyte adhesion [113].

Accordingly, it was shown that antiangiogenesis treatment can overcome the suppression of endothelial adhesion molecules in tumors, leading to the amelioration of leukocyte–vessel wall interactions and to an increased inflammatory infiltrate in orthotropic syngeneic B16F10 melanomas and human LS174T colon carcinoma xenografts [107]. Among the investigated angiostatic agents, the angiogenesis inhibitors anginex, a bpep peptide that elicits angiostatic activity [114], endostatin, the cleaved product of the carboxyl-terminal domain of collagen XVIII, angiostatin, an amino-terminal fragment of plasminogen [115,116] as well as the chemotherapeutic agent paclitaxel with claimed angiostatic properties were found to significantly stimulate leukocyte–vessel wall interactions by circumvention of tumor endothelial anergy in vivo through the upregulation of endothelial adhesion molecules in tumor vessels [107]. The amount of infiltrated leukocytes as well as the number of CD8 + cytotoxic T lymphocytes was markedly enhanced while tumor growth and microvessel density were significantly reduced. Conclusively, these results strongly support the concept of combining modern immunotherapeutic approaches with vessel normalizing therapies.

Interestingly, treatment with low-dose endothelial-targeted cytokines, such as TNFα [117–119] and LIGHT [120,121] were shown to induce adhesion molecule expression and recruitment of immune cells to contribute tumor vessel normalization [70]. The selectively angiogenic tumor endothelium targeting TNFα modification NGR-TNFα, a Cys-Asn-Gly-Arg-Cys peptide-TNF fusion product is already used in phase II/III clinical trials for combination therapy of advanced cancers [122]. In simultaneous combination with the T-cell-directed antibodies anti-CTLA-4 and anti-PD-1, low doses of NGR-TNF altered the endothelial barrier function together with an upregulation of leukocyte-endothelial cell adhesion molecules, the release of pro-inflammatory cytokines, and the infiltration of tumor-specific effector CD8(+) T cells [123]. Immune cell infiltrations correlated nicely with delayed tumor growth and prolonged survival of mice bearing transgenic adenocarcinoma of the mouse prostate cancer and orthotopic B16 melanoma. Thus, immune responses triggered strategies addressing directly the vascular system for vessel normalization might break the immunosuppressive microenvironment to sensitize tumors or improve the therapeutic response of immune checkpoint blocker. In a similar fashion, the TNF superfamily member LIGHT (also known as TNFSF14) was shown to be effective in inducing tumor vessel normalization and therefore ameliorating immunotherapy in breast, lung, pancreatic neuroendocrine cancers, and glioblastoma [124].

Normalization of the tumor vasculature through disruption of the VEGF/VEGFR-2 axis after administration of an anti-VEGF antibody increased extravasation of adoptively transferred T cells into the tumor and improved the overall survival in a B16 melanoma mouse model [125]. Moreover, Huang et al. could show that vascular normalization by antiangiogenic therapy enhanced through activated CD8(+) T cells the anticancer efficacy induced by a whole cancer cell vaccine therapy in both immune-tolerant (MCaP0008) and immunogenic (MMTV-PyVT) murine breast cancer models [53]. An additional antibody-mediated blockade of ANG2 resulted in vascular normalization and facilitated the extravasation and perivascular accumulation of activated, IFNγ-expressing CD8(+) cytotoxic T lymphocytes in the melanoma cancer model as well as in metastatic breast and pancreatic cancer models [126,127]

A positive immunologic synergism of combining anti-VEGF therapy with bevacizumab and immunotherapy in multitreated ovarian cancer patients was recently reported by Napoletano et al. [128]. Within the study peripheral blood mononuclear cells of 20 consecutive recurrent ovarian cancer patients who received bevacizumab or nonbevacizumab-based chemotherapy were analyzed. Clinically responding patients showed a high percentage of nonsuppressive regulatory T cells (Treg) upon bevacizumab treatment and a significant lower interleukin 10 (IL10) production (accounting for a reduced patients' immune suppression) as compared to the nonresponding patients [128]. Similar to the previous observation made in metastatic colorectal cancer patients, responding bevacizumab-treated patients reported a higher percentage of circulating CD4 effector T cells [128,129].

Given that antiangiogenic treatment is supposed to potentiate cancer immunotherapy by promoting immune cell infiltration into tumors and reducing immunosuppression within the tumors, tumor tissue derived from metastatic renal cell carcinoma patients treated with anti-PD-L1 atezolizumab and bevacizumab was recently found to show an increase in intratumoral CD8 T cells as well as an increase in intratumoral MHC-I, Th1, and T-effector cells [130].

Monoclonal antibodies directed against cytotoxic T-lymphocyte associated protein 4 (CTLA-4) and the programmed cell death-1 (PD-1/CD279) T-cell receptor and/or its ligand (programmed death-ligand 1 (PD-L1/B7-H1/CD274)) were central components of immunotherapy for activating therapeutic antitumor immunity [131,132]. However, acquired resistance to immune checkpoint antibody blockades was commonly observed in most cancer patients [133,134]. The success of such therapies again depends on the applied agents, recruited lymphocytes, and/or adoptive transferred cells, that is, tumor-reactive T lymphocytes, in finding their desired place, leaving the blood stream, and subsequently infiltrating the tumor tissues [109,135]. A preclinical study using in polyoma middle T oncoprotein breast cancer and Rip1-Tag2 pancreatic neuroendocrine tumor mouse models revealed that antiangiogenic therapy with sorafenib, a small inhibitor of several tyrosine protein kinases (such as VEGFR, PDGFR, and RAF family kinases) or the anti-VEGFR2 antibody DC101 improved anti-PD-L1 treatment [136]. As shown by reduced microvessel densities, increased diameters and a regular pericyte coverage, promoted lymphocyte infiltration and enhanced cytotoxic T-cell activity was based on vessel normalization [136].

Vessel normalization as determined by increased tumor vessel perfusion could even be used to predict the therapeutic response to immune checkpoint blockade [137]. In two mouse breast tumor models that were either sensitive or resistant to anti-CTLA4 and anti-PD1 treatment, a significantly enhanced vessel perfusion was observed in treatment-sensitive tumors, which was paralleled by increased IFNγ levels and an accumulation of CD8 + T cells, both pointing toward a successful activation of antitumor T-cell immunity [137].

Conclusively these findings strongly argue for new therapeutic approaches including combinations of the antiangiogenic and immunotherapy therapy (Table 9.1).

9.5 Conclusions and further perspectives in therapy

Upon tumor neovascularization, the hierarchical order of normal blood vessels becomes lost: newly formed tumor blood vessels are chaotically organized, run tortuous in the tumor, may end blindly, have arterio-venous shunts, and/or are directed opposite to the blood flow. Together with their structural abnormalities (incomplete endothelial lining, interendothelial gaps, a defective basement membrane, and lack of vascular mural cells), tumor blood vessels contribute a hostile tumor microenvironment that is characterized by instability, an uneven blood stream and pressure finally resulting in areas of apparent excess supply of oxygen and nutrition in addition to areas with an under supply (hypoxia, low pH, and elevated interstitial fluid pressure). In addition, tumor blood vessels cannot actively respond to physiological stimuli. These abnormalities fuel tumor progression, the tumors immune escape, and treatment resistances because this prevents an efficient administration of intravenous anti-cancer drugs. Thus, vessel normalization by antiangiogenic therapy has gained more attention for delivering therapeutic agents by generating mature and regular functioning tumor blood vessels with increased vessel perfusion instead of starving tumors from their blood supply. With vessel normalization, vascular remodeling in tumors takes place in which association with pericytes and SMC stabilizes the immature tumor blood vessels resulting in normalization of the vascular structures. This improves the distribution of circulating blood components, oxygenation, removal of suppressive metabolites, as well as distribution of therapeutically applied drugs. Concomitant, antiangiogenic therapy-mediated vessel normalization reverses endothelial cell anergy resulting in (re)sensitizing tumor blood vessels to inflammatory stimuli by inducing homing molecule expression, and thus an improved T-cell-dependent anticancer immunity. However, the optimal time window of vascular normalization needs to be refined in order to ensure the best possible therapeutic response when combining the different treatment modalities. Until now histological evaluation of vascular morphology and microvessel density have been used to assess vascular normalization. Further effort has to be made for using imaging techniques for the translational approaches to measure vessel permeability, tumor perfusion, and oxygen concentrations in the clinical practice. In addition, predictive biomarkers still need to be identified in order to select patients (potential responder and nonresponder patients), who could benefit from these novel approaches, when angiogenesis inhibitors are combined with other (targeted) therapies to overcome therapy resistance.

9.6 Conflict of interest statement

The author declares that the research was conducted in the absence of any commercial or financial relationships that could be construed as a potential conflict of interest.

Acknowledgments

The author thanks Jürgen Heger for improving the drawing/quality of the presented schemes. Funding: Grants of the DFG (GRK1739/2) and the Brigitte und Dr. Konstanze Wegener-Stiftung.

References

[1] Folkman J. Tumor angiogenesis: a possible control point in tumor growth. Ann Intern Med 1975;82:96—100.

[2] Hanahan D, Weinberg RA. Hallmarks of cancer: the next generation. Cell 2011;144:646—74.

[3] Carmeliet P, Jain RK. Molecular mechanisms and clinical applications of angiogenesis. Nature 2011;473:298—307.

[4] Hanahan D, Folkman J. Patterns and emerging mechanisms of the angiogenic switch during tumorigenesis. Cell 1996;86:353—64.

[5] Weisshardt P, Trarbach T, Durig J, Paul A, Reis H, Tilki D, et al. Tumor vessel stabilization and remodeling by anti-angiogenic therapy with bevacizumab. Histochem Cell Biol 2012;137:391—401.

[6] Klein D, Meissner N, Kleff V, Jastrow H, Yamaguchi M, Ergun S, et al. Nestin(+) tissue-resident multipotent stem cells contribute to tumor progression by differentiating into pericytes and smooth muscle cells resulting in blood vessel remodeling. Front Oncol 2014;4:169.

[7] Klein D. The tumor vascular endothelium as decision maker in cancer therapy. Front Oncol 2018;8:367.

[8] Hida K, Maishi N, Annan DA, Hida Y. Contribution of tumor endothelial cells in cancer progression. Int J Mol Sci 2018;19:1272.

[9] Jain RK. Molecular regulation of vessel maturation. Nat Med 2003;9:685—93.

[10] Ohga N, Ishikawa S, Maishi N, Akiyama K, Hida Y, Kawamoto T, et al. Heterogeneity of tumor endothelial cells: comparison between tumor endothelial cells isolated from high- and low-metastatic tumors. Am J Pathol 2012;180:1294—307.

[11] Dudley AC. Tumor endothelial cells. Cold Spring Harb Perspect Med 2012;2:a006536.

[12] Goel S, Duda DG, Xu L, Munn LL, Boucher Y, Fukumura D, et al. Normalization of the vasculature for treatment of cancer and other diseases. Physiol Rev 2011;91:1071—121.

[13] Folkman J. Tumor angiogenesis: therapeutic implications. N Engl J Med 1971;285:1182—6.

[14] Folkman J. Tumor angiogenesis. Adv Cancer Res 1985;43:175—203.

[15] Ferrara N. The role of VEGF in the regulation of physiological and pathological angiogenesis. EXS 2005;209—31.

[16] Ferrara N, Hillan KJ, Novotny W. Bevacizumab (Avastin), a humanized anti-VEGF monoclonal antibody for cancer therapy. Biochem Biophys Res Commun 2005;333:328—35.

[17] Sandler A, Gray R, Perry MC, Brahmer J, Schiller JH, Dowlati A, et al. Paclitaxel-carboplatin alone or with bevacizumab for non-small-cell lung cancer. N Engl J Med 2006;355:2542—50.

[18] Stevenson JP, Langer CJ, Somer RA, Evans TL, Rajagopalan K, Krieger K, et al. Phase 2 trial of maintenance bevacizumab alone after bevacizumab plus pemetrexed and carboplatin in advanced, nonsquamous nonsmall cell lung cancer. Cancer 2012;118:5580−7.

[19] Erber R, Thurnher A, Katsen AD, Groth G, Kerger H, Hammes HP, et al. Combined inhibition of VEGF and PDGF signaling enforces tumor vessel regression by interfering with pericyte-mediated endothelial cell survival mechanisms. FASEB J 2004;18:338−40.

[20] Jain RK, Duda DG, Clark JW, Loeffler JS. Lessons from phase III clinical trials on anti-VEGF therapy for cancer. Nat Clin Pract Oncol 2006;3:24−40.

[21] Johnson DH, Fehrenbacher L, Novotny WF, Herbst RS, Nemunaitis JJ, Jablons DM, et al. Randomized phase II trial comparing bevacizumab plus carboplatin and paclitaxel with carboplatin and paclitaxel alone in previously untreated locally advanced or metastatic non-small-cell lung cancer. J Clin Oncol 2004;22:2184−91.

[22] Jain RK. Normalizing tumor vasculature with anti-angiogenic therapy: a new paradigm for combination therapy. Nat Med 2001;7:987−9.

[23] Sun J, Wang DA, Jain RK, Carie A, Paquette S, Ennis E, et al. Inhibiting angiogenesis and tumorigenesis by a synthetic molecule that blocks binding of both VEGF and PDGF to their receptors. Oncogene 2005;24:4701−9.

[24] Jain RK. Antiangiogenic therapy for cancer: current and emerging concepts. Oncol (Williston Park) 2005;19:7−16.

[25] Jain RK. Normalization of tumor vasculature: an emerging concept in antiangiogenic therapy. Science 2005;307:58−62.

[26] Xue Y, Religa P, Cao R, Hansen AJ, Lucchini F, Jones B, et al. Anti-VEGF agents confer survival advantages to tumor-bearing mice by improving cancer-associated systemic syndrome. Proc Natl Acad Sci U S A 2008;105:18513−18.

[27] Zhuang HQ, Yuan ZY. Process in the mechanisms of endostatin combined with radiotherapy. Cancer Lett 2009;282:9−13.

[28] Helfrich I, Scheffrahn I, Bartling S, Weis J, von Felbert V, Middleton M, et al. Resistance to antiangiogenic therapy is directed by vascular phenotype, vessel stabilization, and maturation in malignant melanoma. J Exp Med 2010;207:491−503.

[29] Piali L, Fichtel A, Terpe HJ, Imhof BA, Gisler RH. Endothelial vascular cell adhesion molecule 1 expression is suppressed by melanoma and carcinoma. J Exp Med 1995;181:811−16.

[30] Griffioen AW, Damen CA, Blijham GH, Groenewegen G. Tumor angiogenesis is accompanied by a decreased inflammatory response of tumor-associated endothelium. Blood 1996;88:667−73.

[31] Griffioen AW, Damen CA, Martinotti S, Blijham GH, Groenewegen G. Endothelial intercellular adhesion molecule-1 expression is suppressed in human malignancies: the role of angiogenic factors. Cancer Res 1996;56:1111−17.

[32] Griffioen AW, Damen CA, Mayo KH, Barendsz-Janson AF, Martinotti S, Blijham GH, et al. Angiogenesis inhibitors overcome tumor induced endothelial cell anergy. Int J Cancer 1999;80:315−19.

[33] Betsholtz C. Vascular biology: transcriptional control of endothelial energy. Nature 2016;529:160−1.

[34] Rosenberg SA, Restifo NP, Yang JC, Morgan RA, Dudley ME. Adoptive cell transfer: a clinical path to effective cancer immunotherapy. Nat Rev Cancer 2008;8:299−308.

[35] Dougan M, Dranoff G. Immune therapy for cancer. Annu Rev Immunol 2009;27:83−117.

[36] Viallard C, Larrivee B. Tumor angiogenesis and vascular normalization: alternative therapeutic targets. Angiogenesis 2017;20:409−26.

[37] De Sanctis F, Ugel S, Facciponte J, Facciabene A. The dark side of tumor-associated endothelial cells. Semimmunology 2018;35:35−47.

[38] Digklia A, Voutsadakis IA. Combinations of vascular endothelial growth factor pathway inhibitors with metronomic chemotherapy: rational and current status. World J Exp Med 2014;4:58−67.

[39] Jain RK. Antiangiogenesis strategies revisited: from starving tumors to alleviating hypoxia. Cancer Cell 2014;26:605−22.

[40] Deng T, Zhang L, Liu XJ, Xu JM, Bai YX, Wang Y, et al. Bevacizumab plus irinotecan, 5-fluorouracil, and leucovorin (FOLFIRI) as the second-line therapy for patients with metastatic colorectal cancer, a multicenter study. Med Oncol 2013;30:752.

[41] Hurwitz H, Fehrenbacher L, Novotny W, Cartwright T, Hainsworth J, Heim W, et al. Bevacizumab plus irinotecan, fluorouracil, and leucovorin for metastatic colorectal cancer. N Engl J Med 2004;350:2335−42.

[42] Goel S, Wong AH, Jain RK. Vascular normalization as a therapeutic strategy for malignant and nonmalignant disease. Cold Spring Harb Perspect Med 2012;2:a006486.

[43] Martin JD, Seano G, Jain RK. Normalizing function of tumor vessels: progress, opportunities, and challenges. Annu Rev Physiol 2019;81:505−34.

[44] Sorensen AG, Batchelor TT, Zhang WT, Chen PJ, Yeo P, Wang M, et al. A "vascular normalization index" as potential mechanistic biomarker to predict survival after a single dose of cediranib in recurrent glioblastoma patients. Cancer Res 2009;69:5296−300.

[45] Tolaney SM, Boucher Y, Duda DG, Martin JD, Seano G, Ancukiewicz M, et al. Role of vascular density and normalization in response to neoadjuvant bevacizumab and chemotherapy in breast cancer patients. Proc Natl Acad Sci U S A 2015;112:14325−30.

[46] Heist RS, Duda DG, Sahani DV, Ancukiewicz M, Fidias P, Sequist LV, et al. Improved tumor vascularization after anti-VEGF therapy with carboplatin and nab-paclitaxel associates with survival in lung cancer. Proc Natl Acad Sci U S A 2015;112:1547−52.

[47] Giantonio BJ, Catalano PJ, Meropol NJ, O'Dwyer PJ, Mitchell EP, Alberts SR, et al. Bevacizumab in combination with oxaliplatin, fluorouracil, and leucovorin (FOLFOX4) for previously treated metastatic colorectal cancer: results from the Eastern Cooperative Oncology Group Study E3200. J Clin Oncol 2007;25:1539−44.

[48] Goel S, Gupta N, Walcott BP, Snuderl M, Kesler CT, Kirkpatrick ND, et al. Effects of vascular-endothelial protein tyrosine phosphatase inhibition on breast cancer vasculature and metastatic progression. J Natl Cancer Inst 2013;105:1188−201.

[49] Kontos CD, Willett CG. Inhibiting the inhibitor: targeting vascular endothelial protein tyrosine phosphatase to promote tumor vascular maturation. J Natl Cancer Inst 2013;105:1163−5.

[50] Park JS, Kim IK, Han S, Park I, Kim C, Bae J, et al. Normalization of tumor vessels by Tie2 activation and Ang2 inhibition enhances drug delivery and produces a favorable tumor microenvironment. Cancer Cell 2016;30:953−67.

[51] Augustin HG, Koh GY, Thurston G, Alitalo K. Control of vascular morphogenesis and homeostasis through the angiopoietin-Tie system. Nat Rev Mol Cell Biol 2009;10:165−77.

[52] Annese T, Tamma R, Ruggieri S, Ribatti D. Angiogenesis in pancreatic cancer: pre-clinical and clinical studies. Cancers 2019;11::381.

[53] Huang Y, Yuan J, Righi E, Kamoun WS, Ancukiewicz M, Nezivar J, et al. Vascular normalizing doses of antiangiogenic treatment reprogram the immunosuppressive tumor microenvironment and enhance immunotherapy. Proc Natl Acad Sci U S A 2012;109:17561−6.

[54] Mpekris F, Baish JW, Stylianopoulos T, Jain RK. Role of vascular normalization in benefit from metronomic chemotherapy. Proc Natl Acad Sci U S A 2017;114:1994−9.

[55] Awasthi N, Hinz S, Brekken RA, Schwarz MA, Schwarz RE. Nintedanib, a triple angiokinase inhibitor, enhances cytotoxic therapy response in pancreatic cancer. Cancer Lett 2015;358:59−66.

[56] Awasthi N, Zhang C, Schwarz AM, Hinz S, Schwarz MA, Schwarz RE. Enhancement of nab-paclitaxel antitumor activity through addition of multitargeting antiangiogenic agents in experimental pancreatic cancer. Mol Cancer Ther 2014;13:1032−43.

[57] Cabebe E, Fisher GA. Clinical trials of VEGF receptor tyrosine kinase inhibitors in pancreatic cancer. Expert Opin Invest Drugs 2007;16:467−76.

[58] Capdevila J, Carrato A, Tabernero J, Grande E. What could Nintedanib (BIBF 1120), a triple inhibitor of VEGFR, PDGFR, and FGFR, add to the current treatment options for patients with metastatic colorectal cancer? Crit Rev Oncol/Hematol 2014;92:83−106.

[59] Soler A, Serra H, Pearce W, Angulo A, Guillermet-Guibert J, Friedman LS, et al. Inhibition of the p110alpha isoform of PI 3-kinase stimulates nonfunctional tumor angiogenesis. J Exp Med 2013;210:1937−45.

[60] Kim SJ, Jung KH, Son MK, Park JH, Yan HH, Fang Z, et al. Tumor vessel normalization by the PI3K inhibitor HS-173 enhances drug delivery. Cancer Lett 2017;403:339−53.

[61] Murillo MM, Zelenay S, Nye E, Castellano E, Lassailly F, Stamp G, et al. RAS interaction with PI3K p110alpha is required for tumor-induced angiogenesis. The. J Clin Invest 2014;124:3601−11.

[62] Qayum N, Im J, Stratford MR, Bernhard EJ, McKenna WG, Muschel RJ. Modulation of the tumor microvasculature by phosphoinositide-3 kinase inhibition increases doxorubicin delivery in vivo. Clin Cancer Res 2012;18:161−9.

[63] Plate KH, Breier G, Weich HA, Risau W. Vascular endothelial growth factor is a potential tumour angiogenesis factor in human gliomas in vivo. Nature 1992;359:845−8.

[64] Field KM, Jordan JT, Wen PY, Rosenthal MA, Reardon DA. Bevacizumab and glioblastoma: scientific review, newly reported updates, and ongoing controversies. Cancer 2015;121:997−1007.

[65] Yang SB, Gao KD, Jiang T, Cheng SJ, Li WB. Bevacizumab combined with chemotherapy for glioblastoma: a meta-analysis of randomized controlled trials. Oncotarget 2017;8:57337−44.

[66] Mathivet T, Bouleti C, Van Woensel M, Stanchi F, Verschuere T, Phng LK, et al. Dynamic stroma reorganization drives blood vessel dysmorphia during glioma growth. EMBO Mol Med 2017;9:1629−45.

[67] Hellstrom M, Gerhardt H, Kalen M, Li X, Eriksson U, Wolburg H, et al. Lack of pericytes leads to endothelial hyperplasia and abnormal vascular morphogenesis. J Cell Biol 2001;153:543−53.

[68] Meng MB, Zaorsky NG, Deng L, Wang HH, Chao J, Zhao LJ, et al. Pericytes: a double-edged sword in cancer therapy. Future Oncol 2015;11:169−79.

[69] Lu C, Kamat AA, Lin YG, Merritt WM, Landen CN, Kim TJ, et al. Dual targeting of endothelial cells and pericytes in antivascular therapy for ovarian carcinoma. Clin Cancer Res 2007;13:4209−17.

[70] Fan Q, Tao Z, Yang H, Shi Q, Wang H, Jia D, et al. Modulation of pericytes by a fusion protein comprising of a PDGFRβ-antagonistic affibody and TNFα induces tumor vessel normalization and improves chemotherapy. J Control Release 2019;302:63−78.

[71] Lindborg M, Cortez E, Hoiden-Guthenberg I, Gunneriusson E, von Hage E, Syud F, et al. Engineered high-affinity affibody molecules targeting platelet-derived growth factor receptor beta in vivo. J Mol Biol 2011;407:298−315.

[72] Yapp DT, Wong MQ, Kyle AH, Valdez SM, Tso J, Yung A, et al. The differential effects of metronomic gemcitabine and antiangiogenic treatment in patient-derived xenografts of pancreatic cancer: treatment effects on metabolism, vascular function, cell proliferation, and tumor growth. Angiogenesis 2016;19:229−44.

[73] Rahbari NN, Kedrin D, Incio J, Liu H, Ho WW, Nia HT, et al. Anti-VEGF therapy induces ECM remodeling and mechanical barriers to therapy in colorectal cancer liver metastases. Sci Transl Med 2016;8:360ra135.

[74] Jung K, Heishi T, Khan OF, Kowalski PS, Incio J, Rahbari NN, et al. Ly6Clo monocytes drive immunosuppression and confer resistance to anti-VEGFR2 cancer therapy. J Clin Invest 2017;127:3039−51.

[75] Lee JG, Wu R. Erlotinib-cisplatin combination inhibits growth and angiogenesis through c-MYC and HIF-1alpha in EGFR-mutated lung cancer in vitro and in vivo. Neoplasia 2015;17:190−200.

[76] Brand TM, Iida M, Li C, Wheeler DL. The nuclear epidermal growth factor receptor signaling network and its role in cancer. Discov Med 2011;12:419−32.

[77] Krock BL, Skuli N, Simon MC. Hypoxia-induced angiogenesis: good and evil. Genes Cancer 2011;2:1117−33.

[78] Koyama S, Matsunaga S, Imanishi M, Maekawa Y, Kitano H, Takeuchi H, et al. Tumour blood vessel normalisation by prolyl hydroxylase inhibitor repaired sensitivity to chemotherapy in a tumour mouse model. Sci Rep 2017;7:45621.

[79] Mazzone M, Dettori D, de Oliveira RL, Loges S, Schmidt T, Jonckx B, et al. Heterozygous deficiency of PHD2 restores tumor oxygenation and inhibits metastasis via endothelial normalization. Cell 2009;136:839−51.

[80] Ader I, Brizuela L, Bouquerel P, Malavaud B, Cuvillier O. Sphingosine kinase 1: a new modulator of hypoxia inducible factor 1alpha during hypoxia in human cancer cells. Cancer Res 2008;68:8635—42.

[81] Ader I, Gstalder C, Bouquerel P, Golzio M, Andrieu G, Zalvidea S, et al. Neutralizing S1P inhibits intratumoral hypoxia, induces vascular remodelling and sensitizes to chemotherapy in prostate cancer. Oncotarget 2015;6:13803—21.

[82] Kappos L, Radue EW, O'Connor P, Polman C, Hohlfeld R, Calabresi P, et al. A placebo-controlled trial of oral fingolimod in relapsing multiple sclerosis. N Engl J Med 2010;362:387—401.

[83] Gstalder C, Ader I, Cuvillier O. FTY720 (Fingolimod) inhibits HIF1 and HIF2 signaling, promotes vascular remodeling, and chemosensitizes in renal cell carcinoma animal model. Mol Cancer Ther 2016;15:2465—74.

[84] Cartier A, Leigh T, Liu CH, Hla T. Endothelial sphingosine 1-phosphate receptors promote vascular normalization to influence tumor growth and metastasis. bioRxiv 2019;606434.

[85] Cantelmo AR, Pircher A, Kalucka J, Carmeliet P. Vessel pruning or healing: endothelial metabolism as a novel target? Expert Opin Ther Targets 2017;21:239—47.

[86] Cantelmo AR, Conradi LC, Brajic A, Goveia J, Kalucka J, Pircher A, et al. Inhibition of the glycolytic activator PFKFB3 in endothelium induces tumor vessel normalization, impairs metastasis, and improves chemotherapy. Cancer Cell 2016;30:968—85.

[87] Cruys B, Wong BW, Kuchnio A, Verdegem D, Cantelmo AR, Conradi LC, et al. Glycolytic regulation of cell rearrangement in angiogenesis. Nat Commun 2016;7:12240.

[88] Maes H, Kuchnio A, Peric A, Moens S, Nys K, De Bock K, et al. Tumor vessel normalization by chloroquine independent of autophagy. Cancer Cell 2014;26:190—206.

[89] Begg AC, Stewart FA, Vens C. Strategies to improve radiotherapy with targeted drugs. Nat Rev Cancer 2011;11:239—53.

[90] El Alaoui-Lasmaili K, Faivre B. Antiangiogenic therapy: markers of response, "normalization" and resistance. Crit Rev Oncol/Hematol 2018;128:118—29.

[91] Myers AL, Williams RF, Ng CY, Hartwich JE, Davidoff AM. Bevacizumab-induced tumor vessel remodeling in rhabdomyosarcoma xenografts increases the effectiveness of adjuvant ionizing radiation. J Pediatr Surg 2010;45:1080—5.

[92] Myers AL, Orr WS, Denbo JW, Ng CY, Zhou J, Spence Y, et al. Rapamycin-induced tumor vasculature remodeling in rhabdomyosarcoma xenografts increases the effectiveness of adjuvant ionizing radiation. J Pediatr Surg 2012;47:183—9.

[93] Matsumoto S, Batra S, Saito K, Yasui H, Choudhuri R, Gadisetti C, et al. Antiangiogenic agent sunitinib transiently increases tumor oxygenation and suppresses cycling hypoxia. Cancer Res 2011;71:6350—9.

[94] Faivre S, Demetri G, Sargent W, Raymond E. Molecular basis for sunitinib efficacy and future clinical development. Nat Rev Drug Discov 2007;6:734—45.

[95] El Kaffas A, Al-Mahrouki A, Tran WT, Giles A, Czarnota GJ. Sunitinib effects on the radiation response of endothelial and breast tumor cells. Microvasc Res 2014;92:1—9.

[96] El Kaffas A, Giles A, Czarnota GJ. Dose-dependent response of tumor vasculature to radiation therapy in combination with Sunitinib depicted by three-dimensional high-frequency power Doppler ultrasound. Angiogenesis 2013;16:443—54.

[97] Koo HJ, Lee M, Kim J, Woo CW, Jeong SY, Choi EK, et al. Synergistic effect of anti-angiogenic and radiation therapy: quantitative evaluation with dynamic contrast enhanced MR imaging. PLoS One 2016;11:e0148784.

[98] Peterson TE, Kirkpatrick ND, Huang Y, Farrar CT, Marijt KA, Kloepper J, et al. Dual inhibition of Ang-2 and VEGF receptors normalizes tumor vasculature and prolongs survival in glioblastoma by altering macrophages. Proc Natl Acad Sci U S A 2016;113:4470—5.

[99] Solecki G, Osswald M, Weber D, Glock M, Ratliff M, Muller HJ, et al. Differential effects of Ang-2/VEGF-A inhibiting antibodies in combination with radio- or chemotherapy in glioma. Cancers 2019;11:314.

[100] Winkler F, Kozin SV, Tong RT, Chae SS, Booth MF, Garkavtsev I, et al. Kinetics of vascular normalization by VEGFR2 blockade governs brain tumor response to radiation: role of oxygenation, angiopoietin-1, and matrix metalloproteinases. Cancer Cell 2004;6:553—63.

[101] Stupp R, Mason WP, van den Bent MJ, Weller M, Fisher B, Taphoorn MJ, et al. Radiotherapy plus concomitant and adjuvant temozolomide for glioblastoma. N Engl J Med 2005;352:987—96.

[102] McGee MC, Hamner JB, Williams RF, Rosati SF, Sims TL, Ng CY, et al. Improved intratumoral oxygenation through vascular normalization increases glioma sensitivity to ionizing radiation. Int J Radiat Oncol Biol Phys 2010;76:1537—45.

[103] Martini M, De Santis MC, Braccini L, Gulluni F, Hirsch E. PI3K/AKT signaling pathway and cancer: an updated review. Ann Med 2014;46:372—83.

[104] Lee H, Jung KH, Jeong Y, Hong S, Hong SS. HS-173, a novel phosphatidylinositol 3-kinase (PI3K) inhibitor, has anti-tumor activity through promoting apoptosis and inhibiting angiogenesis. Cancer Lett 2013;328:152—9.

[105] Park JH, Jung KH, Kim SJ, Fang Z, Yan HH, Son MK, et al. Radiosensitization of the PI3K inhibitor HS-173 through reduction of DNA damage repair in pancreatic cancer. Oncotarget 2017;8:112893—906.

[106] Wang H, Mu X, He H, Zhang XD. Cancer radiosensitizers. Trends Pharmacol Sci 2018;39:24—48.

[107] Dirkx AE, oude Egbrink MG, Castermans K, van der Schaft DW, Thijssen VL, Dings RP, et al. Anti-angiogenesis therapy can overcome endothelial cell anergy and promote leukocyte-endothelium interactions and infiltration in tumors. FASEB J 2006;20:621—30.

[108] Huang Y, Kim BYS, Chan CK, Hahn SM, Weissman IL, Jiang W. Improving immune-vascular crosstalk for cancer immunotherapy. Nat Rev Immunol 2018;18:195—203.

[109] Ager A, Watson HA, Wehenkel SC, Mohammed RN. Homing to solid cancers: a vascular checkpoint in adoptive cell therapy using CAR T-cells. Biochem Soc Trans 2016;44:377—85.

[110] Al-Soudi A, Kaaij MH, Tas SW. Endothelial cells: from innocent bystanders to active participants in immune responses. Autoimmun Rev 2017;16:951–62.

[111] Taylor CT, Colgan SP. Regulation of immunity and inflammation by hypoxia in immunological niches. Nat Rev Immunol 2017;17:774–85.

[112] Hamzah J, Jugold M, Kiessling F, Rigby P, Manzur M, Marti HH, et al. Vascular normalization in Rgs5-deficient tumours promotes immune destruction. Nature 2008;453:410–14.

[113] Chan EC, Ren C, Xie Z, Jude J, Barker T, Koziol-White CA, et al. Regulator of G protein signaling 5 restricts neutrophil chemotaxis and trafficking. J Biol Chem 2018;293:12690–702.

[114] Griffioen AW, van der Schaft DW, Barendsz-Janson AF, Cox A, Struijker Boudier HA, Hillen HF, et al. Anginex, a designed peptide that inhibits angiogenesis. Biochem J 2001;354:233–42.

[115] Ribatti D. Endogenous inhibitors of angiogenesis: a historical review. Leuk Res 2009;33:638–44.

[116] Javaherian K, Lee TY, Tjin Tham Sjin RM, Parris GE, Hlatky L. Two endogenous antiangiogenic inhibitors, endostatin and angiostatin, demonstrate biphasic curves in their antitumor profiles. Dose Response 2011;9:369–76.

[117] Calcinotto A, Grioni M, Jachetti E, Curnis F, Mondino A, Parmiani G, et al. Targeting TNF-alpha to neoangiogenic vessels enhances lymphocyte infiltration in tumors and increases the therapeutic potential of immunotherapy. J Immunol 2012;188:2687–94.

[118] Curnis F, Gasparri A, Sacchi A, Longhi R, Corti A. Coupling tumor necrosis factor-alpha with alphaV integrin ligands improves its antineoplastic activity. Cancer Res 2004;64:565–71.

[119] Johansson A, Hamzah J, Payne CJ, Ganss R. Tumor-targeted TNFalpha stabilizes tumor vessels and enhances active immunotherapy. Proc Natl Acad Sci U S A 2012;109:7841–6.

[120] Johansson-Percival A, He B, Li ZJ, Kjellen A, Russell K, Li J, et al. De novo induction of intratumoral lymphoid structures and vessel normalization enhances immunotherapy in resistant tumors. Nat Immunol 2017;18:1207–17.

[121] Johansson-Percival A, Li ZJ, Lakhiani DD, He B, Wang X, Hamzah J, et al. Intratumoral LIGHT restores pericyte contractile properties and vessel integrity. Cell Rep 2015;13:2687–98.

[122] Corti A, Curnis F, Rossoni G, Marcucci F, Gregorc V. Peptide-mediated targeting of cytokines to tumor vasculature: the NGR-hTNF example. BioDrugs 2013;27:591–603.

[123] Elia AR, Grioni M, Basso V, Curnis F, Freschi M, Corti A, et al. Targeting tumor vasculature with TNF leads effector T cells to the tumor and enhances therapeutic efficacy of immune checkpoint blockers in combination with adoptive cell therapy. Clin Cancer Res 2018;24:2171–81.

[124] Treps L. EnLIGHTenment of tumor vessel normalization and immunotherapy in glioblastoma. J Pathol 2018;246:3–6.

[125] Shrimali RK, Yu Z, Theoret MR, Chinnasamy D, Restifo NP, Rosenberg SA. Antiangiogenic agents can increase lymphocyte infiltration into tumor and enhance the effectiveness of adoptive immunotherapy of cancer. Cancer Res 2010;70:6171–80.

[126] Schmittnaegel M, De Palma M. Reprogramming tumor blood vessels for enhancing immunotherapy. Trends Cancer 2017;3:809–12.

[127] Schmittnaegel M, Rigamonti N, Kadioglu E, Cassara A, Wyser Rmili C, Kiialainen A, et al. Dual angiopoietin-2 and VEGFA inhibition elicits antitumor immunity that is enhanced by PD-1 checkpoint blockade. Sci Transl Med 2017;9::eaak9679.

[128] Napoletano C, Ruscito I, Bellati F, Zizzari IG, Rahimi H, Gasparri ML, et al. Bevacizumab-based chemotherapy triggers immunological effects in responding multi-treated recurrent ovarian cancer patients by favoring the recruitment of effector T cell subsets. J Clin Med 2019;8::380.

[129] Manzoni M, Rovati B, Ronzoni M, Loupakis F, Mariucci S, Ricci V, et al. Immunological effects of bevacizumab-based treatment in metastatic colorectal cancer. Oncology 2010;79:187–96.

[130] Wallin JJ, Bendell JC, Funke R, Sznol M, Korski K, Jones S, et al. Atezolizumab in combination with bevacizumab enhances antigen-specific T-cell migration in metastatic renal cell carcinoma. Nat Commun 2016;7:12624.

[131] Pardoll DM. The blockade of immune checkpoints in cancer immunotherapy. Nat Rev Cancer 2012;12:252–64.

[132] Li X, Shao C, Shi Y, Han W. Lessons learned from the blockade of immune checkpoints in cancer immunotherapy. J Hematol Oncol 2018;11:31.

[133] Sharma P, Hu-Lieskovan S, Wargo JA, Ribas A. Primary, adaptive, and acquired resistance to cancer immunotherapy. Cell 2017;168:707–23.

[134] Tang H, Wang Y, Chlewicki LK, Zhang Y, Guo J, Liang W, et al. Facilitating T cell infiltration in tumor microenvironment overcomes resistance to PD-L1 blockade. Cancer Cell 2016;30:500.

[135] Nishino M, Ramaiya NH, Hatabu H, Hodi FS. Monitoring immune-checkpoint blockade: response evaluation and biomarker development. Nat Rev Clin Oncol 2017;14:655–68.

[136] Allen E, Jabouille A, Rivera LB, Lodewijckx I, Missiaen R, Steri V, et al. Combined antiangiogenic and anti-PD-L1 therapy stimulates tumor immunity through HEV formation. Sci Transl Med 2017;9:eaak9679.

[137] Zheng X, Fang Z, Liu X, Deng S, Zhou P, Wang X, et al. Increased vessel perfusion predicts the efficacy of immune checkpoint blockade. J Clin Invest 2018;128:2104–15.

Chapter 10

Vascular endothelial growth factor and tumor immune microenvironment

Nabendu Pore

Oncology R&D, AstraZeneca, One MedImmune Way, Gaithersburg, MD, United States

10.1 Introduction

As discussed in detail in the preceding chapter, growth of the vascular network or angiogenesis is a crucial process that drives the proliferation and metastatic spread of cancer cells. This complex process involves the orchestration of numerous signaling pathways but is primarily driven by the vascular endothelial growth factor (specifically the most abundant isoform, known as VEGF-A, also, commonly referred to as VEGF). VEGF signaling has been studied and reviewed extensively in the context of angiogenesis and vasculogenesis associated with the pathogenesis of various cancers. In addition to pathological angiogenesis, VEGF fosters "immune escape," a prerequisite for tumor development [1,2]. Immune escape that is the ability to avoid detection by the body's immune surveillance is pivotal in tumor growth as the capability of the immune system to recognize and eliminate incipient tumor cells needs to be controlled in order to enable tumor progression. VEGF facilitates immune escape by regulating the functions of several immune subpopulations to alter the tumor microenvironment (TME). Given that VEGF's role in tumor growth is vast and includes angiogenesis and other signaling pathways, this review will restrict itself to VEGF signaling in the context of immune escape. Thus the focus of this chapter will be on VEGF driven immune modulation and the cross-talk between VEGF and various immune cells that drive immunosurveillance in the TME.

10.2 The adaptive immune system in tumor growth and suppression

The adaptive immune system consists of cells such as T and B lymphocytes and dendritic cells (DCs) that participate in immune control via various "recognition" processes. With cancer progression, the maturation and activation of these cells are compromised as the TME develops an immunosuppressive environment. A typical tumor includes a core with a surrounding stromal layer. The TME consists of a complex milieu of cells, including an immune infiltrate that contains all immune cell types of the adaptive immune system such as DC, B cells, and T lymphocytes that are embedded within a robust extracellular matrix. Some components of the tumor infiltrate exert a beneficial antitumor effect, while others can downregulate host immunity and promote tumor-immune evasion. The TME supports immune evasion from T lymphocyte recognition [2]. The T lymphocyte family involved in tumor tissue includes T helper 1 (T_H1), T helper 2 (T_H2), regulatory T cells (Tregs), and cytotoxic or effector T cells. While each of these T cells has a specific role, it is VEGF's effect on effector and regulatory T lymphocytes that facilitates immune regulation within the tumor tissue [3]. Adaptive immune systems can be modulated by immunosuppressive cytokine function of VEGF via recruiting myeloid subpopulations, consequently resulting in immune deserts or immune-excluded TME. The effects of VEGF on various immune subpopulations are illustrated in Fig. 10.1 and are summarized in detail in this chapter. Additionally, VEGF also regulates the functions of other immune cells such as NK cells, mast cells, $\gamma\delta T$ cells, and innate lymphoid cells but as this field is still evolving, the next few sections will discuss the effects of VEGF on T cells.

Endothelial Signaling in Vascular Dysfunction and Disease. DOI: https://doi.org/10.1016/B978-0-12-816196-8.00002-3

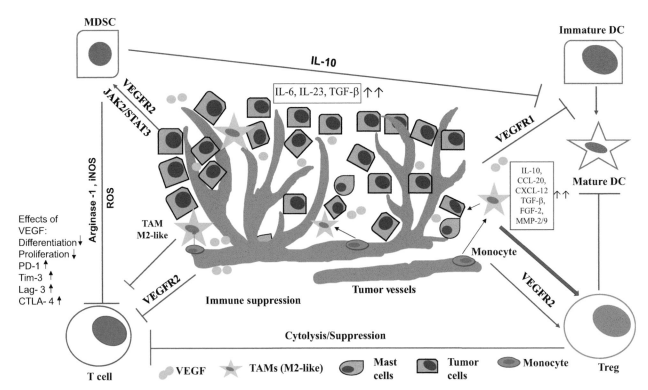

FIGURE 10.1 Schema showing a tumor vessel bed and the interactions between vascular endothelial growth factor (VEGF), its receptor (VEGFR1 and 2), the JAK/STAT3 signaling pathways, and various immune cells on the vascular bed such as T cells, regulatory T cells (Tregs), myeloid-derived suppressor cells (MDSC), tumor-associated macrophages (TAMs), dendritic cells (DC), monocytes, and mast cells [4]. *Adapted with permission from Yang et al. Targeting VEGF/VEGFR to Modulate Antitumor Immunity. Frontiers in immunology 9, 2018, 978.*

10.3 Vascular endothelial growth factor and effector T cells

An important feature of VEGF biology is that VEGF has direct biological effects on T cells. Effector T cell signaling depends upon the expression of surface molecules CD4 and CD8 that play a role in T cell recognition and activation by binding to their respective class II and class I major histocompatibility complex (MHC) ligands on an antigen-presenting cell (APC).

The effect of T cells on tumor cell killing is dependent on two crucial factors: First, CD8 + T cell differentiation and second, the infiltration of CD8 + T cells into the tumor site that occurs by the trafficking of the CD8 + T cells into TME. VEGF is crucial to both these processes. It hinders T cell proliferation by reducing the maturation of hematopoietic progenitor cells and consequently its evolution into functional T effector cells [3]. It also reduces infiltration as observed in the case of ovarian carcinoma, where increased VEGF expression has been observed in the TME and it has been found to inversely modulate the infiltration of T cells [5]. While these point to an immunosuppressive role of VEGF, the exact cross-talk between VEGF and effector T cell function is not very clear. VEGF signals via VEGF receptors (VEGFRs) that have been reported to be expressed on the subpopulations of T cells, such as CD4 + T cells. The VEGF-VEGFRs binding results in the activation of MAPK and PI3K-AKT signaling pathways [6,7].

VEGF signaling acts both as an activator or inhibitor of T cell proliferation. Immunomodulatory cytokines such as interleukin (IL)-2 and interferons γ could be modulated by VEGF [6] and may induce the migration of human memory CD4 + CD45RO + T cells. Conversely, VEGF has been reported elsewhere to reduce the cytotoxic potential of T cells [8].

Blocking the VEGFR-2 signaling axis in T cells through anti-VEGFR-2 therapy reduces the VEGF-induced suppression of T cells [8]. Additionally, VEGF could regulate the function of T cells via cyclooxygenase by increasing Fas Ligand (FasL) expression on endothelial cells [9]. FasL expression on tumor endothelium is associated with reduced CD8 + T cell infiltration into tumor tissue. Thus blocking VEGF signals has been observed to lead to CD8 + T cell dependent tumor growth suppression [10]. High expression of VEGF/VEGFR2 (as in CD47 null mice; CD47 is an integrin-associated protein the lack of which upregulates VEGF/VEGFR2 signaling) modulates T cell proliferation and consequently T cell receptor responses [11].

Thus blocking of VEGF pathways via antiangiogenic therapies is being developed as strategies to regulate T cell functions. Normally T cells recognize and bind to partner proteins (CTLA-4 or cytotoxic T-lymphocyte-associated protein 4, and/or PD-1 or programmed cell death protein-1) on tumor cells, a process called immune checkpoint. Immune checkpoint blockade is a therapeutic approach (e.g., anti-PD-1, anti-PD-L1, or anti-CTLA-4), designed to amplify endogenous antitumor T cell responses. Blocking the costimulatory molecules CTLA-4 and PD-1 (which attenuate T cell activation) has revolutionized cancer treatment. Disinhibition of preexisting immune responses by blocking the immune checkpoint broadly facilitates immune activation significantly extends survival of advanced cancer patients.

Blocking either VEGF or its receptor has been found to be effective in enhancing T cell function and action. Avastin (bevacizumab), a human monoclonal antibody targeting VEGF has been shown to increase both infiltration of B and T cells in tumors in first-line treatment options [12] and augment the cytotoxic potential of T cells in metastatic nonsmall-cell lung carcinoma (NSCLC) [13]. In the preclinical setting, sunitinib, a pan-VEGFR inhibitor and the inhibitor of other tyrosine kinases (such as platelet-derived growth factor or PDGF) demonstrated the reduction of immunosuppressive cytokine, IL-10, and the decrease of immune checkpoint proteins (PD-1 and CTLA-4) on T cells. These led to an increase of T cell infiltration (tumor-infiltrating leucocytes, TILs) in MCA26 (colon carcinoma cells) tumors. Furthermore, sunitinib treatment in MCA26 tumor xenografts resulted in the redistribution of CD4 + and CD8 + T cells of TILs and subsequent enrichment of activated CD8 + cells as evident by increased cytokine interferons γ signature. In summary, above findings suggest that modulation of VEGF signaling axis by broad tyrosine kinase inhibitor, sunitinib favors activation of T cells function through reprogramming of TME via alteration of cytokines and inhibition of immune checkpoint molecules including PD-1 and CTLA-4.

10.4 Modulation of tumor microenvironment architecture

Anti-VEGF therapies can also increase the infiltration of T cells through altering the tumor structure. This occurs via blocking the formation of new blood vessels (neoangiogenesis), which transiently "normalizes" the abnormal structure of tumor vasculature [14,15]. Normalization of the vasculature improves the oxygen concentration, effective drug delivery, and infiltration of CD8 + T cells as well illustrated in various recent reports [16,17]. The extravasation and crawling of T cells in tumor tissues, particularly in tumor endothelial cells, depends upon several intracellular adhesion molecules or vascular adhesion molecules that are often downregulated by VEGF [14,18]. Thus inhibition of VEGF could increase the expression of these adhesion molecules and subsequently increase leucocyte-endothelial interactions and trafficking of TILs in TME [19].

10.5 Vascular endothelial growth factor in modulation of regulatory T cells (signaling via the VEGF/VEGFR axis)

Regulatory T cells (Tregs), defined by markers CD4, CD25, and FOXP3 positive cells, play a critical role in suppression of cytotoxic T cell proliferation and function and consequently influence the immune fitness of the tumors [20]. Overexpression of VEGF in TME drives infiltration of Tregs across various malignancies and consequently scores as a negative prognostic marker for immune therapy [21–23]. Furthermore, researchers have shown that VEGFR2 is overexpressed in Tregs, and VEGF contributes to proliferation and maintenance of Tregs in the TME [23]. Also, neuropilin 1 (NRP-1) expression in FOXP3 + Tregs cells are guided by VEGF; NRP-1 acts as a key mediator of FOXP3 + (+) Treg cell infiltration into the TME, resulting in a dampened antitumor immune response and enhanced tumor progression [24]. Therefore blocking VEGF signaling could potentially improve the immune fitness of TME by ablating Treg functions.

Sunitinib, a multitargeted receptor tyrosine kinase (RTK) has been reported to reduce the Tregs in murine tumors and in patients with metastatic renal cell carcinoma [25–28]. Sunitinib, is a broad RTK, and could potentially target multiple RTK. Although the above studies, however, could not decipher the specific role of VEGF in reducing Treg functions in tumors, nonetheless, treatment of anti-VEGF antibody in murine metastatic colorectal model (CT26) showed reduction of Treg in tumor while no effect was observed with the non-VEGF-targeting RTK, masitinib [29]. Furthermore, proliferation of Tregs was inhibited via blocking of VEGF/VEGFR2 axis [29], which corroborates with the findings reported by Suzuki et al. [30]. Several other studies carried out with sunitinib [29,30], anti-VEGFR2 antibody, DC101 [31] and chimeric antibody blocking both VEGFR1/2 [32] showed a similar outcome.

10.6 Vascular endothelial growth factor and dendritic cells

DCs are specialized hematopoietic cells with a capacity of antigen presentation and play a pivotal role in bridging innate and adaptive immune systems. DCs originate from hematopoietic bone marrow progenitor cells, and are less mature, and exhibit low expression of MHC-I/II proteins, and costimulatory proteins CD80 and CD68. Therefore these immature DCs are unable to present antigen to T cells [33,34]. Various chemokines and cytokines play important roles in differentiation, maturation, and activation of DCs. Earlier research suggests that Flt-1 (VEGFR1) [35] and KDR (VEGFR2), two VEGFRs are expressed on DCs [36]. Follow-up studies conducted for the last two decades showed expression of various VEGF isoforms and VEGFRs in DCs derived from hematopoietic and nonhematopoietic tissues [37,38]. Several preclinical and clinical studies demonstrate that VEGF could impair the differentiation and maturation of DCs via these receptors. Inhibition of DC maturation through the activation of transcription factor NF-κB by the VEGF/VEGFR1 signaling axis has also been reported [39,40]. Similarly, differentiation of DCs from hematopoietic progenitor cells was affected by VEGFR1 and 2 signaling [41]. In clinical settings, increased expression of VEGF in plasma of cancer patients correlate with immature or poor function of DCs isolated from peripheral blood [42]. Likewise, removal of tumor burden through surgical resection led to partial reversal of above phenomenon [43].

Interestingly, targeting VEGF/VEGFR axis to modulate differentiation, maturation, and activation of DCs (as shown in Fig. 10.1) has yielded mixed bag of responses. Several studies carried out with sorafenib, a multikinase inhibitor of REF/MEK/ERK, VEGFR2, VEGFR3, PDGFb, Flt3, and cKIT pathways, led to controversial and ambiguous outcomes [44,45]. This has largely been due to limitation of relevant models and assay method's lack of specificity for sorafenib. However, in preclinical in vitro studies, anti-VEGF antibody, bevacizumab rescued inhibitory effect of VEGF during differentiation of monocytes into DCs [44]. Bevacizumab was reported to enhance the numbers of DCs and their allostimulatory potentials on the peripheral blood of cancer patients, including the patient with metastatic NSCLC [13,46]. In posttreatment biopsies carried out in metastatic melanoma patients, bevacizumab in combination with anti-CTLA antibody (ipilimumab), showed substantial trafficking and infiltration of CD163 + DCs across the tumor vasculature compared to ipilimumab alone [47].

10.7 Vascular endothelial growth factor in regulating myeloid-derived suppressor cells

Myeloid-derived suppressor cells (MDSCs) are known to drive immunosuppression signals that favor tumor growth. About a decade ago, MDSC, were identified and defined by their CD11b + and Gr + signatures in a tumor-bearing murine model [48].

There are two main subtypes of MDSCs: monocytic MDSCs (M-MDSC); polymorphonuclear MDSCs (PMN-MDSCs), which are also known as granulocytic MDSCs (gMDSCs). In murine TME, PMN-MDSCs are the principal cells driving immunosuppressive signals to favor tumor growth while M-MDSCs were reported to dampen human T cell activation under in vitro conditions [48,49]. MDSCs could potentially drive the immunosuppressive effect by (the following) modulating processes as shown in Fig. 10.1: (1) MDSCs could reprogram the tumor-immune metabolism and deplete the key nutrients, necessary for T cell and other lymphocytes' function; (2) generation of oxidative stress; (3) perturbing lymphocytes' trafficking and infiltration and impacting viability thereafter; (4) promoting expansion of regulatory T cells (Tregs) [48]; (5) by elevated inducible nitric oxide synthase (iNOS) expression. MDSCs use nitric oxide (NO) to nitrate both the T cell receptor, thus inhibiting T cell activation and the antitumor immune response [50]. Studies carried out in the late 1990s reported an increase of Gr + 1 cells in tumor free animals following the infusion of VEGF [51]. Likewise, in murine pancreatic ductal carcinoma model, infiltration of MDSCs was shown to be associated with high VEGF expression in the TME [52]. Furthermore, VEGF has been shown to increase accumulation of CD11b + Gr1 + cells in murine tumors via the activation of VEGFR2 and JAK/STAT3 pathway [53]. Therefore targeting VEGF/VEGFR would be a rational approach to block MDSC infiltration and function in tumors.

Sunitinib, a multitargeted RTK described earlier in this chapter, has been shown to reduce the overall MDSCs from spleen, bone marrow, and tumors across various models [27,54]. Similarly, a pan-VEGFR inhibitor, Axitinib, which blocks VEGFR1, R2, and R3, was shown to ablate the immune-suppressive functions in MDSC isolated from spleen and melanoma tumors [55]. MDSCs from axitinib-treated animals demonstrated the capacity to stimulate allogeneic T cells. Thus treatment with axitinib was reported to induce differentiation of monocytic MDSC toward an antigen-presenting phenotype [55]. In clinical settings, sunitinib treatment in renal cell carcinoma patients has demonstrated the reduction of MDSC population and consequently the activation of cytotoxic T cells [56]. Also, recent studies with anti-VEGF monoclonal antibody bevacizumab (Avastin) in NSCLC have shown decrease in granulocytic MDSC in the peripheral blood of the patients compared to non-Avastin arm [57].

10.8 Vascular endothelial growth factor and recruitment of tumor-associated macrophages (TAMs)

VEGF can recruit macrophages into the TME and promote development of TAMs. TAMs exhibit poor antigen-presenting capacity and thus reduce cytotoxic potential due to weak NO production [58]. TAMs could ablate T cell activation and proliferation by secreting various cytokine growth factors including IL-10, tumor growth factor beta, and prostaglandins [21,59]. Emerging studies suggest that high infiltration of macrophages in TME facilitates resistance to anti-VEGF therapy in murine ovarian cancer models. However, such resistance to VEGF therapy did not occur in macrophage-deficient mouse background, suggesting a strong interplay between macrophage function and VEGF signaling [60]. Clinically approved zoledronic acid, a bisphosphonate drug, is currently being used to treat bone metastasis and osteoporosis [61]. The drug is also shown to reduce macrophage populations. The treatment of zoledronic acid, in ovarian murine tumors, has demonstrated effectivity in overcoming resistance to anti-VEGF therapy and increased overall survival of the tumor-bearing mice [60,62]. Additionally, the poor outcome in clinical and preclinical studies carried out with bevacizumab in glioblastomas is associated with high infiltration of TAMs in TME [63,64]. One of the emerging hypotheses in recruiting immune-suppressive cells is the role of stromal-derived factor, SDF-1α (CXCL12) and its receptor, CXCR 4 [65]. Therefore targeting CXCL12/CXCR4 axis could be a potential strategy to block recruitment of TAMs in TME, which could help to overcome anti-VEGF resistance. Preclinical studies carried out with VEGFR and CXCR4 inhibitors have shown positive outcome in glioblastoma tumors [66]. Ongoing clinical trials with small-molecule CXCR4 inhibitor (AMD3100) in combination with bevacizumab are currently being carried out in high-grade glioma. Preliminary data suggest that combination was well tolerated and change in relevant biomarker(s) was consistent with VEGF and CXCR4 inhibition [67].

10.9 Conclusion

VEGF, in addition to its well-established role as a proangiogenic factor, participates in facilitating immune escape via complex cross-talks with various immune-subpopulations in the TME of cancerous tissue. In its role as regulator of "immune subpopulations," VEGF signaling modulates T cell proliferation that plays a role in tumor progression. On the one hand, overexpression of VEGF drives dysfunctional development of DCs and T cells, while on the other it can cause proliferation of T lymphocytes. This chapter showcases how VEGF/VEGFR signaling is crucial in regulation of immune subpopulations across effector T cells, Tregs, MDSCs, DCs, and TAMs.

As mentioned earlier, agents targeting VEGF/VEGFR axis can restore the function and enhance the immune fitness of tumors via (1) infiltration of effector T cells, (2) decreasing the number of immunosuppressive Tregs and TAMs, and (3) blocking the accumulation and immunosuppressive activity of MDSCs. Emerging data from combination trials with anti-VEGF therapy and targeted cancer therapies, including immune-modulatory drugs, point toward promising outcomes, although data across different studies are somewhat inconsistent. Due to these inconsistencies several questions remain unanswered such as the expression and kinetics of VEGF receptors on immune cells and their longitudinal alteration during the progression of cancer. Moreover, a major *caveat* of studies till date is the shortcomings of the murine models used thus far. First, these are not very clinically relevant as they do not represent the human carcinoma pathology. Second, these models cannot recapitulate the intact human immune system and tumor heterogeneity. Simultaneously, ex vivo immune studies using excised tumors are not representative of the in vivo biology and architecture. However, this is not to state that ex vivo tumor and in vivo animal murine models have not been helpful. These models have been successful in providing information on the role of VEGF/VEGFR signaling in invading immune surveillance within the tumor environment. Knowledge of the cross-talk between these signaling processes and immune cells in the TME has broadened our insights into cancer development. However, more studies are needed to decipher various pathways by which VEGF interacts with immune and cancer cells.

References

[1] Hanahan D, Weinberg RA. Hallmarks of cancer: the next generation. Cell 2011;144(5):646−74.

[2] Huang H, Langenkamp E, et al. VEGF suppresses T-lymphocyte infiltration in the tumor microenvironment through inhibition of NF-kappaB-induced endothelial activation. FASEB J 2015;29(1):227−38.

[3] Ohm JE, Gabrilovich DI, et al. VEGF inhibits T-cell development and may contribute to tumor-induced immune suppression. Blood 2003;101 (12):4878−86.

[4] Yang et al. Targeting VEGF/VEGFR to Modulate Antitumor Immunity. Frontiers in immunology 2018;9:978.

[5] Zhang L, Conejo-Garcia JR, et al. Intratumoral T cells, recurrence, and survival in epithelial ovarian cancer. N Engl J Med 2003;348(3): 203−213.

[6] Basu A, Hoerning A, et al. Cutting edge: vascular endothelial growth factor-mediated signaling in human CD45RO + CD4 + T cells promotes Akt and ERK activation and costimulates IFN-gamma production. J Immunol 2010;184(2):545−9.

[7] Zeng H, Dvorak HF, et al. Vascular permeability factor (VPF)/vascular endothelial growth factor (VEGF) peceptor-1 down-modulates VPF/VEGF receptor-2-mediated endothelial cell proliferation, but not migration, through phosphatidylinositol 3-kinase-dependent pathways. J Biol Chem 2001;276(29):26969−79.

[8] Ziogas AC, Gavalas NG, et al. VEGF directly suppresses activation of T cells from ovarian cancer patients and healthy individuals via VEGF receptor type 2. Int J Cancer 2012;130(4):857−64.

[9] Gavalas NG, Tsiatas M, et al. VEGF directly suppresses activation of T cells from ascites secondary to ovarian cancer via VEGF receptor type 2. Br J Cancer 2012;107(11):1869−75.

[10] Motz GT, Santoro SP, Wang LP, et al. Tumor endothelium FasL establishes a selective immune barrier promoting tolerance in tumors. Nat Med. 2014;20(6):607−615. doi:10.1038/nm.3541.

[11] Kaur S, Chang T, et al. CD47 signaling regulates the immunosuppressive activity of VEGF in T cells. J Immunol 2014;193(8):3914−24.

[12] Manzoni M, Rovati B, et al. Immunological effects of bevacizumab-based treatment in metastatic colorectal cancer. Oncology 2010; 79(3-4):187−96.

[13] Martino EC, Misso G, et al. Immune-modulating effects of bevacizumab in metastatic non-small-cell lung cancer patients. Cell Death Discov 2016;2:16025.

[14] Lanitis E, Irving M, et al. Targeting the tumor vasculature to enhance T cell activity. Curr Opin Immunol 2015;33:55−63.

[15] Ramjiawan RR, Griffioen AW, et al. Anti-angiogenesis for cancer revisited: is there a role for combinations with immunotherapy? Angiogenesis 2017;20(2):185−204.

[16] Dirkx AE, oude Egbrink MG, et al. Anti-angiogenesis therapy can overcome endothelial cell anergy and promote leukocyte-endothelium interactions and infiltration in tumors. FASEB J 2006;20(6):621−30.

[17] Tong RT, Boucher Y, et al. Vascular normalization by vascular endothelial growth factor receptor 2 blockade induces a pressure gradient across the vasculature and improves drug penetration in tumors. Cancer Res 2004;64(11):3731−6.

[18] Rodriguez-Ruiz ME, Garasa S, et al. Intercellular adhesion molecule-1 and vascular cell adhesion molecule are induced by ionizing radiation on lymphatic endothelium. Int J Radiat Oncol Biol Phys 2017;97(2):389−400.

[19] Bellone M, Calcinotto A. Ways to enhance lymphocyte trafficking into tumors and fitness of tumor infiltrating lymphocytes. Front Oncol 2013;3:231.

[20] Bettelli E, Carrier Y, et al. Reciprocal developmental pathways for the generation of pathogenic effector TH17 and regulatory T cells. Nature 2006;441(7090):235−8.

[21] Lapeyre-Prost A, Terme M, et al. Immunomodulatory activity of VEGF in cancer. Int Rev Cell Mol Biol 2017;330:295−342.

[22] Sun L, Xu G, et al. Clinicopathologic and prognostic significance of regulatory T cells in patients with hepatocellular carcinoma: a meta-analysis. Oncotarget 2017;8(24):39658−72.

[23] Wada J, Suzuki H, et al. The contribution of vascular endothelial growth factor to the induction of regulatory T-cells in malignant effusions. Anticancer Res 2009;29(3):881−8.

[24] Hansen W, Hutzler M, et al. Neuropilin 1 deficiency on CD4 + Foxp3 + regulatory T cells impairs mouse melanoma growth. J Exp Med 2012;209(11):2001−16.

[25] Adotevi O, Pere H, et al. A decrease of regulatory T cells correlates with overall survival after sunitinib-based antiangiogenic therapy in metastatic renal cancer patients. J Immunother 2010;33(9):991−8.

[26] Finke JH, Rini B, et al. Sunitinib reverses type-1 immune suppression and decreases T-regulatory cells in renal cell carcinoma patients. Clin Cancer Res 2008;14(20):6674−82.

[27] Ozao-Choy J, Ma G, et al. The novel role of tyrosine kinase inhibitor in the reversal of immune suppression and modulation of tumor microenvironment for immune-based cancer therapies. Cancer Res 2009;69(6):2514−22.

[28] Shah S, Lee C, et al. 5-Hydroxy-7-azaindolin-2-one, a novel hybrid of pyridinol and sunitinib: design, synthesis and cytotoxicity against cancer cells. Org Biomol Chem 2016;14(21):4829−41.

[29] Terme M, Pernot S, et al. VEGFA-VEGFR pathway blockade inhibits tumor-induced regulatory T-cell proliferation in colorectal cancer. Cancer Res 2013;73(2):539−49.

[30] Suzuki H, Onishi H, et al. VEGFR2 is selectively expressed by FOXP3high CD4 + Treg. Eur J Immunol 2010;40(1):197−203.

[31] Secondini C, Coquoz O, et al. Arginase inhibition suppresses lung metastasis in the 4T1 breast cancer model independently of the immunomodulatory and anti-metastatic effects of VEGFR-2 blockade. Oncoimmunology 2017;6(6):e1316437.

[32] Li B, Lalani AS, et al. Vascular endothelial growth factor blockade reduces intratumoral regulatory T cells and enhances the efficacy of a GM-CSF-secreting cancer immunotherapy. Clin Cancer Res 2006;12(22):6808−16.

[33] Tan JK, O'Neill HC. Maturation requirements for dendritic cells in T cell stimulation leading to tolerance versus immunity. J Leukoc Biol 2005;78(2):319−24.

[34] Wilson NS, El-Sukkari D, et al. Dendritic cells constitutively present self antigens in their immature state in vivo and regulate antigen presentation by controlling the rates of MHC class II synthesis and endocytosis. Blood 2004;103(6):2187−95.

[35] Hoehn GT, Stokland T, et al. Tnk1: a novel intracellular tyrosine kinase gene isolated from human umbilical cord blood CD34 + /Lin-/CD38-stem/progenitor cells. Oncogene 1996;12(4):903−13.

[36] Katoh O, Tauchi H, et al. Expression of the vascular endothelial growth factor (VEGF) receptor gene, KDR, in hematopoietic cells and inhibitory effect of VEGF on apoptotic cell death caused by ionizing radiation. Cancer Res 1995;55(23):5687−92.

[37] Hamrah P, Zhang Q, et al. Expression of vascular endothelial growth factor receptor-3 (VEGFR-3) in the conjunctiva − a potential link between lymphangiogenesis and leukocyte trafficking on the ocular surface. Adv Exp Med Biol 2002;506(Pt B):851−60.

[38] Mimura T, Amano S, et al. Expression of vascular endothelial growth factor C and vascular endothelial growth factor receptor 3 in corneal lymphangiogenesis. Exp Eye Res 2001;72(1):71−8.

[39] Oyama T, Ran S, et al. Vascular endothelial growth factor affects dendritic cell maturation through the inhibition of nuclear factor-kappa B activation in hemopoietic progenitor cells. J Immunol 1998;160(3):1224−32.

[40] Voron T, Marcheteau E, et al. Control of the immune response by pro-angiogenic factors. Front Oncol 2014;4:70.

[41] Dikov MM, Ohm JE, et al. Differential roles of vascular endothelial growth factor receptors 1 and 2 in dendritic cell differentiation. J Immunol 2005;174(1):215−22.

[42] Della Porta M, Danova M, et al. Dendritic cells and vascular endothelial growth factor in colorectal cancer: correlations with clinicobiological findings. Oncology 2005;68(2-3):276−84.

[43] Almand B, Resser JR, et al. Clinical significance of defective dendritic cell differentiation in cancer. Clin Cancer Res 2000;6(5):1755−66.

[44] Alfaro C, Suarez N, et al. Influence of bevacizumab, sunitinib and sorafenib as single agents or in combination on the inhibitory effects of VEGF on human dendritic cell differentiation from monocytes. Br J Cancer 2009;100(7):1111−19.

[45] Hipp MM, Hilf N, et al. Sorafenib but not sunitinib affects the induction of immune responses. J Clin Oncol 2007;25(18_suppl):3504.

[46] Osada T, Chong G, et al. The effect of anti-VEGF therapy on immature myeloid cell and dendritic cells in cancer patients. Cancer Immunol Immunother 2008;57(8):1115−24.

[47] Hodi FS, Lawrence D, et al. Bevacizumab plus ipilimumab in patients with metastatic melanoma. Cancer Immunol Res 2014;2(7):632−42.

[48] Gabrilovich DI, Ostrand-Rosenberg S, et al. Coordinated regulation of myeloid cells by tumours. Nat Rev Immunol 2012;12(4):253−68.

[49] Mandruzzato S, Solito S, et al. IL4Ralpha + myeloid-derived suppressor cell expansion in cancer patients. J Immunol 2009;182(10):6562−8.

[50] Redd PS, Ibrahim ML, et al. SETD1B activates iNOS expression in myeloid-derived suppressor cells. Cancer Res 2017;77(11):2834−43.

[51] Gabrilovich D, Ishida T, et al. Vascular endothelial growth factor inhibits the development of dendritic cells and dramatically affects the differentiation of multiple hematopoietic lineages in vivo. Blood 1998;92(11):4150−66.

[52] Karakhanova S, Link J, et al. Characterization of myeloid leukocytes and soluble mediators in pancreatic cancer: importance of myeloid-derived suppressor cells. Oncoimmunology 2015;4(4):e998519.

[53] Nefedova Y, Huang M, et al. Hyperactivation of STAT3 is involved in abnormal differentiation of dendritic cells in cancer. J Immunol 2004;172(1):464−74.

[54] Xin H, Zhang C, et al. Sunitinib inhibition of Stat3 induces renal cell carcinoma tumor cell apoptosis and reduces immunosuppressive cells. Cancer Res 2009;69(6):2506−13.

[55] Du Four S, Maenhout SK, et al. Axitinib increases the infiltration of immune cells and reduces the suppressive capacity of monocytic MDSCs in an intracranial mouse melanoma model. Oncoimmunology 2015;4(4):e998107.

[56] Ko JS, Zea AH, et al. Sunitinib mediates reversal of myeloid-derived suppressor cell accumulation in renal cell carcinoma patients. Clin Cancer Res 2009;15(6):2148−57.

[57] Koinis F, Vetsika EK, et al. Effect of first-line treatment on myeloid-derived suppressor cells' subpopulations in the peripheral blood of patients with non-small cell lung cancer. J Thorac Oncol 2016;11(8):1263−72.

[58] Linde N, Lederle W, et al. Vascular endothelial growth factor-induced skin carcinogenesis depends on recruitment and alternative activation of macrophages. J Pathol 2012;227(1):17−28.

[59] Balkwill F, Mantovani A. Inflammation and cancer: back to virchow? Lancet 2001;357(9255):539−45.

[60] Dalton HJ, Pradeep S, et al. Macrophages facilitate resistance to anti-VEGF therapy by altered VEGFR expression. Clin Cancer Res 2017;23(22):7034−46.

[61] Polascik TJ, Mouraviev V. Zoledronic acid in the management of metastatic bone disease. Ther Clin Risk Manag 2008;4(1):261−8.

[62] Rogers TL, Holen I. Tumour macrophages as potential targets of bisphosphonates. J Transl Med 2011;9:177.

[63] Gabrusiewicz K, Liu D, et al. Anti-vascular endothelial growth factor therapy-induced glioma invasion is associated with accumulation of Tie2-expressing monocytes. Oncotarget 2014;5(8):2208−20.

[64] Lu-Emerson C, Snuderl M, et al. Increase in tumor-associated macrophages after antiangiogenic therapy is associated with poor survival among patients with recurrent glioblastoma. Neuro Oncol 2013;15(8):1079−87.

[65] Hambardzumyan D, Gutmann DH, et al. The role of microglia and macrophages in glioma maintenance and progression. Nat Neurosci 2016;19(1):20−7.

[66] Pham K, Luo D, et al. VEGFR inhibitors upregulate CXCR4 in VEGF receptor-expressing glioblastoma in a TGFbetaR signaling-dependent manner. Cancer Lett 2015;360(1):60−7.

[67] Lee EQ, Duda DG, et al. Phase I and biomarker study of plerixafor and bevacizumab in recurrent high-grade glioma. Clin Cancer Res 2018;24(19):4643−9.

Radiation-induced endothelial dysfunction

Chapter 11

Human radiation exposures (occupational, medical, environmental, and radiation incidents) and vascular dysfunction

Andrea L. DiCarlo and Merriline M. Satyamitra

Radiation and Nuclear Countermeasures Program (RNCP), Division of Allergy, Immunology and Transplantation (DAIT), National Institute of Allergy and Infectious Diseases (NIAID), National Institutes of Health (NIH), Rockville, MD, United States

11.1 Introduction

"We are... sitting on the shoulders of giants... they raise us up, and by their great stature add to ours." (John of Salisbury, Metalogicon, 1159). We wish to recognize a 1997 publication that was pivotal in establishing the field's understanding of the impact of radiation exposure on endothelial cells (ECs) and the vasculature [1]. It is not our intent to cover the breadth and depth of information contained in that book, which brought together many experts in the field, but to update some of the scientific references, and focus more on aspects of vascular injuries in different human radiation exposure scenarios.

11.2 Background

11.2.1 Biology of the vasculature and endothelial cell dysfunction

The vascular endothelium is the largest organ in the body. As mentioned in Chapter 1, the endothelial surface area in an adult human is composed of approximately $1-6 \times 10^{13}$ cells, weighs about 1 kg, and covers a surface area of $4000-7000$ m^2 [2]. Normal EC functions include control of vasodilation and vascular growth and maintenance of hemostasis (Fig. 11.1). In addition, these cells have an important role in antioxidant, anti-inflammatory, anticoagulant, anti-atherosclerotic, and fibrinolytic effects [3]. When there is dysfunction in the vasculature, which could be caused by a number of external stressors (such as radiation) or disease states, these normally calming cells can contribute to an overall state of chaos—participating in inflammation and hemorrhage, production of collage, abnormal coagulation, apoptosis, and eventually multiorgan failure (MOF; Fig. 11.1). Juxtapositioned between the flowing blood and the vessel wall, the endothelium plays a role in vaso-regulation and homeostasis, selectively controlling the traffic of hematopoietic cells and nutrients. Descriptions of heterogeneity of the EC phenotype, differences in function, interactions, and cell communication are detailed in recent reviews [4,5].

Initially regarded as an inert "tube" lining of the circulation, providing a conduit for transport of various substances within the bloodstream, the dynamic and tissue-specific nature of the vascular endothelium continues to be of great interest, specifically in relation to its role in the pathologies of radiation injuries. Blood vessels typically have three layers, the tunica intima, media, and adventitia. Capillaries are the most radiosensitive because they have just a single layer of endothelium, the tunica intima, which is extremely susceptible radiation injury [7]. Dimitrievich and colleagues noted that this heightened sensitivity was not due to the abundance of capillaries compared with that of larger vessels. In comparing vessels of different calibers, capillary disruption, extravasation, micropetechiae formation, and inflammatory changes were more pronounced with microvascular irradiation than macrovascular irradiation [7]. A comparable feature in radiation injury to many organs [e.g., liver, heart, kidney, gastrointestinal (GI) tract, skin, brain, and lungs] is the tissues' concentration of highly radiosensitive capillaries lined with ECs.

Endothelial Signaling in Vascular Dysfunction and Disease. DOI: https://doi.org/10.1016/B978-0-12-816196-8.00006-0

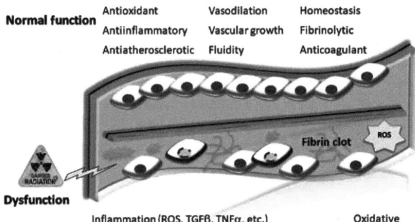

Vascular endothelial cell function

FIGURE 11.1 Overview of radiation injuries to the vascular endothelial cells. *ROS*, reactive oxygen species.
From Satyamitra MM, DiCarlo AL, Taliaferro L. Understanding the pathophysiology and challenges of development of medical counter-measures for radiation-induced vascular/endo-thelial cell injuries: report of a NIAID workshop, August 20, 2015. Radiat Res. 2016 Aug;186(2):99–111. https://doi.org/10.1667/RR14436.1 © 2019 Radiation Research Society [6].

Radiation damage to various organs/tissues is accompanied by perturbations of the vascular endothelium. Just 4 years after the discovery of X-rays, Gassmann [8] published his findings on injury to the endothelium in irradiated skin. Historically, following irradiation, different tissues present with histological evidence of vascular/endothelial damage, lending credence to the pivotal role of vascular injury in tissue toxicity [9]. EC damage and subsequent progressive changes in the vasculature can contribute to chronic lesions in lung, liver, kidney [10,11], heart, and brain [12]. Therefore the radiobiological response of the vascular network, which is still incompletely understood, is of major importance for the medical management of radiation injury.

Several laboratories are studying the role of the vascular niche and ECs in radiation-induced injury and repair. Because hematopoiesis occurs in the bone marrow niche [13], damage to this compartment can impact circulating immune cells. The niche involves a complex interplay of bone marrow, immune, adipocytes, and endothelial-derived cells. ECs are known to have a role in both angiogenesis as well as organ regeneration and repair. The vasculature that feeds different areas of the body is composed of ECs that have distinct, geographical properties unique to the tissue that they perfuse, such as liver, bone marrow, lung, and brain. Within specialized vascular niches such as the bone marrow, vascular ECs interact with other cells via cell-to-cell signaling and provide additional signaling to other niche components. As a result, the blood vessels in each organ system can have differential radiation sensitivities [14]. The vasculature can secrete specific factors at the sites of tissue injury that can influence tissue regeneration, as seen in the lung [15] and liver [16]. This effect has also been noted in other tissues, including pancreatic islet cells [17], adipose stem cells [18], bone [19,20], and cardiac endothelium [21].

Because ECs and the vasculature are a tissue central to all parts of the body, the effects of radiation injury to the vasculature extend beyond the heart and cerebrovascular system. Because the primary target of radiation injury to the vasculature is the EC, radiation-induced apoptosis of ECs and vascular cells contributes to dysfunction and plays an important role in the pathophysiology of acute radiation syndrome (e.g., hematopoietic and GI) and in delayed effects of acute radiation exposure [e.g., lung, renal, and central nervous system (CNS)]. For example, endothelial dysfunction, marked by a loss in thromboresistance and increased inflammatory markers, also inhibits the recovery of the villus epithelium and leads to the breakdown of the GI epithelial barrier [22]. The response of the microvasculature to radiation can contribute to the initiation, progression, and maintenance of damage, both to the vasculature as well as the organs and tissues associated with it. The acute phase of damage occurs within hours to weeks following exposure and is characterized by endothelial swelling, vascular permeability and edema, lymphocyte adhesion and infiltration, and apoptosis [23]. Denudation of ECs is followed by loss of barrier integrity and changes in permeability of the vessel. This phase is often accompanied by inflammation, and migration of leukocytes and platelets as well as cellular debris with fibrin deposition and micro-thrombi formation, and edema (Fig. 11.1). In terms of chronic effects, late vascular injury can occur weeks to months following irradiation and can lead to irreversible damage. Later vascular effects include capillary collapse, thickening of basement membranes, scarring and fibrosis, telangiectasia, and a loss of clonogenic capacity [24]. A common feature in late tissue damage, fibrosis has its roots in the primary damage to ECs. Researchers have attempted to elucidate the differences in the mechanisms of damage to the normal tissue, including the endothelium, in response to different qualities of radiation, protons and gamma rays. Radiation qualities elicit varying responses in EC injury, vascular response and remodeling. In fact, protons and gamma irradiation produced contrasting effects with regards to angiogenesis [25]. These studies present a very insightful overview of response and a means to modulate vascular injury.

ECs undergo radiation-induced death via a variety of mechanisms, such as endothelial apoptosis or senescence. The role of EC senescence in radiation-induced cardiovascular disease (CVD) is relatively undetermined; however, several studies indicate a causal role between EC senescence and CVD. Senescent ECs are incapable of regenerating new cells to maintain the homeostasis of vasculatures and repair damaged blood vessels, which may contribute to the decreased density of cardiac capillaries and small coronary arterioles and to the accelerated atherosclerosis of large blood vessels, including rodent and human coronary arteries [26–28]. Damage to the endothelium plays a key role in development of late radiation-induced CNS injury, and endothelial dysfunction has also been implicated in radiation-induced delayed lung, renal, xerostomia, and spinal cord injuries.

Endothelial changes are involved in early radiation toxicity and may sustain processes leading to fibrosis, and further clinical complications during and following radiation therapy [27]. Clinical outcomes of vascular injury to an organ are specific to its function, the extent of injury, and time elapsed since that injury. Therefore there is a need for greater understanding of the pathophysiology of radiation effects on the micro- and macro-vasculature, particularly to guide strategies to mitigate delayed deleterious effects. These effects could include those observed in patients undergoing clinical radiation procedures, as well as impacts on individuals exposed to radiation during a radiological or nuclear incident (e.g., an accident or attack).

11.3 Human radiation exposures

There are several instances where the human body could be exposed to radiation in a manner that could lead to vascular injuries. These exposures could include environmental, during medical procedures, occupational, or from radiation incidents such as accidental industrial or criticality exposures, large-scale radiation accidents, or nuclear detonations.

11.3.1 Environmental radiation exposures

Consequent to radiation being an integral part of our environment, natural radioactive sources in the soil, water, and air contribute to human exposure to ionizing radiations. Background radiation in the environment is ubiquitous. The sources of environmental radiation are (1) space or cosmogenic radiation including tritium, beryllium-7, carbon-14, and sodium-22 and (2) terrestrial radiation that originates on Earth such as potassium-40, rubidium-87, and radionuclide series resulting from uranium and thorium decay. Radon, a terrestrial radionuclide and important source of background radiation is also a byproduct of uranium and thorium decay.[1] For example, Kerala, a state in India, has a very high rate of background radon exposure due to sands containing thorium, with exposure levels at 70% higher than the global average (70 mSv [29]). More recently, radiation exposures in the areas around nuclear facilities (e.g., the Hanford and Savannah River Sites), nuclear accidents (e.g., Chernobyl, Three Mile Island, Fukushima), and areas contaminated from atmospheric testing, fallout, or other environmental release of radiation (e.g., populations near the Techa River in the Southern Ural region, the Mayak Plutonium Plant, and the Marshall Islands) have been considered for the impact of these environmental exposures [30]. Epidemiological data from extensive research provide no evidence for detrimental health effects below 100 mSv [31] but at higher doses (>0.5 Gy), mortality due to radiation-induced CVDs and pathologies appears to be increased [32]. Inhabitants of nuclear contaminated territories have also been observed to develop other diseases. For example, in 1993–2003 in Belarus, there was a significant increase in blood-based diseases among men and women, including high blood pressure and acute myocardial infarction (in women aged 35–39 years and 55–59 years), cerebrovascular diseases, and atherosclerosis [33]. In addition, young girls (aged 10–15 years) living in regions polluted by ^{137}Cs had substantial impairments in leg blood flow [33].

---

1. https://hps.org/documents/environmental_radiation_fact_sheet.pdf.

11.3.2 Medical radiation exposures

There are a number of well-documented vascular complications that can result from exposure during radiological procedures in the clinic. These include unwanted normal tissue injuries involving the vasculature resulting from standard X-ray therapies such as total body irradiation (TBI) for marrow ablation prior to transplant and focused cancer irradiations [34–39], as well as gamma-knife [40], proton beam, stereotactic radiosurgery [41–43], radioactive iodine use [44], or brachytherapy treatments [45]. These concerns are amplified in children, as they have a longer period of time post-irradiation, during which complications could surface [34,46,47]. Radiation delivered therapeutically to head and neck cancer patients, and to a lesser extent breast cancer and lymphoma patients can be especially problematic, given proximity to the carotid arteries, and their susceptibility to stenosis. For example, a literature review of patients receiving radiotherapy to the head and neck found that the incidence of stenosis of the carotid artery post-irradiation was between 18% and 38% in irradiated patients, compared to 0%–9.2% in the control population [48]. Radiation-induced carotid artery stenosis and atherosclerosis can subsequently lead to a variety of late complications, including strokes and transient ischemia as well as heart attacks [49–51].

Similar vascular complications have been noted in patients undergoing radiotherapy for brain tumors. In one case report, physicians found that patients previously treated with brain irradiation displayed arterial complications both in the short-term and many years posttherapy. These issues included diffuse arteritis and vessel occlusion [52]. However, it is important to note that vascular dysfunction (e.g., changes in tone, vessel stiffness, impaired angiogenesis, etc.) occurs in normal aging in the absence of radiation exposure. These effects are caused by EC senescence and have been linked to changes in structure, inflammation, and oxidative stress [53]. Molecular mechanisms involving NAD + deficiencies in ECs, among other known pathways, have also been linked to age-related vascular complications [54]. These kinds of radiation-induced vascular injuries have also been observed in animal models of exposure [55,56]. Another specific EC-related complication resulting from radiation exposure is the development of late cardiovascular injuries, especially following left-breast radiotherapy treatments. This is because many prevalent exposures involving irradiation of the thorax also lead to cardiac irradiation. Complications from these exposures have been estimated to increase mortality from heart disease by up to 27% [57], with the estimated risk of heart dysfunction even higher for left-sided breast cancer treatments [58]. There is also evidence that heart attack risk is higher in patients receiving left-breast treatments as opposed to right breast [59]. These known complications are reviewed in detail elsewhere [60].

In addition to the obvious exposures attributable to cancer therapies, there are other procedures involving radiation exposure with contributions that can often be quite high in the local microenvironment. These procedures include the use of radiation to prevent restenosis in coronary arteries that have been stented [61–63]. However, retrospective analyses of long-term cardiovascular impacts of these kinds of treatments in patients suggest no adverse outcomes from these exposures [64]. An extensive review of the available literature on the influence of radiation exposure resulting from fluoroscopy procedures, during which X-rays are used continuously to image a tissue over time, yielded only a few studies correlating these kinds of medical procedures with vascular complications. In one, researchers noted that much of the injury to the skin that is sometimes noted after fluoroscopy can be attributed to increased permeability of the vasculature [65]. They also saw a reduction in the density of the vasculature in the skin at 12 weeks post-irradiation in some patients, which might be attributable to necrosis. In addition, a large retrospective study involving more than 77,000 patients in Canada and the United States investigated a possible correlation between radiation exposure during diagnostic fluoroscopy and circulatory disease [66]. Those researchers found that there was evidence for a connection at radiation exposure below 500 cGy, as noted by a trend toward increased incidence of various forms of circulatory dysfunction. Finally, the possibility exists that for radiation exposures associated with diagnostic fluoroscopy, especially those in routine use to guide femoral artery procedures [67,68], unintended vascular dysfunction could result.

11.3.3 Occupational radiation exposures (medical workers, nuclear industry)

In the past, when efforts by the United States and other countries were focused on the production of components to be used in nuclear weapons, there were a number of workers that were exposed to doses of radiation that although not high enough to lead to immediate health effects, were found to be involved in the development of late cardiovascular complications. These occupational exposures have been documented in places like the Mayak Plutonium Plant in the Russian Federation as well as in the United States. CVD incidence has been shown to be statistically higher among the 18,763 workers employed at the Mayak Production Association, who were exposed to doses >0.025 Gy (Fig. 11.2) [69]. Findings of increased risk in nuclear workers as well as other career professionals were also noted in the UK National Registry for Radiation Workers (NRRW) cohort [70], radiation workers in France, the United Kingdom, and the United States [71], the Canadian Registry of Radiation Workers [72], and the 15-country Nuclear Industry Workers study [73].

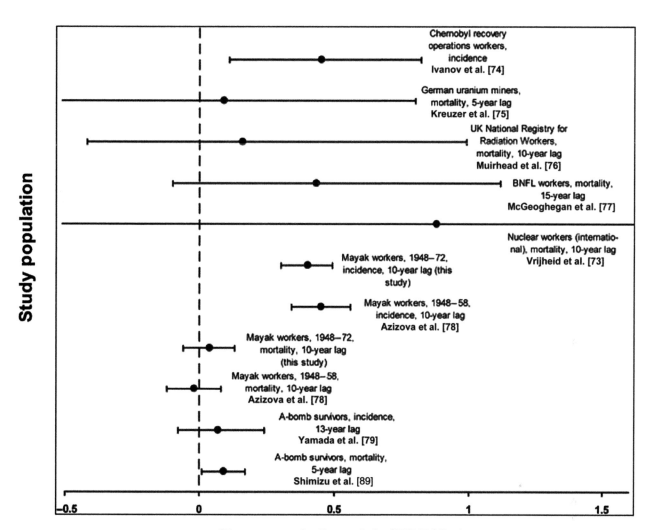

Excess relative risk (ERR/Gy)

FIGURE 11.2 Comparison of risk estimates [excess relative risk (ERR)/Gy—*x*-axis] of CVD incidence and mortality after external radiation exposure. Shown are data collected from Mayak workers, various studies of radiation workers, Chernobyl recovery operation workers, German uranium miners, and Japanese A-bomb survivors. Bars indicate a 95% CI for all, but the British Nuclear Fuel Limited (BNFL) workers study, for which 90% CI is indicated. *CI*, Confidence interval; *MCM*, medical countermeasure. *Reprinted by permission from Azizova TV, Muirhead CR, Moseeva MB, Grigoryeva ES, Sumina MV, O'Hagan J, et al. Cerebrovascular diseases in nuclear workers first employed at the Mayak PA in 1948–1972. Radiat Environ Biophys. 2011;50(4):539. Springer Nature. Ivanov et al. [74] Kreuzer et al. [75]Muirhead et al. [76] McGeoghegan et al. [77] Vrijheid et al. [73] Azizova et al. [78] Yamada et al. [79] Shimizu et al. [89].*

Groups that are occupationally exposed to ionizing radiation, in medicine (radiologists and radiological technicians), nuclear medicine, specialists (dentists and hygienists), industry (nuclear and radiochemical industries, as well as other professions where industrial radiography is used to assess the soundness of materials and structures), defense, research, and even transportation (airline crews as well as workers involved in the maintenance or operation of nuclear-powered vessels) are recognized to be at higher health risks. The type of ionizing radiation exposure varies among occupations, with differing contributions from photons, neutrons, and α- and β-particles [30]. More than two million healthcare workers are exposed to radiation on a daily basis; therefore much attention has been given recently to the analysis of health risks in populations exposed to low or above background radiation doses [80].

11.3.4 Industrial/criticality accidental exposures

The current geopolitical environment heightens the possibility of a mass casualty radiation public health emergency involving an improvised nuclear device or weaponized radiological material. Further, radiological accidents at

Chernobyl (1986), Goiania (1987), Tokaimura (1999), and Fukushima-Daiichi (2011) underscore the pressing need to better understand complex injuries arising from unanticipated radiation exposure. Although the exposures in Goiania, Brazil in 1987 involved four deaths (with over 249 people found to be internally contaminated with ^{137}Cs), one small clinical retrospective study involving individuals exposed during the accident found no correlation between systemic arterial hypertension in victims of the incident and people in the general population [81]. Many of the ^{137}Cs exposures involved the skin, with vascular endotheliitis being the most remarkable vascular complication noted [82]. The same IAEA report found that cardiovascular abnormalities in at least one patient manifested in moderate to light hypertension.

With regards to the industrial criticality accident at the Tokaimura, Japan facility in 1999, two deaths resulted among three exposed workers. Because of the large doses of radiation involved, there were many vascular effects noted in the patients. For example, severe endothelial apoptosis and cell atrophy were noted in the patient that survived for 83 days [83]. Following the Tokaimura incident, Akashi [84] discussed the possible role of inflammation and hemorrhage in MOF. In an elegant review of 110 cases histories of radiation accidents spanning 1945 through 2000, the authors analyzed MOF following TBI and stated that "It also became clear that the symptomatology of organ system involvement could be traced not only to the pathophysiology of the rapidly turning over cell renewal systems but—of equal or more importance—to the vascular system and specifically, to the endothelial components" [85]. Perhaps one of the most well-known accidental exposure incidents was that of Chernobyl. Of the 28 people who died within 98 days of the Chernobyl criticality incident, deaths were attributed to skin, GI, and lung reactions, but most deaths were characterized by circulatory problems with high incidence of edema and focal hemorrhages [31].

11.3.5 Large-scale nuclear detonation exposures (e.g., Hiroshima/Nagasaki)

The vascular/endothelium is intimately linked to both early and late radiation pathologies affecting several major organs. For example, in the aftermath of the Hiroshima and Nagasaki atomic bomb casualties, it is estimated that 60%−100% of the casualties had evidence of hemorrhage at death [86]. In survivors that developed late injury, the vascular tissue was involved in tissue damage, as evidenced from studies on cerebrovascular disease [87], CVDs [88,89], chronic kidney disease [90], and GI diseases [91]. Studies from the Life Span Study of the Japanese atomic bomb survivors demonstrated a statistically increased trend in the incidence of cerebrovascular disease and ensuing mortality [92−98].

Criticality accidents and large-scale radiological incidents, although rare, have been studied for vascular outcomes, especially the role of cardiovascular shock in mortality [99]. A retrospective study was conducted utilizing publicly available autopsy reports, to assess the contribution of radiation-induced bleeding to overall mortality following unintentional radiation exposures, which included nuclear detonations and criticality accidents, clinical overexposures, and industrial incidents [100]. Although focused primarily on the role of thrombocytopenia in radiation deaths, the study found a high incidence of hemorrhage and vascular damage noted in nearly all of the 42 patient files that were reviewed. The vascular endothelium is an organ central to all tissues and is important in acute and delayed radiation injuries. Therefore research focused on the mechanisms of injury and the development of treatment approaches to mitigate these kinds of injuries is necessary.

11.4 Preclinical models to study radiation responses of the vascular endothelium

For any preclinical model of vascular injury selected for study, the mechanism of action of the mitigation/therapy should closely resemble the action of the drug in humans. The most studied model for evaluating radiation injury to the vasculature is the mouse. A number of strains have been used to evaluate direct insult to the vasculature as well as late effect in major organs where the vascular/endothelial dysfunction mediates the delayed injury. Unfortunately, mouse models do not exhibit hemorrhage, while fatalities of the atom bomb incident presented with widespread hemorrhage [101], as did patients from radiation accidents in Norway [102] and Brazil [103]. Given these limitations in the rodent model, work has been done in ferrets and minipigs, as well as other mammals with radiation responses more closely aligned with humans. Few large mammals are currently being studied as models for radiation-induced vascular/endothelial injury, although a study in Yucatan minipigs [104], and anecdotal data for irradiated canines and guinea pigs [105] provide some background information for these species. This paucity of large animal models in this area underscores the need for robust research to identify appropriate models to investigate vascular damage.

11.5 Mechanisms of radiation-induced vascular injuries and biomarkers of damage

There are several excellent references on possible mechanistic drivers of radiation-induced vascular injuries [96,106,107], which can help guide identification of targets for drug development. Pharmaceutical strategies to treat vascular injury often assume intervention in one or more of the pathways from onset of insult to fibrosis/major morbidities. Amid this complex multitude of cross-talk and interactions, key processes and pathways emerge as potential targets. Endothelial apoptosis is the heralding event in injury to the vasculature, accompanied by production of short and long-lived reactive oxygen species (ROS) and inflammation. Although early endothelial and accompanying events precede late tissue damage such as fibrosis, the critical roles of key players and pathways involved in the late effects are as yet unclear.

11.5.1 Apoptosis and senescence of endothelial cells

Apart from the initial DNA damaging effect on the endothelium, radiation also promotes ROS, tissue and systemic senescence [108,109]. Ionizing radiation induces redox-sensitive transcription factors, increases oxidative stress, and triggers the release of damage-associated molecular pathways, with severe consequences to functions such as vascular tone, barrier integrity, fibrogenic dysfunction, inflammation, and coagulation pathways [110]. Similarly, it is well documented that ECs and endothelial progenitor cells undergo senescence when exposed to radiation in vitro [109,111−120] and in cardiac tissue in vivo [121]. Senescent ECs can inhibit the recovery of the vasculature by secreting senescence-associated secretory phenotype that interferes in the angiogenic activity of the endothelial progenitor cells and collectively increases inflammatory cytokines, adhesion molecules, and decrease the levels of thrombomodulin, resulting in prothrombotic and proatherogenic system [118,119,122,123]. Oxidative stress upregulates numerous pathways pertinent to vascular disease, including matrix metalloproteinases, adhesion molecules, pro-inflammatory cytokines, and smooth muscle cell proliferation and apoptosis, while inactivating vasculo-protective NO. Therefore vascular cells can suffer from persistent oxidative stress and inflammation subsequent to irradiation.

11.5.2 Early and delayed expression of biomarkers of vascular/endothelial injury

Although the pathobiology of vascular injury is not completely understood, an abundance of data suggests biomarkers that describe specific aspects of the damage, which can lead to a better understanding of the late effects of radiation. These biomarkers can be signal transducers (e.g., cytokines and chemokines, and exosomes), cells, and cellular products, as well as signaling pathways (e.g., apoptosis, oxidative stress, inflammation, and senolytic). In addition, given that ECs appear to retain a memory of previous exposure, it is likely that there are also epigenetic signals involved. The pathophysiology of aging is thought to be extremely similar to radiation-induced late effects, and monitoring aging-related markers is one potential approach to quantify radiation-induced vascular injury. Tissue damage to the vasculature is an evolving process that can take from seconds to years to manifest in humans; hence some of the processes can take more time than others to reach a level of detection.

The presence of systemic signalers of processes such as the oxidative stress pathway and inflammation is attractive since there is a vast potential for noninvasive or minimally invasive source of biomarkers for early and delayed progression of injury. Published literature is available documenting possible surrogate biomarkers that are predictive of late radiation vascular injuries. For example, fibronectin produced by the endothelium after irradiated appears to contribute to late brain injuries [124], and a marker in residence in the hippocampus predicts late neurocognitive complications associated with vascular damage [125]. In addition, microarray studies have identified other candidate markers involved in microvascular damage [126]. Specific panels of growth factors, matrix metalloproteins, and DNA damage markers have been shown not only to predict damage to a specific major organ, but in some cases are predictive of the progression of the injury (lethality vs survival). Biomarkers may also function as predictors of efficacy of a medical countermeasure (MCM) in mitigating damage, ameliorating major morbidities, and rescuing the organism from lethality. The presence of fibrin in blood can also be an indicator of delayed vascular injury. If fibrin decreases in the organ following antifibrin therapy, this indicates the treatment is efficacious. However, the kinetics of the biomarkers, as well as their multiphasic manifestation, and intimate cross-talk with other signal transducers pose a challenge.

The abundant expression of nitric oxide, thrombomodulin, prostacyclin, endothelin, platelet-activating factor, and cell adhesion molecules by ECs suggest possible signaling cascades that may play a role in mediating anti-inflammatory, antifibrogenic, anticoagulant, antiatherogenic, and vasoactive effects [106]. In addition, radiation-induced damage to the vascular endothelium could shift the balance of pro- and anti-inflammatory cytokines; release a burst of ROS; dysregulate glycolysis, lipid metabolic pathways, or angiogenesis; disrupt telomere function; and/or perturb immunity homeostasis. Like other known off-target effects of radiation (e.g., bystander effect and paracrine and clastogenic effects), these acute effects may trigger long-term vascular dysfunction if unchecked or inadequately compensated for. Activation of the cell death pathway resulting in EC apoptosis is regarded as the initial step in atherosclerosis [127] and is supported by the finding that pro-atherosclerotic factors such as angiotensin II, low-density lipoproteins, ROS, glucose, and inflammatory cytokines induce EC apoptosis. Bombeli et al. demonstrated that apoptotic ECs become procoagulant, with increased expression of phosphatidylserine, and loss of anticoagulant membrane components [128].

Finally, damage caused by irradiation can potentially be monitored in vivo by harnessing signals produced by circulating ECs. Mature ECs are released into the circulation due to the normal turnover and renewal of EC lining of the vasculature. Al-Massarani et al. [129] demonstrated a reduction in circulating ECs following irradiation with acute and fractionated doses of gamma rays in rat peripheral blood, with an incomplete recovery 2 months following exposure. These markers of vascular injury were also observed in patients undergoing chemotherapy [130,131].

11.6   Treatments to address radiation-induced vascular dysfunction

Radiation exposure can cause significant injury to the vasculature in the bone marrow [132]. Previously, infusions of ECs into irradiated mice had been shown to improve survival and lead to renewal of hematopoietic stem and progenitor cells [133,134]. Mouse brain and fetal-derived ECs were both able to rescue the animals; however, mesenchymal cells were not. In addition, compounds that influence and support ECs represent potential approaches to increase tissue survival after radiation injury. Several studies suggest that it is possible to produce organ-specific ECs [135–138]. These ECs expand hematopoietic stem and progenitor cells by deploying angiocrine growth factors [139] and restore hematopoietic recovery in myelo-suppressed mice. ECs can also be generated from embryonic or induced pluripotent stem cells [140]. The problem with cells derived from these sources, however, is that they can still retain some of their fetal characteristics and can be unstable. To address this, mature amniotic cells have been reprogrammed into ECs, with success [141].

Several noncellular approaches, currently under study as mitigators of radiation injury, have been shown to have an impact on the vasculature. Radioprotectants that scavenge free radicals like aminothiols, and sulfhydryl-containing angiotensin-converting enzyme inhibitors such as captopril can attenuate radiation vascular damage [142], statins can limit radiation-induced vascular dysfunction in the skin [143], and dietary inhibitors can modify EC responses in a rodent model of radiation-induced aorta dysfunction [144]. In addition, several pathways targeted by clinically available products are believed to be involved in vascular disease as well as radiation injuries involving the vasculature. Among these are the renin-angiotensin system [145], and the PKCδ pathway [146].

11.7   Conclusion and future directions in vascular therapeutics

Vascular injury has been a recognized complication of exposure since the discovery of radiation; however, research on the role of the vasculature in radiation injury specific to vascular/EC damage is inadequate, and it is clear that research in this area is too premature to consider standardization of models to study the phenomenon. In addition, a better understanding of the cross-talk between ECs and other niche cells needs to be achieved. More studies are required to more fully understand the pathology of vascular damage and modulation of signaling, as well as development of approaches to treat the injuries. Tools continue to become available, including animal models [mouse (e.g., novel genetic knock-in and knock-outs), rabbit, ferret, minipig, and nonhuman primate], state-of-the-art technologies that allow for advanced imaging, and new intervention strategies (e.g., vascular gene editing, and clustered regularly interspaced short palindromic repeats technologies for CVD). If ongoing work continues to yield promising data documenting protective effects gained from modification of these pathways, then drugs targeting the endothelium could represent an important approach to treat clinical complications resulting from radiation exposure. Furthermore, as the vasculature represents a ubiquitous system in the body that likely plays a lead role in radiation-induced multiorgan dysfunction, both basic studies investigating complex intracellular mechanism and cross-talk of ECs with cells in close proximity, and translational work to develop drugs based on the vasculature could provide new approaches for use as medical countermeasures to treat radiation injuries resulting from a public health emergency.

References

[1] The radiation biology of the vascular endothelium. Boca Raton, FL: CRC Press, LLC; 1997.

[2] Sumpio BE, Riley JT, Dardik A. Cells in focus: endothelial cell. Int J Biochem Cell Biol 2002;34(12):1508−12.

[3] Pate M, Damarla V, Chi DS, Negi S, Krishnaswamy G. Endothelial cell biology: role in the inflammatory response. Adv Clin Chem 2010;52:109−30.

[4] Sandoo A, van Zanten JJCSV, Metsios GS, Carroll D, Kitas GD. The endothelium and its role in regulating vascular tone. Open Cardiovasc Med J 2010;4:302−12.

[5] Pugsley MK, Tabrizchi R. The vascular system: an overview of structure and function. J Pharmacol Methods 2000;44(2):333−40.

[6] Satyamitra MM, DiCarlo AL, Taliaferro L. Understanding the pathophysiology and challenges of development of medical countermeasures for radiation-induced vascular/endothelial cell injuries: report of a NIAID workshop, August 20, 2015. Radiat Res. 2016;186(2):99−111. https://doi. org/10.1667/RR14436.1.

[7] Dimitrievich GS, Fischer-Dzoga K, Griem ML. Radiosensitivity of vascular tissue. I. Differential radiosensitivity of capillaries: a quantitative in vivo study. Radiat Res 1984;99(3):511−35.

[8] Gassmann A. Zur histologie der rontgenulcera. Fortschr a d Geb d Röntgenstrahlen 1899;2:199.

[9] Rubin DB, Griem ML. The histopathology of the irradiated endothelium. In: Rubin DB, editor. The radiation biology of the vascular endothelium. Boca Raton, FL: CRC Press LLC; 1998. p. 13−38.

[10] Jaenke RS, Robbins ME, Bywaters T, Whitehouse E, Rezvani M, Hopewell JW. Capillary endothelium. Target site of renal radiation injury. Lab Invest 1993;68(4):396−405.

[11] Juncos LI, Cornejo JC, Gomes J, Baigorria S, Juncos LA. Abnormal endothelium-dependent responses in early radiation nephropathy. Hypertension 1997;30(3 Pt 2):672−6.

[12] Lyubimova N, Hopewell JW. Experimental evidence to support the hypothesis that damage to vascular endothelium plays the primary role in the development of late radiation-induced CNS injury. Br J Radiol 2004;77(918):488−92.

[13] Mendelson A, Frenette PS. Hematopoietic stem cell niche maintenance during homeostasis and regeneration. Nat Med 2014;20(8):833−46.

[14] Nolan DJ, Ginsberg M, Israely E, Palikuqi B, Poulos MG, James D, et al. Molecular signatures of tissue-specific microvascular endothelial cell heterogeneity in organ maintenance and regeneration. Dev Cell 2013;26(2):204−19.

[15] Ding BS, Nolan DJ, Guo P, Babazadeh AO, Cao Z, Rosenwaks Z, et al. Endothelial-derived angiocrine signals induce and sustain regenerative lung alveolarization. Cell. 2011;147(3):539−53.

[16] Ding BS, Cao Z, Lis R, Nolan DJ, Guo P, Simons M, et al. Divergent angiocrine signals from vascular niche balance liver regeneration and fibrosis. Nature. 2014;505(7481):97−102.

[17] Kao DI, Lacko LA, Ding BS, Huang C, Phung K, Gu G, et al. Endothelial cells control pancreatic cell fate at defined stages through EGFL7 signaling. Stem Cell Rep 2015;4(2):181−9.

[18] Tang W, Zeve D, Suh JM, Bosnakovski D, Kyba M, Hammer RE, et al. White fat progenitor cells reside in the adipose vasculature. Science. 2008;322(5901):583−6.

[19] Ramasamy SK, Kusumbe AP, Wang L, Adams RH. Endothelial Notch activity promotes angiogenesis and osteogenesis in bone. Nature. 2014;507(7492):376−80.

[20] Kusumbe AP, Ramasamy SK, Adams RH. Coupling of angiogenesis and osteogenesis by a specific vessel subtype in bone. Nature. 2014;507 (7492):323−8.

[21] Hedhli N, Huang Q, Kalinowski A, Palmeri M, Hu X, Russell RR, et al. Endothelium-derived neuregulin protects the heart against ischemic injury. Circulation 2011;123(20):2254−62.

[22] Wang J, Boerma M, Fu Q, Hauer-Jensen M. Significance of endothelial dysfunction in the pathogenesis of early and delayed radiation enteropathy. World J Gastroenterol 2007;13(22):3047−55.

[23] Paris F, Fuks Z, Kang A, Capodieci P, Juan G, Ehleiter D, et al. Endothelial apoptosis as the primary lesion initiating intestinal radiation damage in mice. Science. 2001;293(5528):293−7.

[24] Milliat F, Sabourin JC, Tarlet G, Holler V, Deutsch E, Buard V, et al. Essential role of plasminogen activator inhibitor type-1 in radiation enteropathy. Am J Pathol 2008;172(3):691−701.

[25] Girdhani S, Sachs R, Hlatky L. Biological effects of proton radiation: what we know and don't know. Radiat Res 2013;179(3):257−72.

[26] Baker JE, Fish BL, Su J, Haworth ST, Strande JL, Komorowski RA, et al. 10 Gy total body irradiation increases risk of coronary sclerosis, degeneration of heart structure and function in a rat model. Int J Radiat Biol 2009;85(12):1089−100.

[27] Stewart FA, Heeneman S, Te Poele J, Kruse J, Russell NS, Gijbels M, et al. Ionizing radiation accelerates the development of atherosclerotic lesions in ApoE − / − mice and predisposes to an inflammatory plaque phenotype prone to hemorrhage. Am J Pathol 2006;168(2):649−58.

[28] Taunk NK, Haffty BG, Kostis JB, Goyal S. Radiation-induced heart disease: pathologic abnormalities and putative mechanisms. Front Oncol 2015;5:39.

[29] UNSCEAR. Sources, effects and risks of ionizing radiation. In: United Nations Scientific Committee on the Efffects of Atomic Radiation [Internet]. New York, NY: United Nations; 1988. p. 49−88.

[30] National Research Council. Health risks from exposure to low levels of ionizing radiation: BEIR VII phase 2. Washington, DC: The National Academies Press; 2006.

[31] Ionizing Radiation: Sources and Effects. Appendix G: early effects in man of high doses of radiation. New York, NY: United Nations Scientific Committee on the Effects of Atomic Radiation (UNSCEAR); 1988.

[32] Baselet B, Rombouts C, Benotmane AM, Baatout S, Aerts A. Cardiovascular diseases related to ionizing radiation: the risk of low-dose exposure (review). Int J Mol Med 2016;38(6):1623−41.

[33] Yablokov AV, Nesterenko VB, Nesterenko AV. 15. Consequences of the chernobyl catastrophe for public health and the environment 23 years later. Ann N Y Acad Sci 2009;1181:318−26.

[34] Maher CO, Raffel C. Early vasculopathy following radiation in a child with medulloblastoma. Pediatr Neurosurg 2000;32(5):255−8.

[35] Mulrooney DA, Blaes AH, Duprez D. Vascular injury in cancer survivors. J Cardiovasc Transl Res 2012;5(3):287−95.

[36] Ye J, Rong X, Xiang Y, Xing Y, Tang Y. A study of radiation-induced cerebral vascular injury in nasopharyngeal carcinoma patients with radiation-induced temporal lobe necrosis. PLoS One 2012;7(8):e42890.

[37] Au KM, Hyder SN, Wagner K, Shi C, Kim YS, Caster JM, et al. Direct observation of early-stage high-dose radiotherapy-induced vascular injury via basement membrane-targeting nanoparticles. Small. 2015;11(48):6404−10.

[38] Tonomura S, Shimada K, Funatsu N, Kakehi Y, Shimizu H, Takahashi N. Pathologic findings of symptomatic carotid artery stenosis several decades after radiation therapy: a case report. J Stroke Cerebrovasc Dis 2018;27(3):e39−41.

[39] Rogers LR. Cerebrovascular complications in patients with cancer. Semin Neurol 2010;30(3):311−19.

[40] Abeloos L, Levivier M, Devriendt D, Massager N. Internal carotid occlusion following gamma knife radiosurgery for cavernous sinus meningioma. Stereotact Funct Neurosurg 2007;85(6):303−6.

[41] Torres-Quinones C, Koch MJ, Raymond SB, Patel A. Left thalamus arteriovenous malformation secondary to radiation therapy of original vermian arteriovenous malformation: case report. J Stroke Cerebrovasc Dis 2019;28(6):e53−9.

[42] Maher CO, Pollock BE. Radiation induced vascular injury after stereotactic radiosurgery for trigeminal neuralgia: case report. Surg Neurol 2000;54(2):189−93.

[43] Song CW, Glatstein E, Marks LB, Emami B, Grimm J, Sperduto PW, et al. Biological principles of stereotactic body radiation therapy (SBRT) and stereotactic radiation surgery (SRS): indirect cell death. Int J Radiat Oncol Biol Phys 2019.

[44] Palumbo B, Palumbo R, Sinzinger H. Radioidine therapy temporarily increases circulating endothelial cells and decreases endothelial progenitor cells. Nucl Med Rev Cent East Eur 2003;6(2):123−6.

[45] Budaus L, Bolla M, Bossi A, Cozzarini C, Crook J, Widmark A, et al. Functional outcomes and complications following radiation therapy for prostate cancer: a critical analysis of the literature. Eur Urol 2012;61(1):112−27.

[46] Partap S. Stroke and cerebrovascular complications in childhood cancer survivors. Semin Pediatr Neurol 2012;19(1):18−24.

[47] Pradhan K, Mund J, Case J, Gupta S, Liu Z, Gathirua-Mwangi W, et al. Differences in circulating endothelial progenitor cells among childhood cancer survivors treated with and without radiation. J Hematol Thromb 2015;1:1.

[48] Fernandez-Alvarez V, Lopez F, Suarez C, Strojan P, Eisbruch A, Silver CE, et al. Radiation-induced carotid artery lesions. Strahlenther Onkol 2018;194(8):699−710.

[49] Gujral DM, Shah BN, Chahal NS, Senior R, Harrington KJ, Nutting CM. Clinical features of radiation-induced carotid atherosclerosis. Clin Oncol (R Coll Radiol) 2014;26(2):94−102.

[50] Gujral DM, Chahal N, Senior R, Harrington KJ, Nutting CM. Radiation-induced carotid artery atherosclerosis. Radiother Oncol 2014;110(1):31−8.

[51] Kang JH, Kwon SU, Kim JS. Radiation-induced angiopathy in acute stroke patients. J Stroke Cerebrovasc Dis 2002;11(6):315−19.

[52] Brant-Zawadzki M, Anderson M, DeArmond SJ, Conley FK, Jahnke RW. Radiation-induced large intracranial vessel occlusive vasculopathy. Am J Roentgenol 1980;134(1):51−5.

[53] Jia G, Aroor AR, Jia C, Sowers JR. Endothelial cell senescence in aging-related vascular dysfunction. Biochim Biophys Acta Mol Basis Dis 2019;1865(7):1802−9.

[54] Csiszar A, Tarantini S, Yabluchanskiy A, Balasubramanian P, Kiss T, Farkas E, et al. Role of endothelial NAD(+) deficiency in age-related vascular dysfunction. Am J Physiol Heart Circ Physiol 2019;316(6):H1253−66.

[55] Li J, De Leon H, Ebato B, Cui J, Todd J, Chronos NA, et al. Endovascular irradiation impairs vascular functional responses in noninjured pig coronary arteries. Cardiovasc Radiat Med 2002;3(3-4):152−62.

[56] Soucy KG, Attarzadeh DO, Ramachandran R, Soucy PA, Romer LH, Shoukas AA, et al. Single exposure to radiation produces early anti-angiogenic effects in mouse aorta. Radiat Environ Biophys 2010;49(3):397−404.

[57] Clarke M, Collins R, Darby S, Davies C, Elphinstone P, Evans V, et al. Effects of radiotherapy and of differences in the extent of surgery for early breast cancer on local recurrence and 15-year survival: an overview of the randomised trials. Lancet 2005;366(9503):2087−106.

[58] Darby S, McGale P, Peto R, Granath F, Hall P, Ekbom A. Mortality from cardiovascular disease more than 10 years after radiotherapy for breast cancer: nationwide cohort study of 90,000 Swedish women. BMJ 2003;326(7383):256−7.

[59] Paszat LF, Mackillop WJ, Groome PA, Boyd C, Schulze K, Holowaty E. Mortality from myocardial infarction after adjuvant radiotherapy for breast cancer in the surveillance, epidemiology, and end-results cancer registries. J Clin Oncol 1998;16(8):2625−31.

[60] Tanteles GA, Whitworth J, Mills J, Peat I, Osman A, McCann GP, et al. Can cutaneous telangiectasiae as late normal-tissue injury predict cardiovascular disease in women receiving radiotherapy for breast cancer? Br J Cancer 2009;101(3):403−9.

[61] Kim EH, Moon DH, Oh SJ, Choi CW, Lim SM, Hong MK, et al. Monte Carlo dose simulation for intracoronary radiation therapy with a rhenium 188 solution-filled balloon with contrast medium. J Nucl Cardiol 2002;9(3):312−18.

[62] Salame MY, Douglas Jr. JS. The restenosis story: is intracoronary radiation therapy the solution? Cardiol Rev 2001;9(6):329−38.

[63] Heckenkamp J, Leszczynski D, Schiereck J, Kung J, LaMuraglia GM. Different effects of photodynamic therapy and gamma-irradiation on vascular smooth muscle cells and matrix: implications for inhibiting restenosis. Arterioscler Thromb Vasc Biol 1999;19(9):2154−61.

[64] Lee SW, Park SW, Hong MK, Kim YH, Lee JH, Park JH, et al. Long-term outcomes after treatment of diffuse in-stent restenosis with rotational atherectomy followed by beta-radiation therapy with a [188]Re-MAG3-filled balloon. Int J Cardiol 2005;99(2):201−5.

[65] Balter S, Hopewell JW, Miller DL, Wagner LK, Zelefsky MJ. Fluoroscopically guided interventional procedures: a review of radiation effects on patients' skin and hair. Radiology. 2010;254(2):326−41.

[66] Tran V, Zablotska LB, Brenner AV, Little MP. Radiation-associated circulatory disease mortality in a pooled analysis of 77,275 patients from the Massachusetts and Canadian tuberculosis fluoroscopy cohorts. Sci Rep 2017;7:44147.

[67] Fairley SL, Lucking AJ, McEntegart M, Shaukat A, Smith D, Chase A, et al. Routine use of fluoroscopic-guided femoral arterial puncture to minimise vascular complication rates in CTO intervention: multi-centre UK experience. Heart Lung Circ 2016;25(12):1203−9.

[68] Bogabathina H, Shi R, Singireddy S, Morris L, Abdulbaki A, Zabher H, et al. Reduction of vascular complication rates from femoral artery access in contemporary women undergoing cardiac catheterization. Cardiovasc Revasc Med 2018;19(6s):27−30.

[69] Azizova TV, Muirhead CR, Moseeva MB, Grigoryeva ES, Sumina MV, O'Hagan J, et al. Cerebrovascular diseases in nuclear workers first employed at the Mayak PA in 1948−1972. Radiat Environ Biophys 2011;50(4):539.

[70] Zhang W, Haylock RGE, Gillies M, Hunter N. Mortality from heart diseases following occupational radiation exposure: analysis of the National Registry for Radiation Workers (NRRW) in the United Kingdom. J Radiol Prot 2019;39(2):327−53.

[71] Gillies M, Richardson DB, Cardis E, Daniels RD, O'Hagan JA, Haylock R, et al. Mortality from circulatory diseases and other non-cancer outcomes among nuclear workers in France, the United Kingdom and the United States (INWORKS). Radiat Res 2017;188(3):276−90.

[72] Zielinski JM, Ashmore PJ, Band PR, Jiang H, Shilnikova NS, Tait VK, et al. Low dose ionizing radiation exposure and cardiovascular disease mortality: cohort study based on Canadian national dose registry of radiation workers. Int J Occup Med Env Health 2009;22(1):27−33.

[73] Vrijheid M, Cardis E, Ashmore P, Auvinen A, Bae JM, Engels H, et al. Mortality from diseases other than cancer following low doses of ionizing radiation: results from the 15-Country Study of nuclear industry workers. Int J Epidemiol 2007;36(5):1126−35.

[74] Ivanov V, Maksioutov MA, Chekin SY, Petrov AV, Biryukov AP, Kruglova ZG, et al. The risk of radiation-induced cerebrovascular disease in Chernobyl emergency workers. Health Phys 2006;90:199−207.

[75] Kreuzer M, Kreisheimer M, Kandel M, Schnelzer M, Tschense A, Grosche B. Mortality from cardiovascular diseases in the German uranium miners cohort study, 1946−1998. Radiat Environ Biophys 2006;45:159−66.

[76] Muirhead CR, O'Hagan JA, Haylock RGE, Phillipson MA, Willcock T, et al. Mortality and cancer incidence following occupational radiation exposure: third analysis of the National Registry for Radiation Workers. Br J Cancer 2009;100:206−12.

[77] McGeoghegan D, Binks K, Gillies M, Jones S, Whaley S. The non-cancer mortality experience of male workers at British Nuclear Fuels plc, 1946−2005. Int J Epidemiol 2008;37:506−18.

[78] Azizova TV, Muirhead CR, Druzhinina MB, Grigoryeva ES, Vlasenko EV, Sumina MV, et al. Cerebrovascular diseases in the cohort of workers first employed at Mayak PA in 1948−1958. Radiat Res 2010;174:851−64.

[79] Yamada M, Wong FL, Fujiwara S, Akahoshi M, Suzuki G. Non-cancer disease incidence in atomic bomb survivors, 1958−1998. Radiat Res 2004;161:622−32.

[80] Cardis E, Vrijheid M, Blettner M, Gilbert E, Hakama M, Hill C, et al. The 15-Country collaborative study of cancer risk among radiation workers in the nuclear industry: estimates of radiation-related cancer risks. Radiat Res 2007;167(4):396−416.

[81] Rodrigues JVR, Pinto MM, Figueredo RMP, Lima H, Souto R, Sacchetim SC. Systemic arterial hypertension in patients exposed to Cesium-137 in Goiania-GO: prevalence study. Arq Bras Cardiol 2017;108(6):533−8.

[82] IAEA. Dosimetry and medical aspects of the radiological accident in Goiania. Vienna: International Atomic Energy Agency; 1998 IAEA TECDOC No. 1009.

[83] Suzuki T, Nishida M, Futami S, Fukino K, Amaki T, Aizawa K, et al. Neoendothelialization after peripheral blood stem cell transplantation in humans: a case report of a Tokaimura nuclear accident victim. Cardiovasc Res 2003;58(2):487−92.

[84] Akashi M. Role of infection and bleeding in multiple organ involvement and failure. Br J Radiol 2005;78(1):69−74.

[85] Fliedner TM, Dörr HD, Meineke V. Multi-organ involvement as a pathogenetic principle of the radiation syndromes: a study involving 110 case histories documented in SEARCH and classified as the bases of haematopoietic indicators of effect. Br J Radiol 2005;27(1):1−8.

[86] Kennedy AR. Biological effects of space radiation and development of effective countermeasures. Life Sci Space Res (Amst) 2014;1:10−43.

[87] Johnson KG, Yano K, Kato H. Cerebral vascular disease in Hiroshima, Japan. J Chron Dis 1967;20(7):545−59.

[88] Wong FL, Yamada M, Sasaki H, Kodama K, Akiba S, Shimaoka K, et al. Noncancer disease incidence in the atomic bomb survivors: 1958−1986. Radiat Res 1993;135(3):418−30.

[89] Shimizu Y, Kodama K, Nishi N, Kasagi F, Suyama A, Soda M, et al. Radiation exposure and circulatory disease risk: Hiroshima and Nagasaki atomic bomb survivor data, 1950−2003. BMJ 2010;340:b5349.

[90] Sera N, Hida A, Imaizumi M, Nakashima E, Akahoshi M. The association between chronic kidney disease and cardiovascular disease risk factors in atomic bomb survivors. Radiat Res 2012;179(1):46−52.

[91] Little MP. Cancer and non-cancer effects in Japanese atomic bomb survivors. J Radiol Prot 2009;29(2A):A43−59.

[92] Preston DL, Shimizu Y, Pierce DA, Suyama A, Mabuchi K. Studies of mortality of atomic bomb survivors. Report 13: solid cancer and noncancer disease mortality: 1950−1997. BIOONE; 2003. p. 381−407, 27 p.

[93] McGale P, Darby SC. Low doses of ionizing radiation and circulatory diseases: a systematic review of the published epidemiological evidence. Radiat Res 2005;163(3):247−57.

[94] Ozasa K, Grant EJ, Kodama K. Japanese legacy cohorts: the life span study atomic bomb survivor cohort and survivors' offspring. J Epidemiol 2018;28(4):162−9.

[95] Little MP, Tawn EJ, Tzoulaki I, Wakeford R, Hildebrandt G, Paris F, et al. A systematic review of epidemiological associations between low and moderate doses of ionizing radiation and late cardiovascular effects, and their possible mechanisms. Radiat Res 2008;169(1):99–109.

[96] Little MP, Tawn EJ, Tzoulaki I, Wakeford R, Hildebrandt G, Paris F, et al. Review and meta-analysis of epidemiological associations between low/moderate doses of ionizing radiation and circulatory disease risks, and their possible mechanisms. Radiat Environ Biophys 2010;49 (2):139–53.

[97] Little MP, Azizova TV, Bazyka D, Bouffler SD, Cardis E, Chekin S, et al. Systematic review and meta-analysis of circulatory disease from exposure to low-level ionizing radiation and estimates of potential population mortality risks. EHP 2012;120(11):1503–11.

[98] Little MP, Lipshultz SE. Low dose radiation and circulatory diseases: a brief narrative review. Cardio Oncol 2015;1(1):4.

[99] Fanger H, Lushbaugh CC. Radiation death from cardiovascular shock following a criticality accident. Report of a second death from a newly defined human radiation death syndrome. Arch Pathol 1967;83(5):446–60.

[100] DiCarlo AL, Kaminski JM, Hatchett RJ, Maidment BW. Role of thrombocytopenia in radiation-induced mortality and review of therapeutic approaches targeting platelet regeneration after radiation exposure. J Radiat Oncol 2016;5(1):19–32.

[101] Liebow AA, Warren S, De CE. Pathology of atomic bomb casualties. Am J Pathol 1949;25(5):853–1027.

[102] Reitan JB. Radiation accidents and radiation disasters. Tidsskr Laegeforen 1993;113(13):1583–8.

[103] Agency IAE. The radiological accident in Goiania Vienna, <http://www-pub.iaea.org/MTCD/publications/PDF/Pub815_web.pdf>; 1988.

[104] Krigsfeld GS, Savage AR, Billings PC, Lin L, Kennedy AR. Evidence for radiation-induced disseminated intravascular coagulation as a major cause of radiation-induced death in ferrets. Int J Radiat Oncol Biol Phys 2014;88(4):940–6.

[105] Krigsfeld GS, Kennedy AR. Is disseminated intravascular coagulation the major cause of mortality from radiation at relatively low whole body doses? Radiat Res 2013;180(3):231–4.

[106] Venkatesulu BP, Mahadevan LS, Aliru ML, Yang X, Bodd MH, Singh PK, et al. Radiation-induced endothelial vascular injury: a review of possible mechanisms. JACC Basic Transl Sci 2018;3(4):563–72.

[107] Flamant S, Tamarat R. Extracellular vesicles and vascular injury: new insights for radiation exposure. Radiat Res 2016;186(2):203–18.

[108] Xu J, Yan X, Gao R, Mao L, Cotrim AP, Zheng C, et al. Effect of irradiation on microvascular endothelial cells of parotid glands in the miniature pig. Int J Radiat Oncol Biol Phys 2010;78:897–903.

[109] Panganiban RAM, Day RM. Inhibition of IGF-1R prevents ionizing radiation-induced primary endothelial cell senescence. PLoS One 2013;8 (10):e78589.

[110] Baselet B, Sonveaux P, Baatout S, Aerts A. Pathological effects of ionizing radiation: endothelial activation and dysfunction. CMLS 2019;76 (4):699–728.

[111] Park H, Kim C-H, Jeong J-H, Park M, Kim KS. GDF15 contributes to radiation-induced senescence through the ROS-mediated p16 pathway in human endothelial cells. Oncotarget 2016;7(9):9634–44.

[112] Dong X, Tong F, Qian C, Zhang R, Dong J, Wu G, et al. NEMO modulates radiation-induced endothelial senescence of human umbilical veins through NF-kappaB signal pathway. Radiat Res 2015;183(1):82–93.

[113] Heo J-I, Kim W, Choi KJ, Bae S, Jeong J-H, Kim KS. XIAP-associating factor 1, a transcriptional target of BRD7, contributes to endothelial cell senescence. Oncotarget 2016;7(5):5118–30.

[114] Kim KS, Kim JE, Choi KJ, Bae S, Kim DH. Characterization of DNA damage-induced cellular senescence by ionizing radiation in endothelial cells. Int J Radiat Biol 2014;90(1):71–80.

[115] Mendonca MS, Chin-Sinex H, Dhaemers R, Mead LE, Yoder MC, Ingram DA. Differential mechanisms of X-ray-induced cell death in human endothelial progenitor cells isolated from cord blood and adults. Radiat Res 2011;176(2):208–16.

[116] Oh CW, Bump EA, Kim JS, Janigro D, Mayberg MR. Induction of a senescence-like phenotype in bovine aortic endothelial cells by ionizing radiation. Radiat Res 2001;156(3):232–40.

[117] Panganiban RA, Mungunsukh O, Day RM. X-irradiation induces ER stress, apoptosis, and senescence in pulmonary artery endothelial cells. Int J Radiat Biol 2013;89(8):656–67.

[118] Sermsathanasawadi N, Ishii H, Igarashi K, Miura M, Yoshida M, Inoue Y, et al. Enhanced adhesion of early endothelial progenitor cells to radiation-induced senescence-like vascular endothelial cells in vitro. J Radiat Res 2009;50(5):469–75.

[119] Ungvari Z, Podlutsky A, Sosnowska D, Tucsek Z, Toth P, Deak F, et al. Ionizing radiation promotes the acquisition of a senescence-associated secretory phenotype and impairs angiogenic capacity in cerebromicrovascular endothelial cells: role of increased DNA damage and decreased DNA repair capacity in microvascular radiosensitivity. J Gerontol A Biol Sci Med Sci 2013;68(12):1443–57.

[120] Yentrapalli R, Azimzadeh O, Barjaktarovic Z, Sarioglu H, Wojcik A, Harms-Ringdahl M, et al. Quantitative proteomic analysis reveals induction of premature senescence in human umbilical vein endothelial cells exposed to chronic low-dose rate gamma radiation. Proteomics 2013;13(7):1096–107.

[121] Azimzadeh O, Sievert W, Sarioglu H, Merl-Pham J, Yentrapalli R, Bakshi MV, et al. Integrative proteomics and targeted transcriptomics analyses in cardiac endothelial cells unravel mechanisms of long-term radiation-induced vascular dysfunction. J Proteome Res 2015;14 (2):1203–19.

[122] Minamino T, Komuro I. Vascular cell senescence: contribution to atherosclerosis. Circ Res 2007;100(1):15–26.

[123] Erusalimsky JD. Vascular endothelial senescence: from mechanisms to pathophysiology. J Appl Physiol (1985) 2009;106(1):326–32.

[124] Andrews RN, Caudell DL, Metheny-Barlow LJ, Peiffer AM, Tooze JA, Bourland JD, et al. Fibronectin produced by cerebral endothelial and vascular smooth muscle cells contributes to perivascular extracellular matrix in late-delayed radiation-induced brain injury. Radiat Res 2018;190(4):361–73.

[125] Farjam R, Pramanik P, Aryal MP, Srinivasan A, Chapman CH, Tsien CI, et al. A radiation-induced hippocampal vascular injury surrogate marker predicts late neurocognitive dysfunction. Int J Radiat Oncol Biol Phys 2015;93(4):908−15.

[126] Kruse JJ, te Poele JA, Russell NS, Boersma LJ, Stewart FA. Microarray analysis to identify molecular mechanisms of radiation-induced microvascular damage in normal tissues. Int J Radiat Oncol Biol Phys 2004;58(2):420−6.

[127] Dimmeler S, Hermann C, Zeiher AM. Apoptosis of endothelial cells. Contribution to the pathophysiology of atherosclerosis? Eur Cytokine Netw 1998;9(4):697−8.

[128] Bombeli T, Karsan A, Tait JF, Harlan JM. Apoptotic vascular endothelial cells become procoagulant. Blood 1997;89(7):2429.

[129] Al-Massarani G, Almohamad K. Evaluation of circulating endothelial cells in the rat after acute and gractionated whole-body gamma irradiation. Nukleonika 2014;59(4):145−51.

[130] Najjar F, Alammar M, Bachour M, Al-Massarani G. Circulating endothelial cells as a biomarker in non-small cell lung cancer patients: correlation with clinical outcome. Int J Biol Markers 2014;29(4):e337−44.

[131] Najjar F, Alammar M, Bachour M, Almalla N, Altahan M, Alali A, et al. Predictive and prognostic value of circulating endothelial cells in non-small cell lung cancer patients treated with standard chemotherapy. J Cancer Res Clin Oncol 2015;141(1):119−25.

[132] Hooper AT, Butler JM, Nolan DJ, Kranz A, Iida K, Kobayashi M, et al. Engraftment and reconstitution of hematopoiesis is dependent on VEGFR2-mediated regeneration of sinusoidal endothelial cells. Cell Stem Cell 2009;4(3):263−74.

[133] Chute JP, Muramoto GG, Salter AB, Meadows SK, Rickman DW, Chen B, et al. Transplantation of vascular endothelial cells mediates the hematopoietic recovery and survival of lethally irradiated mice. Blood 2007;109(6):2365−72.

[134] Salter AB, Meadows SK, Muramoto GG, Himburg H, Doan P, Daher P, et al. Endothelial progenitor cell infusion induces hematopoietic stem cell reconstitution in vivo. Blood 2009;113(9):2104−7.

[135] Rafii S, Dias S, Meeus S, Hattori K, Ramachandran R, Feuerback F, et al. Infection of endothelium with E1(−)E4(+), but not E1(−)E4(−), adenovirus gene transfer vectors enhances leukocyte adhesion and migration by modulation of ICAM-1, VCAM-1, CD34, and chemokine expression. Circ Res 2001;88(9):903−10.

[136] Zhang F, Cheng J, Hackett NR, Lam G, Shido K, Pergolizzi R, et al. Adenovirus E4 gene promotes selective endothelial cell survival and angiogenesis via activation of the vascular endothelial-cadherin/Akt signaling pathway. J Biol Chem 2004;279(12):11760−6.

[137] Seandel M, Butler JM, Kobayashi H, Hooper AT, White IA, Zhang F, et al. Generation of a functional and durable vascular niche by the adenoviral E4ORF1 gene. Proc Natl Acad Sci USA 2008;105(49):19288−93.

[138] Ding BS, Nolan DJ, Butler JM, James D, Babazadeh AO, Rosenwaks Z, et al. Inductive angiocrine signals from sinusoidal endothelium are required for liver regeneration. Nature 2010;468(7321):310−15.

[139] Butler JM, Gars EJ, James DJ, Nolan DJ, Scandura JM, Rafii S. Development of a vascular niche platform for expansion of repopulating human cord blood stem and progenitor cells. Blood 2012;120(6):1344−7.

[140] James D, Nam HS, Seandel M, Nolan D, Janovitz T, Tomishima M, et al. Expansion and maintenance of human embryonic stem cell-derived endothelial cells by TGFbeta inhibition is Id1 dependent. Nat Biotechnol 2010;28(2):161−6.

[141] Ginsberg M, James D, Ding BS, Nolan D, Geng F, Butler JM, et al. Efficient direct reprogramming of mature amniotic cells into endothelial cells by ETS factors and TGFbeta suppression. Cell 2012;151(3):559−75.

[142] Ward WF, Molteni A, Ts'ao CH, Kim YT, Hinz JM. Radiation pneumotoxicity in rats: modification by inhibitors of angiotensin converting enzyme. Int J Radiat Oncol Biol Phys 1992;22(3):623−5.

[143] Holler V, Buard V, Gaugler MH, Guipaud O, Baudelin C, Sache A, et al. Pravastatin limits radiation-induced vascular dysfunction in the skin. J Invest Dermatol 2009;129(5):1280−91.

[144] Soucy KG, Lim HK, Benjo A, Santhanam L, Ryoo S, Shoukas AA, et al. Single exposure gamma-irradiation amplifies xanthine oxidase activity and induces endothelial dysfunction in rat aorta. Radiat Environ Biophys 2007;46(2):179−86.

[145] van Thiel BS, van der Pluijm I, te Riet L, Essers J, Danser AH. The renin-angiotensin system and its involvement in vascular disease. Eur J Pharmacol 2015;763(Pt A):3−14.

[146] Soroush F, Tang Y, Zaidi HM, Sheffield JB, Kilpatrick LE, Kiani MF. PKCdelta inhibition as a novel medical countermeasure for radiation-induced vascular damage. FASEB J 2018;6436−44.

Chapter 12

Flow-adapted vascular systems: mimicking the vascular network to predict clinical response to radiation

Aravindan Natarajan[1,2], Mohan Natarajan[3], Sheeja Aravindan[4] and Sumathy Mohan[3]

[1]Department of Radiation Oncology, University of Oklahoma Health Sciences Center, Oklahoma City, OK, United States, [2]Department of Pathology, University of Oklahoma Health Sciences Center, Oklahoma City, OK, United States, [3]Department of Pathology, University of Texas Health Sciences Center at San Antonio, San Antonio, TX, United States, [4]Stephenson Cancer Center, Oklahoma City, OK, United States

Hemodynamic shear forces have long been realized as the defining factor for vascular assembly, caliber, and remodeling [1−4]. Any events altering hemodynamic forces result in the development of vascular pathology [5−9]. By definition, hemodynamic shear stress (SS) is the blood-flow-inflicted frictional force on the endothelial cell (EC) surface and on the luminal vessel wall [10,11]. Steady-state blood flow with vessel-type defined SS magnitude is critical for the normalcy of vascular function [6,7,12−15]. Recent molecular and cellular studies from diverse settings documented the endothelial response to SS modifications and defined its functional relevance in disease pathogenesis [10,16−20]. Beyond its criticality in cardiovascular normalcy and/or in pathology, the functional significance of the endothelium in other orchestrated complications (e.g., radiation damage) has been indicated. The widespread distribution of endothelium in all tissues places it in the radiation field (regardless of radiation course) of site-directed or whole-body irradiation. The extensive presence of the vascular network increases the probability of exposure of the ECs to radiation; this coupled with the ability of the ECs to mediate local and systemic inflammatory reaction, makes the endothelium a prime player in both acute and delayed radio-response in normal tissue toxicity/organ failure and in tumor biology/therapy response. With such defined functional significance of the endothelium in radiation effects, it is critical to consider the unique physiognomies of the endothelium that influence cell susceptibility to radiation injury and appropriate mirroring for unequivocal conclusive claims. These include the vascular bed, heterogeneity disposition of the endothelium [microvascular vs macrovascular; venous system vs arterial tree; endothelial cell (EC) lining secretory products in the brain vs gut], hemodynamic flow shear stress (HFSS), and associated mechano-transduction signaling [21−24]. In this chapter, we present the progress made in flow-adapted vascular systems that closely reproduce or recreate the vascular environment and discuss the significance of adopting hemodynamic flow shear conditions in models of EC that are used to assess the clinical response to radiation. Further, this chapter highlights the thus-far realized facts and figures of EC function and gene expression by SS, the contribution of the endothelium in radiation injury, and the available tools for mimicking the hemodynamic SS in vitro.

12.1 Endothelium and hemodynamic shear stress

The endothelium is the monolayer tissue that lines the inner surface of the blood vessels and is regarded as the largest organ in the body, as it is disseminated into every tissue [25−27]. ECs do not actively divide (except in areas where there is periodic vascular disruption), yet they critically play functional roles in regulating metabolic, immunologic, and cardiovascular processes, depending on local physiological needs throughout the vascular tree [28−30]. It is now well established that the endothelium regulates inflammatory activity in the vessel walls; limits the adhesion/aggregation of platelets; maintains vascular tone; sustains vascular permeability to hormones, nutrients, macromolecules, and leukocytes; controls the balance between coagulation and fibrinolysis; and supports the composition of the subendothelial matrix in the cardiovascular system. The discrete role of the endothelium in interacting with and influencing the

Endothelial Signaling in Vascular Dysfunction and Disease. DOI: https://doi.org/10.1016/B978-0-12-816196-8.00004-7

129

function of other cell types, including vascular smooth muscle cells, platelets, leukocytes, retinal pericytes, renal mesangial cells, and large artery macrophages, has been well documented [28]. As the prime source of paracrine effectors, the endothelium has a significant role in the preservation of structure and function of every organ in the physiological system. Because ECs are at the strategic interface between the blood stream and tissue, they also contribute to the rapid bidirectional exchange of information [10,31]. Considering the pivotal role of endothelium, dysfunction of the vascular ECs remains an important factor in orchestrated inflammation and in disease pathogenesis [32−34].

The endothelium on the vessel surface is incessantly exposed to hemodynamic forces due to the continuous flow of blood [10,30]. The blood vessels withstand hydrostatic pressure, cyclic stretch, and fluid SS [30]. Because ECs line the inner walls of the vessels and are in direct contact with flowing blood, they are highly exposed to HFSS. The strength of HFSS can be measured in most of the vessels with Poiseuille's law which states, "the HFSS is directly proportional to the blood-flow viscosity and inversely proportional to the third power of the internal radius of the tube" [35]. HFSS in ECs acts in parallel to the vessel wall without influencing the artery dimensions [26] and is unrelated to the perpendicular blood pressure (unsteady pulsatile oscillatory or turbulent laminar flow) that can extend the dimensions of elastic arteries. However, magnification of HFSS is highly vessel-/size-dependent and varies widely in the arteries, capillaries, veins, and venules. Typically, HFSS ranges from 1 to 6 dyne/cm^2 (1 dyne $= 10\,\mu$N) in the venous system [19,31] and between 10 and 70 dyne/cm^2 in the arterial vascular network [19,35]. With specific diameter variations along the artery, achieved through vasoregulatory mechanisms, the mean physiological HFSS level is normally balanced at $15-20$ dynes/cm^2 [30,36]. Researchers have documented that altered HFSS influences vessel wall remodeling [1,2,37]. To that note, sustained HFSS due to chronic increases in blood flow (e.g., radial artery of dialysis patients) concert the luminal radius expansion to harmonize the mean SS to its baseline level [1,38]. Equally, low blood flow or blood viscosity-associated reduction in HFSS has been shown to coordinate a decrease in internal vessel radius [2]. Together, such endothelial-dependent balancing feedback allows the stable maintenance of mean arterial HFSS at the range of $15-20$ dyne/cm^2 [11,38]. However, such an HFSS-stabilizing process is critically dependent on intact EC function and will be negated in the event of prior EC monolayer destruction [2]. In response to inflicted HFSS, ECs endure cytoskeletal rearrangements, signaling specific gene expression modification, and altered function [10,16,26,39−42].

At the molecular level, heightened HFSS has been shown to augment nitric oxide (NO)-dependent cyclic guanosine monophosphate and endothelial nitric oxide synthase (eNOS) transcription/translation/activity, consequently expanding vessel radius [1,43] and normalized shear. Further, transgenic knockout approaches in mouse and nonhuman primate models affirmed the criticality of eNOS in shear-mediated structural remodeling [44,45]. HFSS is associated with the proliferative state of ECs, and studies have shown that reduced HFSS leads to EC cell loss and shedding; morphological changes, including decreased elongation; reduction in actin stress fibers; heightened monocyte adhesion and migration across the endothelium [46]; and robust expression of vascular cell adhesion molecule-1 (VCAM-1) on the EC surface [47]. HFSS increases EC secretion of prostacyclin [48] and NO [48,49], thereby deterring platelet activation [50,51], attenuating smooth muscle proliferation [52], and impeding neointima formation [53,54]. Low-shear forces induce EC loss through a sustained apoptotic process until the harmonization of the shear is realized [55]. The plethora of evidence in humans and animal models [1,7,12] defining the correlations between hemodynamic factors and vascular remodeling paved the way for studies focusing on the in vitro EC response to HFSS [10,16−20,56]. Under in vitro conditions, the morphological transformation of ECs from a polygonal, cobblestone-shape to a fusiform, elongated shape that is aligned in the direction of flow was evident in response to HFSS [25]. HFSS leads to (1) a reduction in EC turnover through regulation of proliferation [57] on the one hand, and inhibition of apoptosis on the other [58,59]; (2) the upsurge of vasodilators [60,61], paracrine growth inhibitors [62], fibronolytics [63−66], and antioxidants [67,68]; and (3) the suppression of vaso-constrictors [69,70], paracrine growth promoters [57,71,72], inflammatory mediators [73], and adhesion molecules [74,75]. These orchestrated events contribute to the HFSS-coordinated phenotype switching and function. Conversely, the oscillatory SS, a weaker inducer of eNOS [76], increases nicotinamide adenine dinucleotide phosphate (NADPH)-dependent endothelial superoxide anion ($O_2{}^-$) [68], suppresses endothelin 1 [77], enhances monocyte adhesion [78], and prompts EC proliferation [79,80]. Likewise, turbulent SS has been shown to induce EC turnover [81] and fails to stimulate eNOS, manganese superoxide dismutase, and cyclooxygenase-2 genes [82]. Together, these findings clearly indicate that the SS type and the magnitude of the shear force are critical components that drive EC response and function.

12.2 Endothelial response to radiation

Unanticipated (and anticipated) radiation exposures is a serious threat that inflicts complex injuries to the personnel involved [83]. In addition to general public (radiological radio-terrorism, radio-accidents), patients (radiotherapy), first responders or military personnel (radio-terrorism, radiation accidents), and radiation workers (nuclear industry workers, interventional medical personnel) are also positioned to receive detrimental radiation doses [83−85]. Radiation-associated perturbations of the vascular endothelium have been realized very early (1899), within years of the discovery of X-rays [86]. Consistently, with the growing list of evidence, it is clear that the vascular endothelium is intimately linked to both early and late radiation pathologies (discussed in detail elsewhere [83]). Conspicuously, radiation (during radiotherapy or nuclear incident) induced endothelial damage and vascular dysregulation contribute to the tissue damage and orchestrated complications (e.g., cerebrovascular, cardiovascular, renal, and gastrointestinal disease) [87−91]. The vascular endothelium exposed to radiation is affected with cellular damage (e.g., apoptosis) within hours, with consequent edema, basal lamina exposure, leakage of plasma proteins, lymphocyte infiltration, and increased vascular permeability [92−96]. Radiation-induced delayed vascular effects that occur after weeks to months postirradiation include capillary collapse, scarring, and fibrosis [92,93,95]. Although radiation exposure-associated damage was realized in both normal and tumor microvasculature [92,93,97,98], the responses heavily relied on the radiation dose, quality, and schedule [99,100]. Utilizing a three-dimensional vessel model study, the radiation quality (LET) and dose-dependent responses of both the developing and mature vessels were recognized [101]. Further, it has been recognized that the radiation fractionation contributed to increased EC activation and vascular inflammatory status [102]. Within the vascular system, the radiation injury to the microvessels is abundant, severe, and qualitatively different from that in larger vessels [98,103]. The micro- and macrovessel difference in radiation response is attributed to having less or no vascular tissue protection (in the form of smooth muscle cushioning) to the capillaries [93,104]. As stated earlier, the presence of the vascular network throughout the body exposes ECs to therapeutic, accidental, intentional, and occupational radiation. ECs are highly responsive to radiation [93,105−111] and their damage can lead to vascular dysfunction [111]. ECs in the vasculature are heterogeneous and are known to exhibit origin-dependent distinctive properties and exclusive responses [112]. Hence, signaling rearrangement in ECs after radiation exposure heavily relies on the organ and vascular bed. For instance, origin-dependent discrepancies in the EC response to radiation were recognized with artery ECs showing actin dissolution [113−116], while dermal microvascular ECs were found to show increased actin polymerization [117]. Consistently, researchers have documented the incidence and significance of acute and late effects in ECs of different origin with low, moderate, and high radiation doses [93,101,102,107−110,113,116−130]. Conceptually, epidemiological data affirmed an association between radiation exposure (low/moderate doses) and delayed cardiovascular complications [131]. Such claims were further corroborated with clinical evidence from cancer patients that develop late cardiovascular complications after radiation therapy [131−134]. These indications from environmental, accidental, diagnostic, therapeutic exposure studies warrants an understanding of the mechanism(s) of EC damage in fitting models that mimic in vivo scenario to interpret the outcome of radiation pathogenesis.

At the cellular level, in microvascular ECs, a two-phase RhoA (Ras homolog A) pathway activation (initial cytoskeletal remodeling and second contractile and hyperpermeable phenotypes) was indicated as the flow-through signaling for the morphologic and functional effects of radiation [117]. Although the mechanism of how radiation activates RhoA signaling is unknown, PI3K [135,136] and reactive oxygen species (ROS) [137] could serve as such effectors. Radiation-induced ROS activated SAPK2/p38 [138] and can play crucial roles in ECs remodeling [113]. Studies from our group recognized that radiation-induced inhibition of EC migration was the translational response of radiation-inhibited SHP-2 phosphatase, and designated SHP2 as the crucial player in EC migration [122]. Interestingly, radiation also could impose a senescence-like phenotype in ECs that are functionally deficient but persist in the vasculature for a prolonged period [127]. In cancer, the sensitivity of tumor vasculature to radiation determines the tumor response to radiotherapy [139]. This is mainly attributed to the explicit dependency of tumors on their blood vessels for survival, and to the secondary cell death inflicted by radiation through tumor-feeding vasculature destruction [140,141]. Consequently, trivial degrees of vascular destruction could translate into measurable tumor kill [142,143]. However, the tumors actively protect themselves from secondary radiation damage (ECs-dependent delayed effects) by endorsing ECs radio-resistance through secreted angiogenic cytokines [144]. Despite the attempts to target angiogenic cytokines [145−148] and/or ECs tyrosine kinase receptors [149] to radiosensitize the tumor vasculature, the outcome is highly limited, mainly because our understanding of how radiation therapy (RT) induces tumors to protect ECs is incomplete. Some studies indicated that binding of secreted cytokines activate ECs' radio-protective signaling through Raf-1 and phosphoinositide 3-kinase/protein kinase B or AKT (PI3K/AKT) [150,151]. In parallel, it was indicated that the secreted cytokines could inhibit EC nonapoptotic cell death [105,152]. Nevertheless, the radio-response of tumor endothelium is beyond the purview this chapter, and herein we discuss the radiation effects on normal vascular endothelium and more importantly the criticality of hemodynamic SS in defining the radiation effects.

Defining the mechanism(s) of EC damage is predominantly dependent on in vitro studies, which provide feasibility to (1) assess contribution of individual cell types, (2) investigate immediate events at the tissue level, (3) adopt application-specific experimental design, and (4) delineate the contribution of specific signaling pathway(s). Till date almost all in vitro studies of the EC radiation response were performed under no-flow conditions, ignoring the crucial HFSS-associated mechanotransduction signaling that plays a major role in EC structure, function and response to stimulus. Relatively few studies incorporated flow conditions while assessing the radiation-related effects [153−156]. Since SS-induced mechanotransduction signaling is a fundamental part of the EC response, outcomes from such no-flow studies of radio-response mechanisms are largely disputed. Essentially, the direction of endothelial radiation response relies on factors that include radiation quality, dose, dose-rate, fractionation, origin of the endothelial cells, and, more importantly, the EC origin-specific fluctuations in HFSS (Fig. 12.1).

12.3 Significance of hemodynamic shear stress in assessing the endothelial cell radiation response

Earlier investigations from our group and from others with the parallel-plate flow model recognized the role of unidirectional steady laminar flow in influencing the structure and function of ECs. [157−162]. Compared with static conditions, shear forces resulted in EC stretching and elongation (oriented with the direction of flow) as a function of exposure time and magnitude of shear force [25]. Unlike static cultures, ECs under HFSS maintain physiological levels of prostacyclin, thrombomodulin, and the release of NO and calcium [163]. In flow-adapted ECs, eNOS activation-dependent maintenance of NO endorses normal homeostasis of vascular function [164]. Our studies, for the first time, showed that human aortic endothelial cells (HAECs) subjected to HFSS at 16 dynes/cm^2 exhibited a consistent higher constitutive level of eNOS than did those maintained under no-flow conditions [25]. Furthermore, our findings demonstrated a significant difference in endothelial response when ECs subjected to shear stress were exposed to radiation; indeed we observed a statistically significant difference in eNOS expression in HAECs subjected to HFSS (16 dynes/cm^2) for 16 hours and exposed to radiation (2 Gy) as compared to irradiated cells under no-flow conditions [25]. Shear forces acting on the ECs' surface are transmitted through the cytoskeleton, allowing the activation of various mechanoreceptors [164,165] that translate mechanical forces into biological signals [10,20,166−168]. Both the integrins connected with cytoskeleton [41,169,170] and the

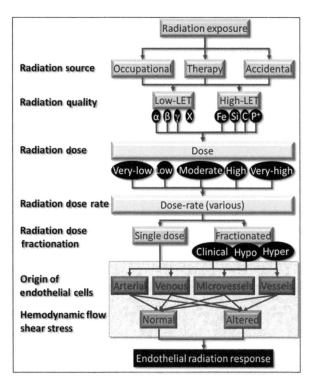

FIGURE 12.1 Schematic representation of complex interplay of factors that play critical roles in deciding the direction and magnification of radiation response in endothelial cells. Irrespective of the source of radiation exposure, the radiation quality, dose, dose-rate, fractionation, type and/or origin of endothelial cells, the vessel-/area-specific hemodynamic flow shear stress critically drives the endothelial response outcome to radiation. As no-flow conditions do not exist in vivo and HFSS-associated mechanotransduction signaling plays a large role in endothelial responses, incorporation of HFSS is critical for assessing the effects of radiation on endothelium in vitro. *HFSS*, Hemodynamic flow shear stress.

nonintegrin mechanoreceptors, including multiple G-protein coupled receptors, the glycocalyx, ion channels, lipid rafts, and receptor tyrosine kinases, have been linked to translate the mechanical stimuli into biological processes [165]. In intact blood vessels, the HFSS-associated mechanotransduction signaling maintains the EC shape, orientation, cytoskeleton reorganization, rate of cell proliferation, expression of adhesion molecules, growth factors and immediate early genes, the release of vasoactive substances, and macromolecular transport [30,171−174]. As this homeostasis is disturbed under no-flow conditions, studies on effects of radiation in static endothelial cultures in which most of these flow mediated signals do not exist, limits our understanding of the real EC responses to radiation. In addition, ECs under no-flow conditions are neither physiologically significant nor relevant, as the endothelium in vivo is never subjected to zero flow. Thus it is safe to conclude that studies assessing the ECs' responses to radiation exposure (both high-LET and low-LET) performed under no-flow conditions may not endothelial responses in vivo (Table 12.1).

In ECs, the steady-state laminar flow leads to the mechanosensors upholding cell-signaling pathways that maintains the cell shape and vessel tone and deters apoptosis or proliferation (Fig. 12.2A). These signaling events include activation of receptor tyrosine kinases, ion channels, NADPH oxidase activation, ROS generation, and eNOS-NO production, that converge on the mitogen-activated protein kinase (MAPK) pathway and orchestrate the homeostasis of the ECs (Fig. 12.2B). Radiation exposure resulting in oxidative stress and inflammatory stimuli drives actin remodeling, apoptosis, cell senescence, and endothelial to mesenchymal transition, leading to endothelial dysfunction. That both HFSS-associated signaling that maintains homeostasis and the radiation-induced endothelial dysfunction signaling that drives damage converge at the "MAPK-signaling pathway" (Fig. 12.2B) indicates a definite interaction between these signaling cascades and signifies the importance of HFSS in augmenting/abridging the radiation response. Hence, it is critical to consider HFSS when assessing radiation damage to ECs for any clinical translation. Our recent gene expression studies comparing the "radiation response" in "no-flow to flow shear conditions" provided a clear insight into the global modifications on (1) endothelial structure and function; (2) inflammation and cell fate function; (3) oxidative stress response; and (4) stemness maintenance. Compared with static conditions, unbranched long aortae simulated HFSS (16 dynes/cm^2) exhibited high baseline levels of EC response molecules (64% of assessed pathway-specific genes showed an increase), early response and inflammatory signaling molecules (57%), oxidative stress response molecules (93%), and stemness maintenance-related molecules (90%) [25]. These altered genes were involved in EC survival, remodeling, thrombosis, cell−cell adhesion, maintenance of vascular tone, and angiogenesis, suggesting that prolonged (24 hours or more) HFSS could allow the cells to maintain a relatively noninflammatory and nonproliferative state compared with no-flow conditions. Independently, a similar HFSS response pattern was documented in HAECs at 12 dynes/cm^2 [40] and in human umbilical vein ECs (HUVECs) at 25 dynes/cm^2 [39] for 24 hours. Conversely, simulating inflammatory-prone branched arteries with low HFSS (2 dynes/cm^2) in HAECs resulted in an intracellular oxidant activity-dependent increase in vascular cell adhesion molecule-1 (VCAM-1) and augmentation of monocyte adhesion [30,157]. Evidently, ECs could sense the variations in hemodynamic forces through unique physiological mechanisms, depending on the type and magnitude of HFSS, to maintain vascular homeostasis. Affirming such criticality of HFSS and in particular the context-specific (vessel/condition) magnitude, it is momentous that the EC response to radiation under no-flow conditions could produce irrelevant equivocal outcomes/claims. Our recent endeavor to assess the differences in radiation response with and without physiological flow SS clearly recognized differences in outcomes (25). Radiation exposure under static conditions activated about 41% of the molecules associated with endothelial architecture and function. However, under prolonged (24 hours) physiological SS (16 dynes/cm^2), only 16% showed significant activation. Our interesting finding revealed an upside response, with one activated (static) gene showing complete inhibition, while four genes were downregulated under static conditions, showing significant activation in the presence of HFSS. These outcomes were indicative of the fact that radiation effects on EC ought to be studied in a HFSS setting and that no flow conditions were insufficient to represent the vasculature in vivo.

Our studies and those of others indicated that radiation induces activation of transcription factor nuclear factor kappa B (NFκB) in ECs [102,109,175,176]. In parallel, our studies under disturbed flow SS (2 dynes/cm^2), pulsatile low-SS (2 ± 2 dynes/cm^2), and under normal physiological HFSS (16 dynes/cm^2) demonstrated NFκB activation in ECs [157−159]. These studies mimicking the in vivo low-shear forces exerted by reversing hemodynamic flow patterns (at arterial branch sites, arterial bifurcations, and inner walls of curved arterial segments) indicated that prolonged activation of NFκB could contribute to recruitment of proatherogenic mediators. It is crucial to accurately mimic the in vivo situation of specific vascular beds in in vitro experiments to assess the radiation response of such vascular beds. In this direction, our studies investigating the NFκB (and signaling pathway) fluctuations between no-flow and flow-adapted conditions in ECs indicated that static cultures are a definite misrepresentation of radiation-associated NFκB signaling. This was evident owing to a stronger radiation induced activation of NFκB signaling pathway in ECs under normal physiological HFSS as compared to ECs kept under no-flow conditions [25].

TABLE 12.1 Brief list of in vitro studies performed under static culture conditions to investigate the ECs' responses to radiation exposure.

Friedman et al. (1986)	Specific disorganization of ECs in terms of reversible increase in permeability, morphology, actin filament distribution, and organization	Main-stem pulmonary ECs of calves	γ-Radiation	[116]
Haimovitz-Friedman et al. (1991)	Radiation-induced de novo synthesis of basic fibroblast growth factor activates potentially lethal damage repair (PLDR) signaling, leading to cells' recovery from radiation damage	Bovine aortic ECs (BAECs)	γ-Radiation	[216]
Haimovitz-Friedman et al. (1994)	Ionizing radiation interaction with cellular membrane to generate ceramide and initiate apoptosis	BAECs	γ-Radiation	[217]
Meeren et al. (1997)	Radiation-induced inflammatory response in vascular endothelium is mediated by IL-6 and IL-8	Human umbilical vein ECs (HUVECs)	γ-Radiation	[218]
Onoda et al. (1999)	Reversible loss of EC cell integrity and permeability barrier function	Pulmonary microvascular ECs (PMVECs)	X-ray	[113]
Oh et al. (2001)	Induction of ECs senescence-like phenotype by radiation	BAEC	γ-Radiation	[127]
Gaugler et al. (2005)	Endothelial activation and blood cells—ECs interaction	HMVEC from lung	γ-Radiation	[219]
Murley et al. (2006)	Delayed radio protection by NFkb-mediated Manganese superoxide (MnSOD) in microvascular ECs with free radical scavengers	Human dermal ECs (HDMECs)	X-ray	[175]
Gabrys et al. (2007)	RhoA/Rho kinase activation after radiation in ECs cytoskeletal remodeling and permeability	HDMECs, HUVECs	X-ray	[117]
Natarajan et al. (2008)	Radiation-induced NFκB-mediated EC survival is dependent on telomerase activity	(BAECs)	γ-Radiation	[109]
Chou et al. (2009)	Radiation-induced p38/NFκb signaling pathway-dependent IL-6 mediates MCL-1- and sIL6-Rα-orchestrated antiapoptosis in ECs	HUVEC	γ-Radiation	[220]
Grabham et al. (2010)	Differential effects on radiation quality and cellular development	HUVEC 3D human vessel models	High LET-Fe-ion, low LET-protons	[101]

(Continued)

TABLE 12.1 (Continued)

Truman et al. (2010)	VEGF and basic fibroblast growth factor inhibit radiation-induced apoptosis via repression of aSMase activation	(BAECs)	γ-Radiation	[221]
Zheng et al. (2011)	Radiation-impaired EC migration is mediated through radiation-induced SHP-2	BAEC, HAEC	γ-Radiation	[122]
Rousseau et al. (2011)	RhoA-ROCK pathway participates in IR-induced adhesion to fibronectin and EC migration	(HDMECs)	X-ray	[128]
Yentrapalli et al. (2013)	PI3K/AKT/MTOR-dependent premature senescence of ECs exposed to chronic radiation	HUVEC	γ-radiation	[222]
Grabham et al. (2013)	Distinct vessel morphology, motile tip activity, and matrix architecture by different radiation qualities	Human brain microvascular ECs (HBMECs)	High LET-Fe-ion, low LET-protons	[119]
Sharma et al. (2013)	Transient and rapid effect of gamma radiation on human endothelial barrier function—PECAM-1 loss dependent	(HBMECs)	γ-Radiation	[118]
Park et al. (2012)	Differences in radiation response of normal ECs derived from diverse organs	HBMECs, HOMECs, HPMECs, HUVECs, HDMECs	γ-Radiation	[223]
Cervelli et al. (2014)	Single and fractionated irradiation effects on EC viability, DNA repair, and ICAM surface expression	HUVECs	X-ray	[102]
Sharma et al. (2014)	Human endothelial barrier structure and function	HUVECs	High LET-Fe-ion, low LET-protons	[224]

Although critical information on how ECs respond to radiation, in terms of modified cellular functions and signaling events was gained, the inclusion of context-/tissue-specific HFSS is warranted to simulate in vivo conditions with intrinsic mechanotransduction signaling that could augment or abridge inflicted radioresponses and for clinical translation of outcomes (arranged in chronological order). ECs, Endothelial cells HDMEC, human dermal microvascular endothelial cell, HMVEC, human microvascular endothelial cell, HUVEC, human umbilical vein endothelial cell.

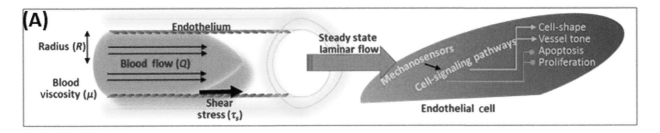

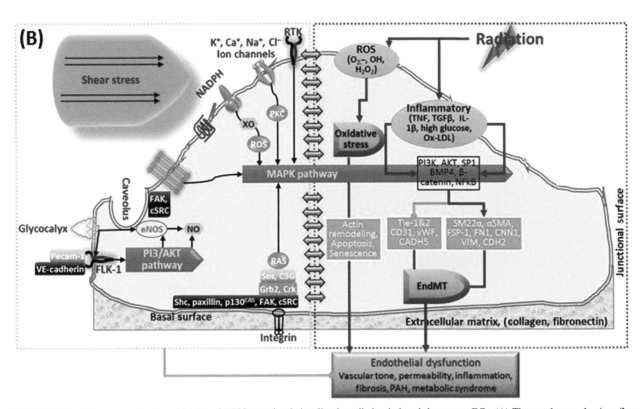

FIGURE 12.2 Schema showing the criticality of HFSS-associated signaling in radiation-induced damage to ECs. (A) The steady-state laminar flow associated shear forces lead to the activation of mechanosensors that uphold the key cell-signaling pathways, maintain cell shape and vessel tone, and prevent apoptosis or proliferation. (B) Steady-state HFSS-associated signaling events in ECs that maintain homeostasis (gold box) and the radiation-associated oxidative stress and inflammatory stimuli signaling events that orchestrate actin remodeling, apoptosis, cell senescence, and endothelial to mesenchymal transition, leading to endothelial dysfunction (red box). The convergence of HFSS-associated homeostasis maintenance signaling and radiation-induced endothelial dysfunction signaling in the "MAPK-signaling pathway" indicates cross-talk signifying that HFSS could augment or abridge the radiation response. Probable other signaling cross-talks within the EC in this setting are indicated with double-sided yellow arrows. *EC*, Endothelial cell; *HFSS*, hemodynamic flow shear stress.

HFSS has been shown to generate ROS species and NO in the vessels [105,159,160] and has been demonstrated to modulate the redox state of the vascular wall by regulating various oxidase systems [177]. ECs exposed to oscillatory SS increased NAD(P)H oxidase activity and ROS production. We showed a relatively high level of ROS generation at low-SS compared with physiological HFSS; the SS-induced ROS-mediated activation of NFκB is NADPH-oxidase-dependent [159]. These observations affirm that different SS conditions could result in specified generation of ROS species and could uniquely dictate (ROS species-specific) downstream intracellular signaling (e.g., NFκB). Our attempt to appreciate the differential EC response with/without flow conditions indicated that ECs are normally primed with heightened (vs static) oxidative response under normal physiological HFSS conditions with widespread activation of oscillatory shear (OS) signaling molecules. With such a primed baseline oxidative response under HFSS, radiation resulted in the activation of only 34% of genes, while no-flow conditions showed about 93% gene activation [25]. Further, the reduced levels of key OS mediators (e.g., NOS2) in HFSS + radiation validate an antioxidant, antiinflammatory phenotype of the vasculature. The relative low OS response in ECs after radiation under HFSS could be attributed to the induction of flow-responsive antioxidant response elements (ARE) [178]. ARE-regulated antioxidant and cytoprotective genes together contribute to defense mechanisms against oxidant-mediated endothelial dysfunction and redox-sensitive inflammatory gene expression [131]. Earlier studies from our group showed that nonlaminar flow prompts greater OS and an inflammatory phenotype [158−161]. Likewise, independent studies have indicated that hemodynamic stressors play a critical role in capillary remodeling [178] and recognized that the exposed ECs endorse the development of stem-like hematopoietic progenitor cells from the endothelium [179] that drive vessel formation. In addition, the induced inflammation that contributes to the vascular endothelial health/survival compromise could promote endothelial turnover [180]. Tissue-specific ECs replication rates can be up to 60% [181] and these ECs retain proliferative potential equal to that of circulating endothelial colony-forming cells (ECFCs) [182]that are derived from vascular endothelium. Our studies assessing the significance of flow conditions in defining the EC response to radiation identified a primed state of ECs stemness with activated stem-cell biology signaling (91% of genes vs static condition) molecules in ECs exposed to HFSS. With that level of baseline priming under HFSS, radiation resulted in significant activation of keratin-18 (KRT-18), which is known to play crucial roles in flexible intracellular scaffolding, resistance to extracellular stress [183], apoptosis, cell cycle progression, and cell signaling [184,185]. These compelling facts emphasize the significance and relevance of incorporating SS for elucidating the EC response to radiation to mimic the in vivo conditions.

Within the scope of radiation damage/response of ECs, only a handful of studies have investigated the orchestrated signaling and/or mechanisms under in vivo relevant shear forces. Peng and colleagues studied UV radiation effects while maintaining the EC morphology at physiological nonreversing steady-state flow and observed morphological changes and total cell detachment under pulse-perfused (7 dynes/cm^2) SS [186]. Further, they identified phosphorylation of AKT and eNOS under pulse-perfused SS conditions and defined the functional role of AKT/eNOS in composing the ECs radiation response [186]. Consistently, under low-SS conditions (1 dyne/cm^2), radiation exposure induced chronic inflammatory response and increased the chemokine-dependent adhesiveness of ECs in vascular endothelium [154]. Radiation increased adhesion of tumor necrosis factor alpha (TNFα)-primed HAECs, while there was no measurable increase in the surface expression of VCAM-1 or intercellular adhesion molecule (ICAM-1) under low SS. However, radiation-induced EC adhesion requires α4, β1, and β2 integrins [154]. An independent study showed neither activation of E-selectin nor an increase in adhesion of the human leukemia cell line (HL-60, a cell type used to monitor immune cell adhesion)) postirradiation (γ-radiation 10 Gy) in HUVECs under SS (parallel-plate flow) ranging from 0.5 to 2.0 dynes/cm^2 [156]. This study indicated that radiation-induced expression of E-selectin and increased HL-60 cell adhesion of human dermal ECs under shear flow conditions, recognizing the ECs' origin-specific radiation response [156]. Recently, Erbeldinger and colleagues in a study measuring leukocyte adhesion to primary ECs after photon (1.5 Gy/min) and charged particle radiation (He-ion, LET of 76Kev/u) reported a twofold increase in binding to human microvascular endothelial cells under static conditions compared with small vessel physiological SS (0.75 dynes/cm^2); significant increase (fivefold) in adhesion under SS (vs 2.5-fold under no-flow) with TNFα stimulation; complete reduction in adhesion with low-dose X-ray exposure under flow-adapted conditions; significant reduction after He-ion exposure of human microvascular endothelial cells under SS; and significant differences in NFκb and adhesion molecules (ICAM-1, VCAM-1, E-selectin) between no-flow and flow-subjected human microvascular endothelial cells exposed to X-ray [187]. Beyond ECs, radiation exposure regulated adhesive interaction between the integrin very late antigen 4 or α4β1-integrin (VLA-4) and its receptor VCAM-1 in an ROS-species-dependent manner under low SS (1.2 dynes/cm^2) in monocytes [188]. Apart from signaling studies, in vivo simulation with SS was also adopted in assessing the effect of radiation and prothrombotic conjugate for stable thrombus formation [189].

12.4   Tools for mimicking the hemodynamic shear stress in vitro

Varying shear forces, which occur at microvasculature/vessels and in different regions within the vessel, can be simulated independently under well-defined experimental conditions. Many in vitro models of shear systems, including parallel-plate flow chambers, cone-and-plate flow chambers, parallel disc viscometers, orbital shakers, and rectangular/tubular capillary tubes, have been developed and characterized to simulate in vivo flow environments and to investigate the signal transduction mechanisms through which hemodynamic forces influence EC structure function and response to external stimuli [30]. However, each of these models comes with its own limitations. The orbital shaker and disc viscometer yield a disorderly HFSS. The rectangular and the round capillary tube models yield low amount of cells that are insufficient for performing endpoint biochemical assays. The parallel-plate and cone-and-plate models are versatile, produce a wide range of uniform shear forces, and are hence exploited for most mechanistic studies.

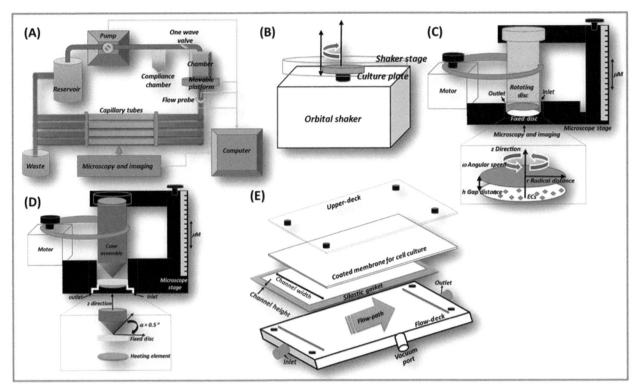

FIGURE 12.3 Schematic representation of the in vitro systems for applying shear flow to ECs. (A) Capillary tube flow chamber; (B) orbital shaker shear force; (C) parallel disc viscometer; (D) cone-and-plate viscometer; and (E) parallel-plate flow chamber. *EC*, Endothelial cells.

12.4.1 Capillary tube flow chamber

The capillary tube flow chamber is an unique bench top system that can be utilized to investigate functional fluctuations of ECs under controlled SS protocols [190,191]. The system allows the growth of ECs inside a rigid glass capillary tube or distensible tubes coupled with the computer-controlled flow in waveforms. Schematically, the culture medium reservoir is hooked up to the compliance chamber through a rate-controlled flow pump (Fig. 12.3A). The flow pump is versatile and could be interchanged for use with capillary tubes or larger distensible tubes. For large distensible tubes, alternatively an inline hydraulic resistor and one-way valve could be adopted. In either case, the fluid passes into a custom sizeable chamber and is then passed through the coated capillary/elastic tubes. The pulse flow can be controlled digitally, while a micromanometer catheter measures luminal pressure. An inline ultrasound flow meter records the phasic and mean flow. Cells can be imaged in real time by fluorescent light microscopy. The system can be regulated by custom-developed pro-portional–derivative–integral feedback control software, and the pump can be controlled by both proportional and integral gain to maintain the flow rate. Tubes that have an incremental elastic modulus mimic the vessels in vivo [163]. The autoclavable tubes are equipped with plastic caps for sterile EC seeding in rotisserie. However, the cell surface area in capillary tubes, representing only <1% of a standard culture dish is a major limitation in this model. Although signaling responses could be assessed in real-time fluorescent imaging [192], this model does not practically allow downstream end-point molecular signaling analysis. However, the distensible tubes may host a large surface area and could provide enough cells for biochemical assays. From an HFSS standpoint, this system is perfect for the study of a diverse range of shear forces from 0.16 dynes to 20 dynes/cm^2. Further, such capillary flow models can also be used to investigate shear forces and cellular interaction (e.g., EC and smooth muscle cells), that precisely simulate the in vivo environment [193]. In this model, a transcapillary EC-smooth muscle cells coculture system was adapted to define the mechanotransduction signaling inflicted by hemodynamically imposed (quantifiable) mechanical stress on ECs.

12.4.2 Orbital shaker shear force

One of the primitive, yet convenient and easily adaptable techniques to induce SS is the use of orbital shaking [194,195]. In this setting, SS can be applied to confluent cultures with a bench top orbital shaker (Fig. 12.3B) [196,197]. The near maximal SS (τ_{max}) can be back calculated with the radius of the orbital rotation (a), density (ρ), viscosity (η) of the medium, and rotation frequency (f), using the formula $\tau_{max} = a\sqrt{\rho\eta(2\pi f)^3}$. For instance, the radius of 1.4-cm rotation with culture medium density of 1.0 g/mL, viscosity of 0.0075, and a frequency of 270 rpm has been indicated to produce a shear force of 14 dynes/cm^2 (adopted from Yun [194]). The advantages of this model include that it (1) can be adopted for diverse culture systems, such as the EC monolayer and coculture; (2) can produce sufficient cells for downstream analysis; (3) allows real-time imaging; and (4) can easily fit radiation settings (X-rays/gamma irradiation) and protocols (SDR/FIR). Despite these advantages, the production of nonuniform laminar SS across the monolayer of cells limits the use of this model.

12.4.3 Parallel disc viscometer

The parallel disc viscometer includes designated directional compartments in a microscopic stage equipped with a motor [198]. The coated flat quartz disc is adhered with a vacuum in a stationary cell plate on a microscopic stage. Atop this cell seeding disc is a cylinder with quartz bottom that can be driven by the motor coupled with a timing belt. The gap distance between the cell seeding plate and parallel top plate can be adjusted at the micrometer level and could be set for the medium to touch the top cylinder and the edges of the chamber (Fig. 12.3C). The cell plate is equipped with injection and ejection ports, and the experiments can be performed at optimal culture temperatures. This model is capable of real-time cellular monitoring and imaging and allows the measurement of molecular endpoints to an extent. It could be useful to investigate the changes in real time with the addition of appropriate antagonist(s) for studying select mechanisms of EC dysfunction. However, the parallel disc viscometer has limitations, including nonuniform SS across the cell monolayer, insufficient yield of exposed cells for biochemical analysis, and the high demand for technical expertise to set the gap distance, the deciding factor for radial SS. With such an intense need for expertise and complexity and disorder in shear forces, this model could be impractical for assessing the radiation damage response in ECs under controlled HFSS.

12.4.4 Cone-and-plate viscometer

The cone-and-plate viscometer is a widely used flow model for studying EC responses to HFSS [81,172,199−202] (Fig. 12.3D). The cone model can produce reproducible flow-induced uniform shear forces. Briefly, the design exhibits a conical surface that rotates relative to a stationary flat plate, where the cells are seeded. The liquid medium between the flat and cone surface is set in motion by the rotation of the cone, leading to a uniform level of shear forces throughout the fluid and over the surface of the cells. The setup can be mounted onto the microscope stage and can be equipped with temperature-maintaining heating elements, similar to the parallel-plate model. The coated surface for cell seeding is fixed onto the plate with vacuum pressure. Additional ports in the diametrically opposed ends are included to adopt experimental interventions. The cone assembly mounted on the dual micrometer-driven translation stages allows the placement of cone gently into the fluid filter environment without disturbing the cells. The rotation of the cone assembly is driven by the timing belt pulley connection to the motor. Control of the dynamic fluid environment can be accomplished by unique motor drive systems coupled with programmed software with which acceleration, deceleration, angular velocity, time duration, and direction can be controlled. The shear force (τ) can be calculated with the viscosity of the fluid (μ), rotational speed (ω), and the angle of the cone (θ) using the formula $\tau = \mu\omega/\theta$. Adding to these advantages, this system can produce shear forces ranging from 0.1 dynes/cm^2, with force rates ranging from 0.1 to 5000 dynes/cm/s. The versatility of the system is the ability to exchange the cone surface with various angles or materials, allowing us to mimic flow conditions to ECs in distinct regions of circulatory system. For instance, Blackman and colleagues reconfigured the cone-and-plate model to a dynamic flow system that can produce pulsatile waveform shear forces simulating the flow in human abdominal aorta [203]. Consistently, studies with this dynamic flow model affirmed its benefit in mimicking actual shear type and force in circumstances such as carotid bifurcations and coronary collateral vessels [204−207].

12.4.5 Parallel-plate flow chamber

The parallel-plate model is the most widely used system for simulating HFSS in vitro and is highly versatile to suit the specifics of vasculatures and the design of experiments [208−213]. With the capability of producing a wide range of uniform shear forces and with the feasibility of collecting sufficient cells for endpoint analysis, this model is aptly used in investigating the physiological and molecular signaling responses in ECs. Although the design of this model has evolved considerably in recent years, conceptually the design utilizes a parallel-plate flow channel created by a gasket with rectangular cutout. This allows a uniform channel height along the path of the flow that is generated between two slits at either end (inlet→outlet) of the rectangular chamber by a constant pressure defined by the pump system (Fig. 12.3E). The shear force (τ) can be calculated with the viscosity of the fluid (μ), width of the channel (w), height of the channel (h), and the flow rate (Q) using the formula $\tau = 6\mu Q/wh^2$. The homogeneity of the force stimulus, ease of adaptability to fit specific experimental design, ease of the access to the cell culture, and ability to allow recovery of sufficient cells for the downstream analysis are the major advantages of this approach.

A number of modifications to the base model of the parallel-plate flow chamber have been adopted in last two decades to meet specific experimental demands. Brown and Larson in 2001 reconfigured the original plate flow chamber to incorporate single-pass and recirculating modalities, minimize tubing and liquid volume, and to permit injection of interventional reagent [214]. The parallel-plate flow model grew popular and was commercially adopted for the development of application-specific assays (e.g., Ibidi flow adhesion assay, Flexcell streamer SS device, and Glycotech Parallel-Plate Flow Chambers). Interestingly, Usami and colleagues designed a parallel-plate flow chamber to study the effect of spatial variation of laminar SS on ECs with a tapered channel that creates a linear variation of shear force—desired (maximum) force at the entrance and near zero at the finish [215]. Such a strategy would allow us to assess EC responses to a spatially arranged range of shear forces in a single run at constant flow. Defined disturbed flow could also be established by installing silicone gaskets to produce backward facing vertical step flow in the channel, allowing assessment of the EC response to both laminar and disturbed flows in the same monolayer (reviewed elsewhere [80] in detail).

12.5 Conclusions and future perspectives

The endothelium being widely distributed and present in every tissue puts the ECs in the field of radiation exposure, irrespective of site-directed or whole-body radiation protocols. The endothelium is known to mediate both systemic and local inflammatory reactions and hence has a functional role in orchestrating the acute or delayed radiation response that actuates normal tissue toxicity and organ failure. Despite considerable attempts to gain insights into the endothelial radiation response and its implications in acute/delayed effects, in vitro mechanistic studies failed to mimic the in vivo environment. This is because shear stress associated with blood flow that drives endothelial phenotype and response has largely been ignored in radiation exposure studies. The criticality of the effect of SS on the physiological function of endothelium cannot be understated; any outcomes reporting the endothelial response to radiation (or beyond) without SS in place may not mimic in vivo conditions. In particular, these studies failed to address the significance of the shear forces-induced mechanotransduction signaling in response to radiation and, hence, are not translatable to health and disease. It is crucial that the studies investigating radiation-induced endothelial response/injury should therefore consider the context-specific shear forces, site- and origin-specific endothelium, and context-specific radiation (quality, source, dose, dose-rate, and fractionation) conditions. Such in vivo environment mimicking, clinically translatable vascular models to study radiation-induced endothelial effects is pivotal as the field is poised to expand considerably with the changing environment (with potential radiological threats and increasing nuclear industry demand) and with newer challenges in the space travel. This chapter highlights the significance of these variables, particularly the hemodynamic flow-SS in experiments to assess the endothelial response to radiation, and beyond.

References

[1] Kamiya A, Togawa T. Adaptive regulation of wall shear stress to flow change in the canine carotid artery. Am J Physiol 1980;239(1):H14−21.

[2] Langille BL, O'Donnell F. Reductions in arterial diameter produced by chronic decreases in blood flow are endothelium-dependent. Science 1986;231(4736):405−7.

[3] Zarins CK, et al. Shear stress regulation of artery lumen diameter in experimental atherogenesis. J Vasc Surg 1987;5(3):413−20.

[4] Gibbons GH, Dzau VJ. The emerging concept of vascular remodeling. N Engl J Med 1994;330(20):1431−8.

[5] Fry DL. Certain histological and chemical responses of the vascular interface to acutely induced mechanical stress in the aorta of the dog. Circ Res 1969;24(1):93−108.

[6] Caro CG, Fitz-Gerald JM, Schroter RC. Atheroma and arterial wall shear. Observation, correlation and proposal of a shear dependent mass transfer mechanism for atherogenesis. Proc R Soc Lond B Biol Sci 1971;177(1046):109−59.

[7] Zarins CK, et al. Carotid bifurcation atherosclerosis. Quantitative correlation of plaque localization with flow velocity profiles and wall shear stress. Circ Res 1983;53(4):502−14.

[8] Kerber CW, et al. Flow dynamics in a fatal aneurysm of the basilar artery. Am J Neuroradiol 1996;17(8):1417−21.

[9] Rossitti S, Svendsen P. Shear stress in cerebral arteries supplying arteriovenous malformations. Acta Neurochir (Wien) 1995;137(3−4):138−45 discussion 145.

[10] Davies PF. Flow-mediated endothelial mechanotransduction. Physiol Rev 1995;75(3):519−60.

[11] LaBarbera M. Principles of design of fluid transport systems in zoology. Science 1990;249(4972):992−1000.

[12] Asakura T, Karino T. Flow patterns and spatial distribution of atherosclerotic lesions in human coronary arteries. Circ Res 1990;66 (4):1045−66.

[13] Gnasso A, et al. In vivo association between low wall shear stress and plaque in subjects with asymmetrical carotid atherosclerosis. Stroke 1997;28(5):993−8.

[14] Motomiya M, Karino T. Flow patterns in the human carotid artery bifurcation. Stroke 1984;15(1):50−6.

[15] Bharadvaj BK, Mabon RF, Giddens DP. Steady flow in a model of the human carotid bifurcation. Part I − flow. Vis J Biomech 1982;15 (5):349−62.

[16] Malek AM, Izumo S. Control of endothelial cell gene expression by flow. J Biomech 1995;28(12):1515−28.

[17] Gimbrone Jr. MA. Vascular endothelium: an integrator of pathophysiologic stimuli in atherosclerosis. Am J Cardiol 1995;75(6):67B−70B.

[18] Traub O, Berk BC. Laminar shear stress: mechanisms by which endothelial cells transduce an atheroprotective force. Arterioscler Thromb Vasc Biol 1998;18(5):677−85.

[19] Nerem RM, et al. The study of the influence of flow on vascular endothelial biology. Am J Med Sci 1998;316(3):169−75.

[20] Chien S, Li S, Shyy YJ. Effects of mechanical forces on signal transduction and gene expression in endothelial cells. Hypertension 1998;31(1 Pt 2):162−9.

[21] Woth K, et al. Endothelial cells are highly heterogeneous at the level of cytokine-induced insulin resistance. Exp Dermatol 2013;22 (11):714−18.

[22] Aird WC. Phenotypic heterogeneity of the endothelium: I. Structure, function, and mechanisms. Circ Res 2007;100(2):158−73.

[23] Aird WC. Mechanisms of endothelial cell heterogeneity in health and disease. Circ Res 2006;98(2):159−62.

[24] Gershon MD, Bursztajn S. Properties of the enteric nervous system: limitation of access of intravascular macromolecules to the myenteric plexus and muscularis externa. J Comp Neurol 1978;180(3):467−88.

[25] Natarajan M, et al. Hemodynamic flow-induced mechanotransduction signaling influences the radiation response of the vascular endothelium. Radiat Res 2016;186(2):175−88.

[26] Chien S. Mechanotransduction and endothelial cell homeostasis: the wisdom of the cell. Am J Physiol Heart Circ Physiol 2007;292(3): H1209−24.

[27] Cines DB, et al. Endothelial cells in physiology and in the pathophysiology of vascular disorders. Blood 1998;91(10):3527−61.

[28] Schalkwijk CG, Stehouwer CD. Vascular complications in diabetes mellitus: the role of endothelial dysfunction. Clin Sci (Lond) 2005;109 (2):143−59.

[29] Chi JT, et al. Endothelial cell diversity revealed by global expression profiling. Proc Natl Acad Sci USA 2003;100(19):10623−8.

[30] Chiu JJ, Chien S. Effects of disturbed flow on vascular endothelium: pathophysiological basis and clinical perspectives. Physiol Rev 2011;91 (1):327−87.

[31] Papaioannou TG, Stefanadis C. Vascular wall shear stress: basic principles and methods. Hellenic J Cardiol 2005;46(1):9−15.

[32] Jaffe EA, et al. Culture of human endothelial cells derived from umbilical veins. Identification by morphologic and immunologic criteria. J Clin Invest 1973;52(11):2745−56.

[33] Gimbrone Jr. MA, Cotran RS, Folkman J. Human vascular endothelial cells in culture. Growth and DNA synthesis. J Cell Biol 1974;60 (3):673−84.

[34] Lewis LJ, et al. Replication of human endothelial cells in culture. Science 1973;181(4098):453−4.

[35] Malek AM, Alper SL, Izumo S. Hemodynamic shear stress and its role in atherosclerosis. JAMA 1999;282(21):2035−42.

[36] Giddens DP, Zarins CK, Glagov S. The role of fluid mechanics in the localization and detection of atherosclerosis. J Biomech Eng 1993;115 (4B):588−94.

[37] Kraiss LW, et al. Shear stress regulates smooth muscle proliferation and neointimal thickening in porous polytetrafluoroethylene grafts. Arterioscler Thromb 1991;11(6):1844−52.

[38] Girerd X, et al. Remodeling of the radial artery in response to a chronic increase in shear stress. Hypertension 1996;27(3 Pt 2):799−803.

[39] McCormick SM, et al. DNA microarray reveals changes in gene expression of shear stressed human umbilical vein endothelial cells. Proc Natl Acad Sci USA 2001;98(16):8955−60.

[40] Chen BP, et al. DNA microarray analysis of gene expression in endothelial cells in response to 24-h shear stress. Physiol Genomics 2001;7(1):55−63.

[41] Shyy JY, Chien S. Role of integrins in endothelial mechanosensing of shear stress. Circ Res 2002;91(9):769−75.

[42] Ohura N, et al. Global analysis of shear stress-responsive genes in vascular endothelial cells. J Atheroscler Thromb 2003;10(5):304−13.

[43] Ben Driss A, et al. Arterial expansive remodeling induced by high flow rates. Am J Physiol 1997;272(2 Pt 2):H851−8.

[44] Rudic RD, et al. Direct evidence for the importance of endothelium-derived nitric oxide in vascular remodeling. J Clin Invest 1998;101 (4):731−6.

[45] Mattsson EJ, et al. Increased blood flow induces regression of intimal hyperplasia. Arterioscler Thromb Vasc Biol 1997;17(10):2245−9.

[46] Walpola PL, Gotlieb AI, Langille BL. Monocyte adhesion and changes in endothelial cell number, morphology, and F-actin distribution elicited by low shear stress in vivo. Am J Pathol 1993;142(5):1392−400.

[47] Walpola PL, et al. Expression of ICAM-1 and VCAM-1 and monocyte adherence in arteries exposed to altered shear stress. Arterioscler Thromb Vasc Biol 1995;15(1):2−10.

[48] Furchgott RF, Zawadzki JV. The obligatory role of endothelial cells in the relaxation of arterial smooth muscle by acetylcholine. Nature 1980;288(5789):373−6.

[49] Palmer RM, Ashton DS, Moncada S. Vascular endothelial cells synthesize nitric oxide from L-arginine. Nature 1988;333(6174):664−6.

[50] Busse R, Hecker M, Fleming I. Control of nitric oxide and prostacyclin synthesis in endothelial cells. Arzneimittelforschung 1994;44 (3A):392−6.

[51] de Graaf JC, et al. Nitric oxide functions as an inhibitor of platelet adhesion under flow conditions. Circulation 1992;85(6):2284−90.

[52] Garg UC, Hassid A. Nitric oxide-generating vasodilators and 8-bromo-cyclic guanosine monophosphate inhibit mitogenesis and proliferation of cultured rat vascular smooth muscle cells. J Clin Invest 1989;83(5):1774−7.

[53] Marks DS, et al. Inhibition of neointimal proliferation in rabbits after vascular injury by a single treatment with a protein adduct of nitric oxide. J Clin Invest 1995;96(6):2630−8.

[54] Janssens S, et al. Human endothelial nitric oxide synthase gene transfer inhibits vascular smooth muscle cell proliferation and neointima formation after balloon injury in rats. Circulation 1998;97(13):1274−81.

[55] Cho A, et al. Effects of changes in blood flow rate on cell death and cell proliferation in carotid arteries of immature rabbits. Circ Res 1997;81 (3):328−37.

[56] Ishida T, et al. Fluid shear stress-mediated signal transduction: how do endothelial cells transduce mechanical force into biological responses? Ann N Y Acad Sci 1997;811:12−23 discussion 23−4.

[57] Malek AM, Izumo S. Molecular aspects of signal transduction of shear stress in the endothelial cell. J Hypertens 1994;12(9):989−99.

[58] Kaiser D, Freyberg MA, Friedl P. Lack of hemodynamic forces triggers apoptosis in vascular endothelial cells. Biochem Biophys Res Commun 1997;231(3):586−90.

[59] Masuda H, et al. Increase in endothelial cell density before artery enlargement in flow-loaded canine carotid artery. Arteriosclerosis 1989;9 (6):812−23.

[60] Fleming I, Bauersachs J, Busse R. Calcium-dependent and calcium-independent activation of the endothelial NO synthase. J Vasc Res 1997;34 (3):165−74.

[61] Chun TH, et al. Shear stress augments expression of C-type natriuretic peptide and adrenomedullin. Hypertension 1997;29(6):1296−302.

[62] Ohno M, et al. Fluid shear stress induces endothelial transforming growth factor beta-1 transcription and production. Modulation potassium channel blockade. J Clin Invest 1995;95(3):1363−9.

[63] Malek AM, et al. Endothelial expression of thrombomodulin is reversibly regulated by fluid shear stress. Circ Res 1994;74(5):852−60.

[64] Diamond SL, Eskin SG, McIntire LV. Fluid flow stimulates tissue plasminogen activator secretion by cultured human endothelial cells. Science 1989;243(4897):1483−5.

[65] Kawai Y, et al. Hemodynamic forces modulate the effects of cytokines on fibrinolytic activity of endothelial cells. Blood 1996;87 (6):2314−21.

[66] Grabowski EF. Thrombolysis, flow, and vessel wall interactions. J Vasc Interv Radiol 1995;6(6 Pt 2 Suppl):25S−9S.

[67] Inoue N, et al. Shear stress modulates expression of Cu/Zn superoxide dismutase in human aortic endothelial cells. Circ Res 1996;79(1):32−7.

[68] De Keulenaer GW, et al. Oscillatory and steady laminar shear stress differentially affect human endothelial redox state: role of a superoxide-producing NADH oxidase. Circ Res 1998;82(10):1094−101.

[69] Sharefkin JB, et al. Fluid flow decreases preproendothelin mRNA levels and suppresses endothelin-1 peptide release in cultured human endothelial cells. J Vasc Surg 1991;14(1):1−9.

[70] Masatsugu K, et al. Physiologic shear stress suppresses endothelin-converting enzyme-1 expression in vascular endothelial cells. J Cardiovasc Pharmacol 1998;31(Suppl. 1):S42−5.

[71] Resnick N, et al. Platelet-derived growth factor B chain promoter contains a cis-acting fluid shear-stress-responsive element. Proc Natl Acad Sci USA 1993;90(10):4591−5.

[72] Kosaki K, et al. Fluid shear stress increases the production of granulocyte-macrophage colony-stimulating factor by endothelial cells via mRNA stabilization. Circ Res 1998;82(7):794−802.

[73] Shyy YJ, et al. Fluid shear stress induces a biphasic response of human monocyte chemotactic protein 1 gene expression in vascular endothelium. Proc Natl Acad Sci USA 1994;91(11):4678−82.

[74] Ando J, et al. Shear stress inhibits adhesion of cultured mouse endothelial cells to lymphocytes by downregulating VCAM-1 expression. Am J Physiol 1994;267(3 Pt 1):C679−87.

[75] Korenaga R, et al. Negative transcriptional regulation of the VCAM-1 gene by fluid shear stress in murine endothelial cells. Am J Physiol 1997;273(5 Pt 1):C1506−15.

[76] Ziegler T, et al. Influence of oscillatory and unidirectional flow environments on the expression of endothelin and nitric oxide synthase in cultured endothelial cells. Arterioscler Thromb Vasc Biol 1998;18(5):686−92.

[77] Malek A, Izumo S. Physiological fluid shear stress causes downregulation of endothelin-1 mRNA in bovine aortic endothelium. Am J Physiol 1992;263(2 Pt 1):C389−96.

[78] Chappell DC, et al. Oscillatory shear stress stimulates adhesion molecule expression in cultured human endothelium. Circ Res 1998;82 (5):532−9.

[79] DePaola N, et al. Vascular endothelium responds to fluid shear stress gradients. Arterioscler Thromb 1992;12(11):1254−7.

[80] Chiu JJ, et al. Effects of disturbed flow on endothelial cells. J Biomech Eng 1998;120(1):2−8.

[81] Davies PF, et al. Turbulent fluid shear stress induces vascular endothelial cell turnover in vitro. Proc Natl Acad Sci USA 1986;83(7):2114−17.

[82] Topper JN, et al. Identification of vascular endothelial genes differentially responsive to fluid mechanical stimuli: cyclooxygenase-2, manganese superoxide dismutase, and endothelial cell nitric oxide synthase are selectively up-regulated by steady laminar shear stress. Proc Natl Acad Sci USA 1996;93(19):10417−22.

[83] Satyamitra MM, DiCarlo AL, Taliaferro L. Understanding the pathophysiology and challenges of development of medical countermeasures for radiation-induced vascular/endothelial cell injuries: report of a NIAID workshop, August 20, 2015. Radiat Res 2016;186(2):99−111.

[84] Al-Abdulsalam A, Brindhaban A. Occupational radiation exposure among the staff of departments of nuclear medicine and diagnostic radiology in Kuwait. Med Princ Pract 2014;23(2):129−33.

[85] Ko S, et al. Occupational radiation exposure and its health effects on interventional medical workers: study protocol for a prospective cohort study. BMJ Open 2017;7(12):e018333.

[86] Gassmann A. Zur histologie der rontgenulcera. Fortschr a d Geb d Röntgenstrahlen 1899;2:199.

[87] Johnson KG, Yano K, Kato H. Cerebral vascular disease in Hiroshima, Japan. J Chronic Dis 1967;20(7):545−59.

[88] Wong FL, et al. Noncancer disease incidence in the atomic bomb survivors: 1958−1986. Radiat Res 1993;135(3):418−30.

[89] Shimizu Y, et al. Radiation exposure and circulatory disease risk: Hiroshima and Nagasaki atomic bomb survivor data, 1950−2003. BMJ 2010;340:b5349.

[90] Sera N, et al. The association between chronic kidney disease and cardiovascular disease risk factors in atomic bomb survivors. Radiat Res 2013;179(1):46−52.

[91] Little MP. Cancer and non-cancer effects in Japanese atomic bomb survivors. J Radiol Prot 2009;29(2A):A43−59.

[92] Roth NM, Sontag MR, Kiani MF. Early effects of ionizing radiation on the microvascular networks in normal tissue. Radiat Res 1999;151 (3):270−7.

[93] Baker DG, Krochak RJ. The response of the microvascular system to radiation: a review. Cancer Invest 1989;7(3):287−94.

[94] Evans ML, et al. Changes in vascular permeability following thorax irradiation in the rat. Radiat Res 1986;107(2):262−71.

[95] Krishnan L, Krishnan EC, Jewell WR. Immediate effect of irradiation on microvasculature. Int J Radiat Oncol Biol Phys 1988;15(1):147−50.

[96] Debbage PL, et al. Vascular permeability and hyperpermeability in a murine adenocarcinoma after fractionated radiotherapy: an ultrastructural tracer study. Histochem Cell Biol 2000;114(4):259−75.

[97] Hopewell JW, et al. Microvasculature and radiation damage. Recent Results Cancer Res 1993;130:1−16.

[98] Fajardo LF. The pathology of ionizing radiation as defined by morphologic patterns. Acta Oncol 2005;44(1):13−22.

[99] Moeller BJ, et al. Pleiotropic effects of HIF-1 blockade on tumor radiosensitivity. Cancer Cell 2005;8(2):99−110.

[100] Ch'ang HJ, et al. ATM regulates target switching to escalating doses of radiation in the intestines. Nat Med 2005;11(5):484−90.

[101] Grabham P, et al. Effects of ionizing radiation on three-dimensional human vessel models: differential effects according to radiation quality and cellular development. Radiat Res 2011;175(1):21−8.

[102] Cervelli T, et al. Effects of single and fractionated low-dose irradiation on vascular endothelial cells. Atherosclerosis 2014;235(2):510−18.

[103] Fajardo LF. Is the pathology of radiation injury different in small vs large blood vessels? Cardiovasc Radiat Med 1999;1(1):108−10.

[104] Krishnaswamy G, et al. Human endothelium as a source of multifunctional cytokines: molecular regulation and possible role in human disease. J Interferon Cytokine Res 1999;19(2):91−104.

[105] Moeller BJ, et al. Radiation activates HIF-1 to regulate vascular radiosensitivity in tumors: role of reoxygenation, free radicals, and stress granules. Cancer Cell 2004;5(5):429−41.

[106] Korpela E, Liu SK. Endothelial perturbations and therapeutic strategies in normal tissue radiation damage. Radiat Oncol 2014;9:266.

[107] Paris F, et al. Endothelial apoptosis as the primary lesion initiating intestinal radiation damage in mice. Science 2001;293(5528):293−7.

[108] Brown CK, et al. Glioblastoma cells block radiation-induced programmed cell death of endothelial cells. FEBS Lett 2004;565(1−3):167−70.

[109] Natarajan M, et al. Induced telomerase activity in primary aortic endothelial cells by low-LET gamma-radiation is mediated through NF-kappaB activation. Br J Radiol 2008;81(969):711−20.

[110] Soucy KG, et al. Dietary inhibition of xanthine oxidase attenuates radiation-induced endothelial dysfunction in rat aorta. J Appl Physiol (1985) 2010;108(5):1250−8.

[111] Cedervall J, Dimberg A, Olsson AK. Tumor-induced local and systemic impact on blood vessel function. Mediators Inflamm 2015;2015:418290.

[112] Ribatti D, et al. Hematopoiesis and angiogenesis: a link between two apparently independent processes. J Hematother Stem Cell Res 2000;9 (1):13−19.

[113] Onoda JM, Kantak SS, Diglio CA. Radiation induced endothelial cell retraction in vitro: correlation with acute pulmonary edema. Pathol Oncol Res 1999;5(1):49−55.

[114] Mooteri SN, et al. WR-1065 and radioprotection of vascular endothelial cells. II. Morphology. Radiat Res 1996;145(2):217−24.

[115] Speidel MT, et al. Morphological, biochemical, and molecular changes in endothelial cells after alpha-particle irradiation. Radiat Res 1993;136(3):373−81.

[116] Friedman M, et al. Reversible alterations in cultured pulmonary artery endothelial cell monolayer morphology and albumin permeability induced by ionizing radiation. J Cell Physiol 1986;129(2):237−49.

[117] Gabrys D, et al. Radiation effects on the cytoskeleton of endothelial cells and endothelial monolayer permeability. Int J Radiat Oncol Biol Phys 2007;69(5):1553−62.

[118] Sharma P, Templin T, Grabham P. Short term effects of gamma radiation on endothelial barrier function: uncoupling of PECAM-1. Microvasc Res 2013;86:11−20.

[119] Grabham P, et al. Two distinct types of the inhibition of vasculogenesis by different species of charged particles. Vasc Cell 2013;5(1):16.

[120] Grabham P, Sharma P. The effects of radiation on angiogenesis. Vasc Cell 2013;5(1):19.

[121] Grabham P, Bigelow A, Geard C. DNA damage foci formation and decline in two-dimensional monolayers and in three-dimensional human vessel models: differential effects according to radiation quality. Int J Radiat Biol 2012;88(6):493−500.

[122] Zheng X, et al. Endothelial cell migration was impaired by irradiation-induced inhibition of SHP-2 in radiotherapy: an in vitro study. J Radiat Res 2011;52(3):320—8.

[123] Milliat F, et al. [Role of endothelium in radiation-induced normal tissue damages]. Ann Cardiol Angeiol (Paris) 2008;57(3):139—48.

[124] Little MP, et al. A systematic review of epidemiological associations between low and moderate doses of ionizing radiation and late cardiovascular effects, and their possible mechanisms. Radiat Res 2008;169(1):99—109.

[125] Nowell CS, et al. Chronic inflammation imposes aberrant cell fate in regenerating epithelia through mechanotransduction. Nat Cell Biol 2016;18(2):168—80.

[126] Rodemann HP, Blaese MA. Responses of normal cells to ionizing radiation. Semin Radiat Oncol 2007;17(2):81—8.

[127] Oh CW, et al. Induction of a senescence-like phenotype in bovine aortic endothelial cells by ionizing radiation. Radiat Res 2001;156 (3):232—40.

[128] Rousseau M, et al. RhoA GTPase regulates radiation-induced alterations in endothelial cell adhesion and migration. Biochem Biophys Res Commun 2011;414(4):750—5.

[129] Bentzen SM. Preventing or reducing late side effects of radiation therapy: radiobiology meets molecular pathology. Nat Rev Cancer 2006;6 (9):702—13.

[130] Adams MJ, et al. Radiation-associated cardiovascular disease. Crit Rev Oncol Hematol 2003;45(1):55—75.

[131] Garlanda C, Dejana E. Heterogeneity of endothelial cells. Specific markers. Arterioscler Thromb Vasc Biol 1997;17(7):1193—202.

[132] Finch W, Shamsa K, Lee MS. Cardiovascular complications of radiation exposure. Rev Cardiovasc Med 2014;15(3):232—44.

[133] Adams MJ, et al. Radiation dose associated with renal failure mortality: a potential pathway to partially explain increased cardiovascular disease mortality observed after whole-body irradiation. Radiat Res 2012;177(2):220—8.

[134] Yeh ET, et al. Cardiovascular complications of cancer therapy: diagnosis, pathogenesis, and management. Circulation 2004;109(25):3122—31.

[135] Zhai GG, et al. Radiation enhances the invasive potential of primary glioblastoma cells via activation of the Rho signaling pathway. J Neurooncol 2006;76(3):227—37.

[136] Tan J, Hallahan DE. Growth factor-independent activation of protein kinase B contributes to the inherent resistance of vascular endothelium to radiation-induced apoptotic response. Cancer Res 2003;63(22):7663—7.

[137] Schmidt-Ullrich RK, et al. Signal transduction and cellular radiation responses. Radiat Res 2000;153(3):245—57.

[138] Kumar P, Miller AI, Polverini PJ. p38 MAPK mediates gamma-irradiation-induced endothelial cell apoptosis, and vascular endothelial growth factor protects endothelial cells through the phosphoinositide 3-kinase-Akt-Bcl-2 pathway. J Biol Chem 2004;279(41):43352—60.

[139] Garcia-Barros M, et al. Tumor response to radiotherapy regulated by endothelial cell apoptosis. Science 2003;300(5622):1155—9.

[140] Folkman J. Role of angiogenesis in tumor growth and metastasis. Semin Oncol 2002;29(6 Suppl. 16):15—18.

[141] Folkman J, Shing Y. Angiogenesis. J Biol Chem 1992;267(16):10931—4.

[142] Denekamp J. Review article: angiogenesis, neovascular proliferation and vascular pathophysiology as targets for cancer therapy. Br J Radiol 1993;66(783):181—96.

[143] Camphausen K, Menard C. Angiogenesis inhibitors and radiotherapy of primary tumours. Expert Opin Biol Ther 2002;2(5):477—81.

[144] Gorski DH, et al. Blockage of the vascular endothelial growth factor stress response increases the antitumor effects of ionizing radiation. Cancer Res 1999;59(14):3374—8.

[145] Geng L, et al. Inhibition of vascular endothelial growth factor receptor signaling leads to reversal of tumor resistance to radiotherapy. Cancer Res 2001;61(6):2413—19.

[146] Hess C, et al. Effect of VEGF receptor inhibitor PTK787/ZK222584 [correction of ZK222548] combined with ionizing radiation on endothelial cells and tumour growth. Br J Cancer 2001;85(12):2010—16.

[147] Kozin SV, et al. Vascular endothelial growth factor receptor-2-blocking antibody potentiates radiation-induced long-term control of human tumor xenografts. Cancer Res 2001;61(1):39—44.

[148] Lund EL, Bastholm L, Kristjansen PE. Therapeutic synergy of TNP-470 and ionizing radiation: effects on tumor growth, vessel morphology, and angiogenesis in human glioblastoma multiforme xenografts. Clin Cancer Res 2000;6(3):971—8.

[149] Ning S, et al. The antiangiogenic agents SU5416 and SU6668 increase the antitumor effects of fractionated irradiation. Radiat Res 2002;157 (1):45—51.

[150] Alavi A, et al. Role of Raf in vascular protection from distinct apoptotic stimuli. Science 2003;301(5629):94—6.

[151] Edwards E, et al. Phosphatidylinositol 3-kinase/Akt signaling in the response of vascular endothelium to ionizing radiation. Cancer Res 2002;62(16):4671—7.

[152] Mauceri HJ, et al. Combined effects of angiostatin and ionizing radiation in antitumour therapy. Nature 1998;394(6690):287—91.

[153] Wang J. Marker-based estimates of relatedness and inbreeding coefficients: an assessment of current methods. J Evol Biol 2014;27 (3):518—30.

[154] Khaled S, Gupta KB, Kucik DF. Ionizing radiation increases adhesiveness of human aortic endothelial cells via a chemokine-dependent mechanism. Radiat Res 2012;177(5):594—601.

[155] Yuan Y, Lee SH, Wu S. The role of ROS in ionizing radiation-induced VLA-4 mediated adhesion of RAW264.7 cells to VCAM-1 under flow conditions. Radiat Res 2013;179(1):62—8.

[156] Prabhakarpandian B, et al. Expression and functional significance of adhesion molecules on cultured endothelial cells in response to ionizing radiation. Microcirculation 2001;8(5):355—64.

[157] Mohan S, Mohan N, Sprague EA. Differential activation of NF-kappa B in human aortic endothelial cells conditioned to specific flow environments. Am J Physiol 1997;273(2 Pt 1):C572—8.

[158] Mohan S, et al. Regulation of low shear flow-induced HAEC VCAM-1 expression and monocyte adhesion. Am J Physiol 1999;276(5): C1100−7.

[159] Mohan S, et al. Low shear stress preferentially enhances IKK activity through selective sources of ROS for persistent activation of NF-kappaB in endothelial cells. Am J Physiol Cell Physiol 2007;292(1):C362−71.

[160] Mohan S, et al. High glucose induced NF-kappaB DNA-binding activity in HAEC is maintained under low shear stress but inhibited under high shear stress: role of nitric oxide. Atherosclerosis 2003;171(2):225−34.

[161] Mohan S, et al. IkappaBalpha-dependent regulation of low-shear flow-induced NF-kappa B activity: role of nitric oxide. Am J Physiol Cell Physiol 2003;284(4):C1039−47.

[162] Hamuro M, et al. High glucose induced nuclear factor kappa B mediated inhibition of endothelial cell migration. Atherosclerosis 2002;162 (2):277−87.

[163] Nichols WW, O'Rourke MF. Contours of pressure and flow waves in arteries. McDonald's blood flow arteries 1998;170−200.

[164] Wang J, et al. Mechanomics: an emerging field between biology and biomechanics. Protein Cell 2014;5(7):518−31.

[165] Gasparski AN, Beningo KA. Mechanoreception at the cell membrane: more than the integrins. Arch Biochem Biophys 2015;586:20−6.

[166] Barakat AI, et al. A flow-activated chloride-selective membrane current in vascular endothelial cells. Circ Res 1999;85(9):820−8.

[167] Grynkiewicz G, Poenie M, Tsien RY. A new generation of Ca^{2+} indicators with greatly improved fluorescence properties. J Biol Chem 1985;260(6):3440−50.

[168] Jufri NF, et al. Mechanical stretch: physiological and pathological implications for human vascular endothelial cells. Vasc Cell 2015;7:8.

[169] Chen W, et al. Role of integrin beta4 in lung endothelial cell inflammatory responses to mechanical stress. Sci Rep 2015;5:16529.

[170] Zakkar M, Angelini GD, Emanueli C. Regulation of vascular endothelium inflammatory signalling by shear stress. Curr Vasc Pharmacol 2016;14(2):181−6.

[171] Veeraraghavan J, et al. Low-dose gamma-radiation-induced oxidative stress response in mouse brain and gut: regulation by NFκB-MnSOD cross-signaling. Mutat Res 2011;718(1−2):44−55.

[172] Dewey Jr. CF, et al. The dynamic response of vascular endothelial cells to fluid shear stress. J Biomech Eng 1981;103(3):177−85.

[173] Paszkowiak JJ, Dardik A. Arterial wall shear stress: observations from the bench to the bedside. Vasc Endovasc Surg 2003;37(1):47−57.

[174] Manevich Y, et al. Oxidative burst and NO generation as initial response to ischemia in flow-adapted endothelial cells. Am J Physiol Heart Circ Physiol 2001;280(5):H2126−35.

[175] Murley JS, et al. Delayed radioprotection by nuclear transcription factor kappaB-mediated induction of manganese superoxide dismutase in human microvascular endothelial cells after exposure to the free radical scavenger WR1065. Free Radic Biol Med 2006;40(6):1004−16.

[176] Coleman MA, et al. Low-dose radiation affects cardiac physiology: gene networks and molecular signaling in cardiomyocytes. Am J Physiol Heart Circ Physiol 2015;309(11):H1947−63.

[177] Cunningham KS, Gotlieb AI. The role of shear stress in the pathogenesis of atherosclerosis. Lab Invest 2005;85(1):9−23.

[178] Chen XL, et al. Laminar flow induction of antioxidant response element-mediated genes in endothelial cells. A novel anti-inflammatory mechanism. J Biol Chem 2003;278(2):703−11.

[179] Adamo L, et al. Biomechanical forces promote embryonic haematopoiesis. Nature 2009;459(7250):1131−5.

[180] Schwartz SM, Gajdusek CM, Selden 3rd SC. Vascular wall growth control: the role of the endothelium. Arteriosclerosis 1981;1(2):107−26.

[181] Yoder MC. Human endothelial progenitor cells. Cold Spring Harb Perspect Med 2012;2(7):a006692.

[182] Ingram DA, et al. Vessel wall-derived endothelial cells rapidly proliferate because they contain a complete hierarchy of endothelial progenitor cells. Blood 2005;105(7):2783−6.

[183] Fuchs E, Cleveland DW. A structural scaffolding of intermediate filaments in health and disease. Science 1998;279(5350):514−19.

[184] Coulombe PA, Wong P. Cytoplasmic intermediate filaments revealed as dynamic and multipurpose scaffolds. Nat Cell Biol 2004;6 (8):699−706.

[185] Weng YR, Cui Y, Fang JY. Biological functions of cytokeratin 18 in cancer. Mol Cancer Res 2012;10(4):485−93.

[186] Peng X, et al. Wall stiffness suppresses Akt/eNOS and cytoprotection in pulse-perfused endothelium. Hypertension 2003;41(2):378−81.

[187] Erbeldinger N, et al. Measuring leukocyte adhesion to (primary) endothelial cells after photon and charged particle exposure with a dedicated laminar flow chamber. Front Immunol 2017;8:627.

[188] Lee SH, et al. Regulation of ionizing radiation-induced adhesion of breast cancer cells to fibronectin by alpha5beta1 integrin. Radiat Res 2014;181(6):650−8.

[189] Subramanian S, et al. Stable thrombus formation on irradiated microvascular endothelial cells under pulsatile flow: pre-testing annexin V-thrombin conjugate for treatment of brain arteriovenous malformations. Thromb Res 2018;167:104−12.

[190] Peng X, et al. In vitro system to study realistic pulsatile flow and stretch signaling in cultured vascular cells. Am J Physiol Cell Physiol 2000;279(3):C797−805.

[191] Moore Jr. JE, et al. A device for subjecting vascular endothelial cells to both fluid shear stress and circumferential cyclic stretch. Ann Biomed Eng 1994;22(4):416−22.

[192] Ziegelstein RC, et al. Initial contact and subsequent adhesion of human neutrophils or monocytes to human aortic endothelial cells releases an endothelial intracellular calcium store. Circulation 1994;90(4):1899−907.

[193] Redmond EM, Cahill PA, Sitzmann JV. Perfused transcapillary smooth muscle and endothelial cell co-culture − a novel in vitro model. Vitro Cell Dev Biol Anim 1995;31(8):601−9.

[194] Yun S, et al. Transcription factor Sp1 phosphorylation induced by shear stress inhibits membrane type 1-matrix metalloproteinase expression in endothelium. J Biol Chem 2002;277(38):34808−14.

[195] Pearce MJ, et al. Shear stress activates cytosolic phospholipase A2 (cPLA2) and MAP kinase in human endothelial cells. Biochem Biophys Res Commun 1996;218(2):500—4.

[196] Ley K, et al. Shear-dependent inhibition of granulocyte adhesion to cultured endothelium by dextran sulfate. Blood 1989;73(5):1324—30.

[197] Kraiss LW, et al. Fluid flow activates a regulator of translation, p70/p85 S6 kinase, in human endothelial cells. Am J Physiol Heart Circ Physiol 2000;278(5):H1537—44.

[198] LaPlaca MC, Thibault LE. An in vitro traumatic injury model to examine the response of neurons to a hydrodynamically-induced deformation. Ann Biomed Eng 1997;25(4):665—77.

[199] Blackman BR, Barbee KA, Thibault LE. In vitro cell shearing device to investigate the dynamic response of cells in a controlled hydrodynamic environment. Ann Biomed Eng 2000;28(4):363—72.

[200] Dewey Jr. CF. Effects of fluid flow on living vascular cells. J Biomech Eng 1984;106(1):31—5.

[201] Hermann C, Zeiher AM, Dimmeler S. Shear stress inhibits H_2O_2-induced apoptosis of human endothelial cells by modulation of the glutathione redox cycle and nitric oxide synthase. Arterioscler Thromb Vasc Biol 1997;17(12):3588—92.

[202] Schnittler HJ, et al. Improved in vitro rheological system for studying the effect of fluid shear stress on cultured cells. Am J Physiol 1993;265 (1 Pt 1):C289—98.

[203] Blackman BR, Garcia-Cardena G, Gimbrone Jr. MA. A new in vitro model to evaluate differential responses of endothelial cells to simulated arterial shear stress waveforms. J Biomech Eng 2002;124(4):397—407.

[204] Dai G, et al. Distinct endothelial phenotypes evoked by arterial waveforms derived from atherosclerosis-susceptible and -resistant regions of human vasculature. Proc Natl Acad Sci USA 2004;101(41):14871—6.

[205] Dai G, et al. Biomechanical forces in atherosclerosis-resistant vascular regions regulate endothelial redox balance via phosphoinositol 3-kinase/Akt-dependent activation of Nrf2. Circ Res 2007;101(7):723—33.

[206] Mack PJ, et al. Biomechanical regulation of endothelium-dependent events critical for adaptive remodeling. J Biol Chem 2009;284 (13):8412—20.

[207] Parmar KM, et al. Integration of flow-dependent endothelial phenotypes by Kruppel-like factor 2. J Clin Invest 2006;116(1):49—58.

[208] Frangos JA, McIntire LV, Eskin SG. Shear stress induced stimulation of mammalian cell metabolism. Biotechnol Bioeng 1988;32 (8):1053—60.

[209] Gallik S, et al. Shear stress-induced detachment of human polymorphonuclear leukocytes from endothelial cell monolayers. Biorheology 1989;26(4):823—34.

[210] Helmlinger G, et al. Effects of pulsatile flow on cultured vascular endothelial cell morphology. J Biomech Eng 1991;113(2):123—31.

[211] Koslow AR, et al. A flow system for the study of shear forces upon cultured endothelial cells. J Biomech Eng 1986;108(4):338—41.

[212] Lawrence MB, McIntire LV, Eskin SG. Effect of flow on polymorphonuclear leukocyte/endothelial cell adhesion. Blood 1987;70(5):1284—90.

[213] Levesque MJ, Nerem RM. The elongation and orientation of cultured endothelial cells in response to shear stress. J Biomech Eng 1985;107 (4):341—7.

[214] Brown DC, Larson RS. Improvements to parallel plate flow chambers to reduce reagent and cellular requirements. BMC Immunol 2001;2:9.

[215] Usami S, et al. Design and construction of a linear shear stress flow chamber. Ann Biomed Eng 1993;21(1):77—83.

[216] Haimovitz-Friedman A, et al. Autocrine effects of fibroblast growth factor in repair of radiation damage in endothelial cells. Cancer Res 1991;51(10):2552—8.

[217] Haimovitz-Friedman A, et al. Ionizing radiation acts on cellular membranes to generate ceramide and initiate apoptosis. J Exp Med 1994;180 (2):525—35.

[218] Meeren AV, et al. Ionizing radiation enhances IL-6 and IL-8 production by human endothelial cells. Mediators Inflamm 1997;6(3):185—93.

[219] Gaugler MH, et al. Pravastatin limits endothelial activation after irradiation and decreases the resulting inflammatory and thrombotic responses. Radiat Res 2005;163(5):479—87.

[220] Chou CH, Chen SU, Cheng JC. Radiation-induced interleukin-6 expression through MAPK/p38/NF-kappaB signaling pathway and the resultant antiapoptotic effect on endothelial cells through Mcl-1 expression with sIL6-Ralpha. Int J Radiat Oncol Biol Phys 2009;75(5):1553—61.

[221] Truman JP, et al. Endothelial membrane remodeling is obligate for anti-angiogenic radiosensitization during tumor radiosurgery. PLoS One 2010;5(8):e12310.

[222] Yentrapalli R, et al. The PI3K/Akt/mTOR pathway is implicated in the premature senescence of primary human endothelial cells exposed to chronic radiation. PLoS One 2013;8(8):e70024.

[223] Park MT, et al. Radio-sensitivities and angiogenic signaling pathways of irradiated normal endothelial cells derived from diverse human organs. J Radiat Res 2012;53(4):570—80.

[224] Sharma P, Guida P, Grabham P. Effects of Fe particle irradiation on human endothelial barrier structure and function. Life Sci Space Res 2014;2:29—37.

Chapter 13

Effects of space radiation on the endovasculature: implications for future human deep space exploration

Melpo Christofidou-Solomidou[1], Thais Sielecki[2] and Shampa Chatterjee[3]

[1]Pulmonary, Allergy and Critical Care Division, Department of Medicine, University of Pennsylvania Perelman School of Medicine, Philadelphia, PA, United States, [2]JT-MeSh Diagnostics LLC, Penn Center of Innovation, Philadelphia, PA, United States, [3]Department of Physiology, Institute for Environmental Medicine, University of Pennsylvania Perelman School of Medicine, Philadelphia, PA, United States

13.1   Introduction

In 1961 the first manned spaceflight was initiated by the launch of Yuri Gagarin on board a Russian Vostok rocket. Since then, numerous manned missions to Earth's orbit have been carried out routinely with the historic moon landing on July 1969 when Neil Armstrong, commander of Apollo 11, set foot upon the moon. Apollo 11 was followed by several lunar missions and the creation and longer term manning of space stations. Over this period of time, NASA has focused on evaluating the health risks to astronauts from space radiation. Research on space radiation and its effects has gathered renewed interest with plans of the first manned mission to Mars in the next decades. Astronauts on a Mars mission will be exposed to radiation and radioactive charged particles in interplanetary space for extended durations. Strategies to protect astronauts from the deleterious physiological effects of exposure to space radiation require identification of the cellular targets of this type of radiation and the subsequent short- and long-term biological effects [1] (Fig. 13.1).

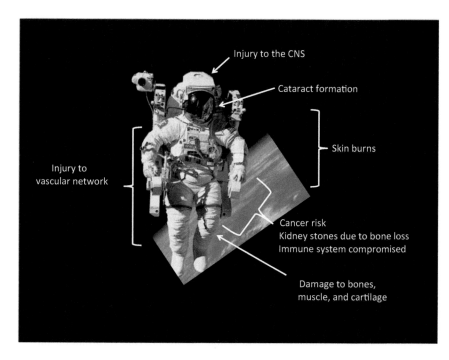

FIGURE 13.1 **Effects on humans due to space radiation exposures.** This figure depicts selected effects on various organs from central nervous system (CNS), eyes (lens), skin burns, damage to bones and bone-associated tissues.

Endothelial Signaling in Vascular Dysfunction and Disease. DOI: https://doi.org/10.1016/B978-0-12-816196-8.00024-2

Space exploration to date has shown that long exposures to space environmental factors, namely, microgravity and space radiation, adversely affect human health. The negative effect would be amplified several fold with interplanetary travel to Mars, a trip that will involve a lengthy period in space, during which astronauts will be exposed to multiple sources of ionizing radiation, including galactic cosmic rays (GCR), solar particle events (SPE), radioactive showers, and trapped radiation in the Van Allen belts [2]. The Van Allen belts are two regions encircling the Earth in which energetic charged particles are trapped inside the Earth's magnetic field. The properties of these charged and radioactive particles vary according to solar activity; additionally, these particles can accelerate to very high energies of several million electron volts within these belts. They represent a significant hazard to any manned mission in space.

All organs, tissues, and cells in the body are susceptible to radiation damage. Prime among these is the radiation-induced damage to cellular DNA via the formation of single and double strand breaks. Endogenous repair processes accompany these strand breaks but these are often ineffective and "disrepairs" lead to mutations, which accumulate to affect cell division and proliferation which then cause cancer. Additionally, ionizing radiation has long been known to cause injury in heart and blood vessels [3]. Radiation leads to atherosclerosis, myocardial fibrosis, and cardiac valve abnormalities; indeed not only astronauts but also radiotherapy patients show manifestations of cardiovascular disease (CVD) as a result of exposure to varying levels of low to high doses of ionizing radiation [4]. NASA conducted the "Twins Study" a year-long study whereby a pair of monozygotic twin astronauts were exposed to the effects of either the spaceflight or the Earth environment [5]. The study identified significant changes associated to spaceflight which, showed among other changes, increased DNA damage and gene expression changes, several of which persisted even several months after returning to earth.

The cardiovascular system comprises of the heart, lungs (cardiopulmonary system) and blood vessels (vascular network). The vascular network, a conduit and connector of blood flow across organs, is a system that integrates biochemical and biophysical signals via systemic transport of blood, nutrients and inflammatory, pathogenic moieties across the body [6]. In addition, the vasculature functions both as an initiator of inflammation and its converging site. Initiation of inflammation occurs via production of moieties such as chemokines, cytokines, and cellular adhesion molecules (CAMs). These moieties transform the vasculature into a converging site of inflammation by facilitating the recruitment and adherence of immune cells (neutrophils, macrophages, and leukocytes) to the vascular wall. Adherence of immune cells is followed by extravasation of these cells into tissue, where these cells release large amounts of radicals (reactive oxygen species, peroxyl radicals, etc.) that damage tissue.

A realistic estimate of the radiation effects on the vascular system can be obtained from the reports of health assessments of atomic bomb survivors [7,8]. These reports show an increased incidence of cardiovascular disease, including ischemic heart disease and stroke, as late as several decades after exposure to doses of gamma radiation as low as 2 Gy [9]. Radiation doses expressed in Gy (Joules per kilogram, J/kg) represent the amount of energy deposited per unit path length for the charged particle. The heavier the charged particle, the greater is the energy absorbed by an astronaut in the path of that radiation. This energy deposited per unit path length is called linear energy transfer (LET).

Furthermore, CVD mortality rates among Apollo lunar astronauts (43%) was four to five times higher than in non-flight and low earth orbit (LEO) flying astronauts [10]. Since vascular dysfunction is the main trigger that drives CVD, it is reasonable to conclude that direct effects of space radiation and weightlessness on the arterial vasculature constitute a major risk for CVD [10].

The NASA Human Integrated Research Program identified cardiovascular disease arising from space radiation as one of the significant risks of spaceflight [11]. The high incidence of cardiovascular disease mortality on astronauts of the Apollo [10] and other missions has been attributed to radiation (proton and gamma) induced downregulation of genes that maintains vascular homeostasis such as vascular growth factor genes [12] and the upregulation of proinflammatory and procoagulant proteins on endothelium [13]. Exposure to gamma radiation increases expression of selectins, integrins, and cellular adhesion molecules including intercellular adhesion molecule (ICAM-1), vascular cell adhesion molecule (VCAM-1), and platelet endothelial cell adhesion molecule (PECAM-1) on the vascular endothelium [14]. These increases have been reported to be both dose- and time-dependent in human endothelial cells, and occur with doses as low as 0.5 Gy [15]. Radiation drives the procoagulant or prothrombotic state of the endothelium via ionizing radiation which besides driving the the production of inflammatory mediators such as tumor necrosis factor alpha (TNF-α) and Interleukin 1 beta (IL-1β) also causes direct oxidative damage to cells and tissues. Inflammatory mediators downregulate thrombomodulin. Thrombomodulin is a cofactor of thrombin and downregulation of thrombomodulin increases free thrombin that activates platelets to create a prothrombotic state. Prothrombotic states cause vascular damage, hemorrhage, etc. and have been recognized as a major cause of death [16]. This chapter reviews the current status of space radiation research with a focus on better understanding the biological targets of space radiation and the associated risks to human health with a specific emphasis on vascular health. In doing so, we provide information on the space environment, the ongoing efforts to simulate the space environment for research via recreating the ion species, radiation doses and energy, and the clinical outcomes of such exposures.

13.2   Space radiation: a major challenge to astronauts

Over the next few decades, space programs will expose humans to different extraterrestrial scenarios as part of the exploration of space. Planned missions to the moon in the next decade will establish a lunar human outpost on the South Pole with constant sunlight illumination and potential resources of water−ice deposits.

An Earth to Mars mission that would cover a distance of 35 million miles away from Earth at the closest point, would involve an estimated cruise time of a 180-day, a 500-day-stay planetary mission, and a 180-day return flight to Earth. This would result in a cumulative radiation dose range between 0.66 and 1.01 Sv based on estimations by NASA's Curiosity Radiation Assessment Detector (RAD) [17].

The two major contributors to health risks in long spaceflights are radiation and microgravity [18]. While microgravity can be simulated in the lab, current models of radiation can only approximate space radiation and do not accurately represent the space radiation environment. To date, studies on astronauts in the low Earth orbit (LEO) or after lunar missions, have served to inform on the effects of space radiation; however, they cannot recreate the conditions that would be encountered in long spaceflights (such as the Mars mission). The difficulty in creating a space radiation model lies in the numerous particle types [proton, iron, silicon, hydrogen (heavy and light), etc.] and varying duration in exposure to each of the radiation sources. An accurate predictive model of space radiation is crucial to obtain the integrated amount of radiation exposure and its biological impact. This will enable to calculate the permissible exposure limits (PELs), and to develop effective countermeasure strategies to enable safe, long-duration space travel.

The space weather environment has three sources of ionizing radiation consisting of both low and high LET. *The first source is the GCR spectrum*, which consists of ionized hydrogen along with heavy charged particles, with high LET. Approximately 85% high-energy protons, 12% alpha particles, and 2% electrons are present in GCR. It also consists of a small fraction of high charge (Z) and energy (E) (HZE) nuclei [19]. Due to their high LET, the HZE component of GCR produces free radicals that can react with biological moieties, ranging from DNA to proteins etc. in cells and cause extensive oxidative modification and cellular damage due to radicals and radionuclides. These can pose a greater danger to astronauts than gamma rays.

The second source is the SPE that are short exposures of high-energy protons from the Sun within regions of solar magnetic instability [20]. SPE are produced either via solar flares or coronal mass ejections. SPE radiation is primarily composed of protons with kinetic energies ranging from 10 MeV up to several GeV (determined by the relativistic speed of particles) and is predicted to produce a heterogeneous dose distribution within an exposed astronaut's body, with a high dose to skin and a lower dose to internal organs. While many SPE show modest energy distributions, there are occasional and unpredictable large SPE events such as in October 1989 which can reportedly deliver doses of 1454 mGy/h to an exposed astronaut in a vehicle traveling in interplanetary space [21]. This is extremely high when compared to the daily dose for long-duration astronauts aboard the International Space Station (ISS), which is approximately 0.282 mGy/day.

The third source is the solar winds, which consist of low-energy protons and electrons. Of these three sources of radiation, the solar wind constitutes the lowest or most negligible risk as it can be easily shielded by modern spacecraft designs.

In addition to these, there is also the cosmic ionizing radiation and secondary radiations produced by interaction of these radiations with spacecraft materials and biological tissues. GCR causes the production of heavy metal ions as it reacts with the spacecraft shielding material; specifically GCR when it collides with the spacecraft, can cause the fragmentation of the spacecraft material into numerous particles of reduced atomic weight [21,22]. This process is called spallation, and produces product ions and additional secondary radiation that are more damaging than the parent particle [21]. Similarly, collision of GCR with human tissue also causes the formation of ions. Average spacecraft shielding is in the order of 20 g/cm^2 [23], whereas the human body has an average tissue thickness of approximately 30 g/cm^2 for internal organ locations [24]. This process of the radiation interacting with spacecraft material and the human body to produce detrimental secondary radiation contributes to the complexity of the radiation environment and is unique to spaceflight.

To model these space radiation environments, NASA has developed the "GCR Simulator" to generate a spectrum of ion beams that approximate the primary and secondary GCR field experienced at human organ locations within a deep space vehicle [25] (Fig. 13.2). In free space, the contribution of the different elements comprising GCR in flux, dose, and dose equivalent is shown in Fig. 13.3. The most abundant GCR particle types include hydrogen (Z = 1), helium (Z = 2), carbon (Z = 6), oxygen (Z = 8), neon (Z = 10), silicon (Z = 14), calcium (Z = 20), and iron (Z = 26). Protons account for 87% of the total flux, helium ions account for approximately 12%, and the remaining heavy ions account

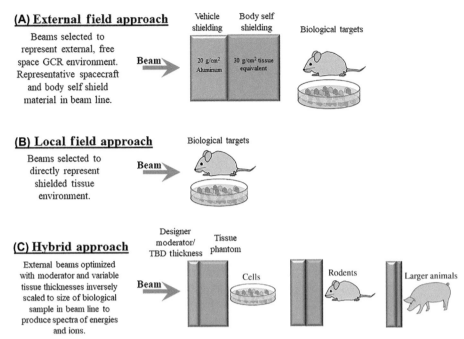

FIGURE 13.2   Three basic strategies for beam selection. (A) Beam selection is representative of the external, free-space GCR spectrum and is approximated by discrete ion and energy beams delivered onto a shielding and tissue equivalent material placed within the beam line, in front of the biological target. (B) Beam selection is representative of the shielded tissue spectrum found in space (e.g., average tissue flux behind vehicle shielding) and is approximated by discrete ion and energy beams delivered directly onto the biological target. (C) Beam selection is representative of energies less than free space with thinner amounts of vehicle shielding and variable thicknesses of tissue equivalent materials to represent the differences in body self-shielding between the physical sizes of species. *GCR*, Galactic cosmic radiation. *From Simonsen LC, et al. NASA's first ground-based galactic cosmic ray simulator: enabling a new era in space radiobiology research. PLoS Biol 2020;18(5):e3000669.*

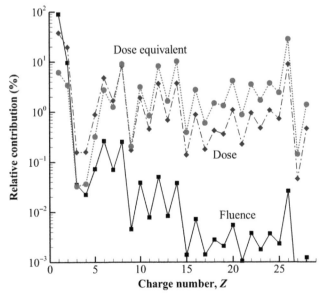

FIGURE 13.3   Relative contribution to fluence (squares), dose (diamonds), and dose equivalent (circles) of different elements in the free-space GCR environment during solar minimum conditions (June 1976) as described by the Badhwar–O'Neill 2010 GCR model. *GCR*, Galactic cosmic radiation. *Adapted from Simonsen LC, Slaba TC, et al. NASA's first ground-based galactic cosmic ray simulator: enabling a new era in space radiobiology research. PLoS Biol 2020;18(5):e3000669.*

for less than 1% of the total flux [26]. The charge squared (Z^2) times abundance, indicates the relative contribution of each particle to the total dose [23]. The radiation is then converted to equivalent whole body dose from the biological effectiveness of the particle producing the dose. The dose equivalent (milliSievert, mSv) represents radiation-induced cancer and genetic damage. However, a challenge with SPE radiation is that it results in nonhomogeneous total body distribution, making it difficult to compare the results of SPE radiation with conventional gamma radiation [27].

NASA's GCR radiation environment simulators represent the current state-of-the-art model of GCR radiation and enable the investigation of the impact of space radiation (and microgravity) elements [25] individually and/or in combination so that risk assessment as well as shielding and protective strategies can be developed. Risk assessment is in the form of effects of simulated space radiation on short- and long-term changes in human tissues and organs and their overall impact of health of astronauts. Protection strategies have been technology development for shielding via spacecraft materials, extravehicular space suits, and other protective gears as well as by countermeasures in the form of nutritional supplements (oral and injected) and topically applied agents. These will be discussed in the next sections.

13.3   Biological impact of space radiation on relevant models of the vasculature

NASA, as part of its Human Research Program (HRP) recognizes four risk areas: (1) Risk of Cardiovascular Disease, (2) Risk of Acute (In-flight) and Late (Post-flight) Central Nervous System, (3) Risk of Acute Radiation Syndromes, and (4) Risk of Radiation Carcinogenesis. This section will review information on the first risk area: that is, space radiation exposure related risks that lead to CVD.

While the effect of ionizing radiation is well established, our understanding of the short- and long-term impact of space radiation is incomplete. Ionizing radiation on earth leads to either acute or delayed responses or both, depending upon the time frame in which the effects are observed. Acute effects (of >1 Sv) appear quite soon after exposure in people that receive high doses in a short period of time (minutes to a few days). Delayed effects (0.7−4 Sv) are in the form of loss of function in cells and pathologies as cancer that generally occurs much (months to years) later. The acute radiation syndrome (ARS) is a sequence of symptoms [28,29] characterized by nausea, vomiting, diarrhea, etc. The magnitude of ARS depends on the type of radiation and the radiation dose absorbed. The delayed effects are in the form of reduced bone marrow function (haematopoietic syndrome), which lead to lymphocyte deprivation, infection, hemorrhage, anemia, and severe fluid loss within 2−6 weeks and death from sepsis [30]. Death rate for this syndrome peaks at 30 days after exposure, but continues up to 60 days. Doses in the range of 6−8 Sv will lead to injury to the vasculature of the submucosa of intestines. Yet higher doses (20−40 Sv) will lead to loss of consciousness with eventual coma and death (neurovascular syndrome). As space radiation comprises of a number of different sources of ionizing radiation, it is reasonable to expect similar short- and long-term impact.

Studying low-dose radiation exposure can help in understanding the health effects from prolonged exposures. For this, the Million Person Study (MPS) was designed to evaluate radiation risks in a cohort (healthy American workers and veterans) that was considered to be more representative of today's populations than Japanese atomic bomb survivors. The comprehensive epidemiologic and data resource repository of 1 million individuals is now being evaluated for CVD, bone marrow function, neurological disorders, cancer, etc. to provide an understanding of prolonged exposure. While these data will be used as guidelines for veterans and military and radiation facility personnel, it can also provide a framework for evaluation of continued use of the linear dose−response model of radiation exposure in the context of space missions [31].

With the Mars mission, astronauts will be subjected to health risks from both the low and high doses of space radiation exposure comprising of SPE and GCR during the flight to Mars and the exposure to gamma during the landing and subsequent colonization. While this environment does not exist on earth, NASA's GCR simulator and other models including that of Chatterjee et al. [32] have shown that among other organs and tissues, the vascular network by permeating every tissue down to the microscopic level, is a prime target for damage from space radiation that passes through the human body. However, (1) there is lack of information on the endothelial cell signaling pathways that would be maximally affected by space radiation and its ramifications on overall health and (2) lack of clarity on space radiation-induced activation and amplification of signaling cascades that lead to long-term injury. An understanding along these lines is crucial in the development of radiological protectors and mitigators for astronaut safety and mission outcome/success.

Potential vascular damage from space radiation includes significant changes in vascular architecture and changes in the blood−brain barrier. Long-term radiation-induced changes in the blood−brain barrier and vascular changes after exposure to 15−30 Gy particle radiation have been observed in the nervous system (recorded by MRI, PET, and CT imaging) [33]. Additionally, recent epidemiological studies have demonstrated increasing risk for cardiovascular disease resulting from low to moderate ionizing radiation exposure [34,35] because ionizing radiation triggers changes in the phenotype of endothelial cells that lead to endothelial activation and potentially to endothelial dysfunction. Risk for cardiovascular disease from exposure to space radiation has long been recognized [36]. Studies of the Apollo lunar astronauts, show, in fact, a higher incidence of cardiovascular disease and possible damage of the vascular endothelium associated with deep space radiation exposure [10]. Endothelial activation is defined as the endothelial expression of adhesion molecules, such as VCAM-1, ICAM-1, and E-selectin [16]. It is usually triggered by proinflammatory cytokines such as tumor necrosis factor (TNF)-α or interleukin (IL)-6. However, these changes in adhesion molecules and proinflammatory cytokines are largely observed following low linear energy transfer (LET) radiation exposures such as X-rays or gamma rays, which have fundamentally different properties from charged HZE particles comprising the GCR. Our studies showed that vascular endothelial cells (kept under flow to mimic the in vivo state) responded to radiation exposure (both gamma and mixed radiation comprising of gamma and protons) by increased expression of inflammation moieties, ICAM-1, and the NLRP3 inflammasome that led to apoptosis [32].

GCR are high-velocity heavy ions traveling through space and the concern is that they may cause potential behavioral and cognitive impairments when acting on the nervous system that might impact the success of the mission [37]. In experiments using rodents exposed to cosmic radiation, a hippocampal and cortical-based performance decrements was observed using six independent behavioral tasks administered between separate cohorts 12 and 24 weeks after irradiation. Radiation-induced impairments in spatial and recognition memory were temporally coincident with deficits in executive function and elevated anxiety [38]. "Space brain", a term used to describe the adverse health effects of the central nervous system, arises from the effect of GCR on reducing cognitive brain function. Cosmic radiation-induced cognitive dysfunction shows little signs of improvement, at least over the time frame of most published studies (3–12 months) [39] [40]. Studies suggested that a potential risk of space radiation exposure may be the induction of Alzheimer's disease [41].

Low-LET proton irradiation significantly downregulates some proangiogenic factors including vascular endothelial growth factor (VEGF), IL-6, IL-8, and hypoxia inducible factor (HIF-1α) in primary human endothelial cells in vitro [42]. Experiments in animal models also indicate that low-LET protons inhibit angiogenesis. Proton radiation also dose dependently reduces blood vessel formation in tissues [43]. The high-LET radiation found in the GCR spectrum can produce excessive free radicals that cause oxidative damage to cell structures and affect endothelial cell integrity. Animal studies using high-LET Fe ions on mouse hippocampal microvessels showed that at low doses (50 cGy) of radiation resulted in a loss of endothelial cells, 1 year post irradiation [44]. Chronic radiation exposure would lead to high oxidative stress and consequent vascular damage thus accelerating aging, cardiovascular disease, retinal damage [45], and the formation of cataracts.

As stated earlier, oxidative damage leads to an increase in inflammatory mediators such as TNF-α and IL-1ß, which activated by radiation can then downregulate thrombomodulin, which increases free thrombin and activates platelets to create a prothrombotic state. These changes mediate acute radiation-induced enteropathy and radiation-induced liver disease that manifests as endothelial injury that in turn drives hepatic sinusoidal occlusion secondary to endothelial damage [46]. Similarly, plasminogen activator inhibitor (PAI)-1 is a fibrinolytic inhibitor activated by radiation, which leads to radiation-induced endothelial cell death [16]. Exposure to radiation increases the von Willebrand factor (vWF), an acute phase reactant constitutively expressed in normal endothelium. Increased vWF leads to a prothrombotic state. The increased expression of cell adhesion molecules ICAM-1 and VCAM-1 enables attachment of macrophages and other immune cells to the endothelium; this characteristic of a prothrombotic state can contribute to a chronic inflammation [47].

While it has been assumed that missions in low earth orbit (LEO) or stays on ISS or short excursions to the moon would not increase the long-term risk for CVD among astronauts, the data show that this group had a mortality rate of approximately 9% due to CVD [48]. When considered as a separate group, the Apollo lunar astronauts, the only group of humans to have traveled outside of the Earth's protective magnetosphere, demonstrate a higher mortality rate due to CVD compared to both the cohort of astronauts that did not travel into space, as well as astronauts who remained in LEO. These data suggest that human travel into deep space may be more hazardous to cardiovascular health than previously estimated [10].

Until now, animals have been observed up to 9 months after exposure to space-relevant irradiation to show poor cardiac function as monitored by noninvasive high-resolution cardiac ultrasound [49]. Additionally, retinal microvasculature has also been observed to be affected during postradiation follow-up of the animals in simulated dosimetry for Yucatan minipigs (using CT-based Monte Carlo dosimetry) [50].

13.4 Potential countermeasures and their efficacy

A concerted action of ground-based studies and space experiments is required to expand radiobiological knowledge about space radiation to improve countermeasure development [51]. Current strategies to reduce exposure to space radiation involve shielding, administration of drugs or dietary supplements to reduce the radiation effects, and selection of crew based on screening for individual radiation sensitivity.

Radiation shielding is the form of passive and active shielding. Passive shielding is the use of materials to block radiation before it reaches the human body., It is inefficient at reducing astronaut exposure levels as increased mass of shielding material can lead to further production of radionuclide fragments from the material [21,52]. Wearable shielding system designed based on Monte Carlo simulations using water or organic compounds as the fabricating material are also under consideration [53]. Active shielding aims to both deflect and to trap portions of the incoming space radiation. It involves electrostatic shielding (i.e., creating an electric field around an astronaut habitat), magnetic shielding (forming a large magnetic field around the spacecraft by the use of superconducting solenoids to deflect the space radiation)and plasma shielding (by which a mass of ionized particles entrapped by electromagnetic fields, swirling around a spacecraft enclosure, serves to deflect charged radiation particles) [54,55]. These strategies are aimed to either provide total protection against radiation exposure within the spacecraft or a radiation shelter within the vehicle. In case of the latter, the crew has to enter the shelter during periods of increasing proton fluxes such as SPE or GCR "events" to avoid high-dose exposure for hours or days.

NASA is also investigating pharmaceutical countermeasures [51,56], which could prove to be more effective than radiation shielding. This involves (1) identification and development of radioprotective agents for protection against tissue damage, (2) evaluation of their stability, integrity, safety, and ease of administration of the countermeasure during prolonged flights and potentially longer planetary stays, and (3) evaluation of their potential to work in a spaceflight environment, including weightlessness and space radiation.

Different radioprotective agents are required for protection against GCR and SPE. For instance, potassium iodide, diethylenetriamine pentaacetic acid (DTPA), and the dye known as "Prussian blue" have been used for decades to treat radiation sickness. Encouraging results were reported in rodent studies with berries or dried plums [57]. Protective effects of soybean-derived Bowman-Birk inhibitor (BBI), BBI concentrate (BBIC), and a combination of ascorbic acid, coenzyme Q10, L-selenomethionine (SeM), and vitamin E succinate against proton and HZE-particle nuclei have been observed (reviewed in [58]). Pyruvate, a natural metabolite of the body, which plays a crucial role in the conversion of food to energy, is an efficient antioxidant and free radical scavenger, and numerous studies have shown that treatment with this compound can protect the collagenized tissue against radiation-induced damage. Our studies on LGM2605, a synthetically derived flaxseed lignan secoisolariciresinol diglucoside (SDG) as well as naturally extracted SDG from whole flaxseed [59], showed this compound to be radioprotective [60,61]. Indeed, we reported that LGM2605, the synthetic version of SDG, protected the vascular network against a space radiation relevant model of mixed gamma and proton beams [32]. Studies by Globus and coworkers on bone loss caused by ionizing radiation identified dried plum as a dietary countermeasure [57]. These studies have highlighted the fact that during long-term missions, astronauts will likely need to take daily doses of such radioprotective agents as meds to mitigate exposure to radiation. These types of radioprotective agents will likely need to be taken during the course of spaceflight as well as potentially for some period prior to and afterward to ensure optimal mitigation of radiation-induced damage. Crew selection needs to be carefully monitored to take into account the lower radiosensitivity of older crew members.

13.5   Conclusions and future perspectives

The space radiation environment is a complex mixture of several radiation species dominated by highly penetrating charged particles from the Sun. From the properties of charged particles and the composition of the environment associated with the vehicles, specific radiation fields would be experienced by astronauts depending upon the space missions. Here we highlighted the vascular damage-related risks of spaceflight. Although extensive studies have not been carried out yet, a major limitation of space radiation research is the lack of models to accurately represent the operational space radiation environment or the complexity of human physiology. Nevertheless, radiation simulators and endothelial cell models that represent, as closely as possible, the vascular network in the space environment, are currently being employed to identify space radiation-induced endothelial signaling and dysfunction [32]. These studies in combination with cardiovascular risks in astronauts and atom bomb and nuclear radiation accident survivors can be employed to build on epidemiological models to estimate health risks and guide safe spaceflight operations.

Studies to date show a disparity between cellular and animal models and the observed empirical effects seen in human astronaut crews [21]. This possibly arises due to the inability in simulation of the complex space environment and the limitations of animal surrogates. However, given the intended future of human spaceflight, there is an urgency to improve upon the understanding of the space radiation risk to the vascular system so as to predict the clinical outcomes of interplanetary radiation exposure, and develop appropriate and effective mitigation strategies for future missions.

References

[1] Cucinotta FA. Review of NASA approach to space radiation risk assessments for Mars exploration. Health Phys 2015;108(2):131−42.

[2] Nachtwey DS, Yang TC. Radiological health risks for exploratory class missions in space. Acta Astronaut 1991;23:227−31.

[3] Adams MJ, Hardenbergh PH, et al. Radiation-associated cardiovascular disease. Crit Rev Oncol/Hematol 2003;45(1):55−75.

[4] Fajardo LF, Berthrong M. Vascular lesions following radiation. Pathol Annu 1988;23(Pt 1):297−330.

[5] Garrett-Bakelman FE, Darshi M, et al. The NASA twins study: a multidimensional analysis of a year-long human spaceflight. Science 2019;364(6436).

[6] Chatterjee S. Endothelial mechanotransduction, redox signaling and the regulation of vascular inflammatory pathways. Front Physiol 2018;9:524.

[7] Little MP, Azizova TV, et al. Systematic review and meta-analysis of circulatory disease from exposure to low-level ionizing radiation and estimates of potential population mortality risks. Environ Health Perspect 2012;120(11):1503−11.

[8] Stewart AM, Kneale GW. A-bomb survivors: reassessment of the radiation hazard. Med Confl Surv 1999;15(1):47−56.

[9] Adams MJ, Grant EJ, et al. Radiation dose associated with renal failure mortality: a potential pathway to partially explain increased cardiovascular disease mortality observed after whole-body irradiation. Radiat Res 2012;177(2):220−8.

[10] Delp MD, Charvat JM, et al. Apollo lunar astronauts show higher cardiovascular disease mortality: possible deep space radiation effects on the vascular endothelium. Sci Rep 2016;6:29901.

[11] Ball JR ECJ, editor. Safe Passage: Astronaut Care for Exploration Missions. Washington (DC): National Academies Press (US); 20012, Risks to Astronaut Health During Space Travel. "Institute of Medicine (US) Committee on Creating a Vision for Space Medicine During Travel Beyond Earth Orbit". Available from: https://www.ncbi.nlm.nih.gov/books/NBK223785/.

[12] Hughson RL, Helm A, et al. Heart in space: effect of the extraterrestrial environment on the cardiovascular system. Nat Rev. Cardiology 2018;15(3):167−80.

[13] Hughson RL. Recent findings in cardiovascular physiology with space travel. Respir Physiol Neurobiol 2009;169(Suppl. 1):S38−41.

[14] Prabhakarpandian B, Goetz DJ, et al. Expression and functional significance of adhesion molecules on cultured endothelial cells in response to ionizing radiation. Microcirculation 2001;8(5):355−64.

[15] Baselet B, Rombouts C, et al. Cardiovascular diseases related to ionizing radiation:the risk of low-dose exposure (Review). Int J Mol Med 2016;38(6):1623−41.

[16] Venkatesulu BP, Mahadevan LS, et al. Radiation-induced endothelial vascular injury: a review of possible mechanisms. JACC Basic Transl Sci 2018;3(4):563−72.

[17] Sobel A, Duncan R. Aerospace environmental health: considerations and countermeasures to sustain crew health through vastly reduced transit time to/from mars. Front Public Health 2020;8:327.

[18] Chancellor JC, Scott GB, et al. Space radiation: the number one risk to astronaut health beyond low earth orbit. Life 2014;4(3):491−510.

[19] Simpson JA. Elemental and isotopic composition of the galactic cosmic rays. Annu Rev Nucl Part Sci. 33, 1983, 323−382.

[20] Wilson JW. Overview of radiation environments and human exposures. Health Phys 2000;79(5):470−94.

[21] Chancellor JC, Blue RS, et al. Limitations in predicting the space radiation health risk for exploration astronauts. NPJ Microgravity 2018;4:8.

[22] Townsend LW, Cucinotta FA, et al. Estimates of HZE particle contributions to SPE radiation exposures on interplanetary missions. Adv Space Res 1994;14(10):671−4.

[23] Sihver L. Transport calculations and accelerator experiments needed for radiation risk assessment in space. Z fur medizinische Phys 2008;18 (4):253−64.

[24] Norbury JW, Schimmerling W, et al. Galactic cosmic ray simulation at the NASA space radiation laboratory. Life Sci Space Res 2016;8:38−51.

[25] Simonsen LC, Slaba TC, et al. NASA's first ground-based galactic cosmic ray simulator: enabling a new era in space radiobiology research. PLoS Biol 2020;18(5):e3000669.

[26] Norbury JW, Slaba TC, et al. Advances in space radiation physics and transport at NASA. Life Sci Space Res 2019;22:98−124.

[27] Hu S, Kim MH, et al. Modeling the acute health effects of astronauts from exposure to large solar particle events. Health Phys 2009;96 (4):465−76.

[28] Cronkite EP. The diagnosis, treatment, and prognosis of human radiation injury from whole-body exposure. Ann N Y Acad Sci 1964;114:341−55.

[29] Chaillet MP, Cosset JM, et al. Prospective study of the clinical symptoms of therapeutic whole body irradiation. Health Phys 1993;64 (4):370−4.

[30] Dainiak N, Waselenko JK, et al. The hematologist and radiation casualties. Hematol Am Soc Hematol Educ Program 2003;473−96.

[31] Boice JD. The Million Person Study relevance to space exploration and mars. Int J Radiat Biol. 4, 2019, 1−9.

[32] Chatterjee S, Pietrofesa RA, et al. LGM2605 reduces space radiation-induced nlrp3 inflammasome activation and damage in in vitro lung vascular networks. Int J Mol Sci 2019;20(1).

[33] Wang YX, King AD, et al. Evolution of radiation-induced brain injury: MR imaging-based study. Radiology 2010;254(1):210−18.

[34] Little MP, Tawn EJ, et al. Review and meta-analysis of epidemiological associations between low/moderate doses of ionizing radiation and circulatory disease risks, and their possible mechanisms. Radiat Environ Biophys 2010;49(2):139−53.

[35] Shimizu Y, Kodama K, et al. Radiation exposure and circulatory disease risk: Hiroshima and Nagasaki atomic bomb survivor data, 1950-2003. BMJ 2010;340:b5349.

[36] Boerma M, Nelson GA, et al. Space radiation and cardiovascular disease risk. World J Cardiol 2015;7(12):882−8.

[37] Cucinotta FA, Cacao E. Predictions of cognitive detriments from galactic cosmic ray exposures to astronauts on exploration missions. Life Sci Space Res 2020;25:129−35.

[38] Parihar V, Allen B, Caressi C, et al. Cosmic radiation exposure and persistent cognitive dysfunction. Sci Rep 6, 2016, 34774.

[39] Acharya MM, Baulch JE, et al. New concerns for neurocognitive function during deep space exposures to chronic, low dose-rate, neutron radiation. eNeuro 2019;6(4).

[40] Jandial R, Hoshide R, et al. Space-brain: the negative effects of space exposure on the central nervous system. Surg Neurol Int 2018;9:9.

[41] Cherry JD, Liu B, et al. Galactic cosmic radiation leads to cognitive impairment and increased abeta plaque accumulation in a mouse model of Alzheimer's disease. PLoS One 2012;7(12):e53275.

[42] Girdhani S, Lamont C, et al. Proton irradiation suppresses angiogenic genes and impairs cell invasion and tumor growth. Radiat Res 2012;178 (1):33−45.

[43] Jang GH, Ha JH, et al. Effect of proton beam on blood vessel formation in early developing zebrafish (Danio rerio) embryos. Arch Pharmacal Res 2008;31(6):779−85.

[44] Mao XW, Favre CJ, et al. High-LET radiation-induced response of microvessels in the Hippocampus. Radiat Res 2010;173(4):486−93.

[45] Mao XW, Boerma M, et al. Acute effect of low-dose space radiation on mouse retina and retinal endothelial cells. Radiat Res 2018;190 (1):45−52.

[46] Sempoux C, Horsmans Y, et al. Severe radiation-induced liver disease following localized radiation therapy for biliopancreatic carcinoma: activation of hepatic stellate cells as an early event. Hepatology 1997;26(1):128−34.

[47] Abderrahmani R, Francois A, et al. PAI-1-dependent endothelial cell death determines severity of radiation-induced intestinal injury. PLoS One 2012;7(4):e35740.

[48] Reynolds RJ, Day SM. Mortality due to cardiovascular disease among apollo lunar astronauts. Aerosp Med Hum Perform 2017;88(5):492−6.

[49] Inoue T, Zawaski JA, et al. Echocardiography differentiates lethally irradiated whole-body from partial-body exposed rats. Front Cardiovascular Med 2018;5:138.

[50] Wilson JM, Sanzari JK, et al. Acute biological effects of simulating the whole-body radiation dose distribution from a solar particle event using a porcine model. Radiat Res 2011;176(5):649−59.

[51] Kennedy AR. Biological effects of space radiation and development of effective countermeasures. Life Sci Space Res 2014;1:10−43.

[52] Wilson JW, Kim M, et al. Issues in space radiation protection: galactic cosmic rays. Health Phys 1995;68(1):50−8.

[53] Vuolo M, Baiocco G, et al. Exploring innovative radiation shielding approaches in space: A material and design study for a wearable radiation protection spacesuit. Life Sci Space Res 2017;15:69−78.

[54] Durante M. Space radiation protection: destination Mars. Life Sci Space Res 2014;1:2−9.

[55] Washburn SA, Blattnig SR, et al. Active magnetic radiation shielding system analysis and key technologies. Life Sci Space Res 2015;4:22−34.

[56] Patel S. The effects of microgravity and space radiation on cardiovascular health: From low-Earth orbit and beyond. Int J Cardiol Heart Vasc 2020;30 100595.

[57] Schreurs AS, Shirazi-Fard Y, et al. Dried plum diet protects from bone loss caused by ionizing radiation. Sci Rep 2016;6:21343.

[58] McLaughlin MF, Donoviel DB, et al. Novel indications for commonly used medications as radiation protectants in spaceflight. Aerosp Med Hum Perform 2017;88(7):665−76.

[59] Christofidou-Solomidou M, Tyagi S, et al. Dietary flaxseed administered post thoracic radiation treatment improves survival and mitigates radiation-induced pneumonopathy in mice. BMC Cancer 2011;11:269.

[60] Mishra OP, Pietrofesa R, et al. Novel synthetic (S,S) and (R,R)-secoisolariciresinol diglucosides (SDGs) protect naked plasmid and genomic DNA From gamma radiation damage. Radiat Res 2014;182(1):102−10.

[61] Velalopoulou A, Chatterjee S, et al. Synthetic secoisolariciresinol diglucoside (lgm2605) protects human lung in an ex vivo model of proton radiation damage. Int J Mol Sci 2017;18(12).

Chapter 14

Role of endothelial cells in normal tissue radiation injury

Marjan Boerma

Division of Radiation Health, Department of Pharmaceutical Sciences, University of Arkansas for Medical Sciences, Little Rock, AR, United States

14.1 Introduction

Radiation therapy is a part of the treatment plan in more than half of all cases of cancer. While radiation therapy is effective in reducing tumor growth, exposure of noncancerous (normal) tissues to ionizing radiation (in this chapter referred to as ionizing radiation or simply radiation) can cause a wide range of short- and long-term adverse effects in many organ systems [1,2]. While radiation therapy planning aims to minimize the exposure of normal tissues to radiation, these exposures cannot be fully avoided in all cases of cancer, leading to sometimes severe side effects. Normal tissue radiation injury also occurs in accidental exposure to ionizing radiation. In scenarios of accidental exposure of a large part of the body, individuals are at risk of developing the acute radiation syndrome within hours after exposure. Those who survive this acute phase of radiation toxicity may experience delayed and chronic dysfunction in multiple organ systems [3,4]. While recent work has shown that low doses of ionizing radiation (several fold lower than those used in radiation therapy) can also cause long-term changes in normal tissues [5,6], this chapter will focus on evidence from high-dose radiation.

Pharmacological interventions that reduce normal tissue radiation injury may make radiation therapy safer. Moreover, such radiation countermeasures are also required to prevent, reduce, or reverse the adverse effects of accidental radiation exposure. However, few such countermeasures are currently available. Insight into the biological mechanisms by which ionizing radiation causes acute and chronic tissue damage may assist in designing new strategies for interventions.

Endothelial cells and the vasculature have long been thought to play a central role in normal tissue radiation injury [7−9]. While most results in this area of research are obtained from animal models, evidence from human subjects also indicates that radiation induces endothelial and vascular dysfunction [10−13]. A large body of research and literature has been devoted to the effects of radiation on endothelial cells and their contribution to adverse remodeling and dysfunction in organs and tissues. Since one chapter cannot describe all available literature, for additional information on the role of vascular injury and endothelial dysfunction in the pathogenesis of normal tissue radiation injury, the reader is referred to prior review articles in this area [14−18].

This chapter describes (1) current knowledge on how radiation alters endothelial cell phenotype and function, (2) insights into biological mechanisms by which radiation-induced changes in endothelial cells contribute to normal tissue radiation injury, and (3) studies that make use of this insight to design new intervention strategies that may reduce normal tissue radiation injury.

14.2 Effects of ionizing radiation on endothelial cell phenotype and vascular function

A large number of studies, many performed in endothelial cells in culture, have addressed the effects of ionizing radiation on endothelial cell phenotype and function. The paragraphs below briefly summarize the results of these prior studies.

Endothelial Signaling in Vascular Dysfunction and Disease. DOI: https://doi.org/10.1016/B978-0-12-816196-8.00020-5

14.2.1 Studies in endothelial cells in culture

Early after exposure to ionizing radiation, endothelial cells show a loss of barrier function and changes in mechanical properties, endothelial cell apoptosis, expression of cytokines, chemokines and endothelial adhesion molecules, and loss of vascular thromboresistance [19−26]. These phenotypic changes are generally referred to as endothelial dysfunction. Recent work in endothelial cells in vitro has identified a wide range of molecular mediators involved in radiation-induced endothelial dysfunction, which include protein kinase C, extracellular signal-regulated kinases, high-mobility group box-1 (HMGB1), the NF-κB pathway, and the nitric oxide (NO) signaling pathway [27−33].

14.2.2 Observations in tissues and organs in vivo

On the tissue level, vascular dysfunction becomes apparent months to years after radiation exposure and includes reduced myocardial capillary density [34], focal loss of endothelial alkaline phosphatase [35,36], and increased expression of von Willebrand factor [37,38], vascular wall thickening, and perivascular fibrosis [13,34]. Imaging of vascular changes is used as a method for early in vivo detection of normal tissue radiation injury [39−41].

14.2.3 Circulating endothelial cells and endothelial progenitor cells

There is a current interest in the contribution of circulating endothelial cells and endothelial progenitor cells in tissue healing from a wide range of injuries [42,43]. Radiation can cause long-term changes in circulating endothelial cells [12] and endothelial progenitor cells that may serve as predictors of the severity of normal tissue radiation toxicity [44]. Infusion of endothelial cells together with hematopoietic stem cells in a mouse model shows the potential of mitigating normal tissue radiation injury [45]. More research is required to determine the exact role of circulating endothelial cells and endothelial progenitor cells in the pathogenesis or healing of vascular radiation injury.

14.3 Mechanisms by which radiation-modified endothelial cells alter tissue/organ function and structure

Changes in endothelial cells and the microvasculature have long been known to precede chronic manifestations of normal tissue radiation injury such as tissue degeneration and fibrosis [7,46−54]. Since then, several studies have addressed mechanisms by which such endothelial changes may play a role in the process of normal tissue radiation injury. The paragraphs below describe some of the known aspects of radiation-induced endothelial dysfunction and its consequences.

14.3.1 Endothelial cell death

Cells that undergo rapid proliferation are particularly sensitive to radiation-induced cell death [55]. Since under normal circumstances most endothelial cells in adult tissues do not rapidly proliferate, they are not considered to undergo massive cell death in irradiated tissues and organs [15,35]. Moreover, a mild to moderate loss of endothelial cells due to radiation may not always lead to a significant loss of vascular density [56]. On the other hand, when endothelial cells are made more susceptible to undergoing radiation-induced cell death, a correlation is seen between endothelial injury and radiation-induced damage in the heart and small intestine [57,58]. Moreover, prevention of endothelial cell apoptosis coincides with reduced gastrointestinal radiation syndrome [59]. Bax, Bak, and plasminogen activator inhibitor type 1 (PAI1) seem required for endothelial cell survival after radiation [60,61]. Bak- and Bax-mediated endothelial cell survival improves hematopoietic stem cell regeneration after irradiation [62]. These results suggest that endothelial cell death may play a role in normal tissue radiation injury in certain organ systems.

14.3.2 Endothelial cell senescence

There is a growing interest in the role of cellular senescence in the pathogenesis of normal tissue radiation injury [63,64]. Similar to other cell types, endothelial cells from a variety of tissue origins and species can undergo senescence after exposure to ionizing radiation in culture [65−78]. In addition, emerging evidence indicates that endothelial cells undergo senescence after exposure to radiation in vivo [79]. Radiation-induced senescence in endothelial cells is associated with changes in cell morphology, reduced ability to form new capillary structures, permanent cell-cycle arrest, increased expression of β-galactosidase, p16 and p21, and increased production of reactive oxygen species [80−85].

Studies suggest that the p53-p21 pathway as well as p16 are required for radiation-induced senescence [86−88]. Future studies need to address whether a decrease in the number of senescent endothelial cells helps decrease chronic structural manifestations of organ radiation injury.

14.3.3 Increased endothelial permeability and deposition of fibronectin

Exposure to ionizing radiation causes an acute increase in endothelial permeability [89,90]. In vivo, this may lead to the deposition of plasma proteins and other macromolecules into the extracellular matrix surrounding the blood vessel [91,92]. Radiation injury in normal tissues is associated with increased levels of fibronectin, derived from the circulation or produced by cells in the vascular wall and thought to accelerate tissue fibrosis [92−94].

14.3.4 The thrombomodulin-protein C pathway

Thrombomodulin (TM) is a transmembrane glycoprotein located on the luminal surface of endothelial cells in most normal blood vessels that regulates coagulation and inflammation. One of the main roles of TM is to bind thrombin. When bound to TM, thrombin loses its procoagulant, proinflammatory, and profibrogenic properties and instead acquires the ability to activate protein C (APC). APC limits further thrombin generation, counteracts thrombin's adverse effects [95], and has additional antiinflammatory and cytoprotective properties [96]. TM itself also has intrinsic antiinflammatory actions by binding to and inhibiting the proinflammatory protein HMGB1 [97]. Loss of TM is a prominent feature of postradiation endothelial dysfunction. In the absence of TM, thrombin acts as a proinflammatory and profibrotic mediator, in part through the activation of the proteinase-activated receptor-1 [98−100]. Various clinical and preclinical studies have confirmed the role of the TM system in the pathophysiology of radiation injury [101−103]. The section of this chapter on radiation countermeasures discusses the potential of the TM-protein C pathway as a target for intervention.

14.3.5 Transforming growth factor-β

Radiation-induced vascular injury and endothelial dysfunction are mediated in part by transforming growth factor-β (TGF-β) [104,105], a pluripotent growth factor that is increased in expression in many irradiated tissues [106] and plays a role in inhibition of angiogenesis [107], adverse vascular remodeling [108], and tissue radiation fibrosis [93,109,110]. As described below, inhibition of TGF-β signaling is under investigation as a potential strategy to intervene in normal tissue radiation injury.

14.3.6 Endothelial to mesenchymal transition

In various disease states, endothelial to mesenchymal transition (EndoMT) is thought to play a role in tissue healing, but when out of balance can also contribute to adverse remodeling [111]. EndoMT has been described to occur in response to radiation [112]. EndoMT may play a role in mucosal damage in radiation proctitis [113,114] and in radiation fibrosis in the lungs [115].

14.3.7 Endothelial extracellular vesicles and microparticles

There is current interest in the role of extracellular vesicles and other microparticles secreted by endothelial cells as carriers of molecular information to other cells and tissues in the body [116,117]. Exposure to radiation is also known to induce the release of extracellular vesicles [118], and endothelial-derived extracellular vesicles may play a role in reducing normal tissue radiation injury [119]. Therefore they are pursued both as a source of biomarkers of radiation exposure and as a novel strategy to intervene in normal tissue radiation toxicity [120].

14.4   Interventions that may reduce normal tissue radiation injury by altering endothelial function

Prevention and mitigation of normal tissue radiation injury is required in cancer radiotherapy and in accidental radiation exposure. A pharmaceutical intervention that may be administered before or during radiation therapy should not reduce tumor growth delay from radiation. Only very few of those interventions are routinely used in the clinic [121,122].

In the case of accidental radiation exposure, since these events are unpredictable, and clinical care may not always be immediately available, countermeasures must be effective when administered 24 hours after radiation. Currently, only one such radiation mitigator, human recombinant granulocyte colony-stimulating factor (G-CSF), has been approved by the FDA [123]. G-CSF enhances the function of acutely suppressed bone marrow. While G-CSF aids in prolonging postradiation survival, it is not likely in itself a mitigator of all types of acute and late radiation toxicity. Therefore additional countermeasures need to be developed.

In light of the central role of endothelial cells in normal tissue radiation injury, endothelial and vascular targeting may serve as promising strategies to develop novel medical radiation countermeasures [124]. Below, some studies that have made use of the insight of the role of endothelial cells in normal tissue radiation injury to design potential new intervention strategies are discussed.

14.4.1 Stimulation of endothelial nitric oxide synthase

Since intermediates in the cholesterol biosynthesis pathway, farnesyl pyrophosphate and geranylgeranyl pyrophosphate serve as donor molecules for certain posttranslational modifications of proteins such as endothelial nitric oxide synthase (eNOS), modifiers of the cholesterol biosynthesis pathway may alter the function of eNOS and other critical enzymes. Administration of geranylgeranyl acetate protects against radiation-induced intestinal injury in a mouse model, at least in part through the prevention of radiation-induced losses of vascular endothelial growth factor and eNOS [125]. The potential involvement of TM was not examined in this study. The vitamin E analogs γ- and δ-tocotrienol inhibit the cholesterol biosynthesis pathway by inducing the intracellular degradation of β-hydroxy β-methylglutaryl-coenzyme A (HMG-CoA) reductase [126,127] and upregulate NO production [128]. In the scenario of exposure to radiation, γ-tocotrienol reduces radiation-induced vascular nitrosative stress [129,130]. While an improvement in eNOS activity may protect against normal tissue radiation injury, an increase in NO levels may in itself not be sufficient to reduce normal tissue radiation injury. An NO releasing aspirin reduced age-related atherosclerosis but not atherosclerosis associated with local exposure to radiation in a mouse model [131].

14.4.2 Modification of the thrombomodulin-protein C pathway

Prior sections have provided evidence for a role of the TM-protein C pathway in normal tissue radiation toxicity. Various strategies have been designed to target pathway in an attempt to reduce normal tissue radiation injury. Administration of soluble TM or APC 24 and 48 hours after total body irradiation improves postradiation survival and bone marrow function [103]. Both statins and the vitamin E analog γ-tocotrienol upregulate endothelial TM [132,133]. γ-Tocotrienol is a strong radiation protector, alone or when administered in combination with a statin [134].

Interestingly, while modulation of the TM-protein C pathway seems to be a potential strategy for intervention, inhibition of coagulation alone may not be as effective. The antiplatelet agent clopidogrel did not modify radiation-induced kidney injury [135] or accelerated atherosclerosis in the carotid artery of ApoE KO mice [136].

14.4.3 Reduction of inflammation and endothelial–leukocyte interactions

Endothelial cells play a clear role in radiation-induced inflammation [137]. Inhibition of chemokine C-C motif ligand 2 reduces radiation-induced lung inflammation [138]. Inhibition of hypoxia-inducible factor-1α in endothelial cells reduces radiation-induced inflammation and improves tissue structure in the intestine [139]. Moreover, the flavonoid baicalein reduces intestinal radiation injury by reducing leukocyte infiltration into the irradiated tissue [140]. Other pharmaceutical modifications of endothelial cell–leukocyte interactions, such as through inhibition of protein kinase Cδ may be an approach to reducing normal tissue radiation injury [141].

14.4.4 Administration of fibroblast growth factor 2

Fibroblast growth factor 2 (FGF2), also known as basic fibroblast growth factor (bFGF) prevents endothelial cells from undergoing cell death and promotes endothelial cell proliferation and angiogenesis [142]. Administration of FGF2 protects against radiation-induced cell death in the endothelium in the mouse small intestine and reduces lethality from intestinal radiation toxicity [51]. Moreover, endothelial targeted FGF2 protects against long-term radiation injury in the lung [143]. An increase in endothelial TM expression may be one of the mechanisms by which FGF2 reduces radiation-induced bladder injury [144]. On the other hand, while not examined in the scenario of radiation exposure, FGF2 has been shown to contribute to unwanted EndoMT [145].

14.4.5 Inhibition of transforming growth factor-β1

Because of the well-known role of TGF-β1 in promoting radiation fibrosis (see also the section on TGF-β1 above), several inhibitors of TGF-β1 signaling have been designed and tested as potential countermeasures against normal tissue radiation injury. The small-molecule inhibitors of the TGF-β type 1 receptor, SM16 and IPW-5371, have been reported to reduce radiation injury in the lung and heart in animal models [146,147] and are under continued investigation as potential radiation countermeasures.

14.5 Future directions

Science is ever developing, and exciting new methods to study endothelial and vascular function are being developed in various research areas, including the radiation sciences [39,148]. In studying endothelial cells, it is important to acknowledge that endothelial cells from different organ or tissue origin have highly different phenotypes and responses to stressors such as ionizing radiation [149]. More research is required to obtain detailed insight into the role of origin on endothelial cell injury from radiation. Lastly, as outlined above, there is a need for countermeasures that mitigate injury from accidental radiation exposure and protect cancer patients from side effects of radiation therapy, and endothelial cells may form a promising target for the development of new countermeasures.

References

[1] Lu L, Sun C, Su Q, et al. Radiation-induced lung injury: latest molecular developments, therapeutic approaches, and clinical guidance. Clin Exp Med 2019;19:417−26.

[2] Ejaz A, Greenberger JS, Rubin PJ. Understanding the mechanism of radiation induced fibrosis and therapy options. Pharmacol Ther 2019;204:107399.

[3] Unthank JL, Ortiz M, Trivedi H, et al. Cardiac and renal delayed effects of acute radiation exposure: organ differences in vasculopathy, inflammation, senescence and oxidative balance. Radiat Res 2019;191(5):383−97.

[4] Medhora M, Gao F, Gasperetti T, et al. Delayed effects of acute radiation exposure (Deare) in Juvenile and old rats: mitigation by Lisinopril. Health Phys 2019;116(4):529−45.

[5] Schollnberger H, Kaiser JC, Jacob P, Walsh L. Dose-responses from multi-model inference for the non-cancer disease mortality of atomic bomb survivors. Radiat Environ Biophys 2012;51(2):165.

[6] Shibamoto Y, Nakamura H. Overview of biological, epidemiological, and clinical evidence of radiation hormesis. Int J Mol Sci 2018;19 (8):2387.

[7] Hopewell JW, Campling D, Calvo W, Reinhold HS, Wilkinson JH, Yeung TK. Vascular irradiation damage: its cellular basis and likely consequences. Br J Cancer Suppl 1986;7:181−91.

[8] Fajardo LF. The unique physiology of endothelial cells and its implications in radiobiology. Front Radiat Ther Oncol 1989;23:96−112.

[9] Denham JW, Hauer-Jensen M. The radiotherapeutic injury — a complex 'wound'. Radiother Oncol 2002;63:129−45.

[10] Rakhypbekov T, Inoue K, Pak L, et al. Endothelial dysfunction in rectal cancer patients chronically exposed to ionizing radiation. Radiat Environ Biophys 2017;56(3):205−11.

[11] Nonoguchi N, Miyatake S, Fukumoto M, et al. The distribution of vascular endothelial growth factor-producing cells in clinical radiation necrosis of the brain: pathological consideration of their potential roles. J Neurooncol 2011;105(2):423−31.

[12] Pradhan K, Mund J, Case J, et al. Differences in circulating endothelial progenitor cells among childhood cancer survivors treated with and without radiation. J Hematol Thromb 2015;1(1):4.

[13] Stewart FA, Hoving S, Russell NS. Vascular damage as an underlying mechanism of cardiac and cerebral toxicity in irradiated cancer patients. Radiat Res 2010;174(6):865−9.

[14] Wang J, Boerma M, Fu Q, Hauer-Jensen M. Significance of endothelial dysfunction in the pathogenesis of early and delayed radiation enteropathy. World J Gastroenterol 2007;13:3047−55.

[15] Hopewell JW, Calvo W, Jaenke R, Reinhold HS, Robbins ME, Whitehouse EM. Microvasculature and radiation damage. Recent Results Cancer Res 1993;130:1−16.

[16] Venkatesulu BP, Mahadevan LS, Aliru ML, et al. Radiation-induced endothelial vascular injury: a review of possible mechanisms. JACC Basic Transl Sci 2018;3(4):563−72.

[17] Baselet B, Sonveaux P, Baatout S, Aerts A. Pathological effects of ionizing radiation: endothelial activation and dysfunction. Cell Mol Life Sci 2019;76(4):699−728.

[18] Soloviev AI, Kizub IV. Mechanisms of vascular dysfunction evoked by ionizing radiation and possible targets for its pharmacological correction. Biochem Pharmacol 2019;159:121−39.

[19] Berends U, Peter RU, Hintermeier-Knabe R, et al. Ionizing radiation induces human intercellular adhesion molecule-1 in vitro. J Invest Dermatol 1994;103(5):726−30.

[20] Dunn MM, Drab EA, Rubin DB. Effects of irradiation on endothelial cell-polymorphonuclear leukocyte interactions. J Appl Physiol 1986;60 (6):1932−7.

[21] Eldor A, Fuks Z, Matzner Y, Witte LD, Vlodavsky I. Perturbation of endothelial functions by ionizing irradiation: effects on prostaglandins, chemoattractants and mitogens. Semin Thromb Hemost 1989;15(2):215−25.

[22] Park JS, Qiao L, Su ZZ, et al. Ionizing radiation modulates vascular endothelial growth factor (VEGF) expression through multiple mitogen activated protein kinase dependent pathways. Oncogene 2001;20(25):3266−80.

[23] Witte L, Fuks Z, Haimovitz-Friedman A, Vlodavsky I, Goodman DS, Eldor A. Effects of irradiation on the release of growth factors from cultured bovine, porcine, and human endothelial cells. Cancer Res 1989;49(18):5066−72.

[24] Baselet B, Belmans N, Coninx E, et al. Functional gene analysis reveals cell cycle changes and inflammation in endothelial cells irradiated with a single X-ray dose. Front Pharmacol 2017;8:213.

[25] Jaillet C, Morelle W, Slomianny MC, et al. Radiation-induced changes in the glycome of endothelial cells with functional consequences. Sci Rep 2017;7(1):5290.

[26] Mohammadkarim A, Mokhtari-Dizaji M, Kazemian A, Saberi H, Khani MM, Bakhshandeh M. Dose-dependent ^{60}Co γ-radiation effects on human endothelial cell mechanical properties. Cell Biochem Biophys 2019;77(2):179−86.

[27] Wang H, Segaran RC, Chan LY, et al. Gamma radiation-induced disruption of cellular junctions in HUVECs is mediated through affecting MAPK/NF-κB inflammatory pathways. Oxid Med Cell Longev 2019;2019:1486232.

[28] Milstone DS, Ilyama M, Chen M, et al. Differential role of an NF-κB transcriptional response element in endothelial versus intimal cell VCAM-1 expression. Circ Res 2015;117(2):166−77.

[29] Sakata K, Kondo T, Mizuno N, et al. Roles of ROS and PKC-βII in ionizing radiation-induced eNOS activation in human vascular endothelial cells. Vasc Pharmacol 2015;70:55−65.

[30] Vu HT, Kotla S, Ko KA, et al. Ionizing radiation induces endothelial inflammation and apoptosis via p90RSK-mediated ERK5 S496 phosphorylation. Front Cardiovasc Med 2018;5:23.

[31] Zhou H, Jin C, Cui L, et al. HMGB1 contributes to the irradiation-induced endothelial barrier injury through receptor for advanced glycation endproducts (RAGE). J Cell Physiol 2018;233(9):6714−21.

[32] Azimzadeh O, Subramanian V, Ständer S, et al. Proteome analysis of irradiated endothelial cells reveals persistent alteration in protein degradation and the RhoGDI and NO signaling pathways. Int J Radiat Biol 2017;93(9):920−8.

[33] Hong CW, Kim YM, Pyo H, et al. Involvement of inducible nitric oxide synthase in radiation-induced vascular endothelial damage. J Radiat Res 2013;54(6):1036−42.

[34] Baker JE, Fish BL, Su J, et al. 10 Gy total body irradiation increases risk of coronary sclerosis, degeneration of heart structure and function in a rat model. Int J Radiat Biol 2009;85:1089−100.

[35] Schultz-Hector S, Balz K. Radiation-induced loss of endothelial alkaline phosphatase activity and development of myocardial degeneration. An ultrastructural study. Lab Invest 1994;71:252−60.

[36] Lauk S. Endothelial alkaline phosphatase activity loss as an early stage in the development of radiation-induced heart disease in rats. Radiat Res 1987;110(1):118−28.

[37] Boerma M, Kruse JJ, Van Loenen M, et al. Increased deposition of von Willebrand factor in the rat heart after local ionizing irradiation. Strahlenther Onkol 2004;180:109−16.

[38] Van Kleef EM, te Poele JA, Oussoren YG, et al. Increased expression of glomerular von Willebrand factor after irradiation of the mouse kidney. Radiat Res 1998;150(5):528−34.

[39] Ashcraft KA, Choudhury KR, Birer SR, et al. Application of a novel murine ear vein model to evaluate the effects of a vascular radioprotectant on radiation-induced vascular permeability and leukocyte adhesion. Radiat Res 2018;190(1):12−21.

[40] Raoufi-Rad N, McRobb LS, Lee VS, et al. In vivo imaging of endothelial cell adhesion molecule expression after radiosurgery in an animal model of arteriovenous malformation. PLoS One 2017;12(9):e0185393.

[41] Medhora M, Haworth S, Liu Y, et al. Biomarkers for radiation pneumonitis using noninvasive molecular imaging. J Nucl Med 2016;57 (8):1296−301.

[42] Guerra G, Perrotta F, Testa G. Circulating endothelial progenitor cells biology and regenerative medicine in pulmonary vascular diseases. Curr Pharm Biotechnol 2018;19(9):700−7.

[43] Erdbruegger U, Dhaygude A, Haubitz M, Woywodt A. Circulating endothelial cells: markers and mediators of vascular damage. Curr Stem Cell Res Ther 2010;5(4):294−302.

[44] Liu Y, Xia T, Zhang W, et al. Variations of circulating endothelial progenitor cells and transforming growth factor-beta-1 (TGF-β1) during thoracic radiotherapy are predictive for radiation pneumonitis. Radiat Oncol 2013;8:189.

[45] Rafii S, Ginsberg M, Scandura J, Butler JM, Ding BS. Transplantation of endothelial cells to mitigate acute and chronic radiation injury to vital organs. Radiat Res 2016;186(2):196−202.

[46] Fajardo LF, Stewart JR. Capillary injury preceding radiation-induced myocardial fibrosis. Radiology 1971;101(2):429−33.

[47] Hallahan DE, Virudachalam S. Ionizing radiation mediates expression of cell adhesion molecules in distinct histological patterns within the lung. Cancer Res 1997;57(11):2096−9.

[48] Hirst DG, Denekamp J, Travis EL. The response of mesenteric blood vessels to irradiation. Radiat Res 1979;77(2):259−75.

[49] Jaenke RS, Robbins ME, Bywaters T, Whitehouse E, Rezvani M, Hopewell JW. Capillary endothelium. Target site of renal radiation injury. Lab Invest 1993;68(4):396−405.

[50] Lyubimova N, Hopewell JW. Experimental evidence to support the hypothesis that damage to vascular endothelium plays the primary role in the development of late radiation-induced CNS injury. Br J Radiol 2004;77(918):488−92.

[51] Paris F, Fuks Z, Kang A, et al. Endothelial apoptosis as the primary lesion initiating intestinal radiation damage in mice. Science 2001;293 (5528):293−7.

[52] Preidl RHM, Möbius P, Weber M, et al. Long-term endothelial dysfunction in irradiated vessels: an immunohistochemical analysis. Strahlenther Onkol 2019;195(1):52−61.

[53] Kalash R, Berhane H, Goff J, et al. Effects of thoracic irradiation on pulmonary endothelial compared to alveolar type-II cells in fibrosis-prone C57BL/6Ntac mice. Vivo 2013;27(3):291−7.

[54] Gabriels K, Hoving S, Seemann I, et al. Local heart irradiation of ApoE(−/−) mice induces microvascular and endocardial damage and accelerates coronary atherosclerosis. Radiother Oncol 2012;105(3):358−64.

[55] Gladstone M, Su TT. Radiation responses and resistance. Int Rev Cell Mol Biol 2012;299:235−53.

[56] Ljubimova NV, Levitman MK, Plotnikova ED, Eidus LKh. Endothelial cell population dynamics in rat brain after local irradiation. Br J Radiol 1991;64(766):934−40.

[57] Lee CL, Moding EJ, Cuneo KC, et al. P53 functions in endothelial cells to prevent radiation-induced myocardial injury in mice. Sci Signal 2012;5(234):ra52.

[58] Lee CL, Daniel AR, Holbrook M, et al. Sensitization of vascular endothelial cells to ionizing radiation promotes the development of delayed intestinal injury in mice. Radiat Res 2019;192(3):258−66.

[59] Rotolo J, Stancevic B, Zhang J, et al. Anti-ceramide antibody prevents the radiation gastrointestinal syndrome in mice. J Clin Invest 2012;122 (5):1786−90.

[60] Rotolo JA, Maj JG, Feldman R, et al. Bax and Bak do not exhibit functional redundancy in mediating radiation-induced endothelial apoptosis in the intestinal mucosa. Int J Radiat Oncol Biol Phys 2008;70(3):804−15.

[61] Abderrahmani R, François A, Buard V, et al. PAI-1-dependent endothelial cell death determines severity of radiation-induced intestinal injury. PLoS One 2012;7(4):e35740.

[62] Doan PL, Russell JL, Himburg HA, et al. Tie2(+) bone marrow endothelial cells regulate hematopoietic stem cell regeneration following radiation injury. Stem Cell 2013;31(2):327−37.

[63] Nguyen HQ, To NH, Zadigue P, Kerbrat, et al. Ionizing radiation-induced cellular senescence promotes tissue fibrosis after radiotherapy. A review. Crit Rev Oncol/Hematol 2018;129:13−26.

[64] He Y, Thummuri D, Zheng G, et al. Cellular senescence and radiation-induced pulmonary fibrosis. Transl Res 2019;209:14−21.

[65] Dong X, Tong F, Qian C, et al. NEMO modulates radiation-induced endothelial senescence of human umbilical veins through NF-κB signal pathway. Radiat Res 2015;183:82−93.

[66] Heo JI, Kim W, Choi KJ, Bae S, Jeong JH, Kim KS. XIAP-associating factor 1, a transcriptional target of BRD7, contributes to endothelial cell senescence. Oncotarget 2016;7 5118-5030.

[67] Kim KS, Kim JE, Choi KJ, Bae S, Kim DH. Characterization of DNA damage-induced cellular senescence by ionizing radiation in endothelial cells. Int J Radiat Biol 2014;90:71−80.

[68] Lowe D, Raj K. Premature aging induced by radiation exhibits pro-atherosclerotic effects mediated by epigenetic activation of CD44 expression. Aging Cell 2014;13:900−10.

[69] Mendonca MS, Chin-Sinex H, Dhaemers R, Mead LE, Yoder MC, Ingram DA. Differential mechanisms of X-ray-induced cell death in human endothelial progenitor cells isolated from cord blood and adults. Radiat Res 2011;176:208−16.

[70] Oh CW, Bump EA, Kim JS, Janigro D, Mayberg MR. Induction of a senescence-like phenotype in bovine aortic endothelial cells by ionizing radiation. Radiat Res 2001;156:232−40.

[71] Panganiban RA, Day RM. Inhibition of IGF-1R prevents ionizing radiation-induced primary endothelial cell senescence. PLoS One 2013;8:e78589.

[72] Panganiban RA, Mungunsukh O, Day RM. X-irradiation induces ER stress, apoptosis, and senescence in pulmonary artery endothelial cells. Int J Radiat Biol 2013;89:656−67.

[73] Park H, Kim CH, Jeong JH, Park M, Kim KS. GDF15 contributes to radiation-induced senescence through the ROS-mediated p16 pathway in human endothelial cells. Oncotarget 2016;7:9634−44.

[74] Petrache I, Petrusca DN, Bowler RP, Kamocki K. Involvement of ceramide in cell death responses in the pulmonary circulation. Proc Am Thorac Soc 2011;8:492−6.

[75] Sermsathanasawadi N, Ishii H, Igarashi K, et al. Enhanced adhesion of early endothelial progenitor cells to radiation-induced senescence-like vascular endothelial cells in vitro. J Radiat Res 2009;50:469−75.

[76] Ungvari Z, Podlutsky A, Sosnowska D, et al. Ionizing radiation promotes the acquisition of a senescence-associated secretory phenotype and impairs angiogenic capacity in cerebromicrovascular endothelial cells: role of increased DNA damage and decreased DNA repair capacity in microvascular radiosensitivity. J Gerontol 2013;68:1443−57.

[77] Yentrapalli R, Azimzadeh O, Sriharshan A, et al. The PI3K/Akt/mTOR pathway is implicated in the premature senescence of primary human endothelial cells exposed to chronic radiation. PLoS One 2013;8:e70024.

[78] Yentrapalli R, Azimzadeh O, Barjaktarovic Z, et al. Quantitative proteomic analysis reveals induction of premature senescence in human umbilical vein endothelial cells exposed to chronic low-dose rate gamma radiation. Proteomics 2013;13:1096−107.

[79] Azimzadeh O, Sievert W, Sarioglu H, et al. Integrative proteomics and targeted transcriptomics analyses in cardiac endothelial cells unravel mechanisms of long-term radiation-induced vascular dysfunction. J Proteome Res 2015;14:1203−19.

[80] Koziel R, Pircher H, Kratochwil M, et al. Mitochondrial respiratory chain complex I is inactivated by NADPH oxidase Nox4. Biochem J 2013;452:231−9.

[81] Frey RS, Ushio-Fukai M, Malik AB. NADPH oxidase-dependent signaling in endothelial cells: role in physiology and pathophysiology. Antioxid Redox Signal 2009;11:791−810.

[82] Donato AJ, Morgan RG, Walker AE, Lesniewski LA. Cellular and molecular biology of aging endothelial cells. J Mol Cell Cardiol 2015;89:122−35.

[83] Muller M. Cellular senescence: molecular mechanisms, in vivo significance, and redox considerations. Antioxid Redox Signal 2009;11:59−98.

[84] Lafargue A, Degorre C, Corre I, et al. Ionizing radiation induces long-term senescence in endothelial cells through mitochondrial respiratory complex II dysfunction and superoxide generation. Free Radic Biol Med 2017;108:750−9.

[85] Marampon F, Gravina GL, Festuccia C, et al. Vitamin D protects endothelial cells from irradiation-induced senescence and apoptosis by modulating MAPK/SirT1 axis. J Endocrinol Inves 2016;39(4):411−22.

[86] Chen J, Huang X, Halicka D, et al. Contribution of p16INK4a and p21CIP1 pathways to induction of premature senescence of human endothelial cells: permissive role of p53. Am J Physiol Heart Circ Physiol 2006;290:H1575−86.

[87] Cho JH, Kim MJ, Kim KJ, Kim JR. POZ/BTB and AT-hook containing zinc finger protein 1 (PATZ1) inhibits endothelial cell senescence through a p53 dependent pathway. Cell Death Differ 2012;19:703−12.

[88] Kim HJ, Cho JH, Kim JR. Downregulation of Polo-like kinase 1 induces cellular senescence in human primary cells through a p53-dependent pathway. J Gerontol 2013;68:1145−56.

[89] Friedman M, Ryan US, Davenport WC, Chaney EL, Strickland DL, Kwock L. Reversible alterations in cultured pulmonary artery endothelial cell monolayer morphology and albumin permeability induced by ionizing radiation. J Cell Physiol 1986;129(2):237−49.

[90] Kabacik S, Raj K. Ionising radiation increases permeability of endothelium through ADAM10-mediated cleavage of VE-cadherin. Oncotarget 2017;8(47):82049−63.

[91] Van der Bijl P, van Eyk AD, Liss J, Böhm L. The effect of radiation on the permeability of human saphenous vein to 17 beta-oestradiol. SADJ 2002;57(3):92−4.

[92] Andrews RN, Caudell DL, Metheny-Barlow LJ, et al. Fibronectin produced by cerebral endothelial and vascular smooth muscle cells contributes to perivascular extracellular matrix in late-delayed radiation-induced brain injury. Radiat Res 2018;190(4):361−73.

[93] Boerma M, Wang J, Sridharan V, Herbert JM, Hauer-Jensen M. Pharmacological induction of transforming growth factor-beta1 in rat models enhances radiation injury in the intestine and the heart. PLoS One 2013;8(7):e70479.

[94] Rosenkrans Jr WA, Penney DP. Cell-cell matrix interactions in induced lung injury. IV. Quantitative alterations in pulmonary fibronectin and laminin following X irradiation. Radiat Res 1987;109(1):127−42.

[95] Esmon CT, Taylor FB, Snow TR. Inflammation and coagulation: linked processes potentially regulated through a common pathway mediated by protein C. Thromb Haemost 1991;66:160−5.

[96] Mosnier LO, Gale AJ, Vegneswaran S, Griffin JH. Activated protein C variants with normal cytoprotective but reduced anticoagulant activity. Blood 2004;104:1740−4.

[97] Abeyama K, Stern DM, Ito Y, et al. The N-terminal domain of thrombomodulin sequesters high-mobility group-B1 protein, a novel antiinflammatory mechanism. J Clin Invest 2005;115(5):1267−74.

[98] Bizios R, Lai L, Fenton JW, Malik AB. Thrombin-induced chemotaxis and aggregation of neutrophils. J Cell Physiol 1986;128(3):485−90.

[99] Chambers RC, Dabbagh K, McAnulty RJ, Gray AJ, Blanc-Brude OP, Laurent GJ. Thrombin stimulates fibroblast procollagen production via proteolytic activation of protease-activated receptor 1. Biochem J 1998;333(Pt 1):121−7.

[100] Wang J, Zheng H, Ou X, Fink LM, Hauer-Jensen M. Deficiency of microvascular thrombomodulin and up-regulation of protease-activated receptor-1 in irradiated rat intestine: possible link between endothelial dysfunction and chronic radiation fibrosis. Am J Pathol 2002;160(6):2063.

[101] Wang J, Albertson CM, Zheng H, Fink LM, Herbert JM, Hauer-Jensen M. Short-term inhibition of ADP-induced platelet aggregation by clopidogrel ameliorates radiation-induced toxicity in rat small intestine. Thromb Haemost 2002;87(1):122−8.

[102] Wang J, Zheng H, Ou X, et al. Hirudin ameliorates intestinal radiation toxicity in the rat: support for thrombin inhibition as strategy to minimize side-effects after radiation therapy and as countermeasure against radiation exposure. J Thromb Haemost 2004;2(11):2027−35.

[103] Geiger H, Pawar SA, Kerschen EJ, et al. Pharmacological targeting of the thrombomodulin-activated protein C pathway mitigates radiation toxicity. Nat Med 2012;18(7):1123−9.

[104] Scharpfenecker M, Kruse JJ, Sprong D, Russell NS, Ten Dijke P, Stewart FA. Ionizing radiation shifts the PAI-1/ID-1 balance and activates notch signaling in endothelial cells. Int J Radiat Oncol Biol Phys 2009;73:506−13.

[105] Rodemann HP, Binder A, Burger A, Guven N, Loffler H, Bamberg M. The underlying cellular mechanism of fibrosis. Kidney Int Suppl 1996;54:S32−6.

[106] Stansborough RL, Bateman EH, Al-Dasooqi N, et al. Vascular endothelial growth factor (VEGF), transforming growth factor beta (TGFβ), angiostatin, and endostatin are increased in radiotherapy-induced gastrointestinal toxicity. Int J Radiat Biol 2018;94(7):645−55.

[107] Imaizumi N, Monnier Y, Hegi M, Mirimanoff RO, Ruegg C. Radiotherapy suppresses angiogenesis in mice through TGFbetaRI/ALK5-dependent inhibition of endothelial cell sprouting. PLoS One 2010;5:e11084.

[108] Kruse JJ, Floot BG, te Poele JA, Russell NS, Stewart FA. Radiation-induced activation of TGF-beta signaling pathways in relation to vascular damage in mouse kidneys. Radiat Res 2009;171:188–97.

[109] Richter KK, Langberg CW, Sung CC, Hauer-Jensen M. Association of transforming growth factor beta (TGF-beta) immunoreactivity with specific histopathologic lesions in subacute and chronic experimental radiation enteropathy. Radiother Oncol 1996;39:243–51.

[110] Barcellos-Hoff MH. How do tissues respond to damage at the cellular level? The role of cytokines in irradiated tissues. Radiat Res 1998;150: S109–20.

[111] Piera-Velazquez S, Jimenez SA. Endothelial to mesenchymal transition: role in physiology and in the pathogenesis of human diseases. Physiol Rev 2019;99(2):1281–324.

[112] Kim M, Choi SH, Jin YB, et al. The effect of oxidized low-density lipoprotein (ox-LDL) on radiation-induced endothelial-to-mesenchymal transition. Int J Radiat Biol 2013;89(5):356–63.

[113] Mintet E, Rannou E, Buard V, et al. Identification of endothelial-to-mesenchymal transition as a potential participant in radiation proctitis. Am J Pathol 2015;185(9):2550–62.

[114] Mintet E, Lavigne J, Paget V, et al. Endothelial Hey2 deletion reduces endothelial-to-mesenchymal transition and mitigates radiation proctitis in mice. Sci Rep 2017;7(1):4933.

[115] Choi SH, Hong ZY, Nam JK, et al. A hypoxia-induced vascular endothelial-to-mesenchymal transition in development of radiation-induced pulmonary fibrosis. Clin Cancer Res 2015;21(16):3716–26.

[116] Vítková V, Živný J, Janota J. Endothelial cell-derived microvesicles: potential mediators and biomarkers of pathologic processes. Biomark Med 2018;12(2):161–75.

[117] Hromada C, Mühleder S, Grillari J, Redl H, Holnthoner W. Endothelial extracellular vesicles-promises and challenges. Front Physiol 2017;8:275.

[118] Gao C, Li R, Liu Y, Ma L, Wang S. Rho-kinase-dependent F-actin rearrangement is involved in the release of endothelial microparticles during IFN-α-induced endothelial cell apoptosis. J Trauma Acute Care Surg 2012;73(5):1152–60.

[119] Piryani SO, Jiao Y, Kam AYF, et al. Endothelial cell-derived extracellular vesicles mitigate radiation-induced hematopoietic injury. Int J Radiat Oncol Biol Phys 2019;104(2):291–301.

[120] Flamant S, Tamarat R. Extracellular vesicles and vascular injury: new insights for radiation exposure. Radiat Res 2016;186(2):203–18.

[121] Devine A, Marignol L. Potential of amifostine for chemoradiotherapy and radiotherapy-associated toxicity reduction in advanced NSCLC: a meta-analysis. Anticancer Res 2016;36(1):5–12.

[122] Kalman NS, Zhao SS, Anscher MS, Urdaneta AI. Current status of targeted radioprotection and radiation injury mitigation and treatment agents: a critical review of the literature. Int J Radiat Oncol Biol Phys 2017;98(3):662–82.

[123] Farese AM, MacVittie TJ. Filgrastim for the treatment of hematopoietic acute radiation syndrome. Drugs Today 2015;51(9):537–48.

[124] Satyamitra MM, DiCarlo AL, Taliaferro L. Understanding the pathophysiology and challenges of development of medical countermeasures for radiation-induced vascular/endothelial cell injuries: report of a NIAID workshop, August 20, 2015. Radiat Res 2016;186(2):99–111.

[125] Han NK, Jeong YJ, Pyun BJ, Lee YJ, Kim SH, Lee HJ. Geranylgeranylacetone ameliorates intestinal radiation toxicity by preventing endothelial cell dysfunction. Int J Mol Sci 2017;18(10):2103.

[126] Parker RA, Pearce BC, Clark RW, Gordon DA, Wright JJ. Tocotrienols regulate cholesterol production in mammalian cells by post-transcriptional suppression of 3-hydroxy-3-methylglutaryl-coenzyme A reductase. J Biol Chem 1993;268(15):11230–8.

[127] Song BL, DeBose-Boyd RA. Insig-dependent ubiquitination and degradation of 3-hydroxy-3-methylglutaryl coenzyme a reductase stimulated by delta- and gamma-tocotrienols. J Biol Chem 2006;281(35):25054–61.

[128] Ali SF, Nguyen JC, Jenkins TA, Woodman OL. Tocotrienol-rich tocomin attenuates oxidative stress and improves endothelium-dependent relaxation in aortae from rats Fed a high-fat western diet. Front Cardiovasc Med 2016;3:39.

[129] Berbée M, Fu Q, Boerma M, Wang J, Kumar KS, Hauer-Jensen M. Gamma-Tocotrienol ameliorates intestinal radiation injury and reduces vascular oxidative stress after total-body irradiation by an HMG-CoA reductase-dependent mechanism. Radiat Res 2009;171(5):596–605.

[130] Berbee M, Fu Q, Boerma M, et al. Reduction of radiation-induced vascular nitrosative stress by the vitamin E analog γ-tocotrienol: evidence of a role for tetrahydrobiopterin. Int J Radiat Oncol Biol Phys 2011;79(3):884–91.

[131] Hoving S, Heeneman S, Gijbels MJ, et al. NO-donating aspirin and aspirin partially inhibit age-related atherosclerosis but not radiation-induced atherosclerosis in ApoE null mice. PLoS One 2010;5(9):e12874.

[132] Fu Q, Wang J, Boerma M, et al. Involvement of heat shock factor 1 in statin-induced transcriptional upregulation of endothelial thrombomodulin. Circ Res 2008;103(4):369–77.

[133] Pathak R, Ghosh SP, Zhou D, Hauer-Jensen M. The vitamin E analog gamma-tocotrienol (GT3) and statins synergistically up-regulate endothelial thrombomodulin (TM). Int J Mol Sci 2016;17(11):1937.

[134] Pathak R, Kumar VP, Hauer-Jensen M, Ghosh SP. Enhanced survival in mice exposed to ionizing radiation by combination of gamma-tocotrienol and simvastatin. Mil Med 2019;184(Suppl. 1):644–51.

[135] Te Poele JA, van Kleef EM, van der Wal AF, Dewit LG, Stewart FA. Radiation-induced glomerular thrombus formation and nephropathy are not prevented by the ADP receptor antagonist clopidogrel. Int J Radiat Oncol Biol Phys 2001;50(5):1332–8.

[136] Hoving S, Heeneman S, Gijbels MJ, et al. Anti-inflammatory and anti-thrombotic intervention strategies using atorvastatin, clopidogrel and knock-down of CD40L do not modify radiation-induced atherosclerosis in ApoE null mice. Radiother Oncol 2011;101(1):100–8.

[137] Boström M, Kalm M, Eriksson Y, et al. A role for endothelial cells in radiation-induced inflammation. Int J Radiat Biol 2018;94(3):259–71.

[138] Wiesemann A, Ketteler J, Slama A, et al. Inhibition of radiation-induced Ccl2 signaling protects lungs from vascular dysfunction and endothelial cell loss. Antioxid Redox Signal 2019;30(2):213–31.

[139]   Toullec A, Buard V, Rannou E, et al. HIF-1α deletion in the endothelium, but not in the epithelium, protects from radiation-induced enteritis. Cell Mol Gastroenterol Hepatol 2017;5(1):15−30.

[140]   Jang H, Lee J, Park S, et al. Baicalein mitigates radiation-induced enteritis by improving endothelial dysfunction. Front Pharmacol 2019;10:892.

[141]   Soroush F, Tang Y, Zaidi HM, Sheffield JB, Kilpatrick LE, Kiani MF. PKCδ inhibition as a novel medical countermeasure for radiation-induced vascular damage. FASEB J 2018;32:6436−44.

[142]   Presta M, Andrés G, Leali D, Dell'Era P, Ronca R. Inflammatory cells and chemokines sustain FGF2-induced angiogenesis. Eur Cytokine Netw 2009;20(2):39−50.

[143]   Guan D, Mi J, Chen X, et al. Lung endothelial cell-targeted peptide-guided bFGF promotes the regeneration after radiation induced lung injury. Biomaterials 2018;184:10−19.

[144]   Zhang S, Qiu X, Zhang Y, et al. Basic fibroblast growth factor ameliorates endothelial dysfunction in radiation-induced bladder injury. Biomed Res Int 2015;2015:967680.

[145]   Lee JG, Ko MK, Kay EP. Endothelial mesenchymal transformation mediated by IL-1β-induced FGF-2 in corneal endothelial cells. Exp Eye Res 2012;95(1):35−9.

[146]   Anscher MS, Thrasher B, Zgonjanin L, et al. Small molecular inhibitor of transforming growth factor-beta protects against development of radiation-induced lung injury. Int J Radiat Oncol Biol Phys 2008;71(3):829−37.

[147]   Rabender C, Mezzaroma E, Mauro AG, Mullangi R, Abbate A, Anscher M. Hart B6, Mikkelsen R. IPW-5371 proves effective as a radiation countermeasure by mitigating radiation-induced late effects. Radiat Res 2016;186(5):478−88.

[148]   Zhao Z, Johnson MS, Chen B, et al. Live-cell imaging to detect phosphatidylserine externalization in brain endothelial cells exposed to ionizing radiation: implications for the treatment of brain arteriovenous malformations. J Neurosurg 2016;124(6):1780−7.

[149]   Park MT, Oh ET, Song MJ, Lee H, Park HJ. Radio-sensitivities and angiogenic signaling pathways of irradiated normal endothelial cells derived from diverse human organs. J Radiat Res 2012;53(4):570−80.

Infection, vascular signaling and injury

Chapter 15

Hemeoxygenase and its metabolites in regulation of vascular endothelial health

Chhanda Biswas

Children's Hospital of Philadelphia, Philadelphia, PA, United States

15.1 Introduction

Vasculogenesis is a systemic process that requires the responses of multiple cell types, including endothelial, mural, inflammatory, and blood-derived cells [1]. Vasculogenesis is essential for many physiological processes such as embryogenesis, tissue repair, and organ regeneration. Angiogenesis or formation of new blood vessels from the preexisting vasculature requires proangiogenic factors, exemplified by vascular endothelial growth factor (VEGF). The expression of VEGF is known to be driven by oxidative signals [2] produced in response to various systemic stimuli, including growth factors and metabolic processes; inflammatory responses-driven cytokines and chemotactic factors; hypoxia (lack of oxygen); and injury-related shear stresses. However, excess exposure to inflammatory reactions and oxidative stress is damaging to vascular endothelial cells (ECs) and rapidly results in various vascular complications. Heme oxygenase 1 (HO-1) is a stress-induced isoform of heme oxygenases, and is induced rapidly after oxidative stress to act as a potent endogenous factor for the resolution of stress-induced inflammatory injury [3]. Therefore maintenance of a healthy endothelial phenotype is pivotal in vascular health, and HO-1 plays a crucial role [4] in this process by attenuating inflammatory and oxidative signals. This chapter will discuss the HO system of enzymes, its interaction with the various endothelial signaling molecules, and highlight its role in endothelial signaling, thusly in vascular health.

Heme oxygenase (HO), heme degradation, and metabolites: HO is the only known metabolic enzyme that catalyzes heme degradation. In a rate-limiting step it adds up 3 mol of oxygen supplied by Nicotinamide adenine dinucleotide phosphate (NADPH) cytochrome p450 reductase leading to oxidation of heme [5]. This catabolic reaction releases equimolar amounts of carbon monoxide (CO), free iron (Fe^{2+}), and biliverdin (BV); the latter is further reduced to bilirubin (BR) by biliverdin reductase (BVR) [6]. Besides releasing heme toxicity, HO-1 is also known for vascular health and antiinflammatory functions attributed by its enzymatic byproducts, perhaps by the promoter activation of interleukin (IL)-10, a known antiinflammatory cytokine (Fig. 15.1). Heme, a primary functional form of iron, is a catalytic factor for the function of many hemoproteins including myoglobin, cytochromes, catalase, peroxidases, soluble guanylate cyclase (sGC), nitric oxide synthase, all of which execute crucial cellular functions. Heme is coordinated in the porphyrin ring, has the ability to transfer electrons and bind diatomic gases such as CO and NO, and characteristically is important for the functional delivery of these hemoproteins. Degradation of these hemoproteins constantly releases heme, which is either recycled to generate fresh hemoproteins, or metabolized by HO; otherwise overload of heme is highly toxic to the cells. Heme is also released in our system by normal process of erythrophagocytosis of senescent red blood cells, or by hemolysis that occurs in many hematologic and nonhematologic diseases. This can wreak oxidative damage in the vasculature and in exposed tissues.

HO isoforms: HO is evolutionarily conserved across the phylogenic kingdoms and functions in various ways precise to the demands [7,8]. To date, two isoforms of the HO (HO-1 and HO-2) enzyme are known in mammals, and a third catalytically inactive isoform, HO-3 has been suggested by Hayashi and his colleagues [9]. HO is membrane anchored with a single C-terminal transmembrane region [10] to endoplasmic reticulum (ER), particularly of smooth ER, and therefore are highly enriched in microsomal fractions [6]. HO-1 (32 kDa) and HO-2 (36 kDa) share high degree of sequence homology, about 45% in total, and 59% in highly conserved regions [10]. The carboxy-terminal membrane

Endothelial Signaling in Vascular Dysfunction and Disease. DOI: https://doi.org/10.1016/B978-0-12-816196-8.00007-2

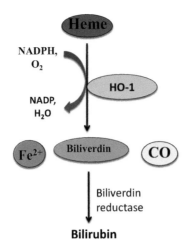

FIGURE 15.1 Heme oxygenase, heme degradation, and metabolites.

Induction of HO-1

FIGURE 15.2 Induction of HO-1.

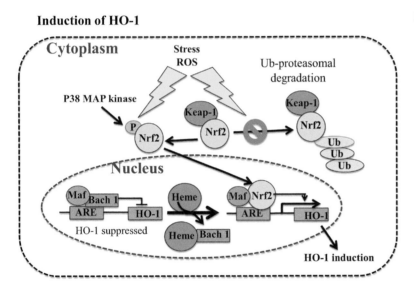

anchor containing 18 amino acids shows similarities but is less conserved ($\sim 15\%$). HO-2 expression is constitutive and predominant in brain, entire neuraxis, and testes [11]. Normally HO-1 is constitutively high in spleen and liver [11,12] with special purpose. Spleen red pulp macrophages are constantly engaged in the clearance of senescent RBCs, which upon degradation release hemoglobin and tons of heme. Likewise, HO-1 as antioxidant helps in maintaining liver homeostasis from high exposure of liver to stressful polar products in processing endogenous compounds and exogenous xenobiotic products and other toxins.

Inducible HO-1: Among different HO isoforms, HO-1 is inducible [13] and has gained major attention in medical research as therapeutic target in various diseases because of its role as a sensor of cellular stress and as a promoter of specific adaptive responses crucial for homeostasis [14]. It can be induced by a plethora of stimuli such as hemorrhage, heme, hypoxia, hyperoxia, heavy metals, reactive oxygen species (ROS), heat shock, ultraviolet irradiation, infection, and inflammation (cytokines). Although HO-1 promoter activation can be achieved by several factors, the most conservative signaling is via binding of nuclear factor erythroid 2-related factor (Nrf2) [8] (Fig. 15.2). Normally constitutive Nrf2 undergoes proteasomal degradation by Keap-1/Cul3-E3 ligase-dependent ubiquitination (Ub), and HO-1 promoter activation remains suppressed by Bach1 [15]. With appropriate stimuli, such as oxidative stress of hypoxia, heavy metal, hyperoxia, or heme itself, Nrf2 is stabilized by MAP kinase-dependent phosphorylation allowing its dissociation from Keap-1 and nuclear delivery. Simultaneously in the nucleus heme replaces Bach1, the suppressor of HO-1, and allows Nrf2/Maf protein complex binds to the antioxidant response element of HO-1 promoter facilitating transcription of HO-1 [16]. HO-1 induction, heme metabolism, and the metabolites further regulate expression of genes coding for antioxidant, antiinflammatory, and detoxifying proteins for detoxification and adaptive responses [14]. Unlike HO-2, HO-1 is vulnerable to tryptic cleavage and the shorter form ($\sim$28 kDa) with noncanonical regulatory function migrates to nuclear compartment in hypoxia, hyperoxia, and in some cancers, and serves in a cellular defense mechanism (discussed elsewhere [17]). HO-1 is quite unique because it requires its own substrate, heme, as a cofactor for heme degradation, and HO-1 is also induced by heme [18] (Fig. 15.2). Recently heme-dependent Nrf2 signaling and HO-1 induction are reported as atheroprotective from intraplaque hemorrhage characteristically by elevating CD163 and IL-10 secretion [19]. Most tissues exhibit robust activation of the highly inducible HO-1 as an adaptive response to environmental insults, and this induction is known to be antioxidative, cytoprotective, and proliferative, and antiinflammatory [14]. The beneficial effects of HO-1 in the mediation of antiinflammatory functions are attributed predominantly to catalytic byproducts including carbon monoxide and BV/BR [29]. Both CO and BV/BR system are referred to as facilitators of promoter activation of IL-10 (Fig. 15.1), which can modulate immune cells of innate and adaptive immune systems to terminate inflammation. However, there is gap in the understanding of the signaling pathways in IL-10 production by the metabolites of heme catabolism. CO predominantly is highly appreciated in vascular health based on many animal and cell-based studies that used loss or gain of HO-1 function, and by demonstrating rescue of cellular homeostasis in HO-1-deficient cells by the use of CO-releasing molecules (CORMs). CORMs are the lipid-soluble tricarbonyldichlororuthenium (II) dimer (CORM-2) [20–22] that when solubilized results in slow release of CO. Notably Nrf2/HO-1 and BR/BVR (biliverdin reductase) are the most conserved systems in mammals perhaps because they mandate physiological functions that are crucial for survival. The focus of this chapter is to elaborate on the beneficial role of HO-1/CO axis and BVR/BR system in vascular health and diseases.

15.2 Heme oxygenase 1 and vascular health

The role of HO in heme degradation was recognized by Tenhunen and colleagues in 1968 [23], and the protective properties of HO-1 were first demonstrated in an animal model of heme protein–induced kidney injury by Nath and colleagues in 1992 [24]. HO-1 deficiency in humans is rare. Yachie et al. [4] reported the first human case of HO-1 deficiency. A 6-year-old male patient was first presented to the medical care with severe growth retardation, recurrent fever, and generalized erythematous rash. The patient showed severe hemolysis, anemia, asplenia, and all symptoms of severe pathophysiology of high inflammatory state. Additionally, an abnormal coagulation/fibrinolysis system, associated with elevated thrombomodulin and von Willebrand factor (vWF), indicated the presence of severe, persistent endothelial damage. In HO-1 targeted mice, growth retardation, anemia, iron deposition, and vulnerability to stressful injury were characteristically observed. Human patients with HO-1 deficiency and HO-1 null mice are both especially prone to oxidative stress and inflammation. Both exhibit similar clinical findings such as low birth rate (embryonic lethal), growth failure, anemia, RBC fragmentation (not known in mice), elevated ferritin, iron deposition, hepatomegaly, splenomegaly (asplenia in human), and lymph node swelling [25]. The patient with HO-1 deficiency though had HO-1 intact died in childhood indicating that HO-1 is an essential survival component. Although HO-1 deficiency is rare in humans, altered expression of HO-1 is observed depending on length polymorphism in microsatellite constituting guanosine-thymidine nucleotide repeats (GT)n in the promoter region of HO-1 gene [26]. The length can range from less than 25 (short) to more than 25 (long) repeats, and the length is inversely proportional to the expression level of HO-1 [27,28]. This polymorphism with longer repeats corresponding to low expression of HO-1 is reported to have association with disease severity and poor treatment outcome [29–33]. Recent clinical data indicate an adverse outcome of this polymorphism on cardiovascular complications in some patients. Particularly the presence of longer (GT)n repeats corresponding to low HO-1 expression has association with an increased risk of arteriovenous fistula failure in people subjected to hemodialysis [34], higher incidence of coronary artery disease in type 2 diabetics or hemodialyzed patients [35,36], clinical restenosis after coronary stent implantation [37], aortic aneurysms [38], and cerebrovascular events [39]. Furthermore, among patients with peripheral artery disease, those carrying longer HO-1 alleles had higher rates of myocardial infarction, percutaneous coronary interventions, and coronary bypass operations [40]. These pathologies are reviewed in detail by Daenen et al. [41] and establish an epidemiological association between the (GT)n polymorphism and cardiovascular disease (CVD), and a presumed protective effect of HO-1 association with shorter repeats (yielding higher HO-1 expression) [42]. Daenen et al. [41] further suggested the possibility of a racial disparity in HO-1 (GT)n repeat length distribution and this might be the framework for the associations of the genotype with CVD status. Taha et al. [31] in support demonstrated that (GT)n promoter allelic variants do exert differential expression and biological effects of HO-1 in primary ECs. Another study by Leaf et al. [42] showed development of postoperative acute kidney injury (AKI) in patients who underwent cardiac surgery with cardiopulmonary bypass. This study further elaborated that longer (GT)n repeats in the *HMOX1* gene promoter associates with increased risk of AKI after cardiac surgery, consistent with heme toxicity as a pathogenic feature of cardiac surgery-associated AKI, and with HO-1 as a potential therapeutic target. Taken together the above reports strongly suggest for a crucial role of HO-1 in vascular health. In order to understand the HO-1-driven mechanism contributing to vascular health, it is important to understand the mediators of two most important physiological processes such as vasculogenesis and angiogenesis.

VEGF and HO-1 in vascular system: Vasculogenesis is dependent on angiogenesis or development of neovascular supply, and both processes serve crucial homeostatic role since blood vessels nourish organs by carrying nutrients and fluid and remove catabolic products. In the embryo, blood vessel formation de novo (neovasculogenesis) and from existing vessels (angiogenesis) results in blood vessels lined by ECs. Judah Folkman hypothesized in 1971 that tumor growth is dependent on angiogenesis, a process by which new blood vessels develop from preexisting vasculature [43,44]. Vasculogenesis and angiogenesis are the two major escalating factors recognized in tumor progression and growth, and targeting VEGF and its receptor VEGF-R has emerged as a breakthrough strategy in cancer treatment. From there the discovery and understanding of VEGF signaling has revolutionized our concept of vasculogenesis and angiogenesis in physiological homeostasis, and in repair of injury and wound healing. Experimentally exogenous VEGF administration was able to acutely improve myocardial functional recovery after ischemia. Increase of heme-dependent HO-1 expression and activity in the glioma cells was able to induce VEGF and brain-derived neurotrophic factor. This finding proposed that the glial cells are capable of recovering brain function from the damage caused by cerebral hemorrhage through the production of neurogenic and angiogenic factors. A recent study [45] further demonstrated that knockdown of the transcription factor Nrf2 in cerebral microvascular ECs under hypoxic conditions was inhibitory to angiogenesis by downregulating VEGF expression through PI3K/Akt signaling pathway. More studies importantly revealed that CO facilitates the beneficial role of HO-1 in vascular health. These studies used different derivatives of CORM capable of regulated release of CO and were able to rescue vascular EC activation and function [46–48] by modulating VEGF. Taken together, HO-1/CO system, via the induction of VEGF, serves a protective role in vascular health and therefore is of potential therapeutic value. The following sections will further elaborate on the role of CO and BR/BV system, in maintaining the physiological processes of vascular system.

CO and VEGF in vascular health: Over the past decade, reports have established that low concentration of CO/CORMs in cell culture or in vivo can modulate several molecular mechanisms that could aid elimination of microbes, regulation of cell death, and protection of organs from damage and dysfunction, and could trigger immune responses. These studies highlighted the important signaling pathways such as HO-1/CO axis [49], activation of p38 mitogen-activated protein kinase (MAPK), enhanced Peroxisome proliferator-activated receptor gamma (PPAR-γ) signaling, and suppression of activation and translocation of toll-like receptors (TLRs) in inflammation [50–52]. Abraham et al. [53] further demonstrated that when human ECs were transduced with a retroviral vector encoded with human HO-1 gene, the transgene expression caused upregulation of several genes associated with cell cycle progression, including cyclin E and D; downregulation of cyclin-dependent kinase inhibitors p21 and p27, cyclin-dependent kinases 2, 5, and 6, and expression of monocyte chemoattractant protein-1, and growth factors, including VEGF, VEGFR1, endothelial growth factor, and hepatic-derived growth factor. These findings identify an array of gene responses to overexpression of human HO-1 and elucidate a new direction of human HO-1 signaling involved in cell growth and proliferation. The beneficial role of CO in VEGF expression was characterized by Choi et al. [54] demonstrating that CO promotes VEGF expression by increasing hypoxia-inducible transcription factor-1 alpha (HIF-1α) protein level via two distinct mechanisms, by translational activation and by stabilization of HIF-1α protein. In this study use of CORM-2 induced not only VEGF, but also HO-1, by a sequential activation of the signaling via c-Src/Pyk2/PKCα/p42/p44 MAPK pathway, thereby upregulating mRNA for the activator protein-1, further described by Lin et al. [55]. Based on findings that regulated supply of CO at a low level can be beneficial for vascular health, studies are underway to understand the differential action of different CORM derivatives and their bioavailability in modulating diseases [56].

CO and angiotensin II in vascular remodeling and hypertension: Angiotensin II (Ang II), a dynamic vasoactive peptide, is a potent mediator of hypertension and organ damage [57]. Angiotensin II (Ang II) acts both as peripheral and central modulator of cardiovascular autonomic function and thus exerts central activation of the sympathetic nerves, increases blood pressure, pro-inflammatory cytokines, and vascular damage [58]. It induces proliferation of splenic lymphocytes and increases cytokine production through its action of angiotensin II receptor type 1 on immune cells [58–60]. Vascular damage and associated pathogenesis of CVDs including cardiomyopathy, coronary artery disease, atherosclerosis, and vascular injury [61] are concurrent with the degradation of the extracellular matrix, through various inflammatory reactions, generation of ROS and matrix metalloproteinase [62,63]. Ang II is known to be the key player in exerting vascular remodeling effects through NADPH oxidase–derived generation of ROS [64–67] leading to the mechanisms in cell proliferation, fibrosis, oxidative stress, and inflammation [68,69]. Ming-Horng Tsai et al. suggested for an inhibitory role of CORM-2 on Ang II-driven ROS generation [70] and human aortic smooth muscle migration.

This study indicated CORM-2-dependent signaling mechanism in the inhibition of ROS/IL-6 generation and matrix metalloproteinase-9 expression [71]. Therefore it is evident that increased Ang II receptor activation by Ang II is capable of exerting various forms of insults that could lead to pro-inflammatory signaling with sustained low-grade inflammation, in turn vascular injury, loss of vascular tone and structure leading to hypertension. And CO plays as an antidote to all this pathophysiology presumably by terminating inflammation.

15.3 Heme oxygenase 1/interleukin-10 signaling in immune suppression and vascular health

Immune dysregulation and chronic inflammatory states are interlinked with arterial hypertension [57,72,73]. Traditionally, arterial hypertension and subsequent end-organ damage have been attributed to hemodynamic factors, but increasing evidence indicates that inflammation dramatically contributes to the deleterious consequences of this disease [73]. Shao et al. [74] suggested that antiinflammatory/immunosuppressive cytokines serve as novel therapeutic targets in inhibiting endothelial dysfunction, vascular inflammation, and cardio- and cerebrovascular diseases. Chronic inflammation can originate from exogenous or sterile infection, and is characterized by occurrence of sustained serum pro-inflammatory cytokines, that is known to trigger endothelial damage and dysfunction and thereby disruption of the blood vessel tone and structure. Our body is constantly exposed to ubiquitous microbial attack that causes the release of pathogen associated molecular patterns (PAMPs); sterile injury drives the release of damage-associated molecular patterns (DAMPs) from damaged tissues. These constantly inflict inflammation through specific pattern recognition receptors (PRRs). PRRs like Toll Like receptors (TLRs) upon activation by PAMPs or DAMPs evoke complex signaling cascades to activate the transcription factor NF-κB. The activated NF-κB is then translocated to the nuclear compartment allowing promoter activation of numerous pro-inflammatory cytokines. These steps in immune responses are necessary for recruitment of immune cells, the elimination of infection, and to ensure, after the clearance of the infectious agents, a well-orchestrated negative regulatory mechanism to restore homeostasis. However, under pathophysiological circumstances lack of such negative regulation fails to pose a break to the inflammatory signals and in a feed-forwarded loop amplifies the inflammatory state. This inflated hyperinflammatory state also known as macrophage activation syndrome (MAS) is typically characterized by high fever, hypercytokinemia, hepatosplenomegaly, pancytopenia, etc. MAS is the underlying cause of many autoimmune diseases and is fatal if not controlled in time. Our studies earlier in murine model of MAS developed by repeat activation of TLR9 have demonstrated expansion of tissue-resident inflammatory monocytes (monocytopoiesis) responsible to amplify hypercytokinemia [75–77]. In this MAS model, blocking IL-10 receptor with anti-IL-10 receptor antibody was fatal to the mice indicating a crucial role for IL-10 in resolving MAS [77]. Currently a role for HO-1 in this IL-10 production is under examination. Nevertheless, IL-10 is long known as a potential terminator of inflammation, and it can regulate immune cells of both innate and adaptive immune systems. Macrophages are the major producers of IL-10, and most literature studies suggest IL-10 signaling–driven HO-1 induction via STAT3 (signal transducer and activator of transcription). It is also suggested that HO-1 induction further amplifies IL-10 induction in a feed-forward loop. The scheme in Fig. 15.3 describes and connects the possible signaling mechanisms in vascular health ensured by HO-1 enzyme activation.

CO-driven IL-10 production: CO-driven cell signaling leading to promoter activation of IL-10 is supported by many literature studies [78,79]. In 2007, Ma et al. in a review discussed how differential sensing of NO, CO, and oxygen is critical for a wide range of physiological processes such as vascular homeostasis, platelet aggregation, host defense, neuronal signaling, and stress responses [80]. However, the antiinflammatory mechanism by IL-10 was not discussed in this study. Here we bring to focus the fact that gaseous CO can only bind to a metal, such as heme, and activate a hemoprotein. An appropriate example is HO-1 itself that requires heme for its activation and therefore the use of CORM drives activation of HO-1. This in turn amplifies production of CO by degradation of heme. However, the precise mechanism in attaining IL-10 synthesis by CO is still obscure. We presumed that in the IL-10 synthesis CO acts on the hemoprotein, sGC [81]. The activated sGC is known to trigger conversion of Guanosine-5′-triphosphate (GTP) to cyclic guanosine monophosphate (cGMP), which further results in the activation of protein kinase G (PKG) phosphotransferases. There are various substrates that are phosphorylated by PKGs [82]. For example, HO-1/CO/CORM-driven sGC/cGMP-mediated signaling is recognized as inhibitory signal for platelet aggregation and thus thrombus formation [83]. Therefore a strategic exploration of phosphoproteins that are activated by HO-1/CO/CORM-driven sGC/cGMP may help to identify specific signaling molecules responsible for the activation of IL-10 promoter.

BR/BV system and activation of aryl hydrocarbon receptor in IL-10 production: Aryl hydrocarbon receptor (AhR) is a ligand-dependent transcriptional factor able to sense a wide range of structurally different exogenous and endogenous molecules. AhR ligands vary in their structure, and their binding affinity can significantly differ between mouse and

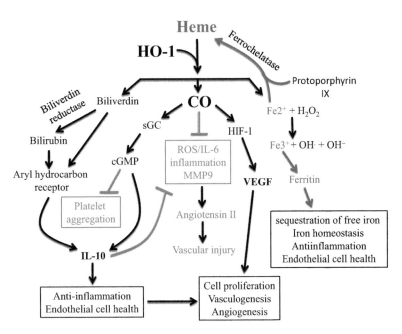

FIGURE 15.3 HO-1 in vascular health.

human AhR [84]. AhR agonists often arise from xenobiotics such as pollutants and indoles primarily derived from tryptophan metabolism occurring in the stomach and intestines, as well as in other organs upon photo-oxidation or oxidative stress. The prototypic high-affinity AhR agonist, the xenobiotic 2,3,7,8-tetrachlorodibenzo-p-dioxin (TCDD), displays a 10-fold higher affinity for mouse AhR compared to human AhR. Conversely, dietary-derived indole metabolites have a better affinity for human AhR. Use of natural AhR ligands as potential therapeutic modalities against inflammatory disorders has already been suggested earlier [85]. BV and BR as endogenous ligands of AHR have gained considerable attention particularly as antiinflammatory mediators. Recently Bock et al. [86] highlighted that human AHR can exert dual function on vascular tissue both as pro-inflammatory and antiinflammatory mediator in a ligand-specific manner, where BR and BV precisely were illustrated as antiinflammatory. Mechanistically AHR is constitutively expressed in the cytoplasm and remains bound to the chaperon protein, HSP90. Binding to appropriate ligand helps AHR to dissociate from HSP90, dimerize, and migrate to nucleus in a complex with its counterpart aryl hydrocarbon receptor nuclear translocator. Circumstances of increased HO-1 function, and thus consequential increased BV/BR is presumed to lead to their binding to AHR and nuclear translocation. The BR/BV-specific trigger to AHR is antiinflammatory, attained particularly via transcription of IL-10 gene that possesses AHR-specific binding element in its promoter region.

15.4 Conclusion

HO-1 is a stress sensor and its induction is crucial in vascular health. HO-1 by increasing heme degradation not only releases heme toxicity, but also by virtue of its metabolites, CO and BV/BR, attains cellular homeostasis by antioxidative and antiinflammatory properties. The third metabolite, free iron (Fe^{2+}) is recycled to generate more heme and hemoproteins, or else is stored as ferritin serving as iron reservoir, and also as a sequestering mechanism of reactive and toxic-free ion from any cellular toxicity (Fig. 15.3). In vascular health, the property of CO in inducing HIF-1-α-driven VEGF is highly recognized for vasculogenesis and angiogenesis, and is considered important for vascular development and also in repair mechanism. Exposure to inflammation either from exogenous or endogenous sources constantly contributes to endothelial injury and vascular damage, and HO-1 plays a crucial role in resolving this, by inducing antioxidative and antiinflammatory signaling. This is attained by two distinct pathways such as (1) CO/sGC/cGMP-driven signaling and (2) BV/BR-driven AHR-dependent transcriptional activity. IL-10 production in resolving antiinflammatory mechanism helps to facilitate the repair of endothelial injury. HO-1 being a major player in cellular homeostasis is thus a major target in several diseases. Notably, aging is a major contributor to much impairment including vascular and CVDs, cancer, diabetes, neurodegenerative diseases, and other causes of death and impairment. Perhaps, this may be attributed to age-related decline in HO-1 expression, and studies support this notion [87].

Taken together, there is a need for therapeutic expedition to modulate HO-1 expression and function in controlling overall vascular and other age-related diseases. A systemic medicinal approach to target Nrf2, the transcriptional regulator of HO-1, is quite relevant and is under exploration [88]. In this context, pharmacologic advantages of dimethyl fumarate (DMF) can be of higher relevance because its signaling is indicated by the enrichment of Nrf2/HO-1 pathway. Additionally DMF is an FDA-approved drug for controlling the inflammatory complications related to multiple sclerosis. Repurposing this drug in vascular health is worth exploring. We also have to consider the disease situation where despite the induction of HO-1, it could undergo profuse degradation by increased immunoproteasomal activity as observed in HIV-related neuropathogenesis [89]. There, more appropriate approach would be to employ regulated release of CO by pharmacologic use of CO-releasing compounds to execute antiinflammatory IL-10 signaling. Likewise, pharmacologic use of bilirubin can also be accounted for antiinflammatory mechanism useful in enforcing AHR-depended transcriptional activation of IL-10. Although this chapter is focused on discussing beneficial effects of HO-1-and HO-1-dependent metabolites, the author further indicates that the advantage of stabilizing Nrf2 is multifaceted. Nrf2 in addition to HO-1 is a master transcriptional regulator of myriads of antioxidative and antiinflammatory genes [88] assuring cellular detoxification and homeostasis. Therefore Nrf2 as a therapeutic target is quite encouraging as preventive mechanism in various degenerative diseases including vascular health.

References

[1] Kim YW, Byzova TV. Oxidative stress in angiogenesis and vascular disease. Blood 2014;123:625−31.

[2] Zhou Y, Yan H, Guo M, Zhu J, Xiao Q, Zhang L. Reactive oxygen species in vascular formation and development. Oxid Med Cell Longev 2013;2013:374963.

[3] Pae HO, Chung HT. Heme oxygenase-1: its therapeutic roles in inflammatory diseases. Immune Netw 2009;9:12−19.

[4] Yachie A, Niida Y, Wada T, Igarashi N, Kaneda H, Toma T, et al. Oxidative stress causes enhanced endothelial cell injury in human heme oxygenase-1 deficiency. J Clin Invest 1999;103:129−35.

[5] Schacter BA, Nelson EB, Marver HS, Masters BS. Immunochemical evidence for an association of heme oxygenase with the microsomal electron transport system. J Biol Chem 1972;247:3601−7.

[6] Tenhunen R, Marver HS, Schmid R. Microsomal heme oxygenase. Characterization of the enzyme. J Biol Chem 1969;244:6388−94.

[7] Kikuchi G, Yoshida T, Noguchi M. Heme oxygenase and heme degradation. Biochem Biophys Res Commun 2005;338:558−67.

[8] Loboda A, Damulewicz M, Pyza E, Jozkowicz A, Dulak J. Role of Nrf2/HO-1 system in development, oxidative stress response and diseases: an evolutionarily conserved mechanism. Cell Mol Life Sci 2016;73:3221−47.

[9] Hayashi S, Omata Y, Sakamoto H, Higashimoto Y, Hara T, Sagara Y, et al. Characterization of rat heme oxygenase-3 gene. Implication of processed pseudogenes derived from heme oxygenase-2 gene. Gene 2004;336:241−50.

[10] Hwang HW, Lee JR, Chou KY, Suen CS, Hwang MJ, Chen C, et al. Oligomerization is crucial for the stability and function of heme oxygenase-1 in the endoplasmic reticulum. J Biol Chem 2009;284:22672−9.

[11] Braggins PE, Trakshel GM, Kutty RK, Maines MD. Characterization of two heme oxygenase isoforms in rat spleen: comparison with the hematin-induced and constitutive isoforms of the liver. Biochem Biophys Res Commun 1986;141:528−33.

[12] Maines MD, Trakshel GM, Kutty RK. Characterization of two constitutive forms of rat liver microsomal heme oxygenase. Only one molecular species of the enzyme is inducible. J Biol Chem 1986;261:411−19.

[13] Trakshel GM, Kutty RK, Maines MD. Purification and characterization of the major constitutive form of testicular heme oxygenase. The noninducible isoform. J Biol Chem 1986;261:11131−7.

[14] Otterbein LE, Soares MP, Yamashita K, Bach FH. Heme oxygenase-1: unleashing the protective properties of heme. Trends Immunol 2003;24:449−55.

[15] Davudian S, Mansoori B, Shajari N, Mohammadi A, Baradaran B. BACH1, the master regulator gene: a novel candidate target for cancer therapy. Gene 2016;588:30−7.

[16] Sun J, Brand M, Zenke Y, Tashiro S, Groudine M, Igarashi K. Heme regulates the dynamic exchange of Bach1 and NF-E2-related factors in the Maf transcription factor network. Proc Natl Acad Sci U S A 2004;101:1461−6.

[17] Biswas C, Shah N, Muthu M, La P, Fernando AP, Sengupta S, et al. Nuclear heme oxygenase-1 (HO-1) modulates subcellular distribution and activation of Nrf2, impacting metabolic and anti-oxidant defenses. J Biol Chem 2014;289:26882−94.

[18] Alam J, Shibahara S, Smith A. Transcriptional activation of the heme oxygenase gene by heme and cadmium in mouse hepatoma cells. J Biol Chem 1989;264:6371−5.

[19] Boyle JJ, Johns M, Lo J, Chiodini A, Ambrose N, Evans PC, et al. Heme induces heme oxygenase 1 via Nrf2: role in the homeostatic macrophage response to intraplaque hemorrhage. Arterioscler Thromb Vasc Biol 2011;31:2685−91.

[20] Wareham LK, Poole RK, Tinajero-Trejo M. CO-releasing metal carbonyl compounds as antimicrobial agents in the post-antibiotic era. J Biol Chem 2015;290:18999−9007.

[21] Qureshi OS, Zeb A, Akram M, Kim MS, Kang JH, Kim HS, et al. Enhanced acute anti-inflammatory effects of CORM-2-loaded nanoparticles via sustained carbon monoxide delivery. Eur J Pharm Biopharm 2016;108:187−95.

[22] Lian S, Xia Y, Ung TT, Khoi PN, Yoon HJ, Kim NH, et al. Carbon monoxide releasing molecule-2 ameliorates IL-1beta-induced IL-8 in human gastric cancer cells. Toxicology 2016;361−362:24−38.

[23] Tenhunen R, Marver HS, Schmid R. The enzymatic conversion of heme to bilirubin by microsomal heme oxygenase. Proc Natl Acad Sci U S A 1968;61:748−55.

[24] Nath KA, Salahudeen AK, Clark EC, Hostetter MK, Hostetter TH. Role of cellular metabolites in progressive renal injury. Kidney Int Suppl 1992;38:S109−113.

[25] Koizumi S. Human heme oxygenase-1 deficiency: a lesson on serendipity in the discovery of the novel disease. Pediatr Int 2007;49:125−32.

[26] Kimpara T, Takeda A, Watanabe K, Itoyama Y, Ikawa S, Watanabe M, et al. Microsatellite polymorphism in the human heme oxygenase-1 gene promoter and its application in association studies with Alzheimer and Parkinson disease. Hum Genet 1997;100:145−7.

[27] Gerbitz A, Hillemanns P, Schmid C, Wilke A, Jayaraman R, Kolb HJ, et al. Influence of polymorphism within the heme oxygenase-I promoter on overall survival and transplantation-related mortality after allogeneic stem cell transplantation. Biol Blood Marrow Transplant 2008;14:1180−9.

[28] Ozaki KS, Marques GM, Nogueira E, Feitoza RQ, Cenedeze MA, Franco MF, et al. Improved renal function after kidney transplantation is associated with heme oxygenase-1 polymorphism. Clin Transplant 2008;22:609−16.

[29] Lee EY, Lee YH, Kim SH, Jung KS, Kwon O, Kim BS, et al. Association between heme oxygenase-1 promoter polymorphisms and the development of albuminuria in type 2 diabetes: a case-control study. Medicine (Baltimore) 2015;94:e1825.

[30] Pechlaner R, Willeit P, Summerer M, Santer P, Egger G, Kronenberg F, et al. Heme oxygenase-1 gene promoter microsatellite polymorphism is associated with progressive atherosclerosis and incident cardiovascular disease. Arterioscler Thromb Vasc Biol 2015;35:229−36.

[31] Taha H, Skrzypek K, Guevara I, Nigisch A, Mustafa S, Grochot-Przeczek A, et al. Role of heme oxygenase-1 in human endothelial cells: lesson from the promoter allelic variants. Arterioscler Thromb Vasc Biol 2010;30:1634−41.

[32] Walther M, De Caul A, Aka P, Njie M, Amambua-Ngwa A, Walther B, et al. HMOX1 gene promoter alleles and high HO-1 levels are associated with severe malaria in Gambian children. PLoS Pathog 2012;8:e1002579.

[33] Wu MM, Chiou HY, Chen CL, Wang YH, Hsieh YC, Lien LM, et al. GT-repeat polymorphism in the heme oxygenase-1 gene promoter is associated with cardiovascular mortality risk in an arsenic-exposed population in northeastern Taiwan. Toxicol Appl Pharmacol 2010;248:226−33.

[34] Lin CC, Yang WC, Lin SJ, Chen TW, Lee WS, Chang CF, et al. Length polymorphism in heme oxygenase-1 is associated with arteriovenous fistula patency in hemodialysis patients. Kidney Int 2006;69:165−72.

[35] Zhang MM, Zheng YY, Gao Y, Zhang JZ, Liu F, Yang YN, et al. Heme oxygenase-1 gene promoter polymorphisms are associated with coronary heart disease and restenosis after percutaneous coronary intervention: a meta-analysis. Oncotarget 2016;7:83437−50.

[36] Chen YH, Chau LY, Chen JW, Lin SJ. Serum bilirubin and ferritin levels link heme oxygenase-1 gene promoter polymorphism and susceptibility to coronary artery disease in diabetic patients. Diabetes Care 2008;31:1615−20.

[37] Gulesserian T, Wenzel C, Endler G, Sunder-Plassmann R, Marsik C, Mannhalter C, et al. Clinical restenosis after coronary stent implantation is associated with the heme oxygenase-1 gene promoter polymorphism and the heme oxygenase-1 +99 G/C variant. Clin Chem 2005;51:1661−5.

[38] Morgan L, Hawe E, Palmen J, Montgomery H, Humphries SE, Kitchen N. Polymorphism of the heme oxygenase-1 gene and cerebral aneurysms. Br J Neurosurg 2005;19:317−21.

[39] Funk M, Endler G, Schillinger M, Mustafa S, Hsieh K, Exner M, et al. The effect of a promoter polymorphism in the heme oxygenase-1 gene on the risk of ischaemic cerebrovascular events: the influence of other vascular risk factors. Thromb Res 2004;113:217−23.

[40] Dick P, Schillinger M, Minar E, Mlekusch W, Amighi J, Sabeti S, et al. Haem oxygenase-1 genotype and cardiovascular adverse events in patients with peripheral artery disease. Eur J Clin Invest 2005;35:731−7.

[41] Daenen KE, Martens P, Bammens B. Association of HO-1 (GT)n promoter polymorphism and cardiovascular disease: a reanalysis of the literature. Can J Cardiol 2016;32:160−8.

[42] Leaf DE, Body SC, Muehlschlegel JD, McMahon GM, Lichtner P, Collard CD, et al. Length polymorphisms in heme oxygenase-1 and AKI after cardiac surgery. J Am Soc Nephrol 2016;27:3291−7.

[43] Folkman J. Tumor angiogenesis: therapeutic implications. N Engl J Med 1971;285:1182−6.

[44] Folkman J, Merler E, Abernathy C, Williams G. Isolation of a tumor factor responsible for angiogenesis. J Exp Med 1971;133:275−88.

[45] Huang Y, Mao Y, Li H, Shen G, Nan G. Knockdown of Nrf2 inhibits angiogenesis by downregulating VEGF expression through PI3K/Akt signaling pathway in cerebral microvascular endothelial cells under hypoxic conditions. Biochem Cell Biol 2018;96:475−82.

[46] Patterson EK, Fraser DD, Capretta A, Potter RF, Cepinskas G. Carbon monoxide-releasing molecule 3 inhibits myeloperoxidase (MPO) and protects against MPO-induced vascular endothelial cell activation/dysfunction. Free Radic Biol Med 2014;70:167−73.

[47] Liu J, Fedinec AL, Leffler CW, Parfenova H. Enteral supplements of a carbon monoxide donor CORM-A1 protect against cerebrovascular dysfunction caused by neonatal seizures. J Cereb Blood Flow Metab 2015;35:193−9.

[48] Parfenova H, Pourcyrous M, Fedinec AL, Liu J, Basuroy S, Leffler CW. Astrocyte-produced carbon monoxide and the carbon monoxide donor CORM-A1 protect against cerebrovascular dysfunction caused by prolonged neonatal asphyxia. Am J Physiol Heart Circ Physiol 2018;315:H978−88.

[49] Chiang N, Shinohara M, Dalli J, Mirakaj V, Kibi M, Choi AM, et al. Inhaled carbon monoxide accelerates resolution of inflammation via unique proresolving mediator-heme oxygenase-1 circuits. J Immunol 2013;190:6378−88.

[50] Ryter SW, Alam J, Choi AM. Heme oxygenase-1/carbon monoxide: from basic science to therapeutic applications. Physiol Rev 2006;86:583−650.

[51] Ryter SW, Choi AM. Targeting heme oxygenase-1 and carbon monoxide for therapeutic modulation of inflammation. Transl Res 2016;167:7−34.

[52] Motterlini R, Otterbein LE. The therapeutic potential of carbon monoxide. Nat Rev Drug Discov 2010;9:728−43.

[53] Abraham NG, Scapagnini G, Kappas A. Human heme oxygenase: cell cycle-dependent expression and DNA microarray identification of multiple gene responses after transduction of endothelial cells. J Cell Biochem 2003;90:1098–111.

[54] Choi YK, Kim CK, Lee H, Jeoung D, Ha KS, Kwon YG, et al. Carbon monoxide promotes VEGF expression by increasing HIF-1alpha protein level via two distinct mechanisms, translational activation and stabilization of HIF-1alpha protein. J Biol Chem 2010;285:32116–25.

[55] Lin CC, Yang CC, Hsiao LD, Chen SY, Yang CM. Heme oxygenase-1 induction by carbon monoxide releasing molecule-3 suppresses interleukin-1beta-mediated neuroinflammation. Front Mol Neurosci 2017;10:387.

[56] Babu D, Leclercq G, Motterlini R, Lefebvre RA. Differential effects of CORM-2 and CORM-401 in murine intestinal epithelial MODE-K cells under oxidative stress. Front Pharmacol 2017;8:31.

[57] Singh MV, Chapleau MW, Harwani SC, Abboud FM. The immune system and hypertension. Immunol Res 2014;59:243–53.

[58] Ganta CK, Lu N, Helwig BG, Blecha F, Ganta RR, Zheng L, et al. Central angiotensin II-enhanced splenic cytokine gene expression is mediated by the sympathetic nervous system. Am J Physiol Heart Circ Physiol 2005;289:H1683–1691.

[59] Nataraj C, Oliverio MI, Mannon RB, Mannon PJ, Audoly LP, Amuchastegui CS, et al. Angiotensin II regulates cellular immune responses through a calcineurin-dependent pathway. J Clin Invest 1999;104:1693–701.

[60] Bataller R, Gabele E, Schoonhoven R, Morris T, Lehnert M, Yang L, et al. Prolonged infusion of angiotensin II into normal rats induces stellate cell activation and proinflammatory events in liver. Am J Physiol Gastrointest Liver Physiol 2003;285:G642–651.

[61] van Thiel BS, van der Pluijm I, te Riet L, Essers J, Danser AH. The renin-angiotensin system and its involvement in vascular disease. Eur J Pharmacol 2015;763:3–14.

[62] McCormick ML, Gavrila D, Weintraub NL. Role of oxidative stress in the pathogenesis of abdominal aortic aneurysms. Arterioscler Thromb Vasc Biol 2007;27:461–9.

[63] Tsai SH, Huang PH, Hsu YJ, Peng YJ, Lee CH, Wang JC, et al. Inhibition of hypoxia inducible factor-1alpha attenuates abdominal aortic aneurysm progression through the down-regulation of matrix metalloproteinases. Sci Rep 2016;6:28612.

[64] Montezano AC, De Lucca Camargo L, Persson P, Rios FJ, Harvey AP, Anagnostopoulou A, et al. NADPH oxidase 5 is a pro-contractile Nox isoform and a point of cross-talk for calcium and redox signaling-implications in vascular function. J Am Heart Assoc 2018;7.

[65] El Assar M, Angulo J, Rodriguez-Manas L. Oxidative stress and vascular inflammation in aging. Free Radic Biol Med 2013;65:380–401.

[66] Nguyen Dinh Cat A, Montezano AC, Burger D, Touyz RM. Angiotensin II, NADPH oxidase, and redox signaling in the vasculature. Antioxid Redox Signal 2013;19:1110–20.

[67] Brandes RP. Role of NADPH oxidases in the control of vascular gene expression. Antioxid Redox Signal 2003;5:803–11.

[68] da Silva AR, Fraga-Silva RA, Stergiopulos N, Montecucco F, Mach F. Update on the role of angiotensin in the pathophysiology of coronary atherothrombosis. Eur J Clin Invest 2015;45:274–87.

[69] Ferrario CM, Strawn WB. Role of the renin-angiotensin-aldosterone system and proinflammatory mediators in cardiovascular disease. Am J Cardiol 2006;98:121–8.

[70] Kobayashi A, Ishikawa K, Matsumoto H, Kimura S, Kamiyama Y, Maruyama Y. Synergetic antioxidant and vasodilatory action of carbon monoxide in angiotensin II-induced cardiac hypertrophy. Hypertension 2007;50:1040–8.

[71] Tsai MH, Lee CW, Hsu LF, Li SY, Chiang YC, Lee MH, et al. CO-releasing molecules CORM2 attenuates angiotensin II-induced human aortic smooth muscle cell migration through inhibition of ROS/IL-6 generation and matrix metalloproteinases-9 expression. Redox Biol 2017;12:377–88.

[72] Rodriguez-Iturbe B, Pons H, Johnson RJ. Role of the immune system in hypertension. Physiol Rev 2017;97:1127–64.

[73] Wenzel U, Turner JE, Krebs C, Kurts C, Harrison DG, Ehmke H. Immune mechanisms in arterial hypertension. J Am Soc Nephrol 2016;27:677–86.

[74] Shao Y, Cheng Z, Li X, Chernaya V, Wang H, Yang XF. Immunosuppressive/anti-inflammatory cytokines directly and indirectly inhibit endothelial dysfunction – a novel mechanism for maintaining vascular function. J Hematol Oncol 2014;7:80.

[75] Weaver LK, Chu N, Behrens EM. TLR9-mediated inflammation drives a Ccr2-independent peripheral monocytosis through enhanced extramedullary monocytopoiesis. Proc Natl Acad Sci U S A 2016;113:10944–9.

[76] Weaver LK, Minichino D, Biswas C, Chu N, Lee JJ, Bittinger K, et al. Microbiota-dependent signals are required to sustain TLR-mediated immune responses. JCI Insight 2019;4.

[77] Behrens EM, Canna SW, Slade K, Rao S, Kreiger PA, Paessler M, et al. Repeated TLR9 stimulation results in macrophage activation syndrome-like disease in mice. J Clin Invest 2011;121:2264–77.

[78] Onyiah JC, Sheikh SZ, Maharshak N, Steinbach EC, Russo SM, Kobayashi T, et al. Carbon monoxide and heme oxygenase-1 prevent intestinal inflammation in mice by promoting bacterial clearance. Gastroenterology 2013;144:789–98.

[79] Sheikh SZ, Hegazi RA, Kobayashi T, Onyiah JC, Russo SM, Matsuoka K, et al. An anti-inflammatory role for carbon monoxide and heme oxygenase-1 in chronic Th2-mediated murine colitis. J Immunol 2011;186:5506–13.

[80] Ma X, Sayed N, Beuve A, van den Akker F. NO and CO differentially activate soluble guanylyl cyclase via a heme pivot-bend mechanism. EMBO J 2007;26:578–88.

[81] Makino R, Obata Y, Tsubaki M, Iizuka T, Hamajima Y, Kato-Yamada Y, et al. Mechanistic insights into the activation of soluble guanylate cyclase by carbon monoxide: a multistep mechanism proposed for the BAY 41–2272 induced formation of 5-coordinate CO-heme. Biochemistry 2018;57:1620–31.

[82] Francis SH, Busch JL, Corbin JD, Sibley D. cGMP-dependent protein kinases and cGMP phosphodiesterases in nitric oxide and cGMP action. Pharmacol Rev 2010;62:525–63.

[83] Brune B, Ullrich V. Inhibition of platelet aggregation by carbon monoxide is mediated by activation of guanylate cyclase. Mol Pharmacol 1987;32:497−504.

[84] Lamas B, Natividad JM, Sokol H. Aryl hydrocarbon receptor and intestinal immunity. Mucosal Immunol 2018;11:1024−38.

[85] Busbee PB, Rouse M, Nagarkatti M, Nagarkatti PS. Use of natural AhR ligands as potential therapeutic modalities against inflammatory disorders. Nutr Rev 2013;71:353−69.

[86] Bock KW. Human AHR functions in vascular tissue: pro- and anti-inflammatory responses of AHR agonists in atherosclerosis. Biochem Pharmacol 2019;159:116−20.

[87] Miyamura N, Ogawa T, Boylan S, Morse LS, Handa JT, Hjelmeland LM. Topographic and age-dependent expression of heme oxygenase-1 and catalase in the human retinal pigment epithelium. Invest Ophthalmol Vis Sci 2004;45:1562−5.

[88] Cuadrado A, Manda G, Hassan A, Alcaraz MJ, Barbas C, Daiber A, et al. Transcription factor NRF2 as a therapeutic target for chronic diseases: a systems medicine approach. Pharmacol Rev 2018;70:348−83.

[89] Kovacsics CE, Gill AJ, Ambegaokar SS, Gelman BB, Kolson DL. Degradation of heme oxygenase-1 by the immunoproteasome in astrocytes: a potential interferon-gamma-dependent mechanism contributing to HIV neuropathogenesis. Glia 2017;65:1264−77.

Chapter 16

Experimental models for identifying target events in vascular injury

Thais Girão-Silva[1], Ayumi Aurea Miyakawa[1] and Silvia Lacchini[2]

[1]Laboratory of Genetics and Molecular Cardiology, Heart Institute (InCor), University of São Paulo Medical School, São Paulo, Brazil, [2]Institute of Biomedical Sciences, University of São Paulo, São Paulo, Brazil

Cardiovascular diseases are the main cause of mortality and morbidity in developed countries, with vascular diseases being an important cause for the onset and progression of tissue damage. The initiation and progression of many of these diseases is related to chronic reduction in tissue blood flow as a consequence of obstruction caused by an increase in arterial perfusion pressure or vascular injury. Thus, inadequate blood supply may be a determinant of tissue injury, such as in myocardial or cerebral infarction, or even amputation of the lower limbs. Therefore, the understanding of the process of initiation and progression of such diseases is fundamental for the development of preventive or therapeutic actions.

Like atherosclerosis, other vascular diseases have an impact on the whole body, as the vascular network is the source of oxygen and nutrients; thus alterations in this network can have ramifications on all tissues. An important strategy for the treatment of atherosclerotic lesion for many years was balloon insufflation angioplasty within the obstructed or stenotic artery. However, the high probability of recurrence of stenosis or restenosis in the treated artery stimulated the development of other approaches, such as the application of endovascular stents and their variations. Individuals in whom restenosis can no longer be treated require the replacement of the vascular segment by arterial or venous graft, which generates a new questioning about how this graft adapts to the arterial conditions. These strategies are important in other common vascular diseases such as aortic and cerebral artery aneurysms, which can lead to sudden death due to unexpected rupture, or even graft vasculopathy, characterized by intimal hyperplasia in arteries of transplanted organs. This chapter will provide an overview of the animal models of vascular disease specifically atherosclerosis and venous graft disease, and showcase how findings from these models play a pivotal role in driving therapy. These last two conditions will not be discussed in this chapter.

16.1 Arterial diseases

Although several causal mechanisms of such diseases have been identified, the pathogenesis of many vascular diseases still needs to be elucidated. Experimental models do not accurately reproduce the known human diseases, but they represent excellent strategies to isolate and identify factors involved with vascular pathophysiology. Considering the experimental advantages, the development of surgical techniques, the constant production of genetically modified animals, and the emergence of new experimental technologies the use of rats and mice is an excellent tool for the study of vascular pathology. Specifically, genetically modified mice represent a powerful tool for the study of the pathogenesis of vascular diseases, addressing immunological, oncological, neural, autoimmune, and congenital processes.

Subsequent sections of this chapter will discuss experimental models that address the processes associated with atherosclerosis and venous graft injury, comprising the changes related to the venous changes during its arterialization.

Currently, we understand the steps involved with the development of atherosclerotic lesion and it is known that this process begins with endothelial damage and dysfunction and culminates with the establishment of atherosclerotic plaque and with large deposition of extracellular matrix (characterizing stable plaques and promoting large vascular stenosis) or with large inflammatory changes (as characterized by unstable plaques with great rupture potential) [1,2]. These two

Endothelial Signaling in Vascular Dysfunction and Disease. DOI: https://doi.org/10.1016/B978-0-12-816196-8.00017-5

important changes (formation and atherosclerotic lesions and the inflammatory response leading to neointima formation) have been well studied and understood in experimental models developed in rabbits and in mice.

16.2 From rabbit to mouse models of atherosclerosis

Although no experimental model combines all the features with humans, rabbits were the first model for the study of atherosclerosis. The pioneering work in the study of atherosclerosis in rabbits was developed by Russian scientists in the early 20th century, where a diet rich in animal proteins showed lesions of the intima accompanied by accumulation of large whitish cells (foam cells) in the aorta of rabbits [3]. This observation was even more evident when the animal received a high cholesterol diet, inducing the formation of human-like atherosclerosis. The foam cells were subsequently identified as macrophages that engulf large amounts of lipid particles. These studies established the basis for the theory of induction of atherosclerosis by dietary cholesterol [4]. Currently, there is evidence suggesting that foam cells, in part, originate from vascular smooth muscle cells (VSMCs or SMCs) [5].

In general, rabbits are able to easily develop atherosclerotic lesions when subjected to cholesterol ingestion [6,7]. As rabbits are sensitive to cholesterol received in the diet, they develop atherosclerosis in the aorta within 12–16 weeks. However, in the late 70s, Watanabe described a spontaneous mutation at the low-density lipoprotein (LDL) receptor in rabbits, which leads to the development of hypercholesterolemia even if the animal receives normal diet, similar to that observed in humans. These animals are called watanabe heritable hyperlipid and develop spontaneous atherosclerotic lesions in the aorta, even without treatment with high cholesterol [6,8].

Atherosclerotic lesions in the rabbit can be evaluated both macroscopically and microscopically. The lesions usually begin in the aortic arch and follow around the orifices of the intercostal arteries in the thoracic aorta. Atherosclerotic lesions in the abdominal aorta are less frequent, different from that observed in humans [6]. As lesions in the aorta of rabbits are large, it is possible to collect fragments of the lesion by extracting mRNA and evaluating gene expression or extracting proteins for the study of protein expression. In addition to genome analysis, histological analysis that can characterize the lesion morphologically as well as study its cellular or extracellular matrix (ECM) components is a valuable tool. Histological analysis can be done on aorta segments embedded in paraffin and stained by different histochemical staining techniques, which permits to define intima, media, and adventitial layers and calculate the proportion of each in the wall vessel. Intima/media ratio is widely used as a synonym for arterial injury, since it allows understanding how much the intima increases compared to the media, providing an idea of vascular stenosis and consequent cardiovascular risk [9,10]. Immunohistochemistry, allows the identification and localization of proteins or cells of interest in the tissue. Although the Western blot method allows measuring protein expression, immunohistochemistry can identify where the protein is located in the tissue or cell and whether it occurs in greater or lesser proportions between groups.

In contrast to rabbits, mice are resistant to dietary cholesterol. As mice have low levels of atherogenic apolipoproteins, unlike rabbits, they do not develop significant lesions when receiving a high cholesterol diet [11]. On the other hand, even wild-type mice show alterations in the vascular wall after receiving a high-fat diet. Although these animals do not present either atherosclerotic lesion or hypertension, it is possible to verify the increase and disorganizing deposition of perivascular collagen fibers in the aorta, as well as the increase of local inflammatory markers, suggesting a vascular stiffening process [14].

If on the one hand, the high-fat diet in wild-type mice leads to the development of vascular changes, on the other hand, when this diet is concomitant with a genetic manipulation that makes such animals sensitive to western diet, the observed lesions approximate human atherosclerotic lesions in complexity. Genetically manipulated mice that are atherosclerosis-prone have been developed and are widely used. These are Atherosclerosis-prone apolipoprotein E-deficient ($Apoe^{-/-}$) mice and Low density lipoprotein receptor ($LDLr^{-/-}$) mice that display poor lipoprotein clearance with subsequent severe hypercholesterolemia, and atherosclerotic lesions even when on a normal diet [13]. In the early 90s, $Ldlr^{-/-}$ or $Apoe^{-/-}$ [14,15]. Mice were reported to present hypercholesterolemia and, on receiving western diet, developed atherosclerotic lesions similar to those seen in humans.

The use of such models allowed to elucidate several stages of the atherosclerotic process, including players involved with initiation, progression, or even the formation of complex plaques. Thus, it became clear that the process involves the migration of smooth muscle cells (SMCs) from the tunica media to the intima, leading to the so-called neointima. There are several triggers for this response from SMCs such as the presence of fat particles in the subendothelial space (lipoprotein particles containing apolipoprotein B) that stimulate the recruitment of monocytes into the intima, and their transformation into macrophages [16]. Both macrophages and endothelial cells are a source of trophic and growth factors, which act on SMCs present in the media layer making SMCs change their contractile phenotype to synthetic phenotype [17]. From this moment, these SMCs begin to produce a great amount of growth factors, cytokines, and

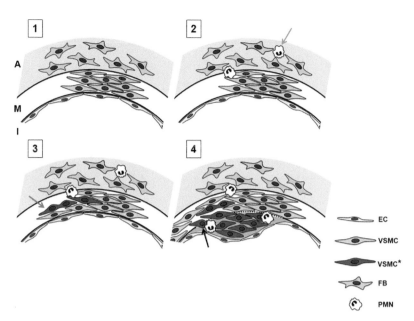

FIGURE 16.1 The most important cells in wall layers: adventitia (A), media (M), and intima (I). Panel 1 shows the structure of a normal artery. After inflammatory stimulus, the acute response starts by PMN cells migration (blue arrow, 2), followed by change of phenotype and proliferation of VSMC (red arrow, 3), and the migration of VSMC from media to intima, originating the neointima (black arrow, 4). *EC*, endothelial cell; *VSMC*, vascular smooth muscle cell; *VSMC**, vascular smooth muscle cell presenting synthetic phenotype; *FB*, fibroblast; *PMN*, polymorphonuclear cell.

components of the ECM proteins, especially collagen (Fig. 16.1). During the atherosclerotic process, this deposition of collagen is responsible for the formation of a fibrous cap on the plaque. Although we know that the size of the fibrous cap is responsible for vascular narrowing or stenosis, it also confers stability to the plaque and prevents its rupture.

As the lesions increase in size, the presence of a necrotic nucleus containing foam cells and, eventually, cholesterol crystals becomes evident. As this structure grows, the VSMCs migrate from the media layer to the intima and synthesize a large amount of ECM components, such as different collagen types and others [16,18]. These findings provided by animal models have been the basis for a better understanding of the atherosclerotic process. Thus, the influence of environmental factors and the importance that phenotypic or genotypic characteristics present on the initiation and/or progression of the atherosclerotic lesion were better understood.

16.3   The study of vascular injury by neointima formation

Although the mechanisms of atherosclerotic lesion development have been studied for more than a century, the first models that allowed us to begin to understand the neointima formation appeared in the 1960s with dogs, and two decades later, with rabbits. The first description of a nonocclusive cuff model around a rabbit artery was made by Japanese researchers where a segment of polyethylene catheter was positioned around the carotid artery [19,20]. These studies showed that the implantation of the cuff around the artery seemed to promote early leukocyte infiltration and VSMC migration. As a consequence, the neointima formation occurred from the first week after the cuff implantation. An alternative to this model was presented by Booth et al. in 1989, where a silicone cuff, positioned around the left carotid artery, enabled the identification of an early infiltration of inflammatory cells, followed by proliferation and migration of VSMC from the media to the intima. Thus, a significant increase in neointima was identified in the first week, which was doubled in the second week [7,21]. Coinciding with this increase in neointima, there was also an increase in the production of von Willebrand factor, which could be related to the deposition of subendothelial ECM and to the modulation of neointimal formation itself [22].

This model helped to understand several initial aspects of vascular injury, since it allowed to interfere with the arterial wall without altering the blood flow. In addition, it used a stimulus that would provide an important mechanism for the atherosclerotic lesion: the inflammatory response. Thus, it was possible to identify three early stages involved with vascular injury: polymorphonuclear cell/neutrophil infiltration 2 hours after cuff implantation, proliferation of VSMCs in the tunica media after 12 hours and its migration to the subendothelial space after 3 days [23]. This initial process resembles that observed at the beginning of the atherosclerotic lesion, as already presented in Fig. 16.1.

The perivascular nonocclusive cuff model was very promising and favored the development of murine models for the study of neointima. The model adapted the protocol for rabbit cuff implantation; however, due to the diameter of the catheter and arteries, it was adapted to the rat femoral artery. Thus, the perivascular cuff model in the rat allowed to confirm the neointima formation after two weeks, coinciding with the increased proliferation of VSMCs in the media [24]. Although the cuff-induced vascular lesion model in rats did not diffuse, it served as the basis for application in mice, a

Elastin staining (Verhoef Van Gieson) **PCNA (proliferating cell nuclear antigen)**

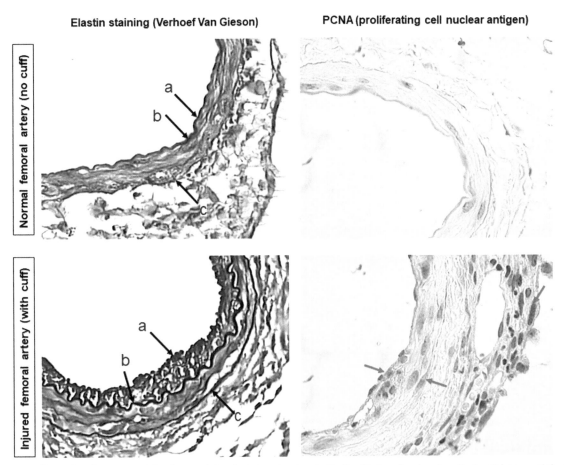

FIGURE 16.2 Comparison between noninjured versus injured arteries, showing the limits of intima (b−a) and media (c−b) layers and the increase of intima in injured artery, characterizing neointima formation. Right panel shows the increase of proliferating cells (Proliferating cell nuclear antigen (PCNA) positive nuclei) in intima, media and adventitia layers (red arrows) in injured artery. a—lumen limit; b—internal elastic lamina; c—external elastic lamina. Magnification 400 × .

model that is now widely used. Thus, from this model developed in rabbits [18,20] and rats [24], the polyethylene-cuff-induced vascular lesion model was originated in mice [25]. This first study permitted to understand the early (5−7 days) increase in the expression of inflammatory markers [interleukins (IL-1β, IL-6), tumor necrosis factor (TNF-α), and interferon (IFN-γ)] in the femoral artery under perivascular cuff stimulation [25]. The morphometric analysis of such lesions showed that VSMC proliferates in the media during the first week and migrates to the intima, participating in the increase of neointimal area, as can be seen in Fig. 16.2.

A major advance in these studies was achieved with the arrival of other technologies, especially the use of genetically modified mice. Just a decade earlier, the description of the first genetically modified mice by the Oliver Smithies group allowed a new tool for understanding the role of specific genes in cardiovascular diseases, since it enabled to study the impact that the presence or absence of a gene of interest would have in vascular injury. The Akishita study [25] made use of genetically modified mice, analyzing knockout animals for the AT2 receptor of angiotensin II (Agt2$^{-/-}$), which made possible to show the importance of this receptor as a vasoprotective mechanism against lesion.

The cuff-induced lesion model is particularly useful in the study of vascular injury, since it enables isolating interest factors to demonstrate the central role of inflammation in this process. This idea contrasts with the models of mechanical deendothelization or stretching of the tunica media by balloon, not only by the absence of mechanical injury but also by the absence of change in blood flow after occlusion of a branch of the studied artery. This is necessary when the lumen of the vessel needs to be accessed by a catheter that will promote mechanical injury. In this way, the cuff model allows to isolate inflammation as a triggering factor, without interference from other factors. Fig. 16.3 shows that any kind of vascular wall change or obstruction to blood flow occurs.

Early studies have shown that placement of perivascular cuff promotes an inflammatory response, resembling particular conditions of atherosclerosis, especially in its early stages. Another specific condition of atherosclerosis to which

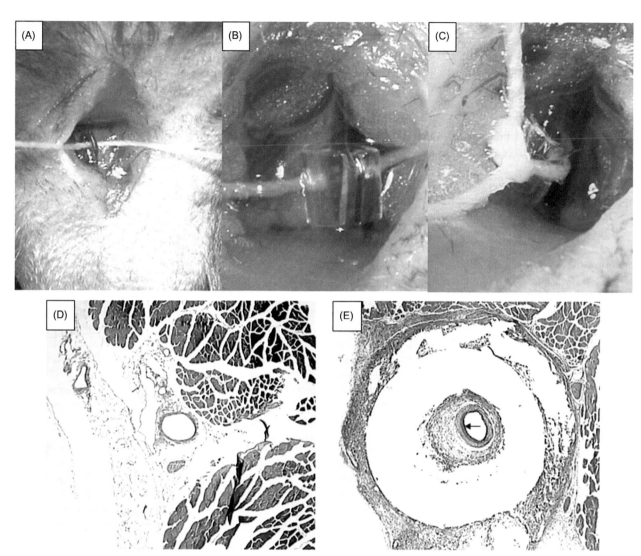

FIGURE 16.3 Different moments of cuff implantation in femoral artery and analysis. (A) isolation of femoral artery; (B) cuff placement, surrounding artery; (C) the cuff opening is closed using cotton thread (notice that cuff does not compress the artery); (D) cross section of noninjured femoral artery (notice the very thin intima); (E) cross section of cuff-injured femoral artery (notice that neointima is easily distinguishable—arrow). (D and E) stained for Verhoeff van Gieson.

the perivascular cuff model resembles is vascular explant disease, since the cuff around the artery would act like a foreign body [26].

It should be noted that the cuff model has been shown to be very effective in providing evidence that the renin-angiotensin system (RAS) plays an important role in the development of neointima and consequently in atherosclerosis. Thus, not only the absence of the AT2 receptor is important to prevent neointima formation [25] but also the presence of the AT1 receptor has been crucial for its formation [27]. However, if we observe an earlier step in the RAS biochemical cascade, it is possible to see that ACE inhibition is able to reduce the formation of vascular injury induced by cuff [26,28]. Another interesting aspect is the fact that increasing or reducing the number of Ace gene copies in mice seems to mimic the genetic polymorphism of Ace observed in humans. In humans, the presence (insertion) or absence (deletion) of a segment in the Ace gene influences enzyme activity and may interfere with cardiovascular parameters [29].

From these studies, a new possibility of investigation arises, where the increase or reduction in the number of copies of the Ace gene is related respectively to the increase or reduction in the tissue ACE activity and also to the neointima formation [28]. This would enable to know if the presence or absence of angiotensin II would be a determinant for the formation of this neointima. This was verified when treating mice with angiotensin II or AT1 receptor blocker. Thus, mice with 3 copies of the Ace gene show lower lesion if treated with AT1 receptor blocker, while animals with 1 copy of the Ace gene develop more lesion if treated with angiotensin II (Fig. 16.4). These data point to a direct causal relationship of angiotensin II via the AT1 receptor on neointima formation [28]. On the other hand, activation of the AT2 receptor inhibits this process [30].

When discussing the role of RAS on vascular injury, it is clear that the ACE/Ang II/AT1 axis plays a pro-proliferative, proinflammatory, and pro-atherosclerotic role. The ACE2/Ang-(1−7)/Mas axis, known for its antagonistic effects on the classical axis, has not yet been fully elucidated in vascular injury. The evaluation of cuff-induced vascular injury in femoral arteries in knockout mice for Mas or AT2 receptors showed that blocking the AT1 receptor axis may increase the activity of the axes involving other receptors (Mas and AT2), inhibiting neointimal formation [31]. Other evidence suggesting a vasoprotective role of Ang-(1−7) was obtained in wild type mice treated with saline overload with implanted perivascular cuff. These animals showed an increase in neointimal formation concomitant with local increase in Ang-(1−7), suggesting that Ang-(1−7) could be a vascular mechanism to counterbalance neointima formation and to maintain a stable vascular structure, despite remodeling induced by cuff [32].

As commented previously, the perivascular cuff model associates the initiation and development of vascular lesion with the inflammatory reaction. Several authors have shown that the inflammatory response can be identified by the release of several mediators. Initially, there is an increase in the expression of vascular monocyte chemoattractant protein-1 (MCP-1), inducing migration of monocytes that infiltrate the tissue. It is not known whether this expression of MCP-1 is a cause or consequence of increased nicotinamide adenine dinucleotide phosphate oxidase activity or superoxide anion production, although all these factors are abolished by AT1 receptor blockade [27]. These data are reinforced by the study which shows that the expression of angiotensin II type 1 receptor-associated protein attenuates the elevation of proliferating cells and reduces the activity of nicotinamide adenine dinucleotide phosphate oxidase and the consequent formation of neointima [33].

The role of the ACE/Ang II/AT1 axis is well known in the pathophysiology of hypertension and in other cardiovascular diseases, including vascular remodeling and atherosclerosis. Likewise, the AT2 receptor of angiotensin II and the ACE2/Ang-(1−7)/Mas axis have also been described as vasoprotective mechanisms. Although it is difficult to isolate the role of inflammation from hypertensive mechanisms in vascular injury and atherosclerosis, some studies have shown

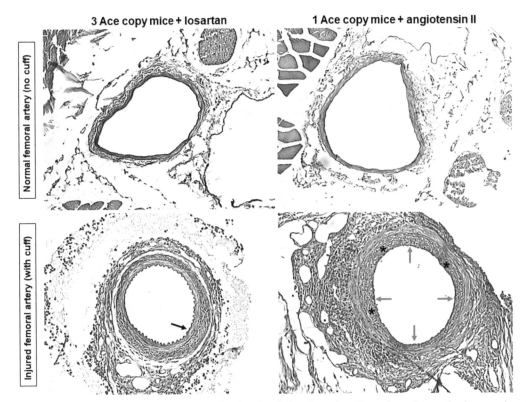

FIGURE 16.4 Effect of angiotensin II and AT1 receptor on cuff-induced neointima formation. Upper photomicrographs show arteries without lesion, whereas the lower photomicrographs are of injured arteries, comparing 3 Ace copy (left) versus 1 Ace copy (right). Note that 3-copy mice, after receiving losartan, did not develop neointima (black arrow); on the other hand, 1-copy mice, receiving angiotensin II, presented important neointima formation (*), narrowing lumen (red arrow). Verhoeff van Gieson staining; magnification 200×.

that angiotensin II can modulate the inflammatory response even in the absence of hypertension. In this case, a discrete increase in angiotensin II can induce an acute inflammatory response in the aorta of normotensive mice (without the presence of perivascular cuff), elevating TGFβ, IL-1β, and iNOS as well as stimulating the migration of CD45-positive cells on the healthy vascular wall [34]. This increase in acute inflammatory response suggests that angiotensin II may facilitate the inflammatory process even in healthy individuals, culminating with the formation of neointima when the vessel is subjected to conditions that induce its remodeling [27,28].

16.4   Vein graft diseases

Coronary artery occlusion reduces or blocks the blood flow inducing myocardial ischemia that evolves to a heart attack. The surgical treatment is the coronary artery bypass graft (CABG), where an autologous vessel (artery or vein) is inserted to bypass the narrow artery and improve blood flow to the heart [35]. In the late 60s, saphenous vein was the first conduit used to revascularize the ischemic heart [35,36]. Although efficient, 50%−60% of the veins thus grafted occlude after 10 years [35]. Nowadays, arterial conduits are commonly used with better patency but saphenous vein is still widely employed in CABG due to its size that allows multiple grafts and easy access because of its superficial location in the leg [35,37].

The main causes in early vein graft occlusion are surgical manipulation and ischemia that promote thrombosis [35]. Later graft failure occurs by intimal hyperplasia and ECM production to compensate the hemodynamic overload [35,38]. The overresponse to the increased mechanical stress provides a suitable soil to the atherosclerosis development and progression [35,37]. It has been demonstrated that hemodynamic differences among grafts beds can explain, at least in part, the saphenous vein worst outcome [38]. Indeed, there is evidence that neointimal thickening occurs when the vein is interposed in arterial circulation, with higher blood flow and pulsatile pressure, whereas such thickening is not observed after vein placement in venous bed [38]. In addition, there is neointimal regression when arterialized graft is placed back into venous circulation [38].

The cardiovascular system is constantly exposed and responsive to the mechanical stimuli associated with pulsative-blood flow and pressure (due to the cardiac cycle) [39,40]. The hemodynamic forces arising from blood flow are shear stress, a force produced by blood flow friction against the inner part of the vessel (endothelium) and cyclic stretch, a force that is tangential to the direction of blood flow [39,41]. Laminar shear stress acts parallel to the vessel and is measured by force per unit area, normally expressed as dyne/cm^2 (or Pa). Endothelial cells by virtue of their location, between blood and tissue, are the major sites for "sensing" of shear stress. Changes in shear regulate release of vasoactive factors from endothelial cells, such as nitric oxide (NO) and prostacyclin, that act in surrounding SMC to control vascular tonus [37,39]. Cyclic stretch results from constant change in vessel diameter as a function of pulsatile blood pressure [37,39]. It creates radial and tangential forces that affect all cell types from tissue and is usually expressed as the percentage of change related to the original dimensions [39,40]. While shear stress regulates vascular tone, cyclic stretch leads to a tension in the wall which controls vessel thickness. Vascular tissue is highly dynamic and is shaped by hemodynamic forces that have a key role in homeostasis maintenance. These forces act differently throughout arterial and venous bed, according to proximity to the heart and physiological needs for an organ's blood perfusion [41].

16.5   Vein graft adaptation to new hemodynamic condition

As is well known, veins and arteries walls are formed by three layers: (1) Intima—monolayer of endothelial cells under a basal lamina, known as endothelium; (2) Media—a muscular layer with SMCs and extracellular matrix; (3) Adventitia—rich in fibroblast, matrix, and capillaries to supply the vessel (vasa vasorum) [42]. While arteries are constantly exposed to high blood pressure and flow, venous system is a pulseless, low pressure, and low flow system [35,43]. Thus, coronary arteries have a greater muscular layer and robust elastic laminas (internal and external) to support and accommodate arterial mechanical stress [36] while saphenous veins have thinner media layer and are less elastic, which increases their capacitance. Also, veins have valves to provide unidirectional flow [42].

A human vein environment has low pressure (5−8 mmHg) and shear stress estimated around 0.2 dynes/cm^2 [44,45]. Thus, post CABG, the saphenous vein graft is suddenly submitted to an arterial hemodynamic condition with great increase in pulsatile pressure (120/80 mmHg), higher cyclic stretch (10%−15%) and shear stress (15−20 dynes) [41,46]. Acute changes in mechanical condition are compensated by vascular tonus control. Long-term mechanical load induces a permanent alteration with modification in the vessel wall structure named as vascular remodeling [47]. The increase in intraluminal pressure induces wall thickness to adjust tangential stress and normalize vessel tension [48]. This is an adaptive process, also termed vein arterialization, which involves endothelium denudation, invasion of inflammatory cells, increase number of SMC like cells for e.g., cells positive to smooth muscle actin (αSMA), cell migration and extracellular matrix production (collagen and elastin) that creates a medial thickening and neointima layer [37,38,49].

Vein graft occlusion in some patients occurs due to an imbalance in this remodeling, which culminates in exacerbated neointima increase. Many questions remain unclear such as: why some grafts have better adaptation than others, the molecular mechanisms responsible for tissue restructuring or what interventions or strategies can be used to modulate the pathway to avoid excessive remodeling.

To answer those questions, experimental models are best suited as they enable the study of molecular pathways associated with a single stimulus. Thus,changes in shear stress and stretching that occur simultaneously in a vascular graft, can be studied individually using in vitro and *in situ* experimental devices. Organ culture (ex vivo) systems have the advantage of closely mimicking human samples in a more controlled condition. In addition, animal models (in vivo) have largely been used to validated data obtained in in vitro and ex vivo models and also to test therapeutic strategies.

16.6 In vitro models

16.6.1 Shear stress

Shear stress intensity is directly proportional to blood velocity and viscosity, and inversely to lumen diameter [45]. Thus, the magnitude and direction of shear stress is not uniform, it changes throughout the bed and vascular geometry [45,50]. In straight vessel segments, there is a steady laminar shear stress (LSS), parallel to vessel wall, which has protector effects on endothelium. Endothelial cells under LSS have activation of genes responsible for quiescent state (e.g., Kruppel like factor 2 or KLF2 and nuclear factor erythroid 2−related factor 2 or Nrf2), allowing a strict permeability control, nonthrombogenic, and antiinflammatory state [51]. However, a low and oscillatory shear stress (OSS) occurs naturally in some regions of arterial bed with disturbed flow such as bifurcations, branches, and curvatures area [45,50,51]. These turbulent flows promote changes in endothelial phenotype leading to proinflammatory activation, an atheroprone region as widely described [50].

After saphenous vein bypass surgery, a great increase in shear stress is experienced by endothelial cells and our understanding was expanded by in vitro studies. Many devices are used to produce shear stress by flowing medium into monolayer of endothelial cells under controlled conditions [52]. Here we are going to discuss two different systems: (1) cone plate viscometer and (2) parallel-plate flow chamber.

1. *Cone plate viscometer:* endothelial cells monolayer is exposed to shear stress achieved by a rotating cone, specifically placed in a certain superficial angle nearby culture plate. Shear stress is calculated through the formula:

$$\tau = \frac{\eta \omega}{\alpha}$$

where
τ = shear stress (dynes/cm^2);
η = viscosity of the fluid (culture medium, dynes.s/cm^2);
ω = rotational velocity (rad/s);
α = cone superficial angle.

This system allows production of different types of shear such as laminar, oscillatory, and also pulsatile, accordingly with the superficial angle and angular velocity.

2. *Parallel-plate flow chamber*: endothelial cells are attached in a chamber slide connected with a unidirectional and continuous flow system [53]. There is a side for inlet and another for the outlet flow. A motor pump controls the system, and shear stress is calculated as the following formula:

$$Q = \frac{\tau w h2}{6\mu}$$

where
Q = desired flow rate;
τ = shear stress;
w = width of flow chamber;
h = height of floe chamber;
μ = viscosity of the fluid.

Currently, chambers are available to mimic all kinds of arterial blood flow such as disturbed flow associated with bifurcation and branches; these chambers allow for studying responses of various regions in the arterial bed [50,54].

Endothelial cells from saphenous vein submitted to higher shear stress have reduction in eNOS, and increased expression of endothelin- 1 and intracellular adhesion molecules, characterizing cell dysfunction [44,54,55]. Also, saphenous vein has larger diameter and compliance than the artery which the bypass is inserted [45]. Thus, besides the high shear stress, the graft also has a turbulent/OSS because of this geometric and compliance mismatch [45,56]. In addition, the presence of valves in saphenous vein may cause even more blood flow disturbances [50]. The importance of this compliance mismatch in neointima formation is still under debate, since there are some discrepant results [45]. Nonetheless is a consensus that increase in flow velocity is associated with endothelium damage at vein graft [56,57].

Higher magnitudes and disturbed flow have been shown to change actin cytoskeleton pattern, leading to a reduction at endothelial barrier function [45,51]. The increase in endothelial permeability occurs by adhesion junction disassembly; OSS decreases the presence of occludin and tight junction protein-1 (TJP1),that are crucial in maintaining cell−cell contacts intact [58,59]. Also, high magnitude of flow or shear stress activates transcription factor NFκB. This leads to induction of adhesion molecules ([vascular adhesion molecules (VCAM) and intracellular adhesion molecules], release of proinflammatory markers (IL8, MCP-1) that recruits surrounding cells and influences SMC proliferation/migration [51,54,60,61]. Cocultured experiments demonstrated that LSS on endothelial cells maintain SMC contractile phenotype [62] whereas increased OSS promotes SMC phenotypic switching for proliferative status [61]. Altogether, these changes can induce neointima formation.

Furthermore, endothelial cells are highly plastic and can undergo endothelial to mesenchymal transition (EndMT) upon certain conditions, which may contribute to neointima layer [63−65]. EndMT is identified by reduction in endothelial markers, gain of mesenchymal markers, loss of cell polarity, increased extracellular matrix production, and migration ability [66]. Initially it was believed that the source of α-SMA-positive cells in neointima was due to SMC proliferation/ migration from media layer. However, experimental data indicated that ∝-SMA positive cells may also come from different sources such as fibroblast differentiation and migration from adventitia, circulating stem cells or from endothelial cells that transformed into ∝-SMA positive cells [67−69]. EndMT could be inhibited by high levels of LSS in the presence of transforming growth factor beta (TGFβ), a potent mesenchymal inducer [70]. TGFβ acts via ERK5 pathway and thus cells with constitutively active mutant for ERK5, did not show EndMT with OSS or disturbed flow. Disturbed flow also facilitates EndMT by induction of bone morphogenic protein 4 (BMP4) and by reactive oxygen species (ROS) production, activate directly TGFβ pathway responsible for mesenchymal transition [71,72].

16.6.2 Stretch

The majority of stretch studies focus on SMC, since these are quite responsive to this hemodynamic force and their proliferation/migration broadly contribute to neointima formation. However, endothelial cells also sense directly this force and many researchers have been showing changes in EC phenotype by enhanced cyclic stretch.

Currently there are several devices capable of recreating cyclic stretch experienced by human vasculature *in vitro* [73,74]. Most of them are based on cultivating cells under a silicone membrane (a malleable surface), connected to a programmable loading system that controls frequency, waveform, and stretch intensity. Mechanical stimulus is often made by vacuum or piston system-based technology, which stretches silicone membranes. Depending on the apparatus is possible to control the type of stretch: uni-, bi- or multiaxial (circumferential) loading.

Arterial cyclic stretch in vivo varies between 2%−20% [75]. Stretch controls cell function and morphology by changing actin cytoskeleton assembly, releasing of vasoactive factors (e.g., NO and prostacyclin), cell proliferation, apoptosis, and antioxidants properties [75]. Endothelial cells from bovine and human aortic source under physiologic stretch (6%− 10%, 1 Hz) are protected against apoptosis, which is not observed in higher cyclic stretch intensity (20%, 1 Hz). The related mechanism involves ROS control [75]. Endothelial cells are constantly in contact with ROS that in physiological level act as second messengers and control several regulatory functions. However, high levels of ROS are deleterious to cardiovascular tissue [76]. Several studies show increases in ROS production by higher cyclic stretch in endothelial cells, thereby impacting actin cytoskeletal mechanotransduction and leading to an increase in vascular permeability [77]. Moreover, induction of mesenchymal markers in HUVEC submitted to higher stretch has also been reported [78,79].

Although the effect of increased stretching on endothelial function has been reported, there is no information on the effect of stretch on endothelial cells from human saphenous vein (hSVEC). Unpublished data from our group identified endothelial dysfunction of hSVEC under arterial stretching for prolonged period. Further experiments are needed to better understand this hemodynamic force during vein graft.

16.6.3 Shear stress and cyclic stretch

As previously mentioned, the best scenario is to evaluate the effects of both forces simultaneously [80]. To do so, some devices are available where shear stress and cyclic stretch can be applied on cells cultured under flexible membranes in controlled systems [81]. To closely represent the human endothelium *in vivo* three-dimensional in vitro models are in current use [82,83]. These equipment are based on cell culture in distensible (silicone) tubes that mimic vessels, connected with an accurate flow/pression perfusion system.

Disadvantages of tridimensional cultures includes difficulties to seed and visualize the cells, to accurately control the strain necessary, to open the tube for following experiments. Furthermore, the tube physical properties are different from native vessels. Nevertheless despite these limitations, 2D and 3D in vitro models have vastly improved our knowledge of how shear stress and cyclic stretch affect EC alignment, cell—cell adhesion, permeability, and oxidative stress.

16.7 Ex vivo models

Organ perfusion systems are an useful model to study hemodynamic changes directly in the vessel used as graft [84]. Several customized devices that mimic grafts have increased our understanding of the molecular mechanisms related to venous arterialization. In our group, we described an equipment that allows the use human saphenous vein segments under a well-controlled perfusion circuit [85]. The pressure and flow can be regulated independently, allowing the vein placement under venous or arterial conditions up to 4 days. Our data demonstrated an increase in apoptosis rate in veins submitted to arterial hemodynamic forces, corroborating previous in vitro and in vivo results [85].

Moreover, ex vivo data also show increased procoagulant state of vein segments cultured in similar flow or shear stress/stretch conditions as those faced by the arterial tree. Human saphenous vein and porcine jugular vein when exposed to coronary arterial shear forces showed increase in expression of tissue factor (TF), the high-affinity receptor and cofactor for factor (F)VII/VIIa. The TF-FVIIa complex is the primary initiator of blood coagulation [86].

These models have the vast advantage of working with human vein used in the grafts. A major disadvantage is that these preps cannot be kept viable (outside the body) for long. Another shortcoming of such models is the absence of endogenous inflammatory cells.

16.8 In vivo models

Rodents were the first and are still the most popular models in use, as they are cost effective, easy to manipulate and care for; besides the availability of a huge body of literature makes the use of rodents very convenient [87,88]. Shunt models that represent venous grafts are easy to be made in rodents; the vein is connected to arterial circulation in an arteriovenous shunt that can be done in several ways between many different beds such as femoral artery into femoral vein, vena cava into abdominal aorta, or carotid artery into jugular vein [87,89]. This model was described by our group in a rat end-to-end anastomose model, which reproduced some morphological changes detected in human vein graft disease [49]. Other rodent models of venous grafts also show endothelial activation with MCP1 and NFκB up regulation, corroborating in vitro data [90]. Furthermore, the possibility of using mouse and rat knockout to target genes and proteins of interest make these models valuable in understanding molecular pathways and testing new therapeutic targets.

Besides rodents, rabbits are also used for vein graft studies. Work on rabbit grafts has shown that gene therapy may help in reduction of graft failure. Indeed intraoperative transfection of rabbit vein grafts with a decoy oligonucleotide that blocks cell-cycle gene transactivation by the transcription factor E2F, has been found to induce long-term stable adaptation that reduces neointimal hyperplasia and atherosclerosiswithout changes in endothelium phenotype [91]. Those exciting data influenced a clinical trial program: Project of Ex Vivo Vein Graft Engineering via Transfection (PREVENT) [43,92,93]. Despite the great results obtained by rabbit's experiments after a 6-month follow-up, data from clinical trials failed to improve human saphenous vein graft [43,92]. This indicates that animal models have their own limitations and that robust models are crucial in translational studies.

Large animal models (e.g., pigs, canines, monkey) too have been in use, since they have vascular anatomy and hemodynamic forces similar to humans [94,95]. Expression of P-selectin and intracellular adhesion molecule-1 was increased in saphenous vein endothelium after arterial anastomosis in pigs [87]. On the other hand, endothelial protective molecules such as PGI2, cAMP, and cGMP were reduced after vein graft [87,96].

Over the years several therapeutic approaches have been used with promising results in neointima reduction: oral drug administration (endothelin 1A receptor antagonist), adenovirus gene delivery (NOS, TIMP-3), internal/external stents, stem cell treatment [87,95,97]. But most studies have been limited to rodents and/or rabbits as large animal models are involve expertise and high costs. Of course the fact still remains that experimental models commonly have a short follow-up (up to 3 months) compared to the long time course for human vein graft disease development (6−24 months) [88,90].

Nevertheless, animal models are essential tools to explore the mechanisms and pathology of vascular diseases and to test preventive strategies and therapies [88]. It is worth noting that majority of animal models are quadrupeds, meaning the venous system remains under constant flow and pressure conditions [98]. Thus, they do not represent the orthostatic pressures difference in bipeds' saphenous vein, which constantly changes according to posture and movement.

16.9 Final considerations

Different approaches to vascular injury have allowed us to understand that the vessel can recognize different stimuli, such as variations in blood pressure or flow as well as deleterious stimuli of agents in blood that lead to injury. Fig. 16.5 summarizes the possible effect of these stimuli, inducing different adaptive responses. Each stimulus is detected by different chemical or mechanical receptors. Activation of such receptors will induce adaptive responses, often mediated via a second messenger in each cell. Communication of vascular cells with each other or with extracellular matrix components or even between different extracellular matrix components will lead to functional and/or structural changes. Depending on the nature of the stimulus, different changes may be observed, altering the ability of the artery or vein to dilate or contract, leading to cell proliferation or death (changing the wall/lumen ratio) or even inducing an inflammatory/healing process. In this context, increased blood pressure will induce responses that will magnify wall thickness, while augmentation of blood flow will expand lumen area. When the vascular system does not adapt to different stimuli, tissue perfusion may be reduced, leading to ischemia or even more severe consequences, such as vascular occlusion.

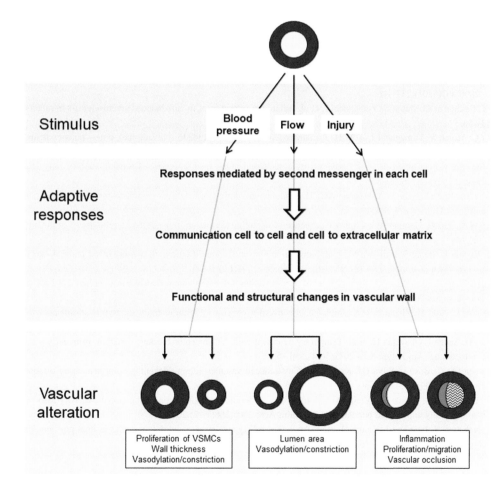

FIGURE 16.5 Possible mechanisms by which the vessel detects various stimuli (mechanical or biochemical stimuli), causing adaptive responses. Vascular responses will include changes in thickness in relation to vascular lumen or alteration in the ability to dilate/constrict, as well as inflammation or proliferation/cell migration processes leading to vessel occlusion. Extracellular matrix or ECM : comprises all substances found in the tissue but outside the cells, such as different types of collagen, hyaluronic acid, proteoglycans, elastic fibers, and others; *VSMCs*, vascular smooth muscle cells.

References

[1] Geovanini GR, Libby P. Atherosclerosis and inflammation: overview and updates. ClinSci (Lond) 2018;132(12):1243−52.

[2] Zhao TX, Mallat Z. Targeting the immune system in atherosclerosis: JACC state-of-the-art review. J Am Coll Cardiol 2019;13(73):1691−706.

[3] Ignatowski AC. Influence of animal food on the organism of rabbits. S Peterb Izviest Imp Voyenno-Med Akad 1908;16:154−73.

[4] Fan J, Chen Y, Yan H, Niimi M, Wang Y, Liang J. Principles and applications of rabbit models for atherosclerosis research. J Atheroscler Thromb 2018;25:213−20.

[5] Owsiany KM, Alencar GF, Owens GK. Revealing the origins of foam cells in atherosclerotic lesions. Arterioscler Thromb Vasc Biol 2019;39(5):836−8.

[6] Fan J, Kitajima S, Watanabe T, Xu J, Zhang J, Liu E, et al. Rabbit models for the study of human atherosclerosis: from pathophysiological mechanisms to translational medicine. Pharmacol Ther 2015;146:104−19.

[7] Booth RFG, Martin JF, Honey AC, Hassall DG, Beesley JE, Moncada S. Rapid development of atherosclerotic lesions in the rabbit carotid artery induced by perivascular manipulation. Atherosclerosis. 1989;16:257−68.

[8] Kobayashi T, Ito T, Shiomi M. Roles of the WHHL rabbit in translational research on hypercholesterolemia and cardiovascular diseases. J Biomed Biotechnol 2011;2011:406473.

[9] Polak JF, Pencina MJ, Pencina KM, O'Donnell CJ, Wolf PA, D'Agostino Sr RB. Carotid-wall intima-media thickness and cardiovascular events. N Engl J Med 2011;365(3):213−21.

[10] Wada S, Koga M, Minematsu K, Toyoda K, Suzuki R, Kagimura T, et al. Baseline carotid intima-media thickness and stroke recurrence during secondary prevention with pravastatin. Stroke. 2019. Available from: https://doi.org/10.1161/STROKEAHA.119.024968.

[11] von Scheidt M, Zhao Y, Kurt Z, Pan C, Zeng L, Yang X, et al. Applications and limitations of mouse models for understanding human atherosclerosis. Cell Metab. 2017;25:248−61.

[12] Santana AB, de Souza Oliveira TC, Bianconi BL, Barauna VG, Santos EW, Alves TP, et al. Effect of high-fat diet upon inflammatory markers and aortic stiffening in mice. Biomed Res Int 2014;2014:914102.

[13] Maeda N. Development of apolipoprotein E-deficient mice. Arterioscler Thromb Vasc Biol 2011;31(9):1957−62.

[14] Ishibashi S, Goldstein JL, Brown MS, Herz J, Burns DK. Massive xanthomatosis and atherosclerosis in cholesterol-fed low density lipoprotein receptor-negative mice. J Clin Invest 1994;93:1885−93.

[15] Plump AS, Smith JD, Hayek T, Aalto-Setälä K, Walsh A, Verstuyft JG, et al. Severe hypercholesterolemia and atherosclerosis in apolipoprotein E-deficient mice created by homologous recombination in ES cells. Cell. 1992;71:343−53.

[16] Veseli BE, Perrotta P, De Meyer GRA, Roth L, Van der Donckt C, Martinet W, et al. Animal models of atherosclerosis. Eur J Pharmacol 2017;816:3−13.

[17] Durham AL, Speer MY, Scatena M, Giachelli CM, Shanahan CM. Role of smooth muscle cells in vascular calcification: implications in atherosclerosis and arterial stiffness. Cardiovasc Res 2018;114(4):590−600.

[18] Lusis AJ. Atherosclerosis. Nature. 2000;407(6801):233−41.

[19] Hirosumi J, Nomoto A, Ohkubo Y, Sekiguchi C, Mutoh S, Yamaguchi I, et al. Inflammatoryresponses in cuff-induced atherosclerosis in rabbits. Atherosclerosis 1987;64(2−3):243−54.

[20] Nomoto A, Hirosumi J, Sekiguchi C, Mutoh S, Yamaguchi I, Aoki H. Antiatherogenicactivity of FR34235 (Nilvadipine), a new potent calcium antagonist. Effect on cuff-induced intimal thickening of rabbit carotid artery. Atherosclerosis. 1987;64(2−3):255−61.

[21] De Meyer GR, Bult H, Herman AG. Early atherosclerosis is accompanied by a decreased rather than an increased accumulation of fatty acid hydroxyderivatives. Biochem Pharmacol 1991;42(2):279−83.

[22] Kockx MM, De Meyer GR, Andries LJ, Bult H, Jacob WA, Herman AG. The endothelium during cuff-induced neointima formation in the rabbitcarotidartery. Arterioscler Thromb 1993;13(12):1874−84.

[23] Kockx MM, De Meyer GR, Jacob WA, Bult H, Herman AG. Triphasic sequence of neointimalformation in the cuffed carotid artery of the rabbit. Arterioscler Thromb 1992;12(12):1447−57.

[24] Akishita M, Ouchi Y, Miyoshi H, Kozaki K, Inoue S, Ishikawa M, et al. Estrogen inhibits cuff-induced intimal thickening of rat femoral artery: effects on migration and proliferation of vascular smooth muscle cells. Atherosclerosis. 1997;130:1−10.

[25] Akishita M, Horiuchi M, Yamada H, Zhang L, Shirakami G, Tamura K, et al. Inflammation influences vascular remodeling through AT2 receptor expression and signaling. Physiol Genomics 2000;2(1):13−20.

[26] Akishita M, Shirakami G, Iwai M, Wu L, Aoki M, Zhang L, et al. Angiotensin converting enzyme inhibitor restrains inflammation-induced vascular injury in mice. J Hypertens 2001;19(6):1083−8.

[27] Inaba S, Iwai M, Furuno M, Kanno H, Senba I, Okayama H, et al. Temporary treatment with AT1 receptor blocker, valsartan, from early stage of hypertension prevented vascular remodeling. Am J Hypertens 2011;5(24):550−6.

[28] Lacchini S, Heimann AS, Evangelista FS, Cardoso L, Silva GJ, Krieger JE. Cuff-induced vascular intima thickening is influenced by titration of the Ace gene in mice. Physiol Genomics 2009;37(3):225−30.

[29] Pereira AC, Morandini Filho AA, Heimann AS, Rabak ET, Vieira AP, Mota GF, et al. Serum angiotensin converting enzyme activity association with the I/D polymorphism in an ethnically admixtured population. Clin Chim Acta 2005;360(1−2):201−4.

[30] Kukida M, Mogi M, Ohshima K, Nakaoka H, Iwanami J, Kanno H, et al. Angiotensin II type 2 receptor inhibits vascular intimal proliferation with activation of PPAR-γ. Am J Hypertens 2016;29(6):727−36.

[31] Ohshima K, Mogi M, Nakaoka H, Iwanami J, Min LJ, Kanno H, et al. Possible role of angiotensin-converting enzyme 2 and activation of angiotensin II type 2 receptorby angiotensin-(1-7) in improvement of vascular remodeling by angiotensin II type 1 receptor blockade. Hypertension. 2014;63(3):e53−9.

[32] Lima CT, Silva JC, Viegas KA, Oliveira TC, Lima RS, Souza LE, et al. Increase in vascular injury of sodium overloaded mice may be related to vascular angiotensin modulation. PLoS One 2015;10(6):e0128141.

[33] Oshita A, Iwai M, Chen R, Ide A, Okumura M, Fukunaga S, et al. Attenuation of inflammatory vascular remodeling by angiotensin II type 1 receptor−associated protein. Hypertension. 2006;48:671−6.

[34] de Lima RS, Silva JCS, Lima CT, de Souza LE, da Silva MB, Baladi MG, et al. Proinflammatory role of angiotensin II in the aorta of normotensive mice. Biomed Res Int 2019;2019:9326896.

[35] Gaudino M, Antoniades C, Benedetto U, Deb S, Di Franco A, Di Giammarco G, et al. Mechanisms, consequences, and prevention of coronary graft failure. Circulation. 2017;136(18):1749−64.

[36] Joviliano EE, Dellalibera-joviliano R, Celotto AC, Capellini VK, Dalio MB, Picconato CE, et al. Pharmacology of the human saphenous vein. Curr Vasc Pharmacol 2011;9:501−20.

[37] Harskamp RE, Lopes RD, Baisden CE, de Winter RJ, Alexander JH. Saphenous vein graft failure after coronary artery bypass surgery. Ann Surg. 2013;257(5):824−33.

[38] Allaire E, Clowes AW. Endothelial cell injury in cardiovascular surgery: the intimal hyperplastic response. Ann Thorac Surg 1997;63 (2):582−91.

[39] Lehoux S, Castier Y, Tedgui A. Molecular mechanisms of the vascular responses to haemodynamic forces. J Intern Med 2006;259(4):381−92.

[40] Hsiai TK, Wu JC. Hemodynamic forces regulate embryonic stem cell commitment to vascular progenitors. Curr Cardiol Rev 2008;4 (4):269−74.

[41] Soulis JV, Farmakis TM, Giannoglou GD, Louridas GE. Wall shear stress in normal left coronary artery tree. J Biomech. 2006;39(4):742−9.

[42] Dilley RJ, McGeachie JK, Prendergast FJ. A review of the histologic changes in vein-to-artery grafts, with particular reference to intimal hyperplasia. Arch Surg. 1988;123(6):691−6.

[43] Conte MS, Bandyk DF, Clowes AW, Moneta GL, Seely L, Lorenz TJ, et al. Results of PREVENT III: a multicenter, randomized trial of edifoligide for the prevention of vein graft failure in lower extremity bypass surgery. J Vasc Surg 2006;43(4):742−51.

[44] Golledge J, Turner RJ, Harley SL, Springall DR, Powell JT. Circumferential deformation and shear stress induce differential responses in saphenous vein endothelium exposed to arterial flow. J Clin Invest 1997;99(11):2719−26.

[45] Vara DS, Punshon G, Sales KM, Hamilton G, Seifalian AM. Haemodynamic regulation of gene expression in vascular tissue engineering. Curr Vasc Pharmacol 2011;9(2):167−87.

[46] Van Andel CJ, Pistecky PV, Borst C. Mechanical properties of porcine and human arteries: implications for coronary anastomotic connectors. Ann Thorac Surg 2003;76(1):58−64.

[47] Epstein FH, Gibbons GH, Dzau VJ. The emerging concept of vascular remodeling. N Engl J Med 2002;330(20):1431−8.

[48] Zwolak RM, Adams MC, Clowes AW. Kinetics of vein graft hyperplasia: association with tangential stress. J Vasc Surg 1987;5(1):126−36.

[49] Borin TF, Miyakawa AA, Cardoso L, de Figueiredo Borges L, Gonçalves GA, Krieger JE. Apoptosis, cell proliferation and modulation of cyclin-dependent kinase inhibitor p21(cip1) in vascular remodelling during vein arterialization in the rat. Int J Exp Pathol 2009;90(3):328−37.

[50] Chiu J-J, Chien S. Effects of disturbed flow on vascular endothelium: pathophysiological basis and clinical perspectives. Physiol Rev. 2011;91 (1):327−87.

[51] Chistiakov DA, Orekhov AN, Bobryshev YV. Effects of shear stress on endothelial cells: go with the flow. Acta Physiol (Oxf) 2017;219 (2):382−408.

[52] Brown TD. Techniques for mechanical stimulation of cells in vitro: a review. J Biomech. 2000;33(1):3−14.

[53] Bacabac RG, Smit TH, Cowin SC, Van Loon JJWA, Nieuwstadt FTM, Heethaar R, et al. Dynamic shear stress in parallel-plate flow chambers. J Biomech. 2005;38(1):159−67.

[54] Sultan S, Gosling M, Abu-Hayyeh S, Carey N, Powell JT. Flow-dependent increase of ICAM-1 on saphenous vein endothelium is sensitive to apamin. Am J Physiol Circ Physiol 2004;287(1):H22−8.

[55] Zhu ZG, Li HH, Zhang BR. Expression of endothelin-1 and constitutional nitric oxide synthase messenger RNA in saphenous vein endothelial cells exposed to arterial flow shear stress. Ann Thorac Surg 1997;64(5):1333−8.

[56] Ward AO, Caputo M, Angelini GD, George SJ, Zakkar M. Activation and inflammation of the venous endothelium in vein graft disease. Atherosclerosis. 2017;265:266−74.

[57] Isaji T, Hashimoto T, Yamamoto K, Santana JM, Yatsula B, Hu H, et al. Improving the outcome of vein grafts: should vascular surgeons turn veins into arteries? Ann Vasc Dis 2017;10(1):8−16.

[58] DeMaio L, Chang YS, Gardner TW, Tarbell JM, Antonetti DA. Shear stress regulates occludin content and phosphorylation. Am J Physiol Circ Physiol 2017;281(1):H105−13.

[59] Conklin BS, Zhong DS, Zhao W, Lin PH, Chen C. Shear stress regulates occludin and VEGF expression in porcine arterial endothelial cells. J Surg Res 2002;102(1):13−21.

[60] Chappell DC, Varner SE, Nerem RM, Medford RM, Alexander RW. Oscillatory shear stress stimulates adhesion molecule expression in cultured human endothelium. Circ Res. 1998;82(5):532−9.

[61] Hastings NE, Simmers MB, McDonald OG, Wamhoff BR, Blackman BR. Atherosclerosis-prone hemodynamics differentially regulates endothelial and smooth muscle cell phenotypes and promotes pro-inflammatory priming. Am J Physiol Cell Physiol 2007;293(6):C1824−33.

[62] Sakamoto N, Kiuchi T, Sato M. Development of an endothelial-smooth muscle cell coculture model using phenotype-controlled smooth muscle cells. Ann Biomed Eng 2011;39(11):2750−8.

[63] Beranek JT. Vascular endothelium-derived cells containing smooth muscle actin are present in restenosis. Lab Invest. 1995;72(6):771.

[64] Frid MG, Kale VA, Stenmark KR. Mature vascular endothelium can give rise to smooth muscle cells via endothelial-mesenchymal transdifferentiation: in vitro analysis. Circ Res. 2002;90(11):1189−96.

[65] Cooley BC, Nevado J, Mellad J, Yang D, Hilaire CS, Negro A, et al. TGF-β signaling mediates endothelial-to-mesenchymal transition (EndMT) during vein graft remodeling. Sci Transl Med 2014;6(227):227ra34.

[66] Pardali E, Sanchez-Duffhues G, Gomez-Puerto MC, Ten Dijke P. TGF-β-induced endothelial-mesenchymal transition in fibrotic diseases. Int J Mol Sci 2017;18(10):2157.

[67] Fogelstrand P, Osterberg K, Mattsson E. Reduced neointima in vein grafts following a blockage of cell recruitment from the vein and the surrounding tissue. Cardiovasc Res. 2005;67(2):326−32.

[68] Hu Y, Zhang Z, Torsney E, Afzal AR, Davison F, Metzler B, et al. Abundant progenitor cells in the adventitia contribute to atheroscleroses of vein grafts in ApoE-deficient mice. J Clin Invest 2004;113(9):1258−65.

[69] Hu Y, Mayr M, Metzler B, Erdel M, Davison F, Xu Q. Both donor and recipient origins of smooth muscle cells in vein graft atherosclerotic lesions. Circ Res. 2002;91(7):e13−20.

[70] Moonen J-RAJ, Lee ES, Schmidt M, Maleszewska M, Koerts JA, Brouwer LA, et al. Endothelial-to-mesenchymal transition contributes to fibro-proliferative vascular disease and is modulated by fluid shear stress. Cardiovasc Res. 2015;108(3):377−86.

[71] Sorescu GP, Sykes M, Weiss D, Platt MO, Saha A, Hwang J, et al. Bone morphogenic protein 4 produced in endothelial cells by oscillatory shear stress stimulates an inflammatory response. J Biol Chem 2003;278(33):31128−35.

[72] Sorescu GP, Song H, Tressel SL, Hwang J, Dikalov S, Smith DA, et al. Bone morphogenic protein 4 produced in endothelial cells by oscillatory shear stress induces monocyte adhesion by stimulating reactive oxygen species production from a Nox1-based NADPH oxidase. Circ Res. 2004;95(8):773−9.

[73] Lee J, Wong M, Smith Q, Baker AB. A novel system for studying mechanical strain waveform-dependent responses in vascular smooth muscle cells. Lab Chip. 2013;13(23):4573−82.

[74] Campos LCG, Miyakawa AA, Barauna VG, Cardoso L, Borin TF, Dallan LA, et al. Induction of CRP3/MLP expression during vein arterialization is dependent on stretch rather than shear stress. Cardiovasc Res 2009;83(1):140−7.

[75] Krenning G, Barauna VG, Krieger JE, Harmsen MC, Moonen J-R. Endothelial plasticity: shifting phenotypes through force feedback. Stem Cells Int 2016;2016:9762959.

[76] Panieri E, Santoro MM. ROS signaling and redox biology in endothelial cells. Cell Mol Life Sci 2015;72(17):3281−303.

[77] Birukov KG. Small GTPases in mechanosensitive regulation of endothelial barrier. Microvasc Res. 2009;77(1):46−52.

[78] Cevallos M, Riha GM, Wang X, Yang H, Yan S, Li M, et al. Cyclic strain induces expression of specific smooth muscle cell markers in human endothelial cells. Differentiation. 2006;74(9−10):552−61.

[79] Mai J, Hu Q, Xie Y, Su S, Qiu Q, Yuan W, et al. Dyssynchronous pacing triggers endothelial-mesenchymal transition through heterogeneity of mechanical stretch in a canine model. Circ J. 2015;79(1):201−9.

[80] Meza D, Abejar L, Rubenstein DA, Yin W. A shearing-stretching device that can apply physiological fluid shear stress and cyclic stretch concurrently to endothelial cells. J Biomech Eng 2016;138(3):031007.

[81] Benbrahim A, L'Italien GJ, Kwolek CJ, Petersen MJ, Milinazzo B, Gertler JP, et al. Characteristics of vascular wall cells subjected to dynamic cyclic strain and fluid shear conditions in vitro. J Surg Res 1996;65(2):119−27.

[82] Casey PJ, Dattilo JB, Dai G, Albert JA, Tsukurov OI, Orkin RW, et al. The effect of combined arterial hemodynamics on saphenous venous endothelial nitric oxide production. J Vasc Surg 2001;33(6):1199−205.

[83] Williams C, Wick TM. Endothelial cell-smooth muscle cell co-culture in a perfusion bioreactor system. Ann Biomed Eng 2005 Jul;33(7):920−8.

[84] Berard X, Déglise S, Alonso F, Saucy F, Meda P, Bordenave L, et al. Role of hemodynamic forces in the ex vivo arterialization of human saphenous veins. J Vasc Surg 2013;57(5):1371−82.

[85] Miyakawa AA, Dallan LAO, Lacchini S, Borin TF, Krieger JE. Human saphenous vein organ culture under controlled hemodynamic conditions. Clinics (Sao Paulo) 2008;63(5):683−8.

[86] Muluk SC, Vorp DA, Severyn DA, Gleixner S, Johnson PC, Webster MW. Enhancement of tissue factor expression by vein segments exposed to coronary arterial hemodynamics. J Vasc Surg 1998;27(3):521−7.

[87] Schachner T, Laufer G, Bonatti J. In vivo (animal) models of vein graft disease. Eur J Cardiothorac Surg 2006;451−63.

[88] Cooley BC. Experimental vein graft research: a critical appraisal of models. Heart Res Open J 2015;2:53−9.

[89] Thomas AC. Animal models for studying vein graft failure and therapeutic interventions. Curr Opin Pharmacol 2012;12(2):121−6.

[90] Owens CD, Gasper WJ, Rahman AS, Conte MS. Vein graft failure. J Vasc Surg 2015;61(1):203−16.

[91] Ehsan A, Mann MJ, Dell'Acqua G, Dzau VJ. Long-term stabilization of vein graft wall architecture and prolonged resistance to experimental atherosclerosis after E2F decoy oligonucleotide gene therapy. J Thorac Cardiovasc Surg 2001;121(4):714−22.

[92] Alexander JH, Hafley G, Harrington RA, Peterson ED, Ferguson TB, Lorenz TJ, et al. Efficacy and safety of edifoligide, an E2F transcription factor decoy, for prevention of vein graft failure following coronary artery bypass graft surgery: PREVENT IV: a randomized controlled trial. JAMA 2005;294(19):2446−54.

[93] Hess CN, Lopes RD, Gibson CM, Hager R, Wojdyla DM, Englum BR, et al. Saphenous vein graft failure after coronary artery bypass surgery: insights from PREVENT IV. Circulation. 2014;130(17):1445−51.

[94] Petrofski JA, Hata JA, Gehrig TR, Hanish SI, Williams ML, Thompson RB, et al. Gene delivery to aortocoronary saphenous vein grafts in a large animal model of intimal hyperplasia. J Thorac Cardiovasc Surg 2004;127(1):27−33.

[95] Wan S, Yim APC, Johnson JL, Shukla N, Angelini GD, Smith FC, et al. The endothelin 1A receptor antagonist BSF 302146 is a potent inhibitor of neointimal and medial thickening in porcine saphenous vein-carotid artery interposition grafts. J Thorac Cardiovasc Surg 2004;127 (5):1317−22.

[96] Jeremy JY, Dashwood MR, Timm M, Izzat MB, Mehta D, Bryan AJ, et al. Nitric oxide synthase and adenylyl and guanylyl cyclase activity in porcine interposition vein grafts. Ann Thorac Surg 1997;63(2):470−6.

[97] George SJ, Wan S, Hu J, MacDonald R, Johnson JL, Baker AH. Sustained reduction of vein graft neointima formation by ex vivo TIMP-3 gene therapy. Circulation 2011;124(11 Suppl):S135−42.

[98] Dashwood M, Loesch A. The saphenous vein as a bypass conduit: the potential role of vascular nerves in graft performance. Curr Vasc Pharmacol 2009;7(1):47−57.

Chapter 17

The endothelium as a target for bacterial infection: challenges at the bedside

Kumkum Ganguly

Bioscience Division, Los Alamos National Laboratory, Los Alamos, NM, United States

17.1 Introduction

The endothelium is the inner lining of the vascular network—a complex system made up of arteries, capillaries, and vein and consists of a single layer of flat cells. Initially considered inert and designated as the "cellophane wrapper" of the vascular tree, with the job of maintaining selective permeability to water and electrolytes [1], the vascular endothelium is now well recognized as an interactive interface which performs a number of important functions in order to maintain adequate blood supply to vital organs. These functions include (1) prevention of coagulation and thrombosis; (2) regulation of vascular tone in the form of vasodilation and vasoconstriction; (3) orchestration of the migration of immune cells or diapedesis by the expression of adhesion molecules; and (4) regulation of permeability and production of chemoattractant compounds. These functions necessitate that the endothelium possess intact machinery for adherence and onset of an inflammation response.

This machinery is used by infectious agents such as bacteria and viruses to bind to the endothelium and trigger an inflammation-immune response. In general, bacterial pathogens colonize mucosal epithelium or skin epithelium, following which they can enter into the circulation. Once the pathogen or pathogen relevant toxins are released into the systemic circulation, these have access to endothelial cells (EC). The large interface afforded by the endothelium by virtue of its surface area (of $4000-7000 \text{ m}^2$ and consisting of 10^{13} endothelial cells in a typical adult human [2]) enables bacteria-induced endothelial dysfunction to pose as a severe health threat.

Bacteria and their toxins can, upon targeting the endothelium, produce severe pathologies, such as sepsis, vascular infections, septicemia, and chronic infections leading to the formation of atherogenic lesions. This targeting occurs via two major mechanisms: (1) direct interaction of the pathogen with the vascular wall and (2) indirect interaction via activation of the innate immune response by the toxins generated by the pathogen.

In this chapter, we will review the aforementioned processes by highlighting the mechanisms by which bacterial pathogens and toxins interact with endothelial cells and activate endothelial inflammatory responses. These responses alter endothelial cell cytoskeleton such that immune cell adherence is facilitated. These events will be discussed in the context of clinical manifestations of the endothelial response to infections.

17.1.1 Regulatory properties of the endothelium and the effect of infection

Thrombosis or blood clot is one of the major complications of infection. Thrombosis is closely associated with inflammation and occurs when the pathogen and/or its products trigger the activation of platelets, which cause endothelial injury leading EC to trigger fibrin formation, platelet adhesion, and aggregation, resulting in thrombus formation [3]). Under physiological conditions, the endothelium prevents thrombosis by providing a surface with antiplatelet and anticoagulant agents that prevents the attachment of cells and clotting proteins [4]. The endothelium regulates clot formation in part via its activation of the intravascular protease-activated receptors (PARs). Endothelial PARs serve as sensors for proteases and initiate a cascade of cell signals upon activation by thrombin, activator protein C (APC), tissue factor (TF/fVIIa/fXa) complex, high concentrations of plasmin, and /or matrix metalloproteases [5–7]. Once thrombin activates PAR-1, the endothelium produces nitric oxide and prostacyclin, which limits platelet activation. Thus PAR-1

Endothelial Signaling in Vascular Dysfunction and Disease. DOI: https://doi.org/10.1016/B978-0-12-816196-8.00003-5

plays an important role in the pro-coagulant response upon stimulation. An intact and healthy endothelium expresses various anticoagulants, such as tissue factor pathway inhibitor, thrombomodulin, endothelial protein C receptor, and heparin-like proteoglycans [8]. Endothelial cells also secrete ecto-nucleotidase CD39/NTPDase1, which metabolizes the platelet agonist adenosine diphosphate (ADP), in maintaining the anticoagulant environment [9]. With severe infection, the endothelial dysfunction leads to reduced antithrombin levels due to reduced synthesis leading to a pro-coagulant state.

Vasodilatory factors such as nitric oxide (NO), prostacyclin (PGI2), and endothelium-derived hyperpolarizing factor or vasoconstrictive factors such as thromboxane (TXA2) and endothelin-1(ET-1) are also produced in concert to maintain endothelial homeostasis. Overall, the balanced production of these vasoactive factors is atheroprotective. In the event of endothelial damage by bacteria or its products, there is a disruption in the production of these factors.

17.1.2   Endothelial activation by infection: access, adhesion, and inflammation

As mentioned earlier, the bloodstream serves as a conduit for dissemination of infectious agents. The endothelium by virtue of its location is an ideal site for these agents have to find ways to entry and exit of pathogens from the blood to distant target organs.

In general, the endothelium is not pervasive to the pathogen or directly accessed by the pathogen unless the pathogen is in the systemic circulation. The involvement of the lymphatics for dissemination is also less common. The vascular system despite well-regulated barrier function allows entry of pathogens via small-scale breaches like the gut-vasculature that allows the entry of *Salmonella typhimurium* [10]. As mentioned in the Introduction section, pathogens attack the endothelium by two major processes, either directly by attaching to the endothelial cell membrane or via release of pore-forming toxins with trigger inflammatory processes.

I. Direct interaction of the pathogen with the vascular wall: Various bacterial surface-bound adhesive molecules called microbial surface components recognizing adhesive matrix molecules (MSCRAMMs) mediate adhesion to tissue components by interacting with cell surface molecules. MSCRAMMs [11] recognize molecules such as fibronectin, fibrinogen, collagen, and von Willebrand factor (vWF). This is the case for bacteria such as *S. aureus* whose adherence to the endothelium of blood vessels is facilitated by vWF, a shear stress-operational protein released from endothelial cells upon activation [12].

Microorganisms such as fungi, bacteria, and viruses also use cellular adhesion molecules (CAMs) to establish contact with host endothelium. Bacteria such as *Mycobacterium tuberculosis*, *Listeria monocytogenes*, *Yersinia* spp, enteropathogenic *Escherichia coli*, *Shigella* spp, *Neisseria* spp, *Bordetella* spp, and *Borrelia burgdorferi*, RTX toxins of *Pasteurella haemolytica*, *Actinobacillus actinomycetemcomitans*, and the superantigen exotoxins of *Staphylococcus aureus* and *Streptococcus pyogenes* use CAMs for entry into cells. Additionally CAMs such as intercellular adhesion molecule (ICAM-1), vascular endothelial cell adhesion molecule (VCAM-1) and endothelial leukocyte adhesion molecule on endothelial cells mediate interactions with infected erythrocytes [13]. Binding to a endothelial cell receptor can trigger cytoskeletal rearrangements that will lead to bacterial internalization. Moreover, signals emanating from the occupied receptors can result in cellular responses such as gene expression events that influence the infected cell.

II. Activation of the Innate immune response by the toxins generated by the pathogen: The endotoxins produced by bacterial pathogens bind to receptors on the endothelium and activate an inflammation cascade. This cascade facilitates recruitment and adherence of leukocytes on the endothelium and alters endothelial phenotype, resulting in barrier dysfunction, increased leukocyte—endothelial interaction, mediator release, and pro-coagulant activity. Depending on the magnitude of the inflammatory response, the endothelium undergoes apoptosis or necrosis or even proliferation, indicating the onset and amplification of a complex interaction between the endotoxin and endothelial layer. These events will be discussed in detail in the next section (Fig. 17.1).

17.1.2.1   Inflammation with pathogenic attack

The central role of the endothelium for the inflammatory process in acute infections is highlighted by the endothelial contribution to the classical signs of infection: calor, dolor, rubor and tumor meaning heat, pain, redness, and swelling. When infection occurs primarily in the lung or in other tissues, endothelial malfunction has fundamental impact on the disease process and contributes in a multitude of ways to the pathologic process as discussed above. One very important manifestation is the loss of barrier function. Overcoming the endothelial barrier may be a key step in the distribution of pathogens into specially protected compartments, like the brain with its tight endothelial blood—brain barrier. Barrier dysfunction also has direct pathophysiologic consequences of clinical significance and is a prerequisite for the perilous

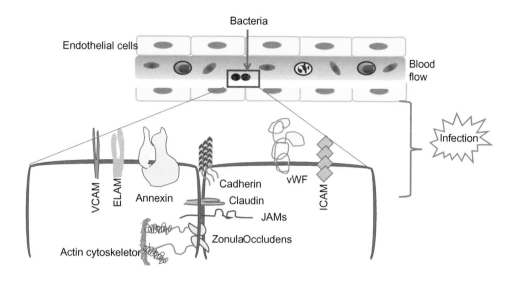

FIGURE 17.1 Schematic representation of the endothelium and bacterial attack. Inset is enlarged to show endothelial receptors for bacterial adhesion and internalization during infection. *VCAM*, vascular cell adhesion molecule; *ICAM*, intercellular adhesion molecule; *ELAM*, endothelial leucocyte adhesion molecule; *vWF*, von Willebrand factor; *JAM*, junctional adhesion molecule.

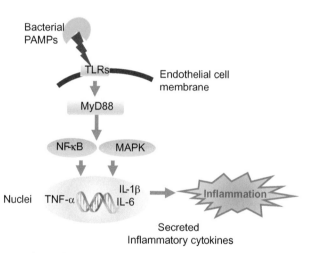

FIGURE 17.2   Schematic of the events associated with onset of inflammation upon pathogenic attack. *PAMPs*, pathogenic associated molecular patterns; *TLR*, toll-like receptors; *MyD88*, myeloid differentiation factor-88; *MAPK*, mitogen activated protein kinase family; *NFκB*, nuclear factor kappa B transcription factor; *TNF-α*, tumor necrosis factor α; *IL*, interleukin.

lung edema formation in diseases like acute respiratory distress syndrome (ARDS), frequently observed in sepsis [14–16]. Injury to endothelial cells in the lungs and brain results in manifestations of disease such as noncardiogenic pulmonary edema, interstitial pneumonia, adult respiratory distress syndrome, meningoencephalitis, and seizures [17]. Other severe manifestations can develop including acute renal failure, hemorrhagic phenomena, focal neurological deficits, peripheral edema, and hypovolemic hypotension due to leakage of intravascular fluid into the extravascular space.

As described earlier, inflammation is a biological response of the immune system that can be triggered by a variety of factors, including pathogens and their products. Microbial structures known as pathogen-associated molecular patterns (PAMPs) can trigger the inflammatory response through activation of germline-encoded pattern-recognition receptors (PRRs) expressed on endothelial and on immune cells [18,19]. Some PRRs also recognize endogenous signals activated during tissue or cell damage and are known as danger-associated molecular patterns (DAMPs) [19]. DAMPs are host biomolecules that can initiate and perpetuate a noninfectious inflammatory response [20] in disrupted cells by recruiting innate inflammatory response [21]. Classes of PRR families include the toll-like receptors (TLRs), C-type lectin receptors, retinoic acid-inducible gene-I-like receptors, and NOD-like receptors [22]. TLRs are a family of highly conserved, mammalian PRRs that participate in the activation of the inflammatory response [23]. Transmission of PAMPs and DAMPs is mediated by myeloid differentiation factor-88 (MyD88) along with TLRs. Signaling through TLRs activates an intracellular signaling cascade [24–26] that leads to nuclear translocation of transcription factors, such as activator protein-1 and NF-κB or interferon regulatory factor 3 (Fig. 17.2).

Inflammatory signaling pathways, most commonly the NF-κB, MAPK, and Janus kinases-signal transducer and activator of transcription proteins (JAK-STAT) pathways, are dysregulated with inflammation [27]. NF-κB activity is induced by a range of stimuli, including pathogen-derived substances, intercellular inflammatory cytokines, and many enzymes [28,29]. Under physiological conditions, IκB proteins present in the cytoplasm inhibit NF-κB. This pathway regulates pro-inflammatory cytokine production and inflammatory cell recruitment, which contribute to the inflammatory responses.

MAPKs are a family of serine/threonine protein kinases that direct cellular responses to a variety of stimuli, including osmotic stress, mitogens, heat shock, and inflammatory cytokines (such as IL-1, TNF-α, and IL-6), which regulate cell proliferation, differentiation, cell survival, and apoptosis [30,31].

The highly conserved JAK-STAT pathway involves diverse cytokines, growth factors, interferons, and related molecules, such as leptin and growth hormone, and is a signaling mechanism through which extracellular factors can control gene expression. Receptor-associated JAKs are activated by ligands and phosphorylate one other, creating docking sites for STATs, which are latent, cytoplasmic transcription factors. Cytoplasmic STATs recruited to these sites undergo phosphorylation and subsequent dimerization before translocation to the nucleus [32]. Tyrosine phosphorylation is essential for STAT dimerization and DNA binding [33]. Therefore, JAK/STAT signaling allows for the direct translation of an extracellular signal into a transcriptional response.

Together, these transcription factors regulate a variety of inflammatory genes, such as IL-1, TNF-α, IL-6, colony stimulating factor, interferons, transforming growth factor, and chemokines [34−36].

17.1.2.2 *Immune regulation by endothelial cells*

Endothelial-target infections represent patho-physiologically relevant models to study two important aspects of the immunological mechanisms of the endothelium, namely antigen presentation and antigen-specific interactions with T lymphocytes. The common patterns of endothelial response as assessed by in vitro studies show the acquisition of a spindled shape with expression of prominent actin stress fibers, increased endothelial cell permeability, activation of the inflammatory response through transcription factors including NFκB, and the consequent expression of cytokines, chemokines, and adhesion molecules like CAMs, ICAM, VCAM, and E-selectin for the recruitment of innate immune cells such as neutrophils. However, neutrophil infiltration is not a prominent feature of infections in vivo. Infected endothelial cells, particularly those in the microvasculature (arterioles, capillaries, and venules) have close contact with circulating T lymphocytes of the adaptive immune system. In general, T lymphocytes interact with antigen presenting cells such as dendritic cells. Antigen presenting cells express major histocompatibility complex (MHC class I and MHC class II) molecules that and can stimulate CD4$^+$ helper T cells as well as cytotoxic T cells to act against pathogens [37]. Normal human constitutive endothelial expression of MHC class I and class II molecules (only class I for mice) makes endothelial cells uniquely positioned for the function of antigen presentation [38]. This aspect of ECs acting as antigen presenting cells, has been investigated almost exclusively from the perspective of organ transplants [39−41]. In addition, the antigen-presenting function of the endothelium mediates antigen-specific translocation of T cells into the interstitium [42]. Endothelial cells express various costimulatory molecules through which they can activate primary immune T-cell responses such as CD58, CD80, CD86, CD2, CD40, CD154, CD134L, and ICOS-L [43−45].

17.1.3 Endothelial membrane reorganization, diapedesis

The leukocytes can find their way between two endothelial cells (the paracellular route). This is facilitated by the disruption of endothelial vascular endothelial (VE)-cadherin contacts, which form a paracellular gap through which the cells migrate. Alternatively, the leukocytes can transmigrate directly through individual endothelial cell (the transcellular route). In the latter case, the endothelial cell junctions remain intact and the membrane of neutrophils and endothelial cells fuse and remodel into a transcellular channel, forming a path for leukocytes, instead. The leukocyte extravasation is a highly regulated process that involves the engagement of complex interactions between the leukocyte and the endothelium, including via selectins, integrins, ICAM, junctional adhesion molecule (JAM), and platelet endothelial cell adhesion molecule (PECAM). These interactions are well coordinated and are known to occur in a sequential manner. Beyond promoting cell−cell interactions, adhesive molecules send bidirectional signaling from the leukocytes to the endothelial cells and vice versa that participate in the establishment of leukocyte polarity, their ability to crawl on the endothelium, and that are instrumental in guiding the mode of leukocyte diapedesis [46,47]. Endothelial (E)- and platelet (P)-selectin that are expressed on the endothelial apical surface upon inflammatory insults capture leukocytes and mediate their rolling onto the endothelium via leukocyte (L)-selectin. Subsequently, firm adhesion is controlled by adhesion receptors of the immunoglobulin family, namely leukocyte integrins [LFA-1 (lymphocyte function-associated antigen-1 also αLβ2 integrin or CD11a/CD18), Mac-1 (macrophage-1 antigen also αMβ2 integrin or CD11b/CD18), and VLA-4 (very late antigen-4 also α4β1 integrin)], which bind to their endothelial ligands, including ICAM (ICAM-1 and ICAM-2), and VCAM-1, respectively. Following firm adhesion, leukocytes adopt a polarized shape and crawl onto the endothelial apical surface in search for a permissive site of extravasation. Locomotion of leukocytes is strictly dependent on β2 integrins [48,49]. Arrest of leukocytes on the endothelium is mediated by a shift from intermediate affinity to high-affinity (HA) β2 integrins [50]. Adhesion molecules including leukocyte integrins and endothelial ICAM redistribute into dense clusters located at the leukocyte−endothelial cell interface and surrounding the cells [50]. These dense clusters stabilize and strengthen leukocyte−endothelial cell interactions. Following arrest and firm anchorage onto apical endothelial surface, the leukocytes flatten and adopt a highly polarized shape enabling their lateral migration or crawling for several microns on the vascular endothelium in search for permissive site of transmigration [48]. Leukocyte motility or crawling depends on asymmetric rearrangement of the leukocyte cytoskeleton in response to chemokines, which is coordinated with a dynamic cycle of assembly and disassembly of adhesive points binding the leukocyte to the endothelium. Interestingly, integrins not only provide dynamic adhesion points, they are also important to regulate the intracellular cytoskeleton and to maintain leukocyte polarization during crawling. Paracellular or junctional diapedesis is itself a multistep process, which is controlled by the sequential involvement of ICAM-1/2, VCAM-1, JAM-1/A/C, PECAM-1, CD99, and ESAM [51,52]. One essential component of the paracellular route is the opening of the endothelial junction. It has been established that leukocyte−endothelial cell interactions via ICAM-β2 integrin trigger the activation of signals to endothelial cells, which lead to the phosphorylation of VE-cadherin—a necessary step for loosening the adherent endothelial cell junctions and facilitating the passage of leukocytes [53]. Then, leukocytes migrate and cross the endothelial junction via sequential interactions with several adhesive molecules. JAM-A/C [54,55] and PECAM [56] are critical for leukocyte diapedesis.

Transcellular diapedesis is a fascinating process enabling leukocytes to cross the endothelial cell barrier away from the endothelial cell junctions. For this, the membrane of leukocytes and endothelial cells fuses to form a transcellular channel between the apical and basal membrane facilitating leukocyte transmigration while leaving the endothelial cell junctions intact [57]. Surprisingly, the adhesive molecules and mechanisms that guide transcellular migration are very similar to those controlling junctional migration. Like for paracellular migration, transcellular diapedesis is always preceded by ICAM-dependent lateral leukocyte crawling onto the endothelial surface during which the cells extend "scanning/invasive" protrusions [57−60]; the formation of a transmigratory cup made of ICAM-1 clusters and of docking structures as well as the recruitment of PECAM-1, CD99, and JAM-A to leukocyte−endothelial cell contact via the lateral border recycling compartment (LBRC) are also necessary for transcellular diapedesis [57,61−63].

The transcellular migration seems to prevail when the endothelial cell junctions are too tight, such as the blood−brain barrier [64,65]. Hence, it has become clear that the transcellular route is a regulated process in vivo. In this regard, several factors have recently been shown to favor transcellular migration, including the stiffness of endothelial cells, the tightness of endothelial cell junctions, or the density of integrin ligands at the endothelial apical surface; these factors will be discussed later [59,66,67].

17.1.4 Direct cytotoxic effects

Several toxins of pathogenic bacteria can hijack host inflammatory responses as well as the endothelial barrier function, inducing direct cytotoxic effects on the endothelial cell membrane specifically the actin cytoskeleton.

Induction of host cell death has been demonstrated in several cases of these bacterial infections. Most successful bacterial pathogens, both Gram-positive like *Staphylococcus* spp, *Bacillus* spp and Gram-negative like *Pseudomonas* spp must at some point interact with endothelium. Usually these pathogens have evolved toxins which can increase vascular permeability by either killing endothelial cells or by interfering with the cytoskeleton or cell−cell junctions. Often the toxins directly kill endothelium, often by forming pores in the plasma membrane of host cells. For example, pneumolysine, a cholesterol dependent cytolysin secreted by *Streptococcus pneunominae*, kills endothelial cells by inducing programmed cell death or apoptosis through p38 MAPK activation. There is increasing evidence that multiple types of programmed cell death play a central role in the complex balance among invading bacteria, the immune system, and host cells leading to inflammation and tissue damage in infections. Endothelial cells undergoing apoptosis typically display shrinkage, membrane blebbing, nuclear condensation, and DNA cleavage resulting in cellular fragmentation, with the cytoplasm retained in subcellular apoptotic bodies [68,69]. Another prominent example is α hemolysin (α-toxin, Hla) which is secreted by *Staphylococcos aureus*. This pore-forming cytotoxin results in barrier disruption and apoptosis [70,71]. *Bacillus anthracis* toxins lethal toxin and edema toxin are primary mediators of disease and cause apoptosis and necrosis of endothelial cells. Necrosis, unlike apoptosis, is not a programmed cell death but premature death of the cells by lack of oxygen or ischemia [72].

Beside simple killing, these bacterial toxins can also manipulate endothelial biology in various ways like (1) inducing calcium flux activating phospholipase A2, which has the potential to produce inflammatory mediators [73] and (2) causing its receptor, ADAM10 [74] to cleave VE-cadherin disrupting cell−cell junctions and contributing to sepsis severity. *Pseudomonas aeruginosa*, a Gram-negative bacterium, produces both type 2 and type 3 secretions systems (T2SS and T3SS). These systems secrete effectors that disrupt endothelial barrier integrity, causing devastating pathology in the lung. The T3SS can secrete different exoenzymes S, T, U, and Y (ExoS, ExoT, ExoU, and ExoY). ExoU directly lyses endothelium through its phospholipase activity, and this is partly achieved through the induction of oxidative stress [75,76]. ExoU can also induce a pro-thrombotic state, by causing ECs to release von Willebrand factor (vWF) and vWF-expressing microparticles, resulting in increased platelet adhesion and disrupting vascular flow [77].

17.2 Pathophysiological consequences of endothelial dysfunction with infection

Diverse pathogens target the endothelium using numerous pathologic mechanisms. The endothelium is both the target and a participant in the inflammation response against the attacking pathogen [78]. As mentioned earlier, the endothelial activation and pro-inflammatory responses are diverse and include mediator release [78], leukocyte recruitment [79], procoagulant activity [80,81], and breakdown of endothelial barrier function [82]. Indeed, markers of endothelial inflammation and damage such as ICAM-1, VCAM, E-selectin, IL-6, procalcitonin, thrombomodulin, and von Willebrand factor correlate with severity of the pathophysiology in sepsis [83]. These responses are general in nature, and a strict pathogen-specific response of the endothelium has not been elucidated thus far. In addition, the response of the endothelium to inflammatory stimuli may differ grossly among different microvasculature sites of the body. Importantly infection may disturb endothelial function, leading to either cell death, often realized by apoptosis, or proliferation. Such a critical location of lining the vasculature allows the endothelium to serve regulatory functions in angiogenesis, hemostasis, permeability and solute exchange, vascular tone, immunity, and inflammation [84−86]. Infection of endothelial cells affects all of these functions. Mechanisms that may participate in the increased vascular permeability observed in humans infected with *Rickettsia* or *Orientia* include endothelial detachment and denudation of vessels, the production of vasoactive prostaglandins as a consequence of increased expression of COX-2 [87], endothelial production of nitric oxide [88], effects of inflammatory cells and their mediators, and possible changes in the interendothelial junctions.

Besides bacteria and fungi a great variety of viruses infect endothelial cells. Viruses causing viral hemorrhagic fevers like Filoviridae (Marburg, Ebola), Arenaviridae (e.g., Lassa fever), Bunyaviridae (e.g., Rift Valley fever), or Flavivirus (Yellow fever), efficiently enter and replicate in endothelial cells [89]. Some of these viruses induce massive direct endothelial damage, for example, Rift Valley fever virus [89]. In all hemorrhagic fevers, the massive host response, with high levels of pro-inflammatory cytokines, contributes significantly to disease pathogenesis and endothelial malfunction [89,90]. Dengue-3 virus causing dengue or dengue hemorrhagic fever infects endothelial cells in vitro although their role as primary targets in vivo is controversial [91]. Interestingly, large-scale gene expression analysis was used to demonstrate that in vitro dengue virus infection resulted in a complex endothelial gene expression pattern, including stress, defense, immune, cell adhesion, wounding, inflammatory, as well as antiviral pathways [92]. Therefore, it has to be considered that dengue virus mediates endothelial activation by both direct infection and cytokine induction.

During shock and cardiovascular disease elevated levels of adhesion molecules could be a sign of endothelial damage. In patients with septic shock, elevated levels of adhesion molecules are found [93]. In peripheral artery disease and coronary artery disease, the levels of ICAM-1 are elevated. These levels could be predictors of thrombotic disorders [94] and ischemic heart disease [95], but results are still inconclusive. The levels of soluble(s) CAMs, such as sVCAM-1 and sPECAM-1 do not predict an adverse outcome in these diseases [96]. Still the finding of increased plasma concentrations of these endothelial surface adhesion molecules is thought to reflect the level of activation and perhaps damage of the endothelial cell [97,98]. In the clinical setting, measurement of E-selectin may be of diagnostic or prognostic value for diseases associated with endothelial pathology, since it is expressed only by vascular endothelial cells and not by any other cell types [99].

17.3 Clinical manifestations of endothelial response to infection

Acute vascular endothelial dysfunction plays the key role in the pathogenesis of infection manifested as sepsis. Sepsis causes acute endothelial dysfunction, inducing a pro-adhesive, pro-coagulant, and antifibrinolytic state in endothelial cells, altering hemostasis, leukocyte trafficking, inflammation, barrier function, and microcirculation [100]. Endothelial cells release the pro-coagulant glycoprotein tissue factor, whereas their synthesis of tissue factor pathway inhibitor is inhibited. The activation of platelets and the coagulation cascade causes microvascular thrombosis [101]. In addition, neutrophil extracellular traps (NETs) which are formed by endothelial membrane rupture provide a scaffold for thrombus formation, promoting hypercoagulability in patients with sepsis [100]. The association of tissue factor with NETs could target thrombin generation and fibrin clot formation at sites of infection/neutrophil activation, with active thrombin leading to increased platelet activation [102]. Acute vascular dysfunction and leakage contribute to hypotension, inadequate organ perfusion, local hypoxia, ischemia and ultimately, to organ failure, acute respiratory distress syndrome, shock and death in the most severe patients [103,104]. The main features of endothelial dysfunction shared by sepsis are (1) increased oxidative stress and systemic inflammation, (2) cell glycocalyx degradation and shedding, (3) disassembly of intercellular junctions, blood−tissue barrier disruption and cell death (4) enhanced leukocyte adhesion and extravasation, (5) induction of a pro-thrombosis and antifibrinolytic state [105]. From a clinical treatment perspective, in elderly patients or those with chronic diseases who suffer an infection, combining antimicrobials with drugs protecting the endothelium could prevent the development of sepsis or improve outcome once sepsis is present. Many potential treatment options like cholesterol lowering, application of antioxidants and angiotensin-converting enzyme inhibition to prevent or treat endothelial dysfunction have been proposed [106−110], but the most successful example of the early administration of intravenous agents has been that of vitamin C, corticosteroids, and thiamine that have been found to prevent progressive organ dysfunction and reduce mortality of patients with sepsis [111]. In that before and after clinical study, Marik et.al. compared the outcome and clinical course of consecutive septic patients treated with intravenous vitamin C, hydrocortisone, and thiamine during a 7-month period (treatment group) with a control group treated in the ICU during the preceding 7 months [111]. While the primary outcome was survival, the long-term effect of vitamin C on vascular function is not clear.

17.4 Summary and conclusions

Infections arising from bacterial and viral pathogens are complex but a common characteristic is their ability to infect the endothelium and lead to endothelial dysfunction. As described in the present chapter, endothelial dysfunction with infectious diseases is a multifactorial pathology that leads to changes in the basic morphology and function of the endothelium. These changes lead to loss of the endothelium barrier and the onset of an inflammatory environment. The endothelium is also involved in the pathogenesis of coagulation disorder following inflammation and is important for regulation of hemostasis. This may lead to an imbalance between platelet function and the regulatory mechanisms of the coagulation cascade and fibrinolysis resulting in bleeding or thrombosis or both and manifest themselves in capillary leakage, loss of endothelial vaso-regulation and thrombocytopenia.

These findings raise the possibility that the cumulative effect of infection could contribute to the development of atherosclerosis. Of course, there would be a role for genetic and environmental predisposing factors as well. Overall these are very premature observations and our current understanding of the short- and long-term impact of infection on the vascular network is very limited. However, the need to understand the events that are triggered on the endothelium postpathogen attack is crucial in informing the therapeutic targeting of the vasculature to restore homeostasis following infection.

A better understanding may open the way to new therapeutic modalities such as vaccination and antibiotic regimens. It is time to start clinical intervention trials to prevent the pro-thrombotic state. Major emphasis is put on the interaction between inflammation and coagulation. Knowledge of the underlying mechanisms leading to thrombosis or bleeding is fundamental for the development of therapeutic strategies. Given the potential role that endothelial injury plays in some of the thrombo hemorrhagic complications of inflammation, it has been hypothesized that intervention in the coagulation pathways may favorably alter the clinical course of these infections.

References

[1] Wilson SH, Lerman A. Function of vascular endothelium. In: Sperelakis N, Kurachi Y, Terzic A, Cohen MV, editors. Heart physiology and pathophysiology. fourth Edition Academic Press; 2001. p. 473−80. Chapter 27.

[2] Aird WC. Spatial and temporal dynamics of the endothelium. J Thrombosis Haemost 2005;3:1392−406.

[3] Beristain-Covarrubias N, Perez-Toledo M, Thomas MR, Henderson IR, Watson SP, Cunningham AF. Understanding infection-induced thrombosis: lessons learned from animal models. Front Immunol 2019;10:2569.

[4] Watson SP. Platelet activation by extracellular matrix proteins in haemostasisand thrombosis. Curr Pharm Des 2009;15:1358−72.

[5] Adams MN, Ramachandran R, Yau MK, Suen JY, Fairlie DP, Hollenberg MD, et al. Structure, function and pathophysiology of protease activated receptors. Pharmacol Ther 2011;130:248−82.

[6] Coughlin SR. Thrombin signalling and protease-activated receptors. Nature 2000;407:258−64.

[7] Coughlin SR. Protease-activated receptors in hemostasis, thrombosis andvascular biology. J Thromb Haemost 2005;3:1800−14.

[8] Esmon CT, Esmon NL. The link between vascular features and thrombosis. Annu Rev Physiol 2011;73:503−14.

[9] Yau JW, Teoh H, Verma S. Endothelial cell control of thrombosis. BMC Cardiovasc Disord 2015;15:130.

[10] Spadoni I, Zagato E, Bertocchi A, Paolinelli R, Hot E, Di Sabatino A, et al. A gut-vascular barrier controls the systemic dissemination of bacteria. Science 2015;350:830−4.

[11] Heying R, van de Gevel J, Que YA, Moreillon P, Beekhuizen H. Fibronectin-binding proteins and clumping factor A in *Staphylococcus aureus* experimental endocarditis: FnBPA is sufficient to activate human endothelial cells. Thromb Haemost 2007;97:617−26.

[12] Claes J, Vanassche T, Peetermans M, Liesenborghs L, Vandenbriele C, Vanhoorelbeke K, et al. Adhesion of *Staphylococcus aureus* to the vessel wall under flow is mediated by von Willebrand factor−binding protein. Blood 2014;124:1669−976.

[13] Ockenhouse CF, Tegoshi T, Maeno Y, Benjamin C, May Ho S, Kan KE, et al. Human vascular endothelial cell adhesion receptors for plasmoclium erythrocytes: roles for endothelial leukocyte adhesion molecule 1 and vascular cell adhesion molecule. J Exp Med 1992;176:1183−9.

[14] Ware LB, Matthay MA. The acute respiratory distress syndrome. N Eng J Med 2000;342:1334−49.

[15] Groeneveld ABJ. Vascular pharmacology of acute lung injury and acute respiratory distress syndrome. Vasc Pharmacol 2003;39:247−56.

[16] MacArthur RD, Miller M, Albertson T, Panacek E, Johnson D, Teoh L, et al. Adequacy of early empiric antibiotic treatment and survival in severe sepsis. Clin Infect Dis 2004;38:284−8.

[17] Walker DH, Valbuena GA, Olano JP. Pathogenic mechanisms of diseases caused by Rickettsia. Ann N Y Acad Sci 2003;990:1−11.

[18] Brusselle G, Bracke K. Targeting immune pathways for therapy in asthma and chronic obstructive pulmonary disease. Ann Am Thorac Soc 2014;11:S322−8.

[19] Gudkov AV, Komarova EA. p53 and the carcinogenicity of chronic inflammation. Cold Spring Harb Perspect Med 2016;6(11):a026161.

[20] Seong SY, Matzinger P. Hydrophobicity: an ancient damage-associated molecular pattern that initiates innate immune responses. Nat Rev Immunol 2004;4:469−478.

[21] Ozinsky A, Underhill DM, Fontenot JD, Hajjar AM, Smith KD, Wilson CB, et al. The repertoire for pattern recognition of pathogens by the innate immune system is defined by cooperation between toll-like receptors. Proc Natl Acad Sci U S A 2000;97:13766−71.

[22] Takeuchi O, Akira S. Pattern recognition receptors and inflammation. Cell 2010;140:805−20.

[23] Janeway Jr CA, Medzhitov R. Innate immune recognition. Annu rev immunol 2002;20:197−216.

[24] Czerkies M, Kwiatkowska K. Toll-like receptors and their contribution to innate immunity: focus on TLR4 activation by lipopolysaccharide. Adv Cell Biol 2014;4:1−23.

[25] Akira S, Takeda K, Kaisho T. Toll-like receptors: critical proteins linking innate and acquired immunity. Nat Immunol 2001;2:675−80.

[26] Chaudhary A, Ganguly K, Cabantous S, Waldo GS, Micheva-Viteva SN. The Brucella TIR-like protein TcpB interacts with the death domain of MyD88. Biochem Biophys Res Commun 2012;417:299−304.

[27] Oeckinghaus A, Hayden MS, Ghosh S. Crosstalk in NF-κB signaling pathways. Nat Immunol 2011;12:695−708.

[28] Pasparakis M, Luedde T, Schmidt-Supprian M. Dissection of the NF-κB signalling cascade in transgenic and knockout mice. Cell Death Differ 2006;13:861−72.

[29] Basak S, Kim H, Kearns JD, Tergaonkar V, O'Dea E, Werner SL, et al. A fourth IκB protein within the NF-κB signaling module. Cell 2007;128:369−81.

[30] Kaminska B. MAPK signalling pathways as molecular targets for anti-inflammatory therapy − from molecular mechanisms to therapeutic benefits. Biochim Biophys Acta Proteins Proteom 2005;1754:253−62.

[31] Pearson G, Robinson F, Beers GT, Xu BE, Karandikar M, Berman K, et al. Mitogen-activated protein (MAP) kinase pathways: regulation and physiological functions. Endocr Rev 2001;22:153−83.

[32] Walker JG, Smith MD. The Jak-STAT pathway in rheumatoid arthritis. J Rheumatol 2005;32:1650−3.

[33] Ivashkiv LB, Hu X. The JAK/STAT pathway in rheumatoid arthritis: pathogenic or protective? Arthritis Rheumatol 2003;48:2092−6.

[34] Iwasaki A, Medzhitov R. Toll-like receptor control of the adaptive immune responses. Nat Immunol 2004;5:987−95.

[35] Opitz B, Van LV, Eitel J, Suttorp N. Innate immune recognition in infectious and noninfectious diseases of the lung. Am J Resp Crit Care Med 2010;81:1294−309.

[36] Rahman I, Adcock IM. Oxidative stress and redox regulation of lung inflammation in COPD. Eur Respir J 2006;28:219−42.

[37] Kreisel D, Krupnick AS, Gelman AE, Engels FH, Popma SH, Krasinskas AM, et al. Non-hematopoietic allograft cells directly activate CD8 + T cells and trigger acute rejection: an alternative mechanism of allorecognition. Nat Med 2002;8:233−9.

[38] Pober JS. Immunobiology of human vascular endothelium. Immunol Res 1999;19:225−32.

[39] Epperson DE, Pober JS. Antigen-presentingfunctionofhumanendothelialcells. Direct activation of resting CD8 T cells. J Immunol 1994;153:5402−12.

[40] Savage CO, Brooks CJ, Harcourt GC, Picard JK, King W, et al. Human vascular endothelial cells process and present autoantigen to human T cell lines. Int Immunol 1995;7:471−9.

[41] Valujskikh A, Heeger PS. Emerging roles of endothelial cells in transplant rejection. Curr Opin Immunol 2003;15:493−8.

[42] Marelli-Berg FM, James MJ, Dangerfield J, Dyson J, Millrain M, Scott D, et al. Cognate recognition of the endothelium induces HY-specific CD8 + T-lymphocyte transendothelial migration (diapedesis) in vivo. Blood 2004;103:3111−16.

[43] Smith ME, Thomas JA. Cellular expression of lymphocyte function associated antigens and the intercellular adhesion molecule-1 in normal tissue. J Clin Pathol 1990;43:893−900.

[44] Karmann K, Hughes CC, Schechner J, Fanslow WC, Pober JS. CD40 on human endothelial cells: inducibility by cytokines and functional regulation of adhesion molecule expression. Proc Natl Acad Sci U S A 1995;92:4342−6.

[45] Choi J, Enis DR, Koh KP, Shiao SL, Pober JS. T lymphocyte-endothelial cell interactions. Annu Rev Immunol 2004;22:683−709.

[46] Herter J, Zarbock A. Integrin regulation during leukocyte recruitment. J Immunol 2013;190:4451−7.

[47] Muller WA. Mechanisms of leukocyte transendothelial migration. Annu Rev Pathol 2011;6:323−44.

[48] Phillipson M, Heit B, Colarusso P, Liu L, Ballantyne CM, Kubes P. Intraluminal crawling of neutrophils to emigration sites: a molecularly distinct process from adhesion in the recruitment cascade. J Exp Med 2006;203:2569−75.

[49] Schenkel AR, Mamdouh Z, Muller WA. Locomotion of monocytes on endothelium is a critical step during extravasation. Nat Immunol 2004;5:393−400.

[50] Shaw SK, Ma S, Kim MB, Rao RM, Hartman CU, Froio RM, et al. Coordinated redistribution of leukocyte LFA-1 and endothelial cell ICAM-1 accompany neutrophil transmigration. J Exp Med 2004;200:1571−80.

[51] Muller WA. Getting leukocytes to the site of inflammation. Vet Pathol 2013;50:7−22.

[52] Nourshargh S, Hordijk PL, Sixt M. Breaching multiple barriers: leukocyte motility through venular walls and the interstitium. Nat Rev Mol Cell Biol 2010;11:366−78.

[53] Vestweber D. VE-cadherin: the major endothelial adhesion molecule controlling cellular junctions and blood vessel formation. Arterioscler Thromb Vasc Biol 2008;28:223−32.

[54] Woodfin A, Reichel CA, Khandoga A, Corada M, Voisin MB, Scheiermann C, et al. JAM-A mediates neutrophil transmigration in a stimulus-specific manner in vivo: evidence for sequential roles for JAM-A and PECAM-1 in neutrophil transmigration. Blood 2007;110:1848−56.

[55] Woodfin A, Voisin MB, Imhof BA, Dejana E, Engelhardt B, Nourshargh S. Endothelial cell activation leads to neutrophil transmigration as supported by the sequential roles of ICAM-2, JAM-A, and PECAM-1. Blood 2009;113:6246−57.

[56] Muller WA, Weigl SA, Deng X, Phillips DM. PECAM-1 is required for transendothelial migration of leukocytes. J Exp Med 1993;178:449−60.

[57] Carman CV, Sage PT, Sciuto TE, de la Fuente MA, Geha RS, Ochs HD, et al. Transcellular diapedesis is initiated by invasive podosomes. Immunity 2007;26:784−97.

[58] Gorina R, Lyck R, Vestweber D, Engelhardt B. β2 integrin-mediated crawling on endothelial ICAM-1 and ICAM-2 is a prerequisite for transcellular neutrophil diapedesis across the inflamed blood-brain barrier. J Immunol, 192. 2014. p. 324−37.

[59] Martinelli R, Zeiger AS, Whitfield M, Sciuto TE, Dvorak A, Van Vliet KJ, et al. Probing the biomechanical contribution of the endothelium to lymphocyte migration: diapedesis by the path of least resistance. J Cell Sci 2014;127(Pt 17):3720−34.

[60] Shulman Z, Shinder V, Klein E, Grabovsky V, Yeger O, Geron E, et al. Lymphocyte crawling and transendothelial migration require chemokine triggering of high-affinity LFA-1 integrin. Immunity 2009;30:384−96.

[61] Carman CV, Jun C, Salas A, Springer TA. Endothelial cells proactively form microvilli-like membrane projections upon intercellular adhesion molecule 1 engagement of leukocyte LFA-1. J Immunol, 171. 2003. p. 6135−44.

[62] Mamdouh Z, Mikhailov A, Muller WA. Transcellular migration of leukocytes is mediated by the endothelial lateral border recycling compartment. J Exp Med 2009;206:2795−808.

[63] Millán J, Hewlett L, Glyn M, Toomre D, Clark P, Ridley AJ. Lymphocyte transcellular migration occurs through recruitment of endothelial ICAM-1 to caveola- and F-actin-rich domains. Nat Cell Biol 2006;8:113−23.

[64] Lossinsky AS, Shivers RR. Structural pathways for macromolecular and cellular transport across the blood−brain barrier during inflammatory conditions. Histol Histopathol 2004;19:535−64.

[65] Wolburg H, Wolburg-Buchholz K, Engelhardt B. Diapedesis of mononuclear cells across cerebral venules during experimental autoimmune encephalomyelitis leaves tight junctions intact. Acta Neuropathol 2005;109:181−90.

[66] Schaefer A, Te Riet J, Ritz K, Hoogenboezem M, Anthony EC, Mul FP, et al. Actin-binding proteins differentially regulate endothelial cell stiffness, ICAM-1 function and neutrophil transmigration. J Cell Sci 2014;127(Pt 20):4470−82.

[67] Yang L, Froio RM, Sciuto TE, Dvorak AM, Alon R, Luscinskas FW. ICAM-1 regulates neutrophil adhesion and transcellular migration of TNF-alpha-activated vascular endothelium under flow. Blood 2005;106:584−92.

[68] Bermpohl D, Halle A, Freyer D, Dagand E, Braun JS, Bechmann I, et al. Bacterial programmed cell death of cerebral endothelial cells involves dual death pathways. J Clin Invest 2005;115:1607−15.

[69] Parthasarathy G, Philipp M. Review: apototic mechanism in bacterial infections of the central nervous system. Front Immunol 2012;3:306.

[70] Berube B, Wardenburg JB. *Staphylococcus aureus* alpha toxin: nearly a century of intrigue. Toxin (Basel) 2013;5:1140−66.

[71] Prevost G, Mourey L, Colin D, Monteil H, Dalla Serra M, Menestrina G. Alpha-helix and beta-barrel pore-forming toxins of *Staphylococcus aureus*. In: Alouf JE, Freer JH, editors. The comprehensive sourcebook of bacterial toxins. London: Academic Press; 2005. p. 590−607.

[72] Kirby JE. Anthrax lethal toxin induces human endothelial cell apoptosis. Infect Immun 2004;72:430−9.

[73] Rubins JB, Mitchell TJ, Andrew PW, Niewoehner DE. Pneumolysin activates phospholipase A in pulmonary artery endothelial cells. Infect Immun 1994;62:3829−36.

[74] Powers ME, Kim HK, Wang Y, Bubeck Wardenburg J. ADAM10 mediates vascular injury induced by *Staphylococcus aureus* alpha-hemolysin. J Infect Dis 2012;206:352−6.

[75] Phillips RM, Six DA, Dennis EA, Ghosh P. In vivo phospholipase activity of the *Pseudomonas aeruginosa* cytotoxin ExoU and protection of mammalian cells with phospholipase A2 inhibitors. J Biol Chem 2003;278:41326−32.

[76] Saliba AM, de Assis MC, Nishi R, Raymond B, Marques Ede A, Lopes UG, et al. Implications of oxidative stress in the cytotoxicity of *Pseudomonas aeruginosa* ExoU. Microbes Infect 2006;8:450−9.

[77] Freitas C, Assis MC, Saliba AM, Morandi VM, Figueiredo CC, Pereira M, et al. The infection of microvascular endothelial cells with ExoU-producing *Pseudomonas aeruginosa* triggers the release of von Willebrand factor and platelet adhesion. Mem Inst Oswaldo Cruz 2012;107:728−34.

[78] Hack CE, Zeerleder S. The endothelium in sepsis: source of and a target for inflammation. Crit Care Med 2001;29:$21−$27.

[79] Granger DN, Kubes, P. The microcirculation and inflammation: modulation of leukocyte-endothelial cell adhesion. J. Leukoe. Biol. 1994;55:662−675.

[80] Keller TT, Mairuhu A, de Kruif M, Klein S, Gerdes V, ten Cate H, et al. Infections and endothelial cells. Cardiovasc Res 2003;60:40−8.

[81] Peters K, Unger R, Brunner J, Kirkpatrick CJ. Molecular basis of endothelial dysfunction in sepsis. Cardiovasc Res 2003;60:49−57.

[82] Stevens T, Garcia JG, Shasby DM, Bhattacharya J, Malik AB. Mechanisms regulating endothelial cell barrier function. Am J Physiol Lung Cell Mol Physiol 2000;279:L419−22.

[83] Reinhart K, Bayer O, Brunkhorst F, Meisner M. Markers of endothelial damage in organ dysfunction and sepsis. Crit CareMed 2002;30: $302−$312.

[84] Cines DB, Pollak ES, Buck CA, Loscalzo J, Zimmerman GA, McEver RP, et al. Endothelial cells in physiology and in the pathophysiology of vascular disorders. Blood 1998;91:3527−61.

[85] Michiels C. Endothelial cell functions. J Cell Physiol 2003;196:430−43.

[86] Danese S, Dejana E, Fiocchi C. Immune regulation by microvascular endothelial cells: directing innate and adaptive immunity, coagulation, and inflammation. J Immunol 2007;78:6017−22.

[87] Rydkina E, Sahni A, Baggs RB, Silverman DJ, Sahni SK. Infection of human endothelial cells with spotted fever group rickettsiae stimulates cyclooxygenase 2 expression and release of vasoactive prostaglandins. Infect Immun 2006;74:5067−74.

[88] Woods ME, Wen G, Olano JP. Nitric oxide as a mediator of increased microvascular permeability during acute rickettsioses. Ann N Y Acad Sci 2005;1063:239−45.

[89] Schnittler HJ, Feldmann H. Molecular pathogenesis of filovirus infections: role of macrophages and endothelial cells. Curr Top Microbiol Immunol 1999;235:175−204.

[90] Peters CJ. Human infection with a renaviruses in the Americas. Curr Top Microbiol Immunol 2002;262:65−74.

[91] Halstead SB. Dengue. Curr Opin Infect Dis 2002;15:471−6.

[92] Warke RV, Xhaja K, Martin K, Fournier M, Shaw S, Brizuela N, et al. Dengue virus induces novel changes in gene expression of human umbilical vein endothelial cells. J Virol 2003;77:11822−32.

[93] Newman W, Beall LD, Carson CW, Hunder GG, Graben N, Randhawa ZI, et al. Soluble E-selectin is found in supernatants of activated endothelial cells and is elevated in the serum of patients with septic shock. J Immunol 1993;150:644−54.

[94] Chong BH, Murray B, Berndt MC, Dunlop LC, Brighton T, Chesterman CN. Plasma P-selectin is increased in thrombotic consumptive platelet disorders. Blood 1994;83:1535−41.

[95] Blann AD, Amiral J, McCollum CN. Prognostic value of increased soluble thrombomodulin and increased soluble E-selectin in is- chaemic heart disease. Eur J Haematol 1997;59:115−120.

[96] Blann AD, Wadley MS, Dobrotova M, Sanders P, Jayson MI, McCollum CN. Soluble platelet endothelial cell adhesion molecule-1 (sPECAM-1) in inflammatory vascular disease, atherosclerotic vascular disease, and in cancer. Blood Coagul Fibrinolysis 1998;9:99−103.

[97] Kayal S, Jais JP, Aguini N, Chaudiere J, Labrousse J. Elevated circulating E-selectin, intercellular adhesion molecule 1, and von Willebrand factor in patients with severe infection. Am J Respir Crit Care Med 1998;157:776−84.

[98] Endo S, Inada K, Kasai T, Takakuwa T, Yamada Y, Koike S, et al. Levels of soluble adhesion molecules and cytokines in patients with septic multiple organ failure. J Inflamm 1995;46:212−19.

[99] McGill SN, Ahmed NA, Christou NV. Endothelial cells: role in infection and inflammation. World J Surg 1998;22:171−8.

[100] Hattori Y, Hattori K, Suzuki T, Matsuda N. Recent advances in the pathophysiology and molecular basis of sepsis-associated organ dysfunction: novel therapeutic implications and challenges. Pharmacol Ther 2017;177:56−66.

[101] Colbert JF, Schmidt EP. Endothelial and microcirculatory function and dysfunction in sepsis. Clin Chest Med 2016;37:263−75.

[102] Gardiner EE, Andrews RK. Neutrophil extracellular traps (NETs) and infection-related vascular dysfunction. Blood Rev 2012;26:255−9.

[103] Pool R, Gomez H, Kellum JA. Mechanisms of organ dysfunction in sepsis. Crit Care Clin 2018;34:63−80.

[104] Crouser ED, Matthay MA. Endothelial damage during septic shock: significance and implications for future therapies. Chest 2017;152:1−3.

[105] Bermejo-Martin JF, Martín-Fernandez M, López-Mestanza C, Duque P, Almansa R. Shared features of endothelial dysfunction between sepsis and its preceding risk factors (aging and chronic disease). J Clin Med 2018;7:400.

[106] Daiber A, Steven S, Weber A, Shuvaev VV, Muzykantov VR, Laher I, et al. Targeting vascular (endothelial) dysfunction. Br J Pharmacol 2017;174:1591−619.

[107] Anderson TJ. Assesment and treatment of endothelial dysfunction in humans. J Am Coll Cardiology 1999;34:631−8.

[108] Darwish I, Liles WC. Emerging therapeutic strategies to prevent infection-related microvascular endothelial activation and dysfunction. Virulence 2013;4:572−82.

[109] Opal SM, van der Poll T. Endothelial barrier dysfunction in septic shock. J Intern Med 2015;277:277−93.

[110] Tarbell JM, Cancel LM. The glycocalyx and its significance in human medicine. J Intern Med 2016;280:97−113.

[111] Marik PE, Khangoora V, Rivera R, Hooper MH, Catravas J. Hydrocortisone, vitamin c, and thiamine for the treatment of severe sepsis and septic shock: a retrospective before-after study. Chest 2017;151:1229−38.

Chapter 18

Coronavirus disease (COVID-19) and the endothelium

Oindrila Paul and Shampa Chatterjee

Department of Physiology, Institute for Environmental Medicine, University of Pennsylvania Perelman School of Medicine, Philadelphia, PA, United States

18.1 Introduction

In late December 2019 an outbreak of a novel coronavirus (severe acute respiratory syndrome coronavirus 2 or SARS-CoV-2) disease causing severe pneumonia was reported in Wuhan, Hubei Province, China [1]. As this article is being written (July 2020), the coronavirus disease (COVID-19) has affected over 11 million people in at least 213 countries worldwide with most of the cases being reported from the United States of America, Brazil, Europe, China, and India [2]. The absolute number of deaths has already surpassed 530,000 globally and is expected to increase further as the disease spreads rapidly. Although the overall mortality rate of COVID-19 is low (1.4%−2.3%), patients with comorbidities are more likely to have severe disease and subsequent mortality [3,4].

COVID-19 has been well established to be associated with a high rate of respiratory tract involvement leading to acute respiratory distress syndrome (ARDS), which requires ventilation in critically ill patients [5]. Pneumonia has been frequently observed in these patients with high mortality rate for those in the intensive care unit (ICU) [6]. In China and the United States, 80% had mild illness, but of the rest that were hospitalized for COVID-19, a quarter of cases needed ICU care. Patients in the ICU showed a mortality rate of ∼49%, whereas those with ARDS had a mortality of 52.4% [4,7].

The pathophysiology of ARDS is characterized by acute inflammation within the alveolar space leading to alveolar edema, which prevents normal gas exchange and causes hypoxemia. Mechanical ventilation is the main therapeutic intervention in the management of ARDS as there are no pharmacologic interventions [8]. Current epidemiological studies on COVID-19 show the emergence of cardiovascular complications including fulminant myocarditis, myocardial injury, heart failure, and arrhythmia [3,9,10]. Additionally, other associated pathologies such as venous and arterial thromboembolism and diffuse intravascular coagulation have been reported [11,12]. These events seem to suggest that the virus is targeting the endothelium.

The vascular endothelium is an active paracrine, endocrine, and autocrine organ. The endothelium that lines the entire vascular network is one of the largest organs in the human body. It plays a pivotal role in maintenance of vascular tone and in vascular homeostasis [13,14]. Furthermore, it is a converging site for inflammation as it can drive recruitment and adherence of immune cells.

Therefore alteration to the endothelium can lead to vasoconstriction that drives ischemia; additionally endothelial damage can lead to a proinflammatory and pro-coagulant phenotype that leads to an aggravated endothelial inflammation response (endotheliitis) and cell death (apoptosis) [15,16]. Reports suggest that SARS-CoV-2 virus directly infects endothelial cells and facilitates the induction of endotheliitis [17]. The presence of viral inclusions in endothelial cells, as reported in small autopsy series of patients with COVID-19, has raised suspicion for endothelial cell injury or activation as a central feature of the pathophysiology of COVID-19, particularly during the inflammatory phase of the disease [17,18].

Mechanistically, the pulmonary complications result from endothelial injury, disintegration of the vascular barrier leading to tissue edema (causing lungs to build up fluid), endotheliitis, and eventual activation of coagulation pathways. Indeed, several studies have also reported evidence of a COVID-19 associated coagulopathy [19]. The severity of this coagulopathy is clear from a cohort study where 71% of patients who eventually died from the disease matched the

Endothelial Signaling in Vascular Dysfunction and Disease. DOI: https://doi.org/10.1016/B978-0-12-816196-8.00008-4

International Society on Thrombosis and Haemostasis (ISTH) criteria for disseminated intravascular coagulation, while this percentage was only 0.6% in patients who survived [20]. As the role of endothelium is central to intravascular coagulation, it seems that endothelial cells play a central role in coagulation associated with COVID-19.

This chapter summarizes molecular pathogenesis of COVID-19 with focus on its interaction with the endothelium; it also provides a reference for contribution of endothelial dysfunction (EDF) to disease progression. Finally we review recent literature for the prevention and drug development of SARS-CoV-2 infection based on both recent research progress of SARS-CoV-2 and past findings on other coronavirus infections (SARS-CoV and middle east respiratory syndrome coronavirus (MERS-CoV)).

18.2 Severe acute respiratory syndrome coronavirus 2 infection

SARS-CoV-2 is transmitted primarily via respiratory droplets. Once in contact with the host, the Coronavirus Spike (S) protein is a significant determinant of virus entry into host cells. A recent study has confirmed that SARS-CoV-2 uses severe acute respiratory syndrome coronavirus (SARS-CoV) receptor angiotensin-converting enzyme 2 (ACE2) for host cell entry [21]. The ACE2 receptor serves as a binding site for the receptor-binding domain (RBD) domain of the S1 protein (one of the two subunits of the S protein). The RBD binding to ACE2 leads to endocytosis of the SARS-CoV-2 virus. Within the endosome, the S1 subunit is cleaved away by exposing the fusion peptide of the S2 subunit that inserts itself into the host cell membrane. ACE2 receptor that facilitates virus entry into host cells is expressed in several organs, including the lung, heart, kidney, and intestine; ACE2 receptors are also expressed by endothelial cells [22,23]. The exact role of endothelial cells in viral infection is not clear; however, in vitro studies have reported that SARS-CoV-2 can directly infect engineered human blood vessel organoids [24].

Although the pathogenesis of COVID-19 is poorly understood, the mechanisms of pathogenesis, action, and clinical manifestations seem to show some similarity with SARS-CoV and MERS-CoV; thus fever, cough, dyspnea, fatigue, decreased leukocyte counts, and radiographic evidence of pneumonia [6,25] which are observed with COVID-19 are similar to the symptoms of SARS-CoV and MERS-CoV infections [26]. However, despite the similarity (there is 72% similarity in the RBD sequence of SARS-CoV and SARS-CoV-2) there is some difference in terms of the enhanced binding ability of SARS-CoV-2 to ACE receptor [27−29] that leads to its increased infectivity as compared to SARS-CoV.

18.3 Endothelium and COVID-19

Injury or damage to the endothelium leads to alteration in its function; EDF refers to a systemic condition in which the endothelium loses its physiological properties, including the tendency to promote vasodilation, control inflammation, and aggregation. EDF is manifested in patients with diabetes, hypertension, and several other comorbidities [30]. In these patients, the balance in the coagulation and fibrinolytic systems as also the inflammatory and antiinflammatory signaling mechanisms is impaired. Therefore maintaining blood fluidity and restoration of vessel wall integrity to avoid conditions such as bleeding are compromised.

Among the distinctive features of COVID-19 are the vascular changes associated with the disease. It is now well established that while the initial symptoms of SARS-CoV-2 infection are due to viral infection, the late-stage disease exaggeration is a consequence of immune-mediated injury induced as a result of the body's response to the virus. While the virus uses ACE2 receptor expressed by epithelial alveolar cells to infect the host, it also infects the ACE2 receptor that is widely expressed on endothelial cells. Indeed SARS-CoV-2 has been reported to directly infect engineered human blood vessel organoids in vitro [24].

Postmortem histology of tissue from COVID-19 patients also revealed endotheliitis (infection to the endothelium) in lung, heart, kidney, and liver [17]. Inflammatory cells were found around the endothelium indicating that either direct viral infection of the endothelium or immune-mediated processes lead to recruitment of immune cells into the vascular wall. Over an extended period this can result in widespread EDF associated with apoptosis. The extravasation of immune cells along with overproduction of early-response proinflammatory cytokines [tumor necrosis factor (TNF), interleukin (IL)-6, and IL-1β] also defined as a "cytokine storm" can lead to EDF via oxidative injury. These events cause vascular leakage and inflammation programmed cell death or pyroptosis. Vascular hyperpermeability leads to leakage of fluid and proteins from blood vessels to the interstitial space in organs causing tissue edema, which often leads to multiple organ failure.

The sequence of events is outlined in Fig. 18.1. First, the virus directly attacks epithelial cells followed by infecting the endothelial cells (were detected in several organs of deceased patients [18]). Second, it (SARS-CoV-2) enters the

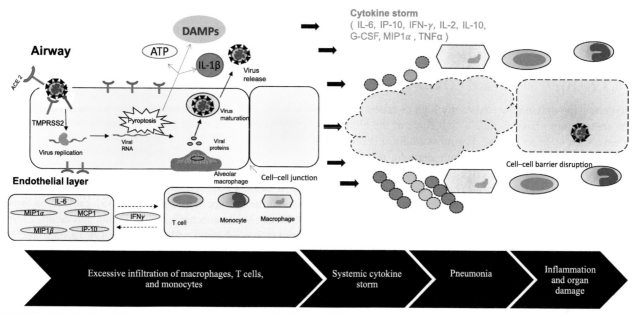

FIGURE 18.1 SARS-CoV-2 infection in the lung: SARS-CoV-2 binds to ACE2 receptor and TMPRSS2 on lung epithelial cells. The viral replication occurs inside the cell. The inflammation programmed cell death (pyroptosis) leads to release of ATP, IL-1β, and damage-associated molecular patterns (DAMPs). These trigger a proinflammatory phenotype by driving a cytokine storm. This causes the recruitment of T cells, monocytes, macrophages, and various other immune cells into the infection site triggering further production of cytokines, chemokines, and oxidants. His feedforward loop eventually damages the lung structure by destroying cell–cell junction and increasing permeability that leads to lung edema. The cytokine storm in the systemic circulation affects other organs as the vascular cells are in contact with systemic blood.

endothelial cells via binding to the ACE2 receptor that is well expressed by the endothelium. Third, in response to the infection, activated neutrophils are recruited, which produce reactive oxygen species (ROS) and inflammatory mediators (cytokines, chemokines). Finally, these inflammatory cytokines enhanced endothelial contractility leading to disruption of the endothelial junctions. This alters vessel barrier integrity causing leakage and edema. This also promotes a procoagulative state leading to formation to mini-thrombi [31]. Furthermore, the endothelial gaps allow for infiltration of immune cells and diffusion of inflammatory mediators.

18.3.1 Inflammation and COVID-19

The major cause for disease severity in patients is the host response to the viral infection (and not the viral infection per se) [31]. The SARS-CoV-2 infection triggers an immune response via viral DNA and other proteins (pathogen-associated molecular patterns or PAMPs). Pattern recognition receptors expressed on alveolar epithelial and endothelial cells detect viral DNA and viral PAMPs and trigger a signaling cascade that initiates an aggressive inflammation response in the form of a "cytokine storm." The cytokine storm is an uncontrolled systemic inflammatory response resulting from the release of large amounts of proinflammatory cytokines [interferons (IFN-α, IFN-γ), interleukins (IL-1β, IL-6, IL-12, IL-18, IL-33), TNF-alpha (TNF-α), transforming growth factor-beta, etc.] and chemokines (CCL2, CCL3, CCL5, CXCL8, CXCL9, CXCL10). The cytokine storm activates a violent attack by the immune system causing recruitment of monocytes, neutrophils, macrophages, and T lymphocytes to the site of infection [32]. The endothelium as the converging site for all these immune cells that are recruited from the circulation to clear the infection via extravasation into tissue is thus subjected to injury. In most cases, the recruited inflammation cells can clear the infection in the lung, as the immune response recedes and patients recover. However, in some patients, a dysfunctional immune response occurs, which triggers a cytokine storm that mediates widespread lung endothelial inflammation and injury.

The endothelial and epithelial proinflammatory phenotype associated with COVID-19 also leads to inflammation programmed cell death (pyroptosis) [33]. Pyroptosis leads to increase in IL-1β, which is associated with vascular leakage. It was observed that patients with severe COVID-19 exhibited higher blood plasma levels of IL-2, IL-7, IL-10, granulocyte colony-stimulating factor (G-CSF), Interferon-inducible protein (IP-10), monocyte chemoattractant protein (MCP-1), macrophage inflammatory protein 1α (MIP1α). Severe COVID-19 patients showed monocyte-derived FCN1 + macrophage population in their bronchoalveolar lavage fluid and a significantly higher percentage of

CD14 + CD16 + inflammatory monocytes in their peripheral blood as compared to patients with mild disease. Thus uncontrolled inflammation can damage the lung via pyroptosis and via inflammatory cell infiltration can damage the lung through excessive secretion of proteases and ROS.

18.3.2   Hypertension and COVID-19

Hypertension defined as persistently high or raised blood pressure, is a condition in which the blood vessels (and vascular endothelium) have to face a high force of the blood against the artery such that it may eventually cause health problems, such as heart disease. For most adults, a healthy blood pressure is usually less than 120 over 80 mm of mercury (systolic pressure reading over your diastolic pressure reading—120/80 mmHg). In general, hypertension is known to be one of the most common diseases and considered a silent killer for worldwide population. Currently, hypertension is emerging as one of the strongest predictors of COVID-19-related death [3,33]. According to available clinical data, $\approx 15\%-30\%$ of the COVID-19 patients have comorbidities in the form of hypertension [6,10,34,35]. Further, in terms of disease severity and a "composite outcome" as assessed by ICU admission, mechanical ventilation, and death, 23.7% of hypertensive patients had disease severity (vs 13.4% of normotensive subjects) and that 35.8% (vs 13.7%) reached the ICU admission, mechanical ventilation, and death [3].

Numerous studies till date show a high rate of hypertensive patients in COVID-19; indeed a prospective analysis on 41 patients admitted to hospital in Wuhan [36] and another study conducted on 138 hospitalized patients with confirmed COVID-19 infection [6] showed that 58% of the COVID-19 patients with the hypertensive state required ICU admission as compared to 21% of COVID-19 affected that were normotensive. The link between hypertension and the worse prognosis in patients with COVID-19 infection could arise from a cause-effect mechanism rather than a casual preexisting association between the two different diseases.

Another concern has been the role that routine drugs against hypertension may have in facilitating COVID-19 infections. As is well known, ACE-2 serves as the receptor for SARS-CoV-2 to gain entry into airway cells. At the same time, patients with hypertension are routinely treated with ACE inhibitors (ACEIs) and angiotensin receptor blockers (ARBs). It has been found that ACEIs and angiotensin II receptor blockers (ARBs) cause an overexpression of ACE2 in patients treated with these drugs [37]. This could increase the possibility of susceptibility to these individuals to COVID-19. However, a recent study from China, on hypertensive patients treated with renin-angiotensin system inhibitors were evaluated for the risk of death from COVID-19 [38], showed that in an overall hospital mortality of 11%, the in-hospital mortality of hypertensive patients was 21.3% but did not differ significantly from those hypertensive patients taking ACEIs/ARBs either in terms of the severity of infections or in survival rates when compared to patients not taking these medications.

Overall, these findings confirm a dual aspect of hypertension during COVID-19 pandemic: first, that hypertension is the most common co-morbidity observed in COVID-19 patients; second, that hypertension is evidenced in patients with worse prognosis and higher rate of death.

18.3.3   Diabetes and COVID-19

Diabetes mellitus (DM) has been associated with ARDS [39]. It has also been reported to be associated with poor prognosis in other viral infections, notably seasonal influenza, pandemic influenza A H1N1 (2009), SARS, and Middle East respiratory syndrome [40−42].

In the context of COVD-19, reports by Guan et al. from China showed that patients with severe disease had a higher prevalence of DM (16.2%) as compared to those with nonsevere disease (5.7%) [3]. In a study using a huge cohort comprising 72,314 cases of COVID-19 (conducted by the Chinese Center for Disease Control and Prevention) patients with DM had higher mortality as compared to nondiabetic patients (7.3% in DM vs 2.3% overall) [4].

Among the causes for this association is hyperglycemia and exaggerated production of inflammatory cytokines such as IL-1β and IL-6 in DM afflicted populations. When the endothelium is exposed to hyperglycemia, it causes EDF so that the endothelium loses its properties of promoting vasodilation, fibrinolysis, and antiaggregation. These lead to a proinflammatory phenotype and apoptosis. The production of inflammatory cytokines in the DM affected causes a dysregulated immune response upon pathogen attack. There is the possibility of a role for ACE2 in the association between DM and COVID-19. ACE2 that breaks down angiotensin-II to smaller peptides, angiotensin (1−7) and angiotensin (1−9); Ang 1−7 exerts a potent vasodilator, antifibrotic, antiproliferation, and antiinflammatory effect, thus protecting the lung against ARDS. ACE2 expression is reduced in patients with DM possibly due to glycosylation; this might explain the increased predisposition to severe lung injury and ARDS with COVID-19 [7].

18.3.4 Coagulation and COVID-19

A major manifestation of COVID-19 infection is a coagulopathy characterized by high D-dimer and fibrinogen concentrations [43]. D-dimers are protein products of cross-linked fibrin degradation, while fibrinogen plays a critical role in hemostasis as it promotes platelet aggregation and, when activated to form fibrin, provides the substrate for red blood cells and platelets to form strong clots.

With COVID-19 infection, the intravascular coagulation and thrombotic disease seem to arise from EDF and damage. Indeed, studies that assessed markers of endothelial cell and platelet activation in critically and noncritically ill patients admitted to the hospital with COVID-19 show that endotheliopathy (disease/infection of the endothelium) is present in COVID-19 and are likely to be associated with critical illness and death [17,18]. In autopsy series, microvascular thrombosis of the pulmonary vasculature has frequently been observed [18]. A biochemical study of 68 noncritically ill and critically ill patients with COVID-19, which assessed markers of endothelial cell and platelet activation [von Willebrand factor (vWF) antigen and thrombomodulin, respectively], reported that elevated levels of plasma vWF antigen and soluble thrombomodulin correlated significantly with mortality. This indicates that endotheliopathy in COVID-19 is likely to be associated with critical illness and death [44].

18.3.5 Thrombosis and COVID-19

Immune response to infection manifests itself in the activation of coagulation pathways. The first step with viral infection is lung inflammation that progresses to a cytokine storm. This is followed by the appearance of an endothelial proinflammatory phenotype and a pro-coagulant phenotype. The lack of pro-coagulant—anticoagulant balance predisposes to the development of micro-thrombosis. Continued inflammation or cytokine storm can cause extension of pulmonary micro-thrombi to become larger thrombi [45]. The lung-specific origin of coagulopathy has been supported by postmortem reports that show thrombi in the lung microvasculature [18]. Thrombin that promotes clot formation by activating platelets and by converting fibrinogen to fibrin also augments inflammation via proteinase-activated receptors (PARs), principally PAR-1. Thrombin generation is tightly controlled by negative feedback loops and anticoagulants, such as antithrombin (AT) III, tissue factor pathway inhibitor, and the protein C system [46]. During inflammation, all three of these control mechanisms can be impaired, with reduced anticoagulant concentrations due to reduced production and increasing consumption as evidenced in severe COVID-19 patients [47].

18.4 Therapeutic approaches to target anticoagulant and antiinflammatory pathways

It is becoming increasingly clear that COVID-19 infection leads to endothelial inflammation and a hypercoagulation state. Further, inflammation in the lungs leads to edema and poor gas exchange. This condition is aggravated by hypoxia, which augments thrombosis by both increasing blood viscosity and hypoxia-inducible transcription factor-dependent signaling pathway [48]. Sepsis-induced coagulopathy with COVID-19 leads to increased mortality. Thus there is an increasing interest for the antiinflammation and anticoagulant therapy for COVID-19.

Heparin therapy has been evaluated by comparing the parameters of coagulation tests and clinical characteristics between survivors and nonsurvivors. In a study conducted at Huazhong University of Science and Technology in Wuhan, heparin treatment appears to be associated with better prognosis in severe COVID-19 patients with coagulopathy [49].

Antithrombin and antifactor Xa direct oral anticoagulants are well established in the prevention and management of venous thromboembolism. Thrombin is targeted as it is the main activator of PAR-1, and coagulation factor Xa can induce production of proinflammatory cytokines via activation of PAR-1 and 2. Thus these drugs might be promising in reducing the severity of COVID-19 although bleeding risk will always be a concern. The effects of nonsteroidal antiinflammatory drugs, such as ibuprofen on COVID-19 patients, are not clear and there have been conflicting reports till date.

18.5 Conclusion

The pandemic by COVID-19 is a serious health issue that has affected billions worldwide. As there are no established therapeutic interventions, the current clinical management is to reduce the virus spread and provide supportive care for diseased patients. However, developing targeted therapies is crucial for reduced disease severity and improving prognosis. Toward that direction, research efforts are focused on detecting the processes that facilitate entry and replication of the virus and also on understanding the cellular signaling pathways that are the targets of the virus and the exaggerated inflammation cascade.

References

[1]   Xiang YT, Li W, et al. Timely research papers about COVID-19 in China. Lancet 2020;395(10225):684−5.

[2]   Khan T, Agnihotri K, Tripathi A, Mukherjee S, Agnihotri N, Gupta G. COVID-19: a worldwide, zoonotic, pandemic outbreak. Alternat Ther Health Med 2020; AT6471.

[3]   Guan WJ, Ni ZY, et al. Clinical characteristics of coronavirus disease 2019 in China. N Engl J Med 2020;382(18):1708−20.

[4]   Wu, Z., McGoogan JM. Characteristics of and important lessons from the coronavirus disease 2019 (COVID-19) outbreak in China: summary of a report of 72314 cases from the Chinese Center for Disease Control and Prevention. JAMA. 2020;323(13):1239−1242.

[5]   Li X, Ma X. Acute respiratory failure in COVID-19: is it "typical" ARDS? Crit Care 2020;24(1):198.

[6]   Wang D, Hu B, et al. Clinical characteristics of 138 hospitalized patients with 2019 novel coronavirus-infected pneumonia in Wuhan, China. JAMA 2020;323:1061−9.

[7]   Wu C, Chen X, et al. Risk factors associated with acute respiratory distress syndrome and death in patients with coronavirus disease 2019 pneumonia in Wuhan, China. JAMA Intern Med 2020;180:1−11.

[8]   Silversides JA, Ferguson ND. Clinical review: acute respiratory distress syndrome − clinical ventilator management and adjunct therapy. Crit Care 2013;17(2):225.

[9]   Ganatra S, Hammond SP, et al. The novel coronavirus disease (COVID-19) threat for patients with cardiovascular disease and cancer. JACC CardioOncol 2020;2:350−5.

[10] Libby P. The heart in COVID19: primary target or secondary bystander? JACC Basic Transl Sci 2020;5(5):537−542.

[11] Klok FA, Kruip M, et al. Incidence of thrombotic complications in critically ill ICU patients with COVID-19. Thromb Res 2020;191:145−7.

[12] Lodigiani C, Iapichino G, et al. Venous and arterial thromboembolic complications in COVID-19 patients admitted to an academic hospital in Milan, Italy. Thromb Res 2020;191:9−14.

[13] Sandoo A, van Zanten JJ, et al. The endothelium and its role in regulating vascular tone. Open Cardiovasc Med J 2010;4:302−12.

[14] Chatterjee S. Endothelial mechanotransduction, redox signaling and the regulation of vascular inflammatory pathways. Front Physiol 2018;9:524.

[15] Samet MM, Lelkes PI. The hemodynamic environment of endothelium in vivo and its stimulation in vitro. In: Lelkes PI, Gimbrone Jr MA, editors. Mechanical forces and the endothelium. London: Harwood Academic; 1999. p. 12.

[16] van Hinsbergh VW. Endothelium − role in regulation of coagulation and inflammation. Semimmunopathology 2012;34(1):93−106.

[17] Varga Z, Flammer AJ, et al. Endothelial cell infection and endotheliitis in COVID-19. Lancet 2020;395(10234):1417−18.

[18] Ackermann M, Verleden SE, et al. Pulmonary vascular endothelialitis, thrombosis, and angiogenesis in Covid-19. N Engl J Med 2020;383:120−8.

[19] Xiong M, Liang X, et al. Changes in blood coagulation in patients with severe coronavirus disease 2019 (COVID-19): a meta-analysis. Br J Haematol 2020;189:1050−2.

[20] Al-Ani F, Chehade S, et al. Thrombosis risk associated with COVID-19 infection. A scoping review. Thromb Res 2020;192:152−60.

[21] Hoffmann M, Kleine-Weber H, et al. SARS-CoV-2 cell entry depends on ACE2 and TMPRSS2 and is blocked by a clinically proven protease inhibitor. Cell 2020;181(2):271−80 e8.

[22] Guo J, Huang Z, et al. Coronavirus disease 2019 (COVID-19) and cardiovascular disease: a viewpoint on the potential influence of angiotensin-converting enzyme inhibitors/angiotensin receptor blockers on onset and severity of severe acute respiratory syndrome coronavirus 2 infection. J Am Heart Assoc 2020;9(7):e016219.

[23] Lovren F, Pan Y, et al. Angiotensin converting enzyme-2 confers endothelial protection and attenuates atherosclerosis. Am J Physiol Heart Circ Physiol 2008;295(4):H1377−84.

[24] Monteil V, Kwon H, et al. Inhibition of SARS-CoV-2 infections in engineered human tissues using clinical-grade soluble human ACE2. Cell 2020;181(4):905−13 e907.

[25] Chan JF, Yuan S, et al. A familial cluster of pneumonia associated with the 2019 novel coronavirus indicating person-to-person transmission: a study of a family cluster. Lancet 2020;395(10223):514−23.

[26] Tang D, Comish P, et al. The hallmarks of COVID-19 disease. PLoS Pathog 2020;16(5):e1008536.

[27] Wrapp D, Wang N, et al. Cryo-EM structure of the 2019-nCoV spike in the prefusion conformation. Science 2020;367(6483):1260−3.

[28] Chen Y, Guo Y. et al. Structure analysis of the receptor binding of 2019-nCoV. Biochem Biophys Res Commun 2020;525(1):135−140.

[29] Coutard B, Valle C, et al. The spike glycoprotein of the new coronavirus 2019-nCoV contains a furin-like cleavage site absent in CoV of the same clade. Antivir Res 2020;176:104742.

[30] Avogaro A, Albiero M, et al. Endothelial dysfunction in diabetes: the role of reparatory mechanisms. Diabetes Care 2011;34(Suppl 2): S285−90.

[31] Huertas A, Montani D, Savale L, et al. Endothelial cell dysfunction: a major player in SARS-CoV-2 infection (COVID-19)? Eur Respir J. 2020;56(1):2001634.

[32] Yang M, Chen S, et al. Pathological findings in the testes of COVID-19 patients: clinical implications. Eur Urol Focus 2020.

[33] Xie J, Tong Z, et al. Clinical characteristics of patients who died of coronavirus disease 2019 in China. JAMA Netw Open 2020;3(4):e205619.

[34] Guan YQ, Zhang M, et al. Medical treatment seeking behaviors and its influencing factors in employed floating population in China. Zhonghua liu xing bing xue za zhi = Zhonghua liuxingbingxue zazhi 2019;40(3):301−8.

[35] Chen N, Zhou M, et al. Epidemiological and clinical characteristics of 99 cases of 2019 novel coronavirus pneumonia in Wuhan, China: a descriptive study. Lancet 2020;395(10223):507−13.

[36] Huang C, Wang Y, et al. Clinical features of patients infected with 2019 novel coronavirus in Wuhan, China. Lancet 2020;395 (10223):497−506.

[37] Ferrario CM, Jessup J, et al. Effect of angiotensin-converting enzyme inhibition and angiotensin II receptor blockers on cardiac angiotensin-converting enzyme 2. Circulation 2005;111(20):2605−10.

[38] Li J, Wang X, et al. Association of renin-angiotensin system inhibitors with severity or risk of death in patients with hypertension hospitalized for coronavirus disease 2019 (COVID-19) infection in Wuhan, China. JAMA Cardiol 2020;5(7):1−6.

[39] Yu S, Christiani DC, et al. Role of diabetes in the development of acute respiratory distress syndrome. Crit Care Med 2013;41(12):2720−32.

[40] Hong KW, Cheong HJ, et al. Clinical courses and outcomes of hospitalized adult patients with seasonal influenza in Korea, 2011−2012: hospital-based influenza morbidity & mortality (HIMM) surveillance. J Infect Chemother 2014;20(1):9−14.

[41] Schoen K, Horvat N, et al. Spectrum of clinical and radiographic findings in patients with diagnosis of H_1N_1 and correlation with clinical severity. BMC Infect Dis 2019;19(1):964.

[42] Yang JK, Feng Y, et al. Plasma glucose levels and diabetes are independent predictors for mortality and morbidity in patients with SARS. Diabet Med 2006;23(6):623−8.

[43] Panigada M, Bottino N, et al. Hypercoagulability of COVID-19 patients in intensive care unit: a report of thromboelastography findings and other parameters of hemostasis. J Thromb Haemost 2020;18(7):1738−42.

[44] Goshua G, Pine AB, Meizlish ML, et al. Endotheliopathy in COVID-19-associated coagulopathy: evidence from a single-centre, cross-sectional study, Lancet Haematol 7(8), 2020, e575−e582.

[45] Connors JM, Levy JH. Thromboinflammation and the hypercoagulability of COVID-19. J Thromb Haemost 2020;18(7):1559−61.

[46] Jose RJ, Williams AE, et al. Proteinase-activated receptors in fibroproliferative lung disease. Thorax 2014;69(2):190−2.

[47] Tang N, Li D, et al. Abnormal coagulation parameters are associated with poor prognosis in patients with novel coronavirus pneumonia. J Thromb Haemost 2020;18(4):844−7.

[48] Gupta N, Zhao YY, et al. The stimulation of thrombosis by hypoxia. Thromb Res 2019;181:77−83.

[49] Tang N, Bai H, et al. Anticoagulant treatment is associated with decreased mortality in severe coronavirus disease 2019 patients with coagulopathy. J Thromb Haemost 2020;18(5):1094−9.

Chapter 19

Endothelial dysfunction and cardiovascular diseases through oxidative stress pathways

Roger Rodríguez-Guzmán, Ela María Céspedes Miranda and Pilar Guzmán-Díaz

Biomedical Sciences Department, Calixto García Faculty, University of Medical Sciences of Havana, Havana, Cuba

19.1 Introduction

Endothelial cells (ECs) line the blood vessels and carry out crucial metabolic, endocrine, and vascular functions. Moreover, they play a significant role in regulating vascular homeostasis [1]. This occurs via endothelial production of several vasoactive substances that maintain vascular tension, normal flow of blood, and endothelial permeability [1,2]. Lack of EC functionality leads to endothelial dysfunction (ED) [2]. ED manifests as a systemic, insidious, and reversible pathological state of the endothelium [3−5] where reduced nitric oxide (NO) bioavailability and imbalance related to other endothelium-derived vasodilator and vasoconstrictor substances is observed.

The etiology of ED is multifactorial and the etiology is influenced by a high uric acid (UA) concentration [6], aldose reductase activity [7], elevated free fatty acid (FFA) levels [8], aging [9], air pollution [10], low-density lipoprotein (LDL) oxidation [11], hyperglycemia [12], serum adipokines levels [13], reactive oxygen species (ROS) [14], epigenetic factors [15], etc. ED plays a central role in the pathogenesis of several vascular and metabolically chronic human diseases such as peripheral vascular diseases [16], stroke [17], heart disease [18], diabetes, and other forms of insulin resistance [19]. Indeed, the adage "Dysfunction of the vascular endothelium is the key to several human diseases" [20] embodies a common observation.

19.2 Endothelial dysfunction through oxidative stress

Oxidative stress (OS) is defined as overproduction of ROS; ROS damages biomolecules such as proteins, nucleic acids, and lipids. Normally, ROS activity is balanced with action of antioxidants; however, when ROS production is uncontrolled and antioxidant activity reduced, this loss of balance leads to ED [21]. As is well known, the main ROS are: superoxide anion $\left(O_2^{\cdot-}\right)$, hydrogen peroxide ($H_2O_2$), hydroxyl radical (OH), and peroxynitrite (ONOO −). The main sources of all ROS are enzymatic systems such as nicotinamide adenine dinucleotide phosphate (NADPH) oxidase (NOX), uncoupled endothelial nitric oxide synthase (eNOS), and xanthine oxidase (XO). ROS and oxidative damage drive endothelial signaling that predisposes the endothelium to ED.

The first step in ED is endothelial activation (EA) that leads ECs to produce chemokines and cytokines [14] such as tumor necrosis factor alpha (TNF-α), interleukin (IL)-6, IL-8, and monocyte chemoattractant protein-1 (MCP-1) [22]. These collectively trigger processes that drive loss of endothelial integrity [23]. With EA, the endothelial surface participates in recruitment and attachment of inflammatory cells [14]. These events occur via adhesion molecules such as cellular adhesion molecules (CAMs), selectins, the immunoglobulin superfamily, and cadherins [24]. The main EC adhesion molecules are the intercellular adhesion molecule-1 (ICAM-1) and the vascular cell adhesion molecule-1 (VCAM-1) that drive the adherence of immune cells and initiate atherogenesis [25,26]. In fact, elevations in the serum levels of adhesion molecules are considered to be an indicator of ED [24].

ROS produced by plasma membrane and/or mitochondria of the endothelium also activate EC; ROS also promote NO reduction by acting as a sink by reacting with NO to form the potent oxidizing species, peroxynitrite [27].

Endothelial Signaling in Vascular Dysfunction and Disease. DOI: https://doi.org/10.1016/B978-0-12-816196-8.00012-6

Mitochondrial antioxidant enzyme systems such as manganese superoxide dismutase (MnSOD) participate in detoxifying mitochondria from ROS [28]. However, under conditions of excessive ROS production, these antioxidants may not be effective. In addition to MnSOD, other antioxidant enzymes such as catalase, glutathione (GSH), peroxiredoxins, and thioredoxins are the scavenger enzymes that act against ROS. These antioxidants, UA [7] and gamma-glutamyl transferase (GGT) act as biomarkers of OS.

19.3 Uric acid and gamma-glutamyl transferase as new etiological elements in endothelial dysfunction through oxidative stress

UA was reported to be a biomarker of cardiovascular diseases for first time in 1879 in Lancet [29]. However, subsequent studies such as the Framingham Heart Studies did not propose it to be an independent risk factor [30,31].

UA drives ED through inflammation and OS. This occurs via expression of high mobility group box chromosomal protein 1 (HMGB1) and its release into the extracellular space by ECs. Extracellular HMGB1 binds to its receptor (receptor for advanced glycation end products or RAGE) on ECs; the HMGB1-RAGE binding leads to EA and recruitment of inflammatory cells [6].

UA activates NOX [7] to produce ROS [30–32]. UA also inhibits endothelial nitric oxide synthase (eNOS) expression and thus NO production [6]. There is an important association between high UA concentrations and ED. Lowering UA concentrations with NOX inhibitor allopurinol is reported to improve ED in asymptomatic hyperuricemia, congestive heart failure, diabetes, chronic kidney disease, obstructive sleep apnea, and smoking. UA reduces NO, and drives vascular constriction, tissue ischemia, and more UA production [33–35]. In addition to UA, ischemia (such as renal ischemia) activates the renin-angiotensin-aldosterone system (RAAS), which in turn stimulates NOX leading to ROS, OS, and ED [36].

GGT is linked to cardiovascular diseases through OS-induced ED [37]. GGT participates in degradation of the antioxidant GSH; during OS, GGT gene expression is increased, and this leads to reduced GSH levels [38] which in turn increase OS [38]. Bradley et al. demonstrated in the multi-ethnic study of atherosclerosis that serum GGT levels was associated with traditional cardiovascular risk factors, OS, immune inflammation, and specifically ED [38,39].

Flow-sensitive microRNAs also play a role in ED. MicroRNAs (miRNAs) are small, noncoding genes that posttranscriptionally regulate gene expression by targeting messenger RNA transcripts [40]. Alteration in blood flow conditions from laminar to disturbed flow regulates expression of miRNAs in ECs. Some of these miRNAs drive smooth muscle cell phenotype and can by virtue of smooth muscle proliferation and vessel remodeling lead to ED induced hypertension [40,41]. miRNAs such as miR-10a, miR-19a, miR-23b, miR-17–92, miR-21, miR-663, miR-92a, miR-143/145, miR-101, miR-126, miR-712, miR-205, and miR-155 have been identified as mechano-miRs that may play a role in ED in conditions of disturbed blood flow as would occur in regions of bifurcation, etc. in the vascular tree in the body [41]. miR-126, an endothelial-specific mRNA, is involved in endothelium integrity and its suppression results in weak vascular cell–cell junction and hemorrhage [41]. miR-217 causes endothelium senescence due to downregulation of eNOS and miR-21 induces endothelial cellular dysfunction [42]. miR-125a-5p and miR-125b-5p affect endothelial vasodilation by downregulate endothelin-1 (ET-1) expression (ET-1, is one of the most potent vasoconstrictors in ECs); miR-125a/b has been involved in the regulation of the potent vasoconstrictor ET-1 especially in animal models of hypertension [43]. miR-155 modifies the expression of endothelial eNOS and angiotensin receptor type 1, which represent the two main players in vascular homeostasis [44]. Collectively miRNAs play a role in regulation of moieties that participate in ED.

19.4 Endothelial dysfunction, cardiovascular diseases and their risk factors

The Framingham Heart Study reported on four "classic" cardiovascular risk factors: hypertension, hypercholesterolemia, diabetes mellitus (DM), and obesity. This study also developed Framingham Risk Scores to predict CVD risk [13,45]. To these factors, the World Health Organization has added five additional risk factors: smoking, sedentary lifestyle, alcohol consumption, fat, and salt diet [46]. All these seem to be linked to ED.

19.4.1 Hypertension

High blood pressure (HBP) is defined as systolic blood pressure values ≥ 140 mmHg and/or diastolic blood pressure (DBP) values ≥ 90 mmHg. Hypertension prevails in 35%–40% of world population [46]. Major complications of hypertension are represented by ischemic heart disease (4.9 million), hemorrhagic stroke (2.0 million), and ischemic stroke (1.5 million). HBP contributes to CVD because of its involvement in ED. The link between HBP and ED is not a direct one but possibly arises from blood flow disturbances that cause mechanical damage to the endothelium [47,48]. Additionally vasoconstriction associated with HBP leads to ischemia or partial ischemia [49,50]. Drugs that reduce HBP are effective in reducing ED. For instance, angiotensin-converting enzyme inhibitors (ACEIs) or angiotensin-receptor blockers significantly decrease ED [51]. Aldosterone antagonists and nebivolol that reduce blood pressure can stimulate NO production and thus improve endothelial function [47]. Other HBP-associated conditions such as blindness, dementia, hearing loss, and stroke could also have underlying ED [52–54].

19.4.2   Hypercholesterolemia

Hypercholesterolemia [45] is defined as increase in total cholesterol ($\geq$240 mg/dL or 6.20 mmol/L), increase in LDL cholesterol (LDL-C; $>$160 mg/dL or 4.13 mmol/L), increase in triglyceride levels ($>$200 mg/dL or 2.25 mmol/L), and decrease in high-density lipoprotein cholesterol (HDL-C; $<$40 mg/dL or 1.03 mmol/L) [54]. The link between LDL-C/HDL-C and CVD possibly arises from ED [45]. There is some evidence to suggest that FFAs drive ED. FFAs are known to facilitate cell death (via apoptosis/necroptosis) of ECs and mediate detrimental effects on endothelial progenitor cells [55]. FFAs also cause generation of ROS via activation of NOX and are therefore involved in ED [26]. ROS and OS from FFA play a role in inflammation [56] and in the metabolic syndrome [8]. FFAs-induced endothelial permeability could play a role in ED [57–59]. Oxidized LDL plays a very important role in ED and atherosclerotic plaque pathophysiology (formation, progression, and destabilization). The mechanism of action of LDL includes induction of EC activation and dysfunction, recruitment of macrophages and formation of foam cells, migration and proliferation of vascular smooth muscle layer [59–61]. In contrast, the protective ability of HDL arises from reverse transport of cholesterol and its antiinflammatory and antioxidants activities, including immunomodulatory activities. Control of cholesterol by statins remarkably reduced ED [62–64]. This seems to arise from statin induced eNOS expression and activation, in addition to statin-driven reduced cholesterol levels.

19.4.3   Diabetes mellitus

DM is defined as a chronic metabolic disease [65] and characterized by chronic hyperglycemia arising from altered glucose metabolism owing to change in insulin quantity or function. DM is a systemic, inflammatory disorder, which involves fat and protein metabolism dysfunction. Epidemiologically, DM is an epidemic worldwide with $>$415 million adults suffering from DM [65]. DM could be classified as type I diabetes due to lack of insulin production from autoimmune (pancreatic) β cell destruction or type 2 diabetes that arises due to a progressive loss of insulin secretion or insulin resistance. More than 95% of people affected by diabetes suffer from DM type 2 [65].

DM is linked to ED via dysregulation of many endothelial cellular mechanisms due to OS, inflammation, endoplasmic reticulum stress, aberrant insulin signaling, accumulation of advanced glycated lipotoxic products, altered signal transduction such as G-protein receptor kinase signaling, RAAS signaling, and β2-adrenergic receptor signaling [66].

DM is also linked to CVD; indeed DM subjects present plaques with a larger necrotic core and significantly greater inflammation consisting mainly of macrophages and T lymphocytes as compared to patients without diabetes [67]. Also subjects with DM show a higher incidence of healed plaque ruptures and positive remodeling, a more active atherogenic process [67].

Additionally, endothelium expresses the insulin transporter via which insulin exerts its endothelium protective functions such as increased production of NO. Insulin resistance is proposed to occur due to peripheral ED [68].

19.4.4   Obesity

Obesity is defined as the hypertrophy of adipocytes or fat cells that have physiological functions such as immune and inflammatory response and endocrinal functions. Obesity is associated with DM and hypercholesterolemia. Sedentary behavior has been associated with adiposity [69,70]. Obesity leads to several pathologies predominantly due to production of cytokines and chemokines by adipocites such as: TNF-α, IL-6 and MCP-1. Adipose tissue is the source of adipokines (leptine, adiponectine, visfatin, vaspin and omentin) that are biologically active molecules of adipose tissue that have local and systematic effects in the body [71]. These inflammation moieties lead to insulin resistance, ED, and hypertension. Obesity leads to ED via ROS generation by adipocytes. ROS signaling drives recruitment of immune cells into the perivascular endothelium. The resultant inflammation causes ED [72]. Overall signaling events associated with metabolic dysregulation play a role in ED. Thus drugs that regulate these signals such as metformin for DM are protective toward the endothelium.

19.5   Summary and recommendations

In conclusion, ROS and OS-induced ED plays a crucial role in CVD. Currently, there are no FDA-approved treatments for ED. Therefore treatment has to be focused on the comorbidities that lead to ED. Thus statins and ACEI have shown benefit in reducing ED. Increased NO production either via statins or L-arginine treatment has been shown to have beneficial effects on both: ED symptoms and Endothelial function too.

Additionally, diet and exercise have both been shown to improve vascular reactivity, and should be encouraged as part of lifestyle behaviors beneficial toward overall CVD health.

References

[1] Meng LB, Chen K, Zhang YM. Gong common injuries and repair mechanisms in the endothelial lining. Chin Med J (Engl) 2018;131 (19):2338−45. Available from: https://doi.org/10.4103/0366-6999.241805. Available from: https://www.ncbi.nlm.nih.gov/pubmed/30246720.

[2] Sorrentino FS, Matteini S, Bonifazzi C, Sebastiani A, Parmeggiani F. Diabetic retinopathy and endothelin system: microangiopathy versus endothelial dysfunction. Eye (Lond) 2018;32(7):1157−63. Available from: https://doi.org/10.1038/s41433-018-0032-4. Available from: https://www.ncbi.nlm.nih.gov/pubmed/29520046.

[3] Sitia S, Tomasoni L, Atzeni F, Ambrosio G, Cordiano C, Catapano A. From endothelial dysfunction to atherosclerosis. Autoimmun Rev 2010;9 (12):830−4. Available from: https://doi.org/10.1016/j.autrev.2010.07.016. Available from: https://www.ncbi.nlm.nih.gov/pubmed/20678595.

[4] Jamwal S, Sharma S. Vascular endothelium dysfunction: a conservative target in metabolic disorders. Inflamm Res 2018;67(5):391−405. Available from: https://doi.org/10.1007/s00011-018-1129-8. Available from: https://www.ncbi.nlm.nih.gov/pubmed/29372262.

[5] Baselet B, Sonveaux P, Baatout S, Aerts A. Pathological effects of ionizing radiation: endothelial activation and dysfunction. Cell Mol Life Sci 2019;76(4):699−728. Available from: https://www.ncbi.nlm.nih.gov/pmc/articles/PMC6514067/.

[6] Cai W, Duan X-M, Liu Y, Yu J, Tang Y-L, Liu Z-L. Uric acid induces endothelial dysfunction by activating the HMGB1/RAGE signaling pathway. Biomed Res Int 2017;2017:4391920. Available from: https://www.ncbi.nlm.nih.gov/pmc/articles/PMC5237466/.

[7] Huang Z, Hong Q, Zhang X. Aldose reductase mediates endothelial cell dysfunction induced by high uric acid concentrations. Cell Commun Signal 2017;15:3. Available from: https://doi.org/10.1186/s12964-016-0158-6. Available from: https://biosignaling.biomedcentral.com/articles/10.1186/s12964-016-0158-6#citeas.

[8] Ghosh A, Gao L, Thakur A, Siu PM, Lai CWK. Role of free fatty acids in endothelial dysfunction. J Biomed Sci 2017;24(1):50. Available from: https://doi.org/10.1186/s12929-017-0357-5. Available from: https://www.ncbi.nlm.nih.gov/pubmed/28750629.

[9] Tesauro M, Mauriello A, Rovella V, Annicchiarico-Petruzzelli M, Cardillo C. Arterial ageing: from endothelial dysfunction to vascular calcification. J Intern Med 2017;281:471−82. Available from: https://doi.org/10.1111/joim.12605. Available from: https://onlinelibrary.wiley.com/doi/epdf/10.1111/joim.12605.

[10] Amadio P, Baldassarre D, Tarantino E, Zacchi E, Gianellini S. Production of prostaglandin E2 induced by cigarette smoke modulates tissue factor expression and activity in endothelial cells. FASEB J 2015;29(9):4001−10. Available from: https://doi.org/10.1096/fj.14-268383. Available from: https://www.fasebj.org/doi/pdf/10.1096/fj.14-268383.

[11] Watt J, Kennedy S, Ahmed N, Hayhurst J, McClure JD, Berry C. The relationship between oxidised LDL, endothelial progenitor cells and coronary endothelial function in patients with CHD. Open Heart 2016;3(1):e000342. Available from: https://doi.org/10.1136/openhrt-2015-000342. Available from: https://www.ncbi.nlm.nih.gov/pubmed/26848395.

[12] Zhu W, Yuan Y, Liao G. Mesenchymal stem cells ameliorate hyperglycemia-induced endothelial injury through modulation of mitophagy. Cell Death Dis 2018;837:9. Available from: https://www.nature.com/articles/s41419-018-0861-x.

[13] Piepoli MF, Hoes AW, Agewall S, Albus C, Brotons C, Catapano AL, et al. 2016 European guidelines on cardiovascular disease prevention in clinical practice. Eur Heart J 2016;37(29):2315−81. Available from: https://doi.org/10.1093/eurheartj/ehw106. Available from: https://www.ncbi.nlm.nih.gov/pmc/articles/PMC4986030/.

[14] Incalza MA, D'Oria R, Natalicchio A, Perrini S, Laviola L, Giorgino F. Oxidative stress and reactive oxygen species in endothelial dysfunction associated with cardiovascular and metabolic diseases. Vasc Pharmacol 2018;100:1−19. Available from: https://doi.org/10.1016/j.vph.2017.05.005. Available from: https://www.ncbi.nlm.nih.gov/pubmed/28579545 > .

[15] Levy E, Spahis S, Bigras JL, Delvin E, Borys JM. The epigenetic machinery in vascular dysfunction and hypertension. Curr Hypertens Rep 2017;19(6):52. Available from: https://doi.org/10.1007/s11906-017-0745-y. Available from: https://www.ncbi.nlm.nih.gov/pubmed/28540644.

[16] Ismaeel A, Brumberg RS, Kirk JS, Papoutsi E, Farmer PJ, Bohannon WT. Oxidative stress and arterial dysfunction in peripheral artery disease. Antioxidants 2018;7(10):145. Available from: https://doi.org/10.3390/antiox7100145 > . Available from: https://www.mdpi.com/2076-3921/7/10/145.

[17] Chung JW, Oh MJ, Cho YH, Moon GJ, Kim GM, Chung CS. Distinct roles of endothelial dysfunction and inflammation in intracranial atherosclerotic stroke. Eur Neurol 2017;77(3−4):211−19. Available from: https://doi.org/10.1159/000460816. Available from: https://www.ncbi.nlm.nih.gov/pubmed/29207392.

[18] Barthelmes J, Nägele MP, Ludovici V, Ruschitzka F, Sudano I, Flammer AJ. Endothelial dysfunction in cardiovascular disease and Flammer syndrome—similarities and differences. EPMA J 2017;8(2):99−109. Available from: https://doi.org/10.1007/s13167-017-0099-1. Available from: https://europepmc.org/articles/PMC5545991;jsessionid = 1308036FC8F12FA3DC442925559072B9.

[19] Shi Y, Vanhoutte PM. Macro- and microvascular endothelial dysfunction in diabetes. J Diabetes 2017;9(5):434−49. Available from: https://doi.org/10.1111/1753-0407.12521. Available from: https://www.ncbi.nlm.nih.gov/pubmed/28044409.

[20] Rajendran P, Rengarajan T, Thangavel J, Nishigaki Y, Sakthisekaran D, Sethi G. The vascular endothelium and human diseases. Int J Biol Sci 2013;9(10):1057−69.

[21] Tokarz P, Kaarniranta K, Blasiak J. Role of antioxidant enzymes and small molecular weight antioxidants in the pathogenesis of age-related macular degeneration (AMD). Biogerontology 2013;14(5):461−82. Available from: https://doi.org/10.1007/s10522-013-9463-2. Available from: https://www.ncbi.nlm.nih.gov/pubmed/24057278.

[22] Abcouwer SF. Angiogenic factors and cytokines in diabetic retinopathy. J Clin Cell Immunol 2013;Suppl 1(11):1−12. Available from: https://doi.org/10.4172/2155-9899. Available from: https://www.ncbi.nlm.nih.gov/pmc/articles/PMC3852182/.

[23] Favero G, Paganelli C, Buffoli B, Rodella LF, Rezzani R. Endothelium and its alterations in cardiovascular diseases: life style intervention. Biomed Res Int 2014;2014:801896. Available from: https://doi.org/10.1155/2014/801896.

[24] Comba A, Çaltepe G, Yank K, Gör U, Yüce Ö, Kalayc AG. Assessment of endothelial dysfunction with adhesion molecules in patients with celiac disease. J Pediatr Gastroenterol Nutr 2016;63(2):247−52. Available from: https://doi.org/10.1097/MPG.0000000000001138. Available from: https://www.ncbi.nlm.nih.gov/pubmed/26835908.

[25] Wu T, McGrath KYC, Death AK. Cardiovascular disease in diabetic nephropathy patients: cell adhesion molecules as potential markers? Vasc Health Risk Manag 2005;1(4):309−16. Available from: https://doi.org/10.2147/vhrm.2005.1.4.309. Available from: https://www.ncbi.nlm.nih.gov/pmc/articles/PMC1993958/.

[26] Scioli MG, D'Amico F, Rodríguez Guzmán R, Céspedes Miranda EM, Orlandi A. Oxidative stress-induced endothelial dysfunction contributes to cardiovascular disease. Rev Cuban Invest Bioméd 2019;38(1). Available from: http://www.revibiomedica.sld.cu/index.php/ibi/article/view/168.

[27] Kluge MA, Fetterman JL, Vita JA. Mitochondria and endothelial function. Circ Res 2013;112(8):1171−88. Available from: https://doi.org/10.1161/CIRCRESAHA.111.300233. Available from: https://www.ncbi.nlm.nih.gov/pubmed/23580773.

[28] Candas D, Li JJ. MnSOD in oxidative stress response-potential regulation via mitochondrial protein influx. Antioxid Redox Signal 2014;20(10):1599−617. Available from: https://doi.org/10.1089/ars.2013.5305. Available from: https://www.ncbi.nlm.nih.gov/pubmed/23581847.

[29] Mohamed FA. On chronic Bright's disease, and its essential symptoms. Lancet 1879;1:399−401.

[30] Feig DI, Kang DH, Johnson RJ. Uric acid and cardiovascular risk. N Engl J Med 2008;359(17):1811−21. Available from: https://doi.org/10.1056/NEJMra0800885.

[31] Snezhkina AV, Kudryavtseva AV, Kardymon OL, Savvateeva MV, Melnikova NV, Krasnov GS. ROS generation and antioxidant defense systems in normal and malignant cells. Oxid Med Cell Longev 2019;2019:17. Available from: https://doi.org/10.1155/2019/6175804 > . Available from: https://www.hindawi.com/journals/omcl/2019/6175804/cta/.

[32] Meitzler JL, Antony S, Wu Y, Juhasz A, Liu H, Jiang G. NADPH oxidases: a perspective on reactive oxygen species production in tumor biology. Antioxid Redox Signal 2014;20(17):2873−89. Available from: https://doi.org/10.1089/ars.2013.5603. Available from: https://www.ncbi.nlm.nih.gov/pubmed/24156355.

[33] Sanchez Lozada LG, Lanaspa MA, Cristóbal-García M, García-Arroyo F, Soto V, Cruz-Robles D. Uric acid-induced endothelial dysfunction is associated with mitochondrial alterations and decreased intracellular ATP concentrations. Nephron Exp Nephrol 2012;121:e71−8. Available from: https://doi.org/10.1159/000345509. Available from: https://www.karger.com/Article/Pdf/345509.

[34] Kang D-H, Han L, Ouyang X, Kahn AM, Kanellis J, Li P. Uric acid causes vascular smooth muscle cell proliferation by entering cells via a functional urate transporter. Am J Nephrol 2005;25:425−33. Available from: https://doi.org/10.1159/000087713.

[35] Feig DI, Madero M, Jalal DI, Sanchez-Lozada LG, Johnson RJ. Uric acid and the origins of hypertension. J Pediatr 2013;162(5):896−902. Available from: https://doi.org/10.1016/j.jpeds.2012.12.078.

[36] Miyata K, Rahman M, Shokoji T, Nagai Y, Zhang GX, Sun GP. Aldosterone stimulates reactive oxygen species production through activation of NADPH oxidase in rat mesangial cells. J Am Soc Nephrol 2016;10:2906−12. Available from: https://doi.org/10.1681/ASN.2005040390.

[37] Céspedes Miranda EM, Rodríguez Guzmán R, Suárez Castillo N. Gamma glutamil transferasa y enfermedad cardiovascular. Arch Hosp Calixto García 2019;7(2):260−73.

[38] Bradley RD, Fitzpatrick AL, Jacobs Jr DR, Lee DH, Swords Jenny N, Herrington D. Associations between γ-glutamyltransferase (GGT) and biomarkers of atherosclerosis: the multi-ethnic study of atherosclerosis (MESA). Atherosclerosis. 2014;233(2):387−93. Available from: https://doi.org/10.1016/j.atherosclerosis.2014.01.010.

[39] Lapenna D, Ciofani G, Giamberardino MA. Glutathione metabolic status in the aged rabbit aorta. Exp Gerontol 2017;91:34−8. Available from: https://doi.org/10.1016/j.exger.2017.02.003.

[40] Wise IA, Charchar FJ. Epigenetic modifications in essential hypertension. Int J Mol Sci 2016;17(4):451. Available from: https://doi.org/10.3390/ijms17040451.

[41] Kumar S, Kim CW, Simmons RD, Jo H. Role of flow-sensitive microRNAs in endothelial dysfunction and atherosclerosis: mechanosensitive athero-miRs. Arterioscler Thromb Vasc Biol 2014;34(10):2206−16. Available from: https://doi.org/10.1161/ATVBAHA.114.303425.

[42] Chamorro-Jorganes A, Araldi E, Suárez Y. MicroRNAs as pharmacological targets in endothelial cell function and dysfunction. Pharmacol Res 2013;75:15−27. Available from: https://doi.org/10.1016/j.phrs.2013.04.002.

[43] Li DJ, Evans RG, Yang Z-W, Song S-W, Wang P, Ma XJ. Dysfunction of the cholinergic anti-inflammatory pathway mediates organ damage in hypertension. Hypertension 2010;57(2):298−307. Available from: https://doi.org/10.1161/hypertensionaha.110.160077.

[44] Santulli G. Micrornas and endothelial (dys) function. J Cell Physiol 2016;231(8):1638−44. Available from: 10.1002/jcp.25276.

[45] Mahmood SS, Levy D, Vasan RS, Wang JJ. The Framingham Heart Study and the epidemiology of cardiovascular disease: a historical perspective. Lancet 2014;383(9921):999−1008. Available from: https://doi.org/10.1016/s0140-6736(13)617523.

[46] Williams B, Mancia G, Spiering W, Rosi EA, Azizi M, Burnier M, et al. 2018 ESC/ESH guidelines for the management of arterial hypertension. Eur Heart J 2018;39:3021−104. Available from: https://doi.org/10.1093/eurheart/ehy339.

[47] Dharmashankar K, Widlansky ME. Vascular endothelial functions and hypertension: insights and directions. Curr Hypertens Rep 2010;12(6):448−55. Available from: https://doi.org/10.1007/s11906-010-0150-2.

[48] Hadi ARH, Carr CS, Swaidi JA. Endothelial dysfunction: cardiovascular risk factors, therapy, and outcome. Vasc Health Risk Manag 2005;1(3):183−98.

[49] Muñoz- Durango N, Fuentes CA, Castillo AE, Gónzalez-Gómez LM, Vecchiola A, Fardella CE. Role of the renin- angiotensin-aldosterone system beyond blood pressure regulation: molecular and cellular mechanisms involved in end-organ damage during arterial hypertension. Int J Mol Sci 2015;17:797. Available from: https://doi.org/10.2290/ijms17070797.

[50] Schiffrin EL, Park JB, Intengan HD, Touzy M. Correction of arterial structure and endothelial dysfunction in human essential hypertension by angiotensin receptor antagonist losartan. Circulation. 2000;101:1653−6. Available from: https://doi.org/10.1161/01cr.101.14.1653.

[51] Panza JA, Epstein SE. Effect of antihypertensive treatment on endothelium- dependant vascular relaxation in patients with essential hypertension. J Am Coll Cardiol 1993;21(5):1145−51.

[52] Cao B, Bary F, Ilbawi A, Soerjomataram I. Effect on longevity of one-third reduction in premature mortality from non-communicable diseases by 2030: a global analysis of the sustainable development goal health target. Lancet 2018;6(12):1288−96. Available from: https://doi.org/10.1016/s2214-109x(18)30411-x.

[53] <https://www.who.int/topics/cerebrovascular_accident/en>.

[54] Kopin L, Lowenstein J. Dyslipidemia. Ann Intern Med 2017;167(119):ITC 81−96. Available from: https://doi.org/10.7326/ATC201712050.

[55] Artwohl M, Roden M, Waldhäusl W, Freudenthaler A, Baumgartner- Parzer SM. Free fatty acids trigger apoptosis and inhibit cell cycle progression in human vascular endothelial cells. FASEB J 2014;18(1):146−8. Available from: https://doi.org/10.1096/fj.03-0301fje.

[56] Jung CH, Park J-Y. Vaspin protects vascular endothelial cells against free fatty acid-induced apoptosis through a phosphatidylinositol 3-kinase/Akt pathway. Biochem Biophys Res Commun 2011;413(2):264−9. Available from: https://doi.org/10.1016/j.bbrc.2011.08.083 > .

[57] Inoguchi T, Li P, Umeda F, Yu HY, Makimoto M, Imamura M. High glucose level and free fatty acid stimulate reactive oxygen species production through protein kinase C-dependent activation of NAD(P)H oxidase in cultured vascular cells. Diabetes 2000;49(11):1939−45. Available from: https://doi.org/10.23337/diabetes.49.11.1939.

[58] Marseglia L, Martin S, D'Angelo G, Nicotera A, Parisi E, Di Rosa G. Oxidative stress in obesity: a critical component in human diseases. Int J Mol Sci 2015;16:378−400. Available from: https://doi.org/10.3390/ijms16010378.

[59] Chen C, Khismatullin DB. Oxidized low-density lipoprotein contributes to atherogenesis via co-activation of macrophages and mast cells. PLoS One 2015;10(3):e0123088. Available from: https://doi.org/10.1371/journal.pone.0123088. Available from: https://www.ncbi.nlm.nih.gov/pmc/articles/PMC4374860/.

[60] Di Pietro N, Formoso G, Pandolfi A. Physiology and pathophysiology of oxLDL uptake by vascular wall cells in atherosclerosis. Vasc Pharmacol 2016;84:1−7. Available from: https://doi.org/10.1016/j.vph.2016.05.013. Available from: https://www.ncbi.nlm.nih.gov/pubmed/27256928.

[61] Masana L. La hipótesis del LDL cero. Hacia concentraciones de LDL extremadamente bajas. Rev Esp Cardiol (Engl Ed) 2018;71(7):591−2. Available from: https://doi.org/10.1016/j.recesp.2017.03.031. Available from: https://www.revespcardiol.org/es-la-hipotesis-del-ldl-cero--articulo-S0300893217302750.

[62] Litwiniuk M, Niemczyk K, Niderla-Bielińska J, Łukawska-Popieluch I, Grzela T. Soluble endoglin (CD105) serum level as a potential marker in the management of head and neck paragangliomas. Ann Otol Rhinol Laryngol 2017;126(10):717−21. Available from: https://doi.org/10.1177/0003489417727548. Available from: https://www.ncbi.nlm.nih.gov/pubmed/28863727.

[63] Pirillo A, Catapano AL, Norata GD. Biological consequences of dysfunctional HDL. Curr Med Chem 2019;26(9):1644−64. Available from: https://doi.org/10.2174/0929867325666180530110543. Available from: https://www.ncbi.nlm.nih.gov/pubmed/29848265 > .

[64] Carr SS, Hooper AJ, Sullivan DR, Burnett JR. Non-HDL-cholesterol and apolipoprotein B compared with LDL-cholesterol in atherosclerotic cardiovascular disease risk assessment. Pathology. 2019;51(2):148−54. Available from: https://doi.org/10.1016/j.pathol.2018.11.006. Available from: https://www.ncbi.nlm.nih.gov/pubmed/30595507.

[65] Unnikrishnan R, Anjana RM, Mohan V. Diabetes mellitus and its complications in India. Nat Rev Endocrinol 2016;12(6):357−70. Available from: https://doi.org/10.1038/nrendo.2016.53. Available from: https://www.ncbi.nlm.nih.gov/pubmed/27080137.

[66] Kenny HC, Abel ED. Heart failure in type 2 diabetes mellitus. Circ Res 2019;124(1):121−41. Available from: https://doi.org/10.1161/CIRCRESAHA.118.311371. Available from: https://www.ncbi.nlm.nih.gov/pubmed/30605420.

[67] Yahagi K, Kolodgie FD, Lutter C. Pathology of human coronary and carotid artery atherosclerosis and vascular calcification in diabetes mellitus. Arterioscler Thromb Vasc Biol 2017;37(2):191−204. Available from: https://doi.org/10.1161/ATVBAHA.116.306256.

[68] Muniyappa R, Sowers JR. Role of insulin resistance in endothelial dysfunction. Rev Endocr Metab Disord 2013;14(1):5−12. Available from: https://doi.org/10.1007/s11154-012-9229-1. Available from: https://www.ncbi.nlm.nih.gov/pmc/articles/PMC3594115/.

[69] Kolka CM, Bergman RN. The endothelium in diabetes: its role in insulin access and diabetic complications. Rev Endocr Metab Disord 2013;14(1):13−19. Available from: https://doi.org/10.1007/s11154-012-9233-5.

[70] Fletcher E, Leech R, McNaughton SA, Dunstan DW, Lacy KE, Salmon J. Is the relationship between sedentary behaviour and cardiometabolic health in adolescents independent of dietary intake? A systematic review. Obes Rev 2015;16:795−805. Available from: https://doi.org/10.1111/obr.12302. Available from: https://onlinelibrary.wiley.com/doi/pdf/10.1111/obr.12302.

[71] Antuna-Puente B, Feve B, Fellahi S, Bastard JP. Adipokines: the missing link between insulin resistance and obesity. Diabetes Metab 2008;34 (1):2−11. Available from: https://www.ncbi.nlm.nih.gov/pubmed/18093861.

[72] Boden G. Obesity and free fatty acids. Endocrinol Metab Clin North Am 2008;37(3):635−46. Available from: https://doi.org/10.1016/j.ecl.2008.06.007. Available from: https://www.ncbi.nlm.nih.gov/pubmed/18775356.

Approaches in understanding endothelial function

Chapter 20

Mathematical modeling of endothelial network

Sourav Roy

Department of Microbiology and Immunology, Brody School of Medicine, East Carolina University, Greenville, NC, United States

20.1 Introduction

Endothelial cells comprise the innermost lining of all blood and lymphatic vessels. Typically, in vertebrates, the vasculature extends to all corners of the body. The endothelium is not a mere layer of cells like a "parchment paper," rather it actively participates in a multitude of physiological functions such as modulation of vasomotor tone, monitoring blood fluidity, regulating transfer of water, nutrients and leukocytes across the vascular wall, regulating innate and acquired immunity, and angiogenesis. The endothelium plays a crucial role in a plethora of diseases both as a primary determinant of pathophysiology or in a secondary role as a result of collateral damage [1]. Although much is known about how the endothelium functions, there exists a considerable gap in a "bench-to-bedside" understanding in the field of endothelial biomedicine. While endothelium is well established to be a major participant in vascular disease, a clear and concrete understanding of endothelium in health and diseases extends beyond its role in atherosclerosis [2]. A major factor in bridging the bench-to-bedside gap is the heterogeneity of the endothelium so that it can be treated both as giant monopoly of identical cells or a large collection of disparate cells. This makes it difficult to design mathematical models that can recapitulate endothelial functions. Because the endothelium is an enormous consortium of different phenotypes within the same cell each with its own identity, identification of components and events associated with health and disease is challenging [3,4]. Structural and functional heterogeneity of the endothelium reflects its role in meeting the diverse demands of underlying tissues as well as the need to adapt to and survive in distinct environments across the body. At a molecular level, mapping of endothelial-cell phenotypes to obtain vascular bed-type-specific expression profiles or "vascular post boxes" has often been employed as a strategy to model the endothelium [5–7]. The endothelium plays a major role in angiogenesis and vasculogenesis. Angiogenesis, or new blood vessel formation, is typically a multiscale process, initiated by the activation of endothelial cells (ECs, the main bricks of the capillary walls) via appropriate biochemical stimuli released both by various cells [8–10]. Vasculogenesis consists of formation of a primitive vascular network, assisted by autonomous migration, aggregation, and organization of the endothelial cells.

Vasculogenesis is a major event accompanying the onset of vertebrate life (in the embryo), with the assembly of mesoderm-derived precursors of ECs [11] in polygons with well-determined topological characteristics, guided by the oxygen transport cues to the tissues. After remodeling, these geometrical properties are conserved in the adult body, where the capillary network embedded in the tissues and stemmed by the vascular tree has the same shape as the minimal unit participating in the formation of the embryo vascular network [12–14].

Angiogenesis, on the other hand, is the formation of new vessels from existing vessels and occurs during adult life of a vertebrate as a response to various physiological and pathological processes. Myriad pathophysiological complexities like chronic inflammatory diseases such as rheumatoid arthritis and psoriasis, vasculopathies like diabetic microangiopathy, degenerative disorders like atherosclerosis and cirrhosis, tissue injury occurring in ischemia, are only a few of them.

Angiogenic development plays a pivotal role in cancer where new vessels are a source of nutrients and oxygen to malignant cells, thus allowing these cells to proliferate and remain viable, and, eventually, to cause metastases and invasion into the circulatory system [9]. Moreover, it drives the transition of cancer from dormant metastases to an aggressive status [8]. Transition into angiogenic phenotype therefore dictates a fast progression to a potentially fatal stage cancer and thus presents an important target for therapeutic interventions in most types of malignancies for researchers

Endothelial Signaling in Vascular Dysfunction and Disease. DOI: https://doi.org/10.1016/B978-0-12-816196-8.00016-3

and clinicians [15]. Thus deciphering the molecular mechanisms that facilitate angiogenesis and/or vasculogenesis is of prime importance in cancer therapeutics, for development of antiangiogenic therapies, and for the designing and optimization of drug delivery and the efficacies of these drugs to the sites of malignancies. Computational models may therefore be employed to predict optimal therapies in the context of antiangiogenic therapeutics. This chapter will describe the various computational models that are currently in use to study the vascular endothelium and summarize their ability to simulate the vasculature.

20.2 Models of the vascular system

Several experimental studies on vascular systems have established migration, proliferation, and chemotactic growth as the pivotal events in vascular endothelial function, both in physiological conditions and in pathological situations. These studies can be further refined in the context of drug delivery by use of computational models to examine the effects of spatiotemporal scales, interstitial and vessel pressure, etc., on drug delivery. This strategy helps to circumvent extremely difficult and practically nonfeasible laboratory-based biological methods and assays to study the vasculature. In that direction, in silico models can play a major role in replicating and simulating selected features of the experimental system in question and in gaining further insights.

Broadly there are two types of models that have evolved over the past two decades for modeling the flow of blood, namely, (1) the continuum and (2) the discrete model to simulate the complex processes of vasculogenesis and angiogenesis.

In continuum-based modeling, the system is approximated as a continuous series of entities or events. The individual parts are modeled very similar to their nearest neighbor, and variation across the system is accounted for by gradual transitions lacking discontinuities. In the context of angiogenesis modeling, continuum models ignore details of the constituents (e.g., cell-level details) so that new capillary growth is often modeled as changes in vascular density at the network level [16]. Continuum models are usually implemented by solving systems of differential equations that describe physical phenomena as being a continuous spread in space and/or time.

In the discrete model, the entities being modeled (e.g., endothelial cells) and their unique behaviors (e.g., proliferation) are explicitly represented for an accurate reconstruction of the system. Discrete models consider increments of time and space as distinct entities and the objects within the environment as individual, unique units. For example, behaviors of individual endothelial cells are accounted for [17,18]. Discrete models are often implemented by solving systems of differential equations at discrete locations [i.e., points on a two-dimensional (2D) grid] or by using computational algorithms where individual behaviors are explicitly modeled according to logic algorithms using discrete event simulators (e.g., agent-based modeling). Stochastic modeling approaches use probabilities to define biological phenomena. These models explicitly accommodate the randomness associated with biological processes by using probability distributions to dictate the probable outcomes of simulated events [19]. Stochastic modeling assumes that random variation and fluctuation in the system dominates the overall behavior of the system. In this chapter/review, we would primarily focus on the mathematical models of angiogenesis and its implications.

20.2.1 Continuous or continuum model of angiogenesis

Classically, the continuous or continnum model of angiogenesis (neo-vascularization) involves solving numerically coupled flux-density partial differential equations with appropriate parameters and boundary conditions. Due to physiological and clinical importance of tumor-induced angiogenesis, the focus would be on modeling for that using continuous model of angiogenesis. Secretion of tumor angiogenic factors (TAF) into surrounding tissues from cancerous cells of solid tumors are one of the first steps involved in the process [20].

In tumor tissue, the chemical gradient that is created between the tumor and existing vasculature by the diffusion of these factors in the tissue space reaches neighboring blood vessels to activate the endothelial cells to degrade the parent venule basement membranes and then migrate through the disrupted membrane toward the tumor. This process of aided by angiogenic factors like vascular endothelial growth factor (VEGF), fibroblast growth factor acidic (aFGF), and basic (bFGF), etc. Additionally, endothelial-cell receptors for these proteins combined with direct experimental evidence have shown that growth factors and disruption of these receptors have direct influence on the final structure of the capillary network [21−25].

Typically motion of endothelial cells (at or near a capillary sprout tip) is dictated by three factors: "random motility" (analogous to molecular diffusion), "chemotaxis" in response to TAF gradients [26−28], and haptotaxis in response to fibronectin gradients [29−33].

To derive the partial differential equation governing endothelial-cell motion, we first consider the total cell flux and then use the conservation equation for cell density. The three contributions to the endothelial-cell flux J_n, can be conceived to be,

$$J_n = J_{random} + J_{chemo} + J_{hapto}. \tag{20.1}$$

Description of random motility of the endothelial cells at or near the sprout tip can be approximated by the equation of the form $J_{random} = D_n \nabla n$ where D_n is a positive constant, the cell random-motility coefficient. The chemotactic flux can be modeled as $J_{chemo} = \chi(c)n\nabla c$, where $\chi(c)$ is a chemotactic function. In previous models of tumor-induced angiogenesis, $\chi(c)$ is often assumed to be a constant, meaning that endothelial cells always respond to a chemosensory stimulus (e.g., TAF) in the same manner, regardless of the stimulus concentration. A receptor kinetic law of the form

$$x(C) = x_0 \frac{k_1}{k_1 + C} \tag{20.2}$$

is the best form of modeling in this case, that reflects the more realistic assumption that chemotactic sensitivity decreases with increased TAF concentration, where x_0, the chemotactic coefficient, and k_1 are positive constants [34–38].

The influence of basement matrix fibronectin on the endothelial cells is modeled by the haptotactic flux, $J_{hapto} = \rho_0 n \nabla f$, where $\rho_0 > 0$ is the (constant) haptotactic coefficient. As we are focusing attention on the endothelial cells at the sprout tips (where there is no proliferation) and given that endothelial cells have a long half-life, in the order of months [39], we omit any birth and death terms associated with the endothelial cells. The conservation equation for the endothelial-cell density n is therefore given by

$$\frac{\partial n}{\partial t} + \nabla \cdot J_n = 0 \tag{20.3}$$

and hence the partial differential equation governing endothelial-cell motion (in the absence of cell proliferation) is,

$$\frac{\partial n}{\partial t} = Dn\nabla^2 n - \nabla \cdot (\chi(c)n\nabla c) - \nabla \cdot (\rho_0 n \nabla f). \tag{20.4}$$

To derive the TAF equation, we first of all consider the initial event of tumor-induced angiogenesis which is the secretion of TAF by the tumor cells. Once secreted, TAF diffuses into the surrounding corneal tissue and extracellular matrix and sets up a concentration gradient between the tumor and any preexisting vasculature such as the nearby limbal vessels. During this initial stage, where the TAF diffuses into the surrounding tissue (with some natural decay), we assume that the TAF concentration c satisfies an equation of the form

$$\frac{\partial c}{\partial t} = Dc\nabla^2 c - \theta c \tag{20.5}$$

where Dc is the TAF diffusion coefficient and θ is the decay rate. We will assume that the steady state of this equation establishes the TAF gradient between the tumor and the nearby vessels and provides us with the initial conditions for the TAF concentration profile. As the endothelial cells migrate through the extracellular matrix in response to this steady-state gradient [19], there is some uptake and binding of TAF by the cells [24,40].

We model this process by a simple uptake function, resulting in the following equation for the TAF concentration form:

$$\frac{\partial c}{\partial t} = -\lambda n c \tag{20.6}$$

where λ is a positive constant, and the initial TAF concentration profile is obtained from the steady state of Eq. (20.5).

Fibronectin is known to be present in most mammalian tissue and has been identified as a component of the tissue of the cornea [41–44]. In addition to this preexisting fibronectin, it is known that the endothelial cells themselves produce and secrete fibronectin [45–51] which then becomes bound to the extracellular matrix and does not diffuse [46,52].

Therefore the equation for fibronectin does not account for diffusion There is also some uptake and binding of fibronectin to the endothelial cells as they migrate toward the tumor [52]. These production and uptake processes are modeled by the following equation:

$$\frac{\partial f}{\partial t} = \omega n - \mu n f \tag{20.7}$$

where ω, μ are positive constants.

Hence the complete system of equations describing the interactions of the endothelial cells, TAF and fibronectin as detailed in the previous paragraphs is

$$\frac{\partial n}{\partial t} = Dn\nabla^2 n - \nabla\left(\frac{\chi_0 k1}{k1 + C} n\nabla c\right) - \nabla \cdot \left(\rho_0 n\nabla f\right). \tag{20.8}$$

$$\frac{\partial f}{\partial t} = \omega n - \mu n f$$

$$\frac{\partial c}{\partial t} = -\lambda n c$$

The above equations constitute the most generalized version of a continuum models, and all the models in recent years have their basis in these following equations. Parameters pertaining to these equations have been mostly obtained from experimental data [27,53]. This simplistic continuum model was employed in a seminal work [54]. All of the numerical solutions presented in this section were obtained from a finite difference approximation of the system Eq. (20.8) with appropriate boundary, initial conditions and parameter values, and elucidated a wide range of spatiotemporal behavior. Important conclusions from this modeling are as follows:

1. In the absence of haptotaxis, the regions of endothelial-cell density were observed to migrate directly across the extracellular matrix to the tumor.
2. With haptotaxis, the regions of endothelial-cell density seemed to migrate more slowly, with lateral movement between the clusters clearly visible.
3. With an appropriate choice of parameters, the endothelial cells do not connect with the tumor. It was also noted in this study [54] that the geometry of the tumor, and consequently the TAF concentration profile, clearly plays a role in influencing cell migration and therefore the shape of the capillary network.

Experimental angiogenesis in a corneal assay with implanted cornea of an animal such as a rabbit or a mouse also corroborates the findings of continuum modeling. Due to the transparency of the cornea, the growing blood vessels can be easily observed and these show the migration of tips and vascularization behind the tips (Fig. 20.1) shows tips migrating with increasing density toward the corneal tumor.

Vascularization occurs behind the evolving tips this model is also popularly known as a snail-trail model owing to the graphs of the numerical simulations as shown in Fig. 20.2.

Continuum models generally examine the response of endothelial cells to angiogenic cytokines through the process of chemotaxis. They are also important in elucidating the interactions of endothelial cells with extracellular matrix molecules (fibronectin) through haptotaxis. Numerical simulations of the continuum model [54] demonstrated that a sufficiently intense chemotactic response is crucial for the initial outgrowth of the capillary network. It also demonstrated the necessity of interaction between the endothelial cells and extracellular matrix which leads to accurately modeling of local gradients around areas of high cell density permitting lateral migration. In the absence of interactions between the endothelial cells and the extracellular matrix, cells move directly to the tumor without any significant lateral movement. These results show that the important large-scale features of angiogenesis can be captured qualitatively using a continuum model. However, important processes on a smaller scale, such as sprout branching, are not captured. This requires the development of a discrete model applicable at the level of a single individual endothelial cell.

20.2.2 Discrete model of angiogenesis

Discrete models employ cellular automata to study branching and sprouting of capillary networks in a qualitative way using coupled map lattice models, fractal models, diffusion limited aggregation models and L-systems [55−68]. Typically, cellular automata models have finite dimensions with respect to time space and state. These models have a distinct advantage of being computationally inexpensive and robust. They do not generally require exact parametrization and can provide qualitative insight into the deployment and outcome of a particular model in question. Such models lay the framework for further sophisticated discrete model for greater accuracy.

A major bottleneck with cellular automata models is the accurate description and descriptors of the state the system. Thus the transitions between the various states within the biological tissue are often unclear. Spatial movements in cellular automata models are based on nearest neighbor interactions directed by phenomenological laws. In the context of tumor-induced angiogenesis, the early and seminal work of Anderson et al. [54] forms the cornerstone and foundation for many discrete models till date. This model is a discretized form of the continuum model as elaborated by certain Eq. (20.8). An assumption of this

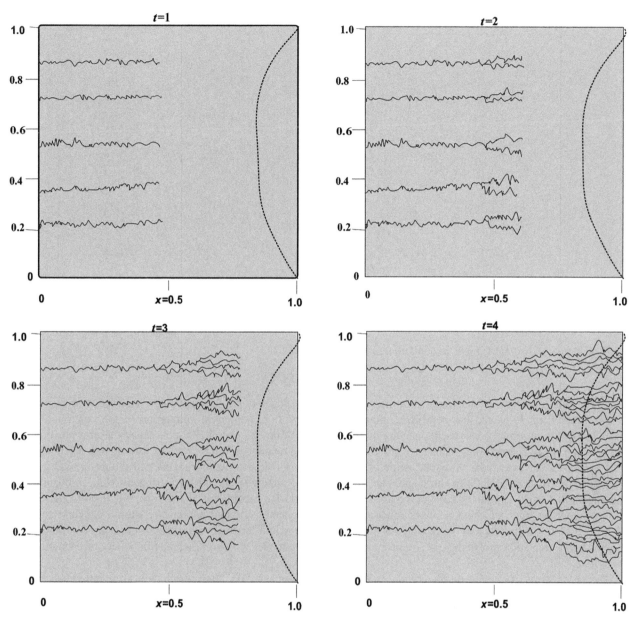

FIGURE 20.1   Progression of endothelial capillary network from a numerical simulation. The figure shows the endothelial cells at the capillary sprout tips migrating from the parent vessel ($x = 0$) toward a corneal tissue ending in ($x = 1$) in the absence of haptotaxis ($\rho = 0$). It is observed that there is very little lateral movement (in the y direction) or branching, no anastomosis and by $t = 4$ the sprouts connect with the tumor.

model is (1) that the movement a single endothelial cell located at the tip of a capillary sprout is responsible for the movement of the whole sprout (2) that the other endothelial cells lining the capillary sprout wall are contiguous [39,69].

A successful initiation method in discretizing Eq. (20.8) is to follow standard finite difference approach. This method would allow to monitor the path of a single endothelial sprout tip during the course of tumor angiogenesis under chemotactic and haptotactic stimuli. The coefficients resulting from the finite difference approach would then give probabilities of movement of a single endothelial cell in its local environment. There is an element of stochastic events in the model discussed and it is comparable to similar models [19,70,71] and different in foundation from other forms of discrete models [72]. This particular discrete model is a biased random walk describing the motion of single endothelial cells based on the system of partial differential equations. Formulation of the discrete model led to the approximation of the otherwise continuous domain of coordinates to a set of discrete points of finite dimensions "h" (mesh size) and time (t) with discrete increments in steps of "k." This was in effect achieved by Euler finite difference approximation [73].

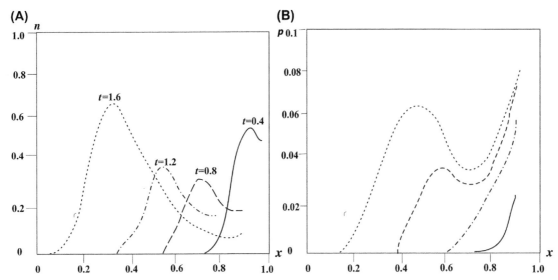

FIGURE 20.2 A tumor acting as a source for an angiogenic factor (AF) on the left boundary ($x = 0$), leading to the migration of the tip to the right boundary on 2D grid box ($x = 1$). It also shows the proliferation of tip and vessel cells: (A) spatial tip profile evolving in time and (B) spatial vessel profile evolving in time. Here "n" denotes the endothelial-cell density and "ρ" haptotactic concentration.

A generalized discrete form of the endothelial-cell equation is

$$n_{l,m}^{q+1} = n_{l,m}^{q} P_0 + n_{l+1,m}^{q} P_1 + n_{l-1,m}^{q} P_2 + n_{l,m+1}^{q} P_3 + n_{l,m-1}^{q} P_4 \tag{20.9}$$

where the subscripts specify the location on the 2D grid and the superscripts the time steps. That is $x = lh$, $y = mh$, and $t = qk$ where l, m, k, q, and h are positive integer parameters. In a numerical simulation of the continuous model [Eq. (20.8)], the purpose of the discrete equation (14) is to determine the endothelial-cell density at grid position (l,m), and time $q + 1$, by averaging the density of the four surrounding neighbors at the previous time step q. The five coefficient's $P_0 - P_4$ are essentially probabilities defining the motion of the single endothelial cell within the 2D grid box. These probability coefficient's in the context of tumor-induced angiogenesis can be attributed to the probabilities of endothelial cell being stationary P_0, moving left P_1, moving right P_2, moving up P_3, or down P_4. Each of these probabilities comprises of three components of random movement, chemotactic and haptotactic stimuli which in turn reinforces that both the discrete and the continuous models are closely knit together. Stationary probability coefficient of P_0 is similar to a transition probability of a reinforced random walk model with an inherent component of chemotaxis and haptotaxis [74].

The exact forms of $P_0 - P_4$ involve functions of the fibronectin and TAF concentrations near an individual endothelial cell. Therefore if there were no fibronectin or TAF the values of $P_1 - P_4$ would be equal, with P_0 smaller (or larger, depending on the precise values chosen for the space and time steps) which defines that there is no bias in any one direction and the endothelial cell is less (more) likely to be stationary—approximating an unbiased random walk. However, if there is a TAF gradient and no fibronectin gradient, chemotaxis dominates and the coefficients $P_0 - P_4$ will become biased toward the tumor source.

To start solving the discrete model of equations through numerical simulations for the coefficient of $P_0 - P_4$, it is essential to first formulate and integrate the branching rules. Initially it is prudent to assume that only new sprouts should progress from existing sprouts. These new sprouts are unlikely to mature and show progression in branching immediately and that there should be enough number of endothelial cells at the sprout tips for their progression. Since the new sprout becomes stunted as the tumor invades, it is evident that the endothelial-cell density required for sprouting is inversely proportional to TAF concentrations.

These assumptions lead to three conditions, which must be satisfied before a capillary sprout can branch at its tip and generate a new sprout.

1. Age of current sprout must be greater than a threshold branching age ψ, which implies that new sprouts must progress for a time period equal to ψ before branching.
2. Sufficient space for the formation of a sprout locally and nonoverlap into an existing sprout.
3. The endothelial-cell density is greater than a threshold level denoted by n_b.

It is also important to note that these rules are not exclusive for the formation of new sprouts and other equally opportune choices for new sprout formation and branching may exist. On the basis of the above three conditions, it is assumed that a singular sprout tip has a probability P_b of branching (generating new sprouts) which in turn is dependent on local TAF concentrations and gradients.

Solving the discrete equation involves numerical simulation for each time step to generate the five probabilistic coefficients of $P_0 - P_4$. Ranges involving the probability coefficients are then calculated by summation to produce five ranges $R_0 = 0 - P_0$ and $R_j = \sum_{i=0}^{j-1} P_i - \sum_{i=0}^{j} P_i$, where $j = 1,4$. To preserve the inherent probabilistic nature of the model, a random number between 0 and 1 is generated. The range of this random number in turn decides the motion of the individual endothelial cell in question, as to whether it would remain stationary (R_0) move left (R_1), right (R_2), up (R_3), or down (R_4). Depending on the strength of these ranges, the corresponding probability coefficient would be selected which in turn would dictate the movement of the individual endothelial cell into its neighboring orthogonal grid boxes or remain stationary. The basic framework of the numerical simulations entailing to the discrete model also took into account the effect of initial fibronectin gradient and the presence or absence of haptotactic effects.

Numerical simulation results emerging from the discrete model permitted the tracking of an individual endothelial cell seated at the tip of the capillary sprout thus enabling researchers to obtain the trajectory of the complete capillary sprout. Rules for the formation of new sprouts (branching), fusion of sprouts to form loops (anastomosis), and extension of sprouts through endothelial-cell proliferation are also guided by the discrete model. Discrete model simulations also reiterate the findings from the continuum model that both haptotaxis and chemotaxis are crucial to the formation of a capillary network in a large scale.

20.2.2.1 Implications of using models for vascular studies

As stated earlier, the continuum model comprises of a system of nonlinear partial differential equations. The model elucidated responses of endothelial cells to angiogenic cytokines (VEGF) via chemotaxis and essential interactions with extracellular matrix molecules (fibronectin) via haptotaxis. Continuum model simulations demonstrate that (1) for capillary network formation, a sufficiently strong chemotactic response is crucial for initial outgrowth and (2) interactions between endothelial cells and the extracellular matrix molecules are critical for capillary formation. Inclusion of uptake terms in the continuum simulations for TAF and fibronectin by the endothelial cells generate local gradients around high cell density which facilitate lateral migration. In the absence of such local gradients, endothelial cells move toward the tumor without any significant lateral motion. Such results qualitatively show the macroscopic features of angiogenesis modeled by continuum nonlinear partial differential equations. However, microscopic phenomena (like sprout branching) at the level of individual endothelial cells could not be modeled through such a model and ultimately led to the development of discrete mathematical models at cellular level. The discrete model which was derived from finite difference approximation of Eq. (20.8) and included realistic physiological phenomena like branching, anastomosis, and cell proliferation bore similarities with previous stochastic discrete models [70,71,74]. Time scales of capillary network formation as predicted by discrete models also corroborated very well with experimental dataset [40,75,76].

Both the discrete and continuum model have been employed with some accuracy in studies using transmembrane receptor tyrosine kinase (RTK). Experimental data suggest that disrupting RTK has direct impact on the structural morphogenesis of the capillary network in the form of immature capillary network that lacks complete branching structure [21,23,24]. The scope of this model can be extended by including the interaction of heparin with angiogenic cytokines like VEGF, aFGF, bFGF, and angiogenin in the bound form as well as soluble form in the extracellular matrix (ECM) [77−81]. Including interaction of heparin with ECM into these models would involve modeling the distribution of heparin and its subsequent modes of binding with TAF. Discrete and continuum used in tandem can account for the modeling of such a complex physiological phenomena. This can be best achieved by the improvised technique of using partial differential equations as a basis for discrete models to generate movements of individual cells based on a continuum model of a population of cells.

20.3 Conclusions

These computational models are a powerful means of employing microscale events into a macroscopic projection. Since these models are easily parametrizable, they can be tinkered with to modify the spatial configuration and the environment of the endothelial medium. This is especially important in angiogenesis, to emulate a realistic situation of an obstruction by a cartilage or the inclusion of a spatially heterogeneous medium which realistically mimic ECM makeup. Other factors that when introduced seamlessly into the current framework of the discrete and continuum models could make the models more robust include (1) blood flow through capillary network, (2) involvement of oxygen concentration and gradients, and (3) invasion of macrophages [82−85]. Recent theoretical studies on model development and extensions based on the framework of these models have addressed these physiological factors with more rigor and detail. The above aspects are of special importance to angiogenesis with relevance to wound healing and tissue repair. A practical relevance, importance, and application of mathematical modeling of endothelial network has proven to be crucial in developing and perfecting antiangiogenesis strategies. These strategies include preferential destruction of endothelial cells [86], inhibition of endothelial-cell proliferation via angiostatin [87], development of antichemotactic and antihaptotactic drugs [88,89]. These strategies seem to be hold immense promise in therapeutics, especially in treatment of cancer patients [90]. In particular use of these strategies as an adjuvant chemotherapy has now been recognized as an effective way to treat secondary tumors (metastases). Mathematical models of angiogenesis pave the way for perfecting, testing, application, and in further development of these therapeutic strategies in drug delivery and effective cancer treatment.

References

[1] Cines DB, Pollak ES, Buck CA, Loscalzo J, Zimmerman GA, McEver RP, et al. Endothelial cells in physiology and in the pathophysiology of vascular disorders. Blood 1998;91:3527−61.

[2] Hwa C, Sebastian A, Aird W. Endothelial biomedicine: its status as an interdisciplinary field, its progress as a basic science, and its translational bench-to-bedside gap. Endothelium 2005;12:139−51.

[3] Aird WC. Phenotypic heterogeneity of the endothelium, I: structure, function, and mechanisms. Circ Res 2007;100:158−73.

[4] Aird WC. Phenotypic heterogeneity of the endothelium, II: representative vascular beds. Circ Res 2007;100:174−90.

[5] Ruoslahti E. Vascular zip codes in angiogenesis and metastasis. Biochem Soc Trans 2004;32:397−402.

[6] Ruoslahti E, Rajotte D. An address system in the vasculature of normal tissues and tumors. Annu Rev Immunol 2000;18:813−27.

[7] Narasimhan K. Zip codes: deciphering vascular addresses. Nat Med 2002;8:116.

[8] Bussolino F, Arese M, Audero E, Giraudo E, Marchio S, Mitola S, et al. Biological aspects in tumor angiogenesis. In cancer modeling and simulation. In: Preziosi L, editor. Mathematical biology and medicine sciences. Chapman and Hall/CRC; 2003. p. 1−16.

[9] Carmeliet P, Jain RK. Angiogenesis in cancer and other diseases. Nature 2000;407:249−57.

[10] Carmeliet P. Angiogenesis in life, disease and medicine. Nature 2005;438:932−6.

[11] Risau W, Flamme I. Vasculogenesis. Annu Rev Cell Dev Biol 1995;11:73−91.

[12] Chilibeck PD, Paterson DH, Cunningham DA, Taylor AW, Noble EG. Muscle capillarization O_2 diffusion distance, and VO_2 kinetics in old and young individuals. J Appl Physiol 1997;82:63−9.

[13] Guyton A, Hall J. Textbook of medical physiology. St. Louis: W.B. Saunders; 2000.

[14] Krogh A. The number and distribution of capillaries in muscle with calculations of the oxygen pressure head necessary for supplying the tissue. J Physiol 1919;52:409−15.

[15] Straume O, Salvesen HB, Akslen LA. Angiogenesis is prognostically important in vertical growth phase melanomas. Int J Oncol 1999;15:595−9.

[16] Hogea CS, Murray BT, Sethian JA. Simulating complex tumor dynamics from avascular to vascular growth using a general level-set method. J Math Biol 2006;53(1):86−134.

[17] Levine HA, Sleeman BD, Nilsen-Hamilton M. A mathematical model for the roles of pericytes and macrophages in the initiation of angiogenesis. I. The role of protease inhibitors in preventing angiogenesis. Math Biosci 2000;168(1):77−115.

[18] Mac Gabhann F, Ji JW, Popel AS. Multi-scale computational models of pro-angiogenic treatments in peripheral arterial disease. Ann Biomed Eng 2007;35(6):982−94.

[19] Stokes CL, Lauffenburger DA. Analysis of the roles of microvessel endothelial cell random motility and chemotaxis in angiogenesis. J Theor Biol 1991;152(3):377−403.

[20] Folkman J, Klagsbrun M. Angiogenic factors. Science 1987;235:442−7.

[21] Dumont DJ, Gradwohl G, Fong GH, Puri MC, Gertsenstein M, Auerbach A, et al. Dominant-negative and targeted null mutations in the endothelial receptor tyrosine kinase, TEK, reveal a critical role in vasculogenesis of the embryo. Genes Dev 1994;8:1897−909.

[22] Fong GH, Rossant J, Gertsenstein M, Breitman ML. Role of the FLT-1 receptor tyrosine kinase in regulating the assembly of vascular endothelium. Nature 1995;376:66−70.

[23] Sato TN, Tozawa Y, Deutsch U, Wolburgbuchholz K, Fujiwara Y, Gendronmaguire M, et al. Distinct roles of the receptor tyrosine kinases TIE-1 and TIE-2 in blood-vessel formation. Nature 1995;376:70−4.

[24] Hanahan D. Signaling vascular morphogenesis and maintenance. Science 1997;227:48−50.

[25] Scalerandi M, Sansone BC, Condat CA. Diffusion with evolving sources and competing sinks: development of angiogenesis. Phys Rev E Stat Nonlin Soft Matter Phys 2002;65.

[26] Terranova VP, Diflorio R, Lyall RM, Hic S, Friesel R, Maciag T. Human endothelial cells are chemotactic to endothelial cell growth factor and heparin. J Cell Biol 1985;101:2330−4.

[27] Stokes CL, Rupnick MA, Williams SK, Lauffenburger DA. Chemotaxis of human microvessel endothelial cells in response to acidic fibroblast growth factor. Lab Invest 1990;63:657–68.

[28] Stokes CL, Lauffenburger DA, Williams SK. Migration of individual microvessel endothelial cells: stochastic model and parameter measurement. J Cell Sci 1991;99:419–30.

[29] Schor SL, Schor AM, Brazill GW. The effects of fibronectin on the migration of human foreskin fibroblasts and Syrian hamster melanoma cells into three-dimensional gels of lattice collagen fibres. J Cell Sci 1981;48:301–14.

[30] Bowersox JC, Sorgente N. Chemotaxis of aortic endothelial cells in response to fibronectin. Cancer Res 1982;42:2547–51.

[31] Quigley JP, Lacovara J, Cramer EB. The directed migration of B-16 melanoma-cells in response to a haptotactic chemotactic gradient of fibronectin. J Cell Biol 1983;97:A450–1.

[32] Lacovara J, Cramer EB, Quigley JP. Fibronectin enhancement of directed migration of B16 melanoma cells. Cancer Res 1984;44:1657–63.

[33] McCarthy JB, Furcht LT. Laminin and fibronectin promote the directed migration of B16 melanoma cells in vitro. J Cell Biol 1984;98:1474–80.

[34] Lapidus IR, Schiller R. Model for the chemotactic response of a bacterial population. Biophys J 1976;16:779–89.

[35] Lauffenburger D, Aris R, Kennedy CR. Travelling bands of chemotactic bacteria in the context of population growth. Bull Math Biol 1984;46:19–40.

[36] Sherratt JA. Chemotaxis and chemokinesis in eukaryotic cells: the Keller–Segel equations as an approximation to a detailed model. Bull Math Biol 1994;56:129–46.

[37] Woodward DE, Tyson R, Myerscough MR, Murray JD, Budrene EO, Berg HC. Spatio-temporal patterns generated by *Salmonella typhimurium*. Biophys J 1995;68:2181–9.

[38] Olsen L, Sherratt JA, Maini PK, Arnold F. A mathematical model for the capillary endothelial cell-extracellular matrix interactions in wound-healing angiogenesis. J Math Appl Med Biol 1997;14:261–81.

[39] Paweletz N, Knierim M. Tumor-related angiogenesis. Crit Rev Oncol Hematol 1989;9:197–242.

[40] Ausprunk DH, Folkman J. Migration and proliferation of endothelial cells in preformed and newly formed blood vessels during tumour angiogenesis. Microvasc Res 1977;14:53–65.

[41] Kohno T, Sorgente N, Ishibashi T, Goodnight R, Ryan SJ. Immunofluorescent studies of fibronectin and laminin in the human eye. Invest Opthamol VisSci 1987;28:506–14.

[42] Kohno T, Sorgente N, Patterson R, Ryan SJ. Fibronectin and laminin distribution in bovine eye. Jpn J Opthamol 1983;27:496–505.

[43] Ben-Zvi A, Rodrigues MM, Krachmer JH, Fujikawa LS. Immunohistochemical characterisation of extracellular matrix in the developing human cornea. Curr Eye Res 1986;5:105–17.

[44] Sramek SJ, Wallow IH, Bindley C, Sterken G. Fibronectin distribution in the rat eye. An immunohistochemical study. Invest Opthamol Vis Sci 1987;28:500–5.

[45] Birdwell CR, Gospodarowicz D, Nicolson GL. Identification, localisation and role of fibronectin in cultured endothelial cells. Proc Natl Acad Sci USA 1978;75:3273–7.

[46] Birdwell CR, Brasier AR, Taylor LA. Two-dimensional peptide mapping of fibronectins from bovine aortic endothelial cells and bovine plasma. Biochem Biophys Res Commun 1980;97:574–81.

[47] Jaffee EA, Mosher DF. Synthesis of fibronectin by cultured endothelial cells. J Exp Med 1978;147:1779–91.

[48] Macarak EJ, Kirby E, Kirk T, Kefalides NA. Synthesis of cold-insoluble globulin cultured by calf endothelial cells. Proc Natl Acad Sci USA 1978;75:2621–5.

[49] Monaghan P, Warburton MJ, Perusinghe N, Rutland PS. Topographical arrangement of basement membrane proteins in lactating rat mammary gland: comparison of the distribution of Type IV collagen, laminin, fibronectin and Thy-1 at the ultrasructural level. Proc Natl Acad Sci USA 1983;80:3344–8.

[50] Rieder H, Ramadori G, Dienes HP, Meyer zum Buschenfelde KH. Sinusoidal endothelial cells from guinea pig liver synthesize and secrete cellular fibronectin in vitro. Hepatology 1987;7:856–86.

[51] Sawada H, Furthmayr H, Konomi H, Nagai Y. Immuno-electron microscopic localization of extracellular matrix components produced by bovine corneal endothelial cells in vitro. Exp Cell Res 1987;171:94–109.

[52] Hynes RO. Fibronectins. New York, NY: Springer-Verlag; 1990.

[53] Rupnick MA, Stokes CL, Williams SK, Lauffenburger DA. Quantitative analysis of human microvessel endothelial cells using a linear under-agarose assay. Lab Invest 1988;59:363–72.

[54] Anderson ARA, Chaplain MAJ. Continuous and discrete mathematical models of tumor-induced angiogenesis. Bull Math Biol 1998;60:857–900.

[55] Bell AD, Roberts D, Smith A. Branching patterns: the simulation of plant architecture. J Theor Biol 1979;81:351–75.

[56] Bell AD. The simulation of branching patterns in modular organisms. Philos Trans R Soc Lond 1986;B313:143–59.

[57] Gottlieb ME. Modelling blood vessels: a deterministic method with fractal structure based on physiological rules. In: Proceedings of the 12th international conference of IEEE EMBS. New York, NY: IEEE Press; 1990. p. 1386–1387.

[58] Gottlieb ME. The VT model: a deterministic model of angiogenesis and biofractals based on physiological rules. In: Proceeding of IEEE 17th annual northeast bioengineering conference. New York, NY: IEEE Press; 1991a. p. 38–39.

[59] Gottlieb ME. Vascular networks: fractal anatomies from non-linear physiologies. IEEE Eng Med Bio Mag 1991;13:2196–7.

[60] Düchting W. Tumor growth simulation. Comput Graph 1990;14:505–8.

[61] Düchting W. Computer simulation in cancer research. In: Meöller DPF, editor. Advanced simulation in biomedicine. New York, NY: Springer-Verlag; 1990. p. 117–39.

[62] Düchting W. Simulation of malignant cell growth. In: Encarncao JL, Peitgen H-O, Sakas G, Englert G, editors. Fractal geometry and computer graphics. New York, NY: Springer-Verlag; 1992. p. 135–43.

[63] Prusinkiewicz P, Lindenmayer A. The algorithmic beauty of plants. New York, NY: Springer-Verlag; 1990.

[64] Kiani M, Hudetz A. Computer simulation of growth of anastomosing microvascular networks. J Theor Biol 1991;150:547–60.

[65] Landini G, Misson G. Simulation of corneal neo-vascularization by inverted diffusion limited aggregation. Invest Opthamol Vis Sci 1993;34:1872–5.

[66] Indermitte C, Liebling Th M, Cĺemonçon H. Culture analysis and external interaction models of mycelial growth. Bull Math Biol 1994;56:633–64.

[67] Düchting W, Ulmer W, Ginsberg T. Cancer: a challenge for control theory and computer modelling. Eur J Cancer 1996;32A:1283–92.

[68] Nekka F, Kyriacos S, Kerrigan C, Cartilier L. A model of growing vascular structures. Bull Math Biol 1996;58:409–24.

[69] Pettet G, Chaplain MAJ, McElwain DLS, Byrne HM, On the rôle of angiogenesis in wound healing. Proc. R Soc Lond **B**.1996;263:1487–1493.

[70] Weimar JR, Tyson JJ, Watson LT. Diffusion and wave-propagation in cellular automaton models of excitable media. Physica 1992; D55:309–27.

[71] Weimar JR, Tyson JJ, Watson LT. Third generation cellular automaton for modelling excitable media. Physica 1992;D55:327–39.

[72] Dallon JC, Othmer HG. A discrete cell model with adaptive signalling for aggregation of *Dictyostelium discoideum*. Philos Trans R Soc Lond 1997;B352:391–417.

[73] Mitchell AR, Griffiths DF. The finite difference method in partial differential equations. Chichester: Wiley; 1980.

[74] Othmer H, Stevens A. Aggregation, blowup and collapse: the ABCs of taxis and reinforced random walks. SIAM J Appl Math 1997;57:1044–81.

[75] Gimbrone MA, Cotran RS, Leapman SB, Folkman J. Tumor growth and neovascularization: an experimental model using the rabbit cornea. J Natl Cancer Inst 1974;52:413–27.

[76] Muthukkaruppan VR, Kubai L, Auerbach R. Tumor-induced neovascularization in the mouse eye. J Natl Cancer Inst 1982;69:699–705.

[77] D'Amore PA, Klagsbrun M. Endothelial cell mitogens derived from retina and hypothalamus—biochemical and biological similarities. J Cell Biol 1984;99:1545–9.

[78] Gospodarowicz D, Cheng J, Lui GM, Baird A, Bohlen P. Isolation of brain fibroblast growth-factor by heparin-Sepharose affinity-chromatography—identity with pituitary fibroblast growth-factor. Proc Natl Acad Sci USA 1984;81:6963–7.

[79] Lobb RR, Fett JW. Purification of two distinct growth-factors from bovine neural tissue by heparin affinity-chromatography. Biochemistry 1984;23:6295–9.

[80] Maciag T, Mehlman T, Friesel R, Schreiber AB. Heparin binds endothelial cell-growth factor, the principal endothelial cell mitogen in bovine brain. Science 1984;225:932–5.

[81] Sullivan R, Klagsbrun M. Purification of cartilage-derived growth-factor by heparin affinity-chromatography. J Biol Chem 1985;260:2399–403.

[82] Knighton DM, Hunt TK, Scheuenstuhl H, Halliday BJ. Oxygen tension regulates the expression of angiogenesis factor by macrophages. Science 1983;221:1283–5.

[83] Polverini PJ, Cotran RS, Gimbrone Jr. MA, Unanue ER. Activated macrophages induce vascular proliferation. Nature 1977;269:804–6.

[84] Knighton DM, Silver IA, Hunt TK. Regulation of wound-healing angiogenesis—effect of oxygen gradients and inspired oxygen concentration. Surgery 1981;90:262–70.

[85] Lewis CE, Leek R, Harris A, McGee JOD. Cytokine regulation of angiogenesis in breast-cancer—the role of tumour-associated macrophages. J Leukoc Biol 1995;57:747–51.

[86] Brooks PC, Montgomery AMP, Rosenfeld M, Reisfled RA, Hu T, Klier G, et al. Integrin $\alpha v\beta 3$ antagonists promote tumor regression by inducing apoptosis of angiogenic blood vessels. Cell 1994;79:1157–64.

[87] O'Reilly MS, Holmgren L, Shing Y, Chen C, Rosenthal RA, Moses M, et al. Angiostatin: a novel angiogenesis inhibitor that mediates the suppression of metastases by a Lewis lung carcinoma. Cell 1994;79:315–28.

[88] Bussolino F, Di Renzo MF, Ziche M, Bocchietto E, Olivero M, Naldini L, et al. Hepatocyte growth factor is a potent angiogenic factor which stimulates endothelial cell motility and growth. J Cell Biol 1992;119:629–41.

[89] Yamada KM, Olden K. Fibronectin-adhesive glycoproteins of cell surface and blood. Nature 1978;275:179–84.

[90] Harris AL. Anti-angiogenesis for cancer therapy. Lancet 1997;349(Suppl. II):13–15, 81.

Chapter 21

Comparative assessment of electrocardiographic parameters of some birds—an essential diagnostic tool in veterinary practice

Oyebisi Mistura Azeez[1], Sirajo Garba[2], Afisu Basiru[1], Adakole Sylvanus Adah[1], Folashade Helen Olaifa[1], Soliu Akanni Ameen[3], Hauwa Motunrayo Ambali[3], Moshood Bolaji[4] and Rashidat Bolanle Balogun[5]

[1]*Department of Veterinary Physiology and Biochemistry, University of Ilorin, Ilorin, Nigeria,* [2]*Department of Veterinary Medicine, Usmanu Danfodiyo University, Sokoto, Nigeria,* [3]*Department of Veterinary Medicine, University of Ilorin, Ilorin, Nigeria,* [4]*Department of Veterinary Pathology, University of Ilorin, Ilorin, Nigeria,* [5]*Veterinary Teaching Hospital, University of Ilorin, Ilorin, Nigeria*

21.1   Introduction

21.1.1   General consideration

Birds have an efficient cardiovascular systems to meet the metabolic demands of flight. Like mammals, birds have a 4-chambered heart consisting of 2 atria & 2 ventricles. The right ventricle pumps blood to the lungs, while the left ventricle pumps blood to the rest of the body. Early in the course of cardiac disease, birds often do not show any signs of cardiovascular disease (CVD) and may often present acute death without symptoms or history. Among the currently employed diagnostics, electrocardiography has emerged as an efficient tool of avian medicine.

Electrocardiography in clinical practice is the recording at the body surface of electric fields generated by the heart. Specific waveforms represent stages of myocardial depolarization and repolarization. The electrocardiogram (ECG) is basic and valuable diagnostic tool in veterinary medicine and relatively easy to acquire. It is the initial test of choice in the diagnosis of cardiac arrhythmia and may also yield information regarding chamber dilation and hypertrophy.

When arrhythmia is detected during physical examination, such bird should be sent for electrocardiographic assessment. This may include bradycardia, tachycardia, or irregularity in rhythm that is not secondary to respiratory sinus arrhythmia. Animals with a history of syncope or episodic weakness may have cardiac arrhythmias, and an electrocardiography is indicated in these cases. Arrhythmias in such cases may be transient; a normal ECG result does not rule out transient arrhythmias. In some cases, long-term electrocardiographic monitoring (Holter monitor or cardiac event recorder) is warranted. Arrhythmias often accompany significant heart disease and may significantly affect the clinical status of the patient. An electrocardiograph assessment should be advised in animals with significant heart disease, which can be combined with echo cardiogram. The electrocardiogram is also used to monitor efficacy of antiarrhythmic therapy and determine whether arrhythmias may have developed secondary to cardiac medications (e.g., digoxin). Significant arrhythmias may also occur in animals with systemic disease, including those diseases associated with electrolyte abnormalities (hyperkalemia, hyponatremia, hypercalcemia, and hypocalcemia), neoplasia (particularly splenic neoplasia), gastric dilatation-volvulus, and sepsis [1].

Cardiac chamber enlargement: Changes in waveforms may provide indirect evidence of cardiac chamber enlargement. The ECG result may be normal, however, in cases with chamber enlargement. Right ventricular hypertrophy most consistently results in waveform changes. As heart disease progresses, waveform changes may indicate progressive chamber enlargement. Thoracic radiography and, ideally, echocardiography should be performed for definitive assessment of chamber enlargement. The ECG may provide evidence of pericardial effusion (electrical alternans, low-

Endothelial Signaling in Vascular Dysfunction and Disease. DOI: https://doi.org/10.1016/B978-0-12-816196-8.00005-9

amplitude complexes). Electrocardiographic abnormalities are often present with hypothyroidism and hyperthyroidism. A pronounced sinus arrhythmia may be present in animals with elevated vagal tone (often seen with diseases affecting the respiratory tract, central nervous system, and gastrointestinal tract) [1].

Recording the electrocardiogram: The ECG should be recorded in an area as quiet and as free of distraction as possible. Noises from clinical activity and other animals may significantly affect rate and rhythm. Any use of electrically operated equipment, such as clippers, may cause interference and should be minimized during recording of electrocardiogram. In some cases, fluorescent lighting may result in electrical interference. The patient should ideally be placed in right lateral recumbency. Electrocardiographic reference values were obtained from animals in right lateral recumbency. Limbs should be held perpendicular to the body. Each pair of limbs should be held parallel, and limbs should not be allowed to contact one another. The animal should be held as still as possible during the recording of ECG. When possible, panting should be avoided. When dyspnea or other factors prevent standard positioning, the ECG may be recorded while the animal is standing or, less ideally, sitting. Alligator clips or adhesive electrodes may be used. For reduction of discomfort, teeth of alligator clips should be blunted, and the spring should be relaxed. Limb electrodes are placed either distal or proximal to the elbow (caudal surface) and over the stifle. Electrodes placed proximal to the elbow may increase respiratory artifact. Each electrode should be wetted with 70% isopropyl alcohol to ensure electrical contact [1]. On the other hand ultrasound gel can be applied to the recording sight before clipping for the same purpose.

Cardiac conduction and genesis of waveforms: The function of the cardiac conduction system is to coordinate the contraction and relaxation of the four cardiac chambers. For each cardiac cycle, the initial impulse originates in the sinoatrial (SA) node located in the wall of the right atrium near the entrance of the cranial vena cava. This impulse is rapidly propagated through the atrial myocardium, with a resulting depolarization of the atria. Depolarization of the atria results in the P wave and atrial contraction. The initial S-A nodal impulse is small and does not produce an electrocardiographic change on the body's surface. Immediately after atrial depolarization, the impulse travels through the atrioventricular (AV) node, located near the base of the right atrium. Conduction is slow here, which allows atrial contraction to be completed before ventricular depolarization occurs. As the impulse travels through the AV node, there is no electrocardiographic activity on the body's surface—rather, the PR-interval is generated. On leaving the AV node, conduction velocity increases significantly, and the impulse is rapidly spread through the bundle of His, bundle branches, and Purkinje system. This results in rapid and nearly simultaneous depolarization of the ventricles. Depolarization of the ventricles results in the prominent QRS complex and causes ventricular contraction. The Q wave represents initial depolarization of the interventricular septum and is defined as the first negative deflection following the P wave and occurring before the R wave. A 'Q' wave may not be identified in all animals. The R wave represents depolarization of the ventricular myocardium from the endocardial surface to the epicardial surface. The R wave is the first positive deflection following the P wave and is usually the most prominent waveform. The S wave represents depolarization of the basal sections of the ventricular posterior wall and interventricular septum. The S wave is defined as the first negative deflection following the R wave in the QRS complex. Ventricular repolarization quickly follows ventricular depolarization and results in the T wave. The delay in repolarization results in the ST-segment on the surface electrocardiograph [1].

The electrocardiogram should be evaluated from left to right. Areas of artifact should be identified and avoided in the evaluation. Calculate the heart rate (HR). Determine the number of R waves (or R-R intervals) within 3 seconds and multiply by 20 to obtain beats per minute (beats/min) (for an ECG recorded at 50 mm/s, vertical timing marks above the gridlines occur every 1.5 seconds). If the rhythm is regular, the HR may be derived by determining the number of small boxes in one R-R interval and dividing 3000 by that number (for paper speed of 25 mm/second, use 1500). The method is also useful for determining the rate of paroxysmal ventricular tachycardia and other arrhythmias lasting less than 3 seconds [1].

The avian cardiovascular system is highly developed to accommodate the specialized requirements of the abilities to fly, run, and/or swim [2,3]. The avian cardiovascular system is also designed to match the high metabolic demand of birds and it comprises of a large heart, high HR, high cardiac output, and high blood pressure. This chapter summarizes the current state of the field of ECG in avian medicine with a focus on the clinical manifestations and their association with interpretation of ECG data.

21.1.2 Anatomical description

The anatomy and physiology of the avian heart allow for highly efficient blood circulation and oxygen delivery [2]. The avian heart is situated cranially within the coelomic cavity on the ventral midline. There is no diaphragm in birds, the right and left lobes of the liver enclose the apex of the heart on each side [4]. Avian cardiac muscle cells are one-fifth to one-tenth the diameter of mammalian cardiac myocytes and lack the M-band and transverse tubules (T-tubules) found in mammalian cardiac muscle [2]. The avian heart is proportionally larger than mammals relative to body mass [4]. The proportion of body weight taken up by the heart increases as the size of the bird decreases [4].

These unique high-performance features of the avian heart enable birds to fly, run, or dive. However, pet birds often are compromised in their cardiovascular capabilities by restricted exercise, poor diet, and abnormal climate. These factors may predispose for cardiovascular disease in captive birds [5]. Many times backyard poultry and companion birds are presented for evaluation of poor performance. Avian practitioners have given relatively little attention to electrocardiography probably because of scarcity of reference values in companion, zoo and wild birds [6]. Although the avian cardiovascular system shares many similarities with those of mammals, several unique and specific details exist. The mass of the avian heart is nearly twice as large as that of a mammal of comparable size and it is designed to meet high-performance demands [7]. Stroke volume and cardiac output in avian species are higher than those of mammals. One of the most useful tools for evaluation of cardiac function of animals in both infectious and noninfectious condition is electrocardiography [6].

The mean deflection of QRS complex in birds is negative and not positive as in mammals. It was demonstrated in 1949 that this mean electrical axis in ventricular depolarization in birds occurs because the depolarization wave begins epicardially and then spreads through the myocardium toward the endocardium. The ECG may be of help in diagnosing diseases that cause signs of lethargy, fatigue seizure, fever, and collapse. Metabolic, cardiac, neurologic, and systemic diseases that cause toxemia can cause one or all of these clinical changes. Avian HR are so rapid that inspecting and measuring the tracing are less accurate at lower speed. ECG machine with a speed above the 50 mm/s needs to be selected. The machine should be standardized at 1 cm = 1 mV. The calibration and the paper speed should always be marked on the ECG paper with date, time, name, and case number of the patient [8].

In veterinary medicine, avian cardiology is one of the previously neglected and presently developing areas. It is always advised to carry out the electrocardiography without any anesthesia which will be facilitated during the interpretation. The electrocardiograph of most birds show inverted QRS wave in lead II, which is indicative of negative mean electrical axis. To record the regular parameters standard bipolar limb lead II is considered as the standard lead for analysis. The morphology of P wave, PR-interval, QrS complexes, ST-segment, T wave, QT-interval, and U wave was analyzed in lead II.

21.1.3 Clinical presentation

Heart disease should be suspected when patients are presented with dyspnea, coughing, weakness, lethargy, exercise intolerance, collapse/syncope, or coelomic distention and more [4,9]. Early in the course of cardiac disease, birds may present without any obvious signs. Birds may also be presented for acute death singly or in flock with no history to suggest the presence of heart disease [4,9,10]. Detailed history is important to help determine risk factors that may contribute to heart disease.

21.1.4 Predisposing factors

Predisposing factors to heart disease in birds may include species, age, gender, and diet in the development of heart disease in companion avian species [6,11]. For example broilers (*Gallus domesticus*) often develop pulmonary hypertension and ascites syndrome, a spontaneous cardiomyopathy thought to develop as a result of increased demand placed on the cardiovascular system by the large breast muscle mass and rapid growth [4,12]. Copper deficiency may be linked to the formation of dissecting aortic aneurysms in turkeys and ostriches [4].

21.1.5 Methods of examination

- The physical examination is an important diagnostic tool in avian medicine. Birds should first be evaluated in the cage at a close look to assess them at rest. Auscultation of the heart is best performed on the left and right ventral thorax at the base of the sternum [6]. Muffled heart sounds may indicate pericardial effusion or hepatomegaly [13].
- Complete blood cell count and biochemical analysis.
- Radiography—there is need for understanding normal avian radiographic anatomy to give accurate interpretation of the radiographs.

- Ultrasound can help to evaluate the coelomic cavity for effusion and soft-tissue masses. Ultrasound is a useful tool, but poses some difficulty in birds because internal organs are surrounded by air sacs.
- Electrocardiography demonstrates the summed electrical activity of the heart and can be utilized to measure HR and detect arrhythmias, cardiac chamber enlargement, and electrical conductance abnormalities [6]. In birds, the mean electrical axis is negative (and thus the QrS wave is inverted in lead II); however, in many other respects the avian ECG is similar to mammals [2]. The standard bipolar limb lead II is commonly used to evaluate ECG waveforms in birds [6]. Avian ECG differs somewhat from its mammalian counterpart in that the rS wave is inverted in most species as a norm [14]; it showed that the avian ECG, unlike the human ECG, had a deep inverted S wave but no R wave [15]. To evaluate morphologic features of the avian ECG, it is important to have an ECG machine that has a speed of 100 mm/s or greater (may need 200 mm/s in very small patients). At slower speeds the ECG waveforms will be too close together.
- This work was designed to assess the electrocardiographic parameters of various birds as alternative/additional means of clinical diagnosis. The objective of this study was to identify every aspect of the Lead II ECG waveform. The electrocardiogram is a useful tool in avian medicine as it can be utilized to measure HR and detect arrhythmias, cardiac chamber enlargement, and electrical conductance abnormalities [16].

21.2 Materials and method

EDAN 10 veterinary electrocardiographic equipment made in China, with a 200 mm/s paper speed and a sensitivity of 100 mm/mV, was used to measure the ECG. The five alligator clip electrodes were fixed directly to the skin under the feather—on the forearms (muscular part of the wing), on the hind limbs above the stifle joint, and the heart as described earlier by Azeez et al. [13]. Birds were placed on dorso-lateral recumbency. The EDAN was connected to the laptop and information about each bird was recorded and saved. Birds considered include broilers, domestic duck, white geese, Chinese geese, laying birds (chicken), point of lay birds, and turkey. They were all carefully restrained. Five birds from each group were used.

21.3 Our findings

21.3.1 Results

The ECG exhibited positive P wave, inverted (Q)rS, and positive T wave in all of them. S-S interval was regular in turkey and duck, and irregular in chicken and Chinese geese. The PR-interval in the laying birds and broilers was very long with overlap by QRS. The (Q)rS was shorter (29−44 ms) in the chicken with very short amplitude, and longer (50−65 ms) in turkey and duck with longer amplitude. No significant difference was observed in the (Q)rS within the groups. QT-interval was longer in turkey, geese, and duck (297−456 ms) but shorter in chicken (Table 21.1).

21.3.1.1 ECG patterns in different birds
Figs. 21.1−21.7

TABLE 21.1 ECG parameters of birds.

	Broilers	Domestic duck	White geese	Chinese geese	Point of lay	Laying birds	Turkey
HR (bpm)	394 ± 6.4	158 ± 4.1	125 ± 2.1	87 ± .97	375 ± 8.97	294 ± 3.6	198 ± 2.4
P (ms)	70 ± 1.9	92 ± 2.31	61 ± 1.54	61 ± 1.1	67 ± 1.34	21 ± 0.3	88 ± 1.2
PR (ms)	76 ± 1.7	120 ± 3.17	95 ± 3.21	118 ± 3.3	84 ± 1.74	30 ± 0.67	98 ± 1.6
(Q)rS (ms)	50 ± 1.0	33 ± 1.24	32 ± 0.95	37 ± 0.48	25 ± 0.57	31 ± 1.0	50 ± 0.45
QT (ms)	108 ± 2.3	343 ± 11.23	456 ± 11.4	212 ± 5.1	107 ± 1.9	116 ± 2.3	297 ± 6.1
QTc (ms)	276 ± 3.9	556 ± 12.94	658 ± 14.62	255 ± 7.4	267 ± 6.3	256 ± 5.4	539 ± 8.1

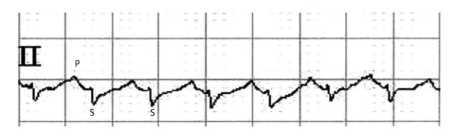

FIGURE 21.1 Electrocardiographic record of waves in lead II of normal parameters in broilers showing inverted S. S-S interval is irregular.

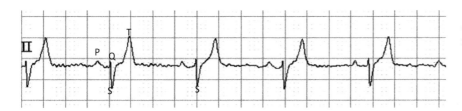

FIGURE 21.2 Electrocardiographic record of waves in lead II of domestic duck showing inverted S. S-S interval is regular.

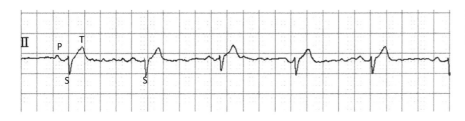

FIGURE 21.3 Electrocardiographic record of waves in lead II of white geese showing inverted S. S-S interval is regular.

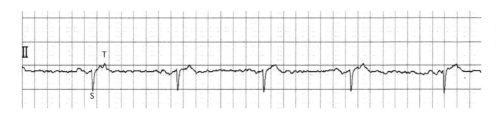

FIGURE 21.4 Electrocardiographic record of waves in lead II of Chinese geese showing inverted S. S-S interval is regular.

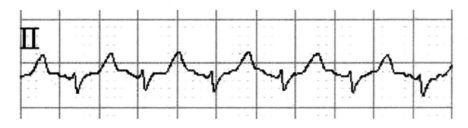

FIGURE 21.5 Electrocardiographic record of waves in lead II of point of lay birds showing inverted S. S-S interval is regular.

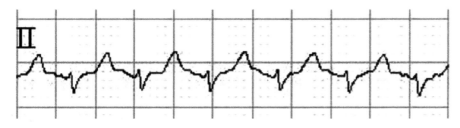

FIGURE 21.6 Electrocardiographic record of waves in lead II of laying birds showing inverted S. S-S interval is regular.

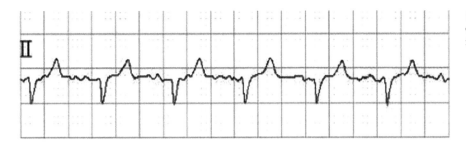

FIGURE 21.7 Electrocardiographic record of waves in lead II of turkey showing inverted S. S-S interval is regular.

21.3.2 Discussion

In birds, the mean electrical axis is negative (and thus the QrS wave is inverted in lead II); however, in many other respects the avian ECG is similar to mammals. From our findings, HR in broilers, point of lay birds, and laying birds was significantly higher than in geese, turkey, and duck. We suspect that broiler, laying birds, and point of lay birds are kept intensively where there is restricted movement. They have lighter body weight compared with geese, duck, or turkey. Boredom, accumulation of fat as they age is common with broilers and the laying birds. This could be a contributing factor to increased HR. There is no significant difference in the P and PR-interval of all the groups. However, P and PR-interval in the laying birds are significantly lower compared with others. There is no significant difference in the QrS complex timing of all the groups. Amplitude of S is significantly higher in geese, duck, and turkey compared with laying birds, point of lay birds, and broilers. QT and QTc are significantly higher in the duck, geese, and turkey. For this birds, reference Electrocardiographic assessment for individual species are not available.

The avian ECG also starts with a P wave, but in some groups of birds it is followed by a slight depression of the initial part of the PR-interval. This depression, referred to as the Ta wave, represents atrial repolarization and appears to be a physiologic phenomenon as seen in almost all the birds sampled above. The second difference is the form of the QrS complex, which should more accurately be described as an rS complex as described by Lumeij and Ritchie [6] and Yogeshpriya et al. [15]. This is due to the fact that the ventricular depolarization in the avian heart starts subepicardially spreading via the myocardium to the endocardial surface, in contrast to the mammalian heart, where it spreads from the endocardial side to the epicardial side. The ST-segment is rather short or absent. When present, as found in the turkey, laying birds, and broilers it is often elevated or shows ST-slurring (S wave directly merging into the T wave). The ST-segment is absent in the geese and duck. This is similar to the finding of Yogeshpriya et al. [15] in pigeons or psittacine birds. In mammals, an elevated ST-segment or ST-slurring is associated with cardiac disease. The T wave is opposite to the direction of the ventricular complex and always positive in lead II. The P-on-T phenomenon (the P wave is superimposed on the T wave) is a normal finding as described by Yogeshpriya et al. [15].

Mechanisms: There are three types of cells that provide the histologic components of the conduction system. The pacemaker cells or P-cells are small, spherical cells in the nodes that have repetitive spontaneous depolarizations. Transitional cells or T-cells are intermediate in morphology as they have smaller numbers of myofibrils than cardiac cells and are much smaller. Purkinje cells are large, elongated cells that are more rectangular in shape. They transmit the electrical impulse through the substance of the myocardium or heart muscle. In mammals, the Purkinje cells conduct the electrical impulse at a faster rate than the myocytes (muscle cells). The size of birds' Purkinje cells is greater than their myocytes, so the conduction velocity of birds is greater than that of mammals because bird hearts beat much faster than mammals. High levels of ionized serum calcium concentration were suspected to be the cause of shortened ST-segment in the birds. The bird hearts, even if fast beating (e.g., in finch and hummingbird), have no T-tubules, despite fiber sizes comparable to those of mammalian ventricle, but are rich in EjSR/corbular SR. It is known that cardiac myocytes contain three categories of calcium release units all bearing arrays of Ryanodine Receptor 2 (RyR2): peripheral couplings, constituted of an association of the junctional SR (jSR) with the plasmalemma. Perni et al. [17] found out that chicken heart has a spatial relationship between RyR2 clusters in jSR of peripheral couplings and clusters of intra-membrane particles identifiable as voltage-sensitive calcium channels (CaV1.2) in the adjacent plasmalemma. This provides the structural basis for initiation of the heartbeat.

Conclusion: Electrocardiography is a useful diagnostic tool in birds. The unique high-performance features of the avian heart enable birds to fly, run, or dive. However, pet birds often are compromised in their cardiovascular capabilities by restricted exercise, poor diet, and abnormal climate. In birds, severe cardiac histopathology is not always reflected in ECG abnormalities. While interpreting ECG, clinicians should always consider history, clinical findings, and laboratory results before final diagnosis. More emphasis should be placed on use of ECG by veterinarians and clinicians in handling cases of cardiovascular issues in birds.

21.4 Perspectives in therapy

This includes vaccination regimen and medications for the poultry in general.

The under-mentioned vaccination regimen should be adhered to avoid outbreak of the associated disease conditions (Tables 21.2 and 21.3).

21.4.1 Turkeys

The most common diseases are Fowl pox, Fowl cholera, Fowl typhoid (FT), mycoplasmosis, Blue comb/blackhead, and roundworm infections. Blackhead is an asymptomatic disease in chicken but can cause mortality in turkey flocks. Hence, it is generally recommended that turkey poults and chicks as well as adult turkeys and chicken, should not be

TABLE 21.2 Vaccination regime of poultry.

S. no.	Age	Type of vaccine	Route of administration
1	Day old	Newcastle disease (ND) i/o vaccine	Intra-ocular/eye drop
2	Day old	Mareks disease vaccine	Subcute (SC)
3	10–14 days	Infectious bursal disease (IBD/gumboro)	Drinking water/oral
4	3 weeks	New castle disease Lasota vaccine	Drinking water
5	4–5 weeks	Gumboro (IBD) vaccine	Drinking water
6	6 weeks	Fowl pox vaccine	Intra-dermal (I/d) or wing web
7	7 weeks	New castle disease Komarov vaccine	Intramuscular
8	8 weeks	Fowl cholera vaccine	SC
9	9 weeks	Fowl typhoid vaccine	SC
10	16 weeks	ND Komarov vaccine	Intramuscular (I/M)
		Repeat ND Komarov as booster at every 3 months because it is a mesogenic strain	I/M
		Repeat ND La Sota as booster at every 2 months because of its lentogenic/weaker strain	Drinking water

TABLE 21.3 The common diseases and the treatment accorded to them.

S. no.	Disease	Medications	Breeds commonly affected
1	Coccidiosis	Amprolium, sulfadimidine, sulfadiazine	Turkeys, layers, cockrel
2	Salmonellosis Foul typhoid/pullorum disease	Gentamicin, erythromycin, chloramphenicol	Chicken, turkey, and other poultry
3	Lousiness	Cypermethrin powder, oral ivermectin	Common in depending on management
4	Scaly leg	Benzyl benzoate lotion	Chicken and other birds
5	Viral diseases	(e.g., IBD, ND, IB, etc.) iodine oral solution, virucine solution, vitamin C powder, multivitamin	All birds can be affected
6	Helminthosis	Levamisole solution	All poultry considered

housed together. Protozoa *Histomonas meleagridis*, tiny, single-celled organisms are spread to the bird by the roundworm *Heterakis gallinarum*, which invades the cecal mucosa and spreads, via blood, to the liver. Lesions occur in the cecum and liver. Clinical signs include anorexia, depression, and yellow droppings. Mortality rates may be very high and reach a peak 1 week after the onset of the first clinical signs. Historically dimetridazole was used for prophylaxis or treatment in turkeys and game birds, but it has been withdrawn from use in food-producing poultry [18].

Coccidiosis: is a disease that can cause diarrhea and lack of "thrift" or good growth in poults. Keeping litter dry is also important, as this organism spreads and grows in the wet, dirty litter. Also getting poults out onto pasture by 8 weeks of age, and moving roosts to fresh ground frequently, will help prevent coccidiosis. Some hatcheries will vaccinate against coccidiosis for a small fee per chick.

Fowl pox: Fowl pox is a viral infection caused by large DNA virus (an avipoxvirus in the Poxviridae family) that may survive in the environment for extended periods in dried scabs. Nodular lesions on unfeathered skin are common in the cutaneous form. In the diphtheritic form, which affects the upper gastrointestinal (GI) and respiratory tracts, lesions occur from the mouth to the esophagus and on the trachea. Diagnosis is by observing characteristic gross and microscopic lesions and polymerase chain reaction (PCR). Vaccination can prevent the disease and limit spread in affected flocks.

Airsacculitis: This is an untreatable respiratory disease that affects the turkeys' air sacs. The main prevention is to make sure to apriori test for the disease. The disease is transmitted in the egg, so the poults get it before they are even born.

Salmonellosis: Fowl thyphoid (FT) and pullorum disease (PD) are septicemic diseases, primarily of chickens and turkeys, caused by Gram-negative bacteria, *Salmonella gallinarum* and *Salmonella Pullorum*, respectively. Clinical signs in chicks and poults include anorexia, diarrhea, dehydration, weakness, and high mortality. In mature fowl, FT and PD are manifested by decreased egg production, fertility, hatchability, and anorexia, and increased mortality. Gross and microscopic lesions due to FT and PD in chicks and poults include hepatitis, splenitis, typhlitis, omphalitis, myocarditis, ventriculitis, pneumonia, synovitis, peritonitis, and ophthalmitis. In mature fowl, lesions include oophoritis, salpingitis, orchitis, peritonitis, and perihepatitis. Transovarian infection resulting in infection of the egg and subsequently the chick or poult is one of the most important modes of transmission of these two diseases.

21.4.2 The geese and duck

Common diseases in the duck and geese include: aspergillosis, avian adenovirus, chlamydiosis, coccidiosis, cryptosporidiosis, Derzsy's disease, erysipelas, and flukes.

Using the ECG to monitor any of the disease conditions will depend on the clinician handling and understanding of ECG interpretations. For example, the diphtheritic form of fowl pox affect the respiratory tract leading to respiratory distress, difficult breathing and hypoxia. if presented early, ECG recording will show ventricular hypertrophy, decrease in amplitude of P, R, S and T waves. There is variation in ST segment in different birds. If the increase in coronary blood flow is insufficient for the increase in myocardial oxygen demand, the myocardium becomes hypoxic, particularly in the ventricular subendocardium. The patient may experience symptoms of exertional angina and at the same time display an abnormal ECG. The resulting cardiac shock is the major cause of death in the bird.

Conflict of interest

There is no conflict of interest in relation to the research presented.

References

[1] Smith FWK, Tilley LP, Oyama MA, Sleeper MM. Manual of canine and felie cardiology. 5th edition Elsevier; 2016. p. 49–56.

[2] Smith FM, West NH, Jones DR. The cardiovascular system. In: Whittow GC, editor. Sturkie's avian physiology. 5th edition San Diego, CA: Academic Press; 2000. p. 141–231.

[3] Butler PJ, Bishop CM. Flight. In: Whittow GC, editor. Sturkie's avian physiology. 5th edition San Diego, CA: Academic Press; 2000. p. 391–435.

[4] Strunk A, Wilson GH. Avian cardiology. Vet Clin Exot Anim 2003;6:1–28.

[5] de Wit M, Schoemaker NJ. Clinical approach to avian cardiac disease. Semin Avian Exot Pet Med. 2005;14:6–13.

[6] Hassanpour H, Shamsabadi MG, Dehkordi IN, Dehkordi MM. Normal electrocardiogram of the laughing dove (*Spilopelia senegalensis*). J Zoo Wildl Med 2014;45(1):41–6.

[7] Pees M, Straub J, Krautwald-Junghanns ME. Echocardiographic examinations of 60 African grey parrots and 30 other psittacine birds. Vet Rec. 2004;155:73−6.

[8] Lumeij J, Ritchie B. Cardiology. In: Ritchie B, Harrison G, Harrison L, editors. Avian medicine: principles and applications. Lake Worth, FL: Wingers Publishing; 1994. p. 695−722.

[9] Rosenthal K, Miller M, Orosz S, et al. Cardiovascular system. In: Altman RB, Clubb SL, Dorrestein GM, editors. Avian medicine and surgery. 1st edition Philadelphia, PA: W.B. Saunders; 1997. p. 489−500.

[10] Schmidt RE. Sudden death in pet birds. In: Association of Avian Veterinarians Annual Conference Proceedings, Reno; 1995. p. 473−8.

[11] Krautwald-Junghanns ME, Straub J. Avian cardiology: part I. In: Association of Avian Veterinarians Annual Conference Proceedings, Orlando; 2001. p. 323−330.

[12] Riddell C. Developmental, metabolic, and other non-infectious disorders. In: Calnek B, Barnes H, Beard C, et al., editors. Diseases of poultry. 10th edition Ames, IA: Iowa State University Press; 1997. p. 913−50.

[13] Azeez OM, Adah SA, Olaifa FH, Basiru A, Abdulbaki R. The ameliorative effects of *Moringa oleifera* leaf extract on cardiovascular functions and osmotic fragility of Wistar rats exposed to petrol vapour. Sokoto J Vet Sci 2017;15(2):36−42.

[14] Machida N, Aohagi Y. Electrocardiography, heart rates, and heart weights of free-living birds. J Zoo Wildl Med 2001;32(1):47−54.

[15] Yogeshpriya S, Selvaraj P, Ramkumar PK, Veeraselvam M, Saravanan M, Venkatesan M, et al. Review on avian electrocardiogram. Int J Curr Microbiol App Sci 2018;7(8):1389−95.

[16] Reddy BS, Sivajothi S. Avian electrocardiography: a simple diagnostic tool. Int J Avian Wildl Biol 2017;2(5):166−7.

[17] Perni S, Iyer VR, Franzini-Armstrong C. Ultrastructure of cardiac muscle in reptiles and birds: optimizing and/or reducing the probability of transmission between calcium release units. J Muscle Res Cell Motil 2012;33(2):145−52.

[18] Pattison M, McMullin P, Bradbury J, Alexander D. Poultry diseases. 6th Edition Edinburgh: Butterworth Heinemann; 2008. ISBN: 9780702028625 Pattison M, McMullin P, Bradbury J, Alexander D. Poultry Diseases 6th Edition; Elsevier, ISBN: 9780702037269. 2007.

Therapeutic interventions to limit vascular disease

Chapter 22

Percutaneous coronary interventions: scaffolds versus stents

Syed Raza Shah

University of Louisville Hospital, Louisville, KY, United States

22.1 Introduction to percutaneous coronary intervention

Atherosclerosis causes coronary artery disease (CAD) leading to obstruction of vessels, which affects approximately one in every 13 American adults. It is the leading cause of mortality and morbidity, which imposes a major health and economic burden on both the developed and developing nations. In the United States alone it is estimated to cause about 800,000 heart attacks each year with an estimated cost of $89 billion as of 2016 that is expected to increase to $215 billion by 2035 [1,2]. In the past decade, improved therapies and treatment options have decreased the mortality accompanying CAD while increasing survival rates. Despite this decrease in mortality, the prevalence of CAD is expected to continue to increase due to the increase in the aging population [2].

Percutaneous transluminal coronary arteriography is the most reliable means of investigating and quantifying the severity of coronary artery occlusive disease. It utilizes a contrast dye that travels through the vessels and shows up on fluoroscopy as a depiction of the inner blood vessels. This procedure was first performed in the 1920s, and it ultimately led to the development of percutaneous transluminal coronary angioplasty, also known as balloon angioplasty, which is a transcatheter treatment procedure that restores blood flow in the coronary arteries identified by arteriography. It is also commonly known as percutaneous coronary intervention (PCI). In other words, PCI is a nonsurgical procedure that uses a catheter (a thin flexible tube) to place a small structure called a stent to open up blood vessels in the heart that have been narrowed by plaque buildup. The first milestone in treating CAD was achieved by the introduction of PCI performed by Andreas Grüntzig in 1977. Since then, the treatment of CAD has been transformed by the introduction of various treatment modalities, which remains the focus of intensive research and development [3,4]. This chapter summarized clinical achievements of PCI and outlines the drawbacks faced by this technology.

22.2 Bioresorbable vascular scaffold versus metallic stents

PCI has major drawbacks: Thrombosis and acute occlusion caused by vascular elastic recoil was observed to occur immediately after the procedure, and development of neointimal proliferation with restenosis occurs within the first 6 months. In an effort to combat the shortcomings of elastic recoil, pioneering work performed by Sigwart et al. led to the development and implantation of the first self-expanding bare-metal stent (BMS), and in 1987 the BMS was the first Food and Drug Administration (FDA)-approved stent in the United States [5]. Although this new technology reduced early elastic recoil, it was accompanied by two major problems: stent thrombosis (ST) and in-stent restenosis (ISR) [6]. Despite the potentially serious complications associated with the procedure, BMS implantation became the standard of care following the publication of the results from two landmark trials in 1993, the STRESS and the BENESTENT, which indicated that BMS implantations were superior to balloon angioplasty alone [7].

Current BMS have thin struts and adopt biocompatible or biodegradable polymer coating that improved their safety profile, making them the device of first choice for the treatment of CAD. However, permanent caging of the vessel with a metallic implant runs the risk of impairing endothelial function, decreasing positive lumen remodeling, and the risk of side branches occlusion due to neointimal hyperplasia [7]. Hence, the concept of using bioresorbable vascular scaffolds (BVS) is an attractive strategy and has been credited as the fourth revolution in interventional cardiology. The

Endothelial Signaling in Vascular Dysfunction and Disease. DOI: https://doi.org/10.1016/B978-0-12-816196-8.00001-1

BVS is expected to provide the necessary temporal support to the vessel and disappear in due time, allowing the vessel to return to a more natural and healed state. Theoretically, after the healing period, the BVS will degrade and be resorbed completely, leaving the vessel with a healthy endothelium, normal vasomotion, and free of caging.

22.3 Stent thrombosis and in-stent restenosis

ST is an uncommon but serious complication that carries with it significant mortality and morbidity. ST results from occlusion of the endoprosthetic lumen by thrombus and is an entity with a wide chronological spectrum that can occur anywhere from intraprocedurally to years after implantation. It is one of the most serious complications of PCI, so its incidence and prevalence have been followed very closely. The rates of ST have also paralleled the evolution of improved stents and antiplatelet agents. The majority of ST occurs within the first 30 days after PCI. In general clinical practice, the expected rate of early ST is approximately 1%, and beyond 30 days is 0.2%−0.6% per year [8,9].

Restenosis is defined as a reduction in lumen diameter after PCI. In-stent restenosis (ISR) occurs after the lumen is obstructed post stent introduction. While the exact incidence of ISR depends on a number of factors such as single or multivessel occlusion, the overall expected state is 17−41% [10]. ISR rate appears to be higher when the patient has a multivessel disease rather than a single-vessel disease, as reported by Zhao et al.; this study showed that the occurrence of ISR was significantly higher in patients with two-vessel (OR: 2.922; 95% CI: 1.266−6.745; P = 0.012) or three-vessel disease (OR: 2.574; 95% CI: 1.128−5.872; P = 0.025) when compared with those with one-vessel disease [11].

22.4 Bioresorbable vascular scaffolds stents in percutaneous coronary intervention

As discussed previously, revascularization with PCI is associated with benefits such as angina relief, reduction in the rate of myocardial infarction, improved left ventricular function, less need for subsequent coronary artery bypass grafting, and better patient survival regardless of the presence of collateral circulation [8,9,12]. After successful recanalization of the obstructed blood vessel, stenting is mandatory, preferably with BMS, to ensure long-term vessel patency [13,14]. The data have heavily favored the use of BMS implantation; however, there is always an augmented risk of restenosis and thrombosis, and impairment of vasomotion. Also, frequently multiple sequential long stents are required (termed vessel "caging") to treat a chronically occluded vessel that hinders the placement of these stents in the long run, especially excluding the possibility of future bypass graft anastomosis within these segments [15−17]. To overcome these shortcomings, BVS were developed.

BVS stents were initially made to overcome some of the disadvantages of the metallic stents. Hence, the U.S. FDA Circulatory System Device Panel members reviewed the ABSORB III (A Clinical Evaluation of Absorb BVS, the Everolimus Eluting Bioresorbable Vascular Scaffold in the Treatment of Subjects with de Novo Native Coronary Artery Lesions) trial outcomes and additional data with subsequent analyses presented by the sponsor and the FDA. [18,19] BVS was confirmed by FDA advisory panel of experts who recommended approval of the device based on the analysis of its risks and rewards in July 2016 [20]. Since then, BVS offers an alternative treatment option. Their unique properties of potentially promoting vessel healing and avoiding late lumen enlargement by permitting vascular remodeling have led to widespread success [21,22]. BVS stents also restore the normal vasomotion of the vessels that theoretically avoids the vessel caging issues and risk of late thrombosis [21,22]. This precludes the need for further metal layers and preserves the natural healing process of the coronary artery. In addition, complete resorption by struts within 3 years preserves the possibility of further interventions by percutaneous or surgical means [23]. This is important in terms of future complications in these set of patients: the idea of "leaving nothing behind" after PCI is a very exiting concept in modern interventional cardiology and also gives the chance of the possibility of future graft anastomosis in case of the need for coronary artery bypass graft.

Conversely, there are also many limitations of BVS use in a subset of lesions: severely calcified vessels may be poorly accessible for bulky devices, their low radial strength can bear the risk of vessel recoil, and underestimation of vessel size raise the risk of malapposition. Furthermore, efficacy of BVS use in tortuous vessels is still undefined. Theoretically, due to the thick profile increases coronary tortuosity may limit the use of the BVS due to the difficulty of deliverability and predilatation for lesion preparation. In addition, BVS use in the treatment of chronic total occlusion is currently unknown as patients with these lesions were generally excluded from large BVS randomized trials. Many techniques have been described to overcome these challenges, including the use of stiffer wires, anchor balloon technique, and deep seating of the guide catheter; however, data are still limited.

22.5 Conclusion—past and future prospectives

There have been recent controversies regarding the efficacy of various types of stents to maintain vessel structure and prevent obstruction. Conventional stents, which are metallic, cage-like structures, have some limitations; notably, they leave a permanent implant and come with the risk of restenosis or regrowth such that the lumen diameter is reduced. Recently, PCI with BVS has emerged as an interesting alternative since the presence of the prosthesis in the coronary artery is transient. The preliminary results of the Amsterdam Investigator-initiateD Absorb Strategy All-Comers (AIDA) trial were released. The study included 1845 patients undergoing PCI to receive either a BVS (924 patients) or a metallic stent (921 patients). The primary end point of the study was target-vessel failure that included target-vessel myocardial infarction, target-vessel revascularization, or cardiac death. The preliminary results revealed no significant difference of target-vessel failure when BVS was compared with metallic stenting [24]. However, during the 2 years of follow-up, BVS was associated with a higher rate of device thrombosis. This is seen as an important development in the trial. Although the full results of the study are yet to be published, the safety monitoring board recommended early reporting of the study results because of safety concerns.

The development of coronary-artery stents was a major advance in interventional cardiology. Although the preliminary results for BVS are not very encouraging, it is still too early to make conclusions. The AIDA trial is a landmark in the cardiovascular world. These preliminary results raise some concerns regarding thrombosis and the real vessel functionality restoration at long-term observation. However, the AIDA trial is path-breaking as it introduces newer concepts in interventional vascular surgery by advancing the cause of thinner devices that might minimize coronary blood flow perturbations as well as this can lead to the decreased thrombogenicity of such devices. Devices with technical improvements that ensure high shear stress and decrease blood flow separation with subsequent reduced platelet activation should be used.

Despite an impressive list of promising technological advances accompanied by serious research efforts expended in the field of stent therapy over the past two decades, restenosis remains the principal factor contributing to stent-associated morbidity and mortality rates. In parallel to the technological advances in the stent design, individualized treatment that takes into consideration the individual's genetic make up and lifestyle to tailor healthcare decisions and treatments to the individual patient are required. The Human Genome Project laid the groundwork for the development of precision medicine, also known as personalized medicine. Genome-wide association studies have identified a multitude of single nucleotide polymorphisms that increase the susceptibility to develop CAD and ISR, while others were associated with response to antiplatelet therapy. Therefore genetic testing in the clinical setting will help physicians when making medical decisions to improve the outcomes associated with PCI and lower healthcare cost. Nevertheless, even with these improvements still there are some concerns regarding ST and the real vessel functionality restoration at long-term observation. For the same reason, metallic stents remain the treatment of choice for PCI. However, larger studies with long-term follow-up are needed to adequately address the safety and efficacy of BVS use in such settings.

References

[1] Benjamin E.J., Blaha M.J., Chiuve S.E., Cushman M., Das S.R., Deo R., et al. Heart disease and stroke statistics-2017 update: a report from the American Heart Association. Circulation. 2017;135:e146−e603. Available from: https://doi.org/10.1161/CIR.0000000000000485.

[2] American Heart Association (AHA). Projections of cardiovascular disease prevalence and costs: 2015−2035. Dallas, TX: American Heart Association; 2016. p. 1−54.

[3] Stefanini GG, Holmes Jr. DR. Drug-eluting coronary-artery stents. N Engl J Med 2013;368:254−65. Available from: https://doi.org/10.1056/NEJMra1210816.

[4] Gruntzig A. Transluminal dilatation of coronary-artery stenosis. Lancet. 1978;311:1093. Available from: https://doi.org/10.1016/S0140-6736(78)90500-7.

[5] Sigwart U, Puel J, Mirkovitch V, Joffre F, Kappenberger L. Intravascular stents to prevent occlusion and restenosis after transluminal angioplasty. N Engl J Med 1987;316:701−6. Available from: https://doi.org/10.1056/NEJM198703193161201.

[6] Serruys PW, Strauss BH, Beatt KJ, Bertrand ME, Puel J, Rickards AF, et al. Angiographic follow-up after placement of a self-expanding coronary-artery stent. N Engl J Med 1991;324:13−17. Available from: https://doi.org/10.1056/NEJM199101033240103.

[7] Fischman DL, Leon MB, Baim DS, Schatz RA, Savage MP, Penn I, et al. A randomized comparison of coronary-stent placement and balloon angioplasty in the treatment of coronary artery disease Stent restenosis study investigatorsN Engl J Med 1994;331:496−501. Available from: https://doi.org/10.1056/NEJM199408253310802.

[8] Christakopoulos GE, Christopoulos G, Carlino M, et al. Meta-analysis of clinical outcomes of patients who underwent percutaneous coronary interventions for chronic total occlusions. Am J Cardiol 2015;115:1367−75.

[9] Chung C-M, Nakamura S, Tanaka K, et al. Effect of recanalization of chronic total occlusions on global and regional left ventricular function in patients with or without previous myocardial infarction. Catheter Cardiovasc Interv 2003;60:368−74.

[10] García Del Blanco B, Rumoroso Cuevas JR, Hernández Hernández F, et al. Spanish cardiac catheterization and coronary intervention registry. 22nd official report of the Spanish Society of Cardiology Working Group on Cardiac Catheterization and Interventional Cardiology (1990-2012). Rev Esp Cardiol (Engl Ed) 2013;66:894−904.

[11] Zhao LP, Xu WT, Wang L, et al. Influence of insulin resistance on in-stent restenosis in patients undergoing coronary drug-eluting stent implantation after long-term angiographic follow-up. Coron Artery Dis 2015;26:5−10.

[12] Jang WJ, Yang JH, Choi S-H, et al. Long-term survival benefit of revascularization compared with medical therapy in patients with coronary chronic total occlusion and well-developed collateral circulation. JACC Cardiovasc Interv 2015;8:271−9.

[13] Roffi M, Iglesias JF. CTO PCI in patients with diabetes mellitus: sweet perspectives. JACC Cardiovasc Interv 2017;10:2182−4. Available from: https://doi.org/10.1016/j.jcin.2017.09.008.

[14] Dinesch V, Buruian M. Drug-eluting stent failure: a complex scenario. Int J Cardiol 2017;247:26. Available from: https://doi.org/10.1016/j.ijcard.2017.01.058.

[15] Fam JM, Ojeda S, Garbo R, et al. Everolimus-eluting bioresorbable vascular scaffolds for treatment of complex chronic total occlusions. EuroIntervention 2017;13:355−63.

[16] Suh J, Park D-W, Lee J-Y, et al. The relationship and threshold of stent length with regard to risk of stent thrombosis after drug-eluting stent implantation. JACC Cardiovasc Interv 2010;3:383−9.

[17] Shirai S, Kimura T, Nobuyoshi M, et al. Impact of multiple and long sirolimus-eluting stent implantation on 3-year clinical outcomes in the j-Cypher Registry. JACC Cardiovasc Interv 2010;3:180−8.

[18] Chakraborty R, Patra S, Banerjee S, et al. Outcome of everolimus eluting bioabsorbable vascular scaffold (BVS) compared to non BVS drug eluting stent in the management of ST-segment elevation myocardial infarction (STEMI) − a comparative study. Cardiovasc Revasc Med 2016;17:151−4. Available from: https://doi.org/10.1016/j.carrev.2016.01.004.

[19] Jamshidi P, Nyffenegger T, Sabti Z, et al. A novel approach to treat in-stent restenosis: 6- and 12-month results using the everolimus-eluting bioresorbable vascular scaffold. EuroIntervention 2016;11:1479−86. Available from: https://doi.org/10.4244/EIJV11I13A287.

[20] Steinvil A, Rogers T, Torguson R, Waksman R. Overview of the 2016 U.S. Food and Drug Administration circulatory system devices advisory panel meeting on the absorb bioresorbable vascular scaffold system. JACC Cardiovasc Interv 2016;9(17):1757−64. Available from: https://doi.org/10.1016/j.jcin.2016.06.027.

[21] Serruys PW, Ormiston JA, Onuma Y, et al. A bioabsorbable everolimus-eluting coronary stent system (ABSORB): 2-year outcomes and results from multiple imaging methods. Lancet. 2009;373:897−910.

[22] Serruys PW, Ormiston J, Van Geuns R-J, et al. A polylactide bioresorbable scaffold eluting everolimus for treatment of coronary stenosis: 5-year follow-up. J Am Coll Cardiol 2016;67:766−76.

[23] Indolfi C, De Rosa S, Colombo A. Bioresorbable vascular scaffolds − basic concepts and clinical outcome. Nat Rev Cardiol 2016;13:719−29. Available from: https://doi.org/10.1038/nrcardio.2016.151.

[24] Wykrzykowska JJ, Kraak RP, Hofma SH, van der Schaaf RJ, Arkenbout EK, IJsselmuiden AJ, et al. Bioresorbable scaffolds versus metallic stents in routine PCI AIDA investigatorsN Engl J Med 2017;376(24):2319−28. Available from: https://doi.org/10.1056/NEJMoa1614954 Epub 2017 Mar 29.

Chapter 23

Statins in venous thrombosis: biochemical approaches to limiting vascular disease

Willem M. Lijfering[1], Suzanne C. Cannegieter[1,2] and Frits R. Rosendaal[1,2]

[1]Department of Clinical Epidemiology, Leiden University Medical Center, Leiden, The Netherlands, [2]Department of Thrombosis and Haemostasis, Leiden University Medical Center, Leiden, The Netherlands

23.1 Introduction

Venous thrombosis (deep vein thrombosis or pulmonary embolism) is a common and potentially lethal disease that occurs each year in about 1−2 of 1000 people [1]. The condition can be prevented and treated with anticoagulants, but as a side effect, bleeding often occurs [2]. Currently, the duration of treatment of venous thrombosis with anticoagulants depends on whether the event was provoked or not [3]. Provoking risk factors, such as surgery, immobilization, and use of oral contraceptives, are transient causes. These temporarily increase the "thrombotic potential" of a person which explains, for example, why recurrence risk is low (<1%) in patients who developed their first event after surgery. Patients with provoking risk factors are usually treated for 3−6 months only, while patients with unprovoked thrombosis are prescribed treatment for a longer period. This decision is based on the high incidence of venous recurrence in patients with unprovoked events (30% within 5 years after the 3−6 months of oral anticoagulation) [4,5]. Only 40%−50% of patients can be classified as patients with provoked events, which leads to a dilemma in the other 50%−60% of patients: discontinuing treatment may lead to a new venous thrombotic event, while continuing oral anticoagulant treatment overtreats the majority of patients and is accompanied with a yearly 1%−3% risk of major bleeding [2,6]. Therefore, novel therapeutic strategies that are not associated with bleeding complications need to be sought.

23.2 The classic triad of Virchow

Although medical textbooks consider venous thrombosis and arterial cardiovascular disease (CVD) as different disease entities [7], Virchow's triad (1865) postulates that the pathophysiology of thrombosis, either venous or arterial, is an interplay between (1) stasis of the blood, (2) hypercoagulability, and (3) vessel wall injury [8]. Provoking factors for venous thrombosis that are currently used to classify a patient at low recurrence risk are those that transiently lead to a hypercoagulable state (oral contraceptives), stasis (immobilization) or both (pregnancy/puerperium/surgery). The vascular component of Virchow's triad has largely been neglected for venous thrombosis. Vessel wall injury, due to atherosclerosis or kidney disease, is chronically progressing but can be slowed with drugs such as statins [9]. Recent studies have suggested substantial risk reductions (20%−50%) for venous thrombosis with statin use [10,11], of which the pathophysiology till date is not fully understood. The basis of this book chapter is the hypothesis that arterial vessel wall injury leads to a systemic prothrombotic state and that the risk of venous thrombosis can be reduced by pharmacologic agents that target vessel wall injury (i.e., statins).

23.3 Atherosclerosis as a procoagulant disease

In 1973, Russell Ross proposed that atherosclerosis begins with an injury to the lining of the artery [12], and he demonstrated the role of inflammation in the formation of atherosclerotic plaques, thereby showing that atherosclerosis does

Endothelial Signaling in Vascular Dysfunction and Disease. DOI: https://doi.org/10.1016/B978-0-12-816196-8.00009-6

TABLE 23.1 Risk of venous thrombosis according to levels of hsCRP adjusted for factor VIII.

hsCRP percentile (mg/L)	FVIII (U/dL)[a]	Case	Control	Overall OR[b] (CI95)	Adjusted for FVIII OR[b] (CI95)
<25th (<0.68)	103 (38)	371	684	1.0 (reference)	1.00 (reference)
25−50th (0.68−1.43)	112 (39)	496	701	1.32 (1.11−1.57)	1.16 (0.96−1.39)
50−75th (1.43−3.15)	121 (43)	563	697	1.51 (1.28−1.79)	1.17 (0.97−1.40)
75−90th (3.15−6.22)	127 (46)	395	418	1.77 (1.46−2.13)	1.19 (0.97−1.46)
90−95th (6.22−9.26)	126 (41)	153	139	2.05 (1.58−2.67)	1.30 (0.97−1.73)
95−97.5th (9.26−13.48)	125 (45)	99	69	2.68 (1.92−3.74)	1.49 (1.04−2.16)
>97.5th (13.48)	147 (67)	128	69	3.47 (2.52−4.78)	1.55 (1.10−2.21)

CI 95, 95% confidence interval; *FVIII*, factor VIII; *OR*, odds ratio.
[a]*Mean level and SD derived from MEGA control population.*
[b]*Odds ratio adjusted for age and sex.*

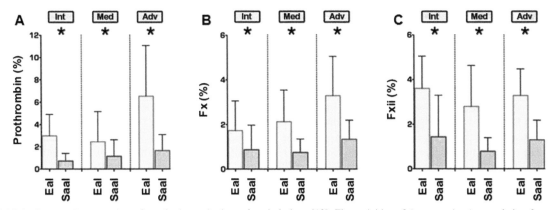

FIGURE 23.1 Procoagulant state in early and advanced atherosclerosis lesions [18]. The activities of (among others) coagulation factors II (pro-thrombin), X (FX), and XII (FXII) were tested in tunica intima, media, and adventitia. *Statistical significance ($P < .05$). *INT*, tunica intima; *MED*, tunica media; *ADV*, tunica adventitia; *EAL*, early atherosclerosis; *SAAL*, stable advanced atherosclerosis.

not result simply from the accumulation of lipids [13]. In 1980, DeWood showed that thrombus formation on erupted plaques is the final step toward myocardial infarction, closely linking atherosclerosis and thrombosis in arterial disease [14]. Although atherosclerosis does not affect the venous vessel walls, the same mechanistic links also exist in venous disease: several risk factors are common to both atherosclerosis and venous thrombosis: obesity, chronic kidney disease, aging, and male sex [1,15−17]. In the MEGA study, we observed that chronic inflammation (measured by high sensi-tive C-reactive protein [hsCRP]) is tightly linked with procoagulant factor VIII, and that the relation between increasing levels of hsCRP and venous thrombosis risk is at least partly mediated by concurrently raised factor VIII levels [15−17] (Table 23.1).

We also showed that the risk of venous thrombosis in obese individuals (in whom an heightened inflammatory state is present) is related to concurrently raised factor VIII levels [15]. Furthermore, in another study, we showed that chronic kidney disease leads to a dose dependent (up to >3-fold) increased risk of venous thrombosis, which was explained by raised levels of factor VIII and von Willebrand factor [16].

These findings illustrate that conditions as chronic kidney disease and obesity may lead to venous thrombosis through vessel wall injury and atherosclerosis primarily via inflammation and increased factor VIII and von Willebrand factor levels. Interestingly, a histopathological study in human patients showed that early atherosclerotic specimens were associated with a prothrombotic state, whereas advanced atherosclerosis was not (Fig. 23.1) [18]. This is important because the hemostatic system not only regulates thrombosis but also has the potential to acceler-ate atherosclerosis [19].

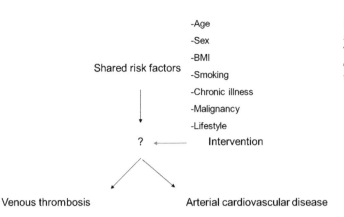

FIGURE 23.2 The relationship between venous thrombosis and arterial cardiovascular disease is explained by shared risk factors [22]. The underlying mechanism, yet unknown and identified with the question mark, might reveal interventions that could both decrease the risk of venous thrombosis and arterial cardiovascular disease.

23.3.1   The relationship between venous thrombosis and arterial cardiovascular disease

Several studies have shown that patients with venous thrombosis are at a long-term increased risk of subsequent arterial thrombosis [20−22]. This result can be explained by underlying shared risk factors, like age, sex, smoking, unhealthy lifestyle that are related to both venous thrombosis and arterial cardiovascular disease (Fig. 23.2) [22]. Since these shared risk factors cannot in itself explain the relationship, it is reasonable to conclude that there must be an underlying biologic mechanism. Understanding the mechanism that links venous thrombosis to arterial CVD could lead to interventions that could both decrease the risk of both.

23.3.2   The link between arterial cardiovascular disease and subsequent venous thrombosis

In contrast to the link between venous thrombosis and subsequent arterial CVD, the increase in incidence of venous thrombosis after arterial CVD is confined to the first three months only [23,24]. One direct causal explanation for this is that patients with arterial CVD are, obviously, subjected to hospitalization which is generally accompanied by periods of immobilization that offers an alternative (indirect causal) explanation of venous thrombosis due to stasis.However, other causal factors such as a temporary procoagulant or inflammatory response due to an acute cardiovascular event etc. cannot be ruled out [24]. Irrespective of the cause, what is clear is that the increase in risk is only temporary.

23.3.3   Why is the relationship from arterial cardiovascular disease to subsequent venous thrombosis different as compared to venous thrombosis to subsequent arterial cardiovascular disease?

One answer could be that early atherosclerosis, identified as question mark in Fig. 23.2, acts as the underlying culprit in the relationship. According to Borissoff et al., only the early stages of atherosclerosis are prothrombotic, while the later stages are not [18,19]. Why coagulation factors are more abundantly present within early atherosclerotic vessels than in advanced atherosclerosis is as yet unknown, but it may be attributable to primary protective mechanisms against vascular injury [25]. Atherosclerosis is usually measured by ultrasound measures, but ultrasound can only distinguish atherosclerosis in its more advanced stages [26], which are not prothrombotic [18]. This could explain why atherosclerosis as measured with ultrasound is not associated with venous thrombosis in most studies [27−29], while it clearly is a predictor for arterial CVD [29]. In contrast, early atherosclerosis can be determined in vivo by MRI [30], but whether early atherosclerosis increases the risk of venous thrombosis is currently unknown. As shown in Fig. 23.3, early atherosclerosis could explain that venous thrombosis does increase the risk of arterial CVD, while the vice versa case does not hold (i.e. arterial CVD does not increase the risk of venous thrombosis except for the time that patients are immobilized). According to Fig. 23.3, the underlying early atherosclerosis is procoagulant, which leads to an increased risk of onset of venous thrombosis. In contrast, patients with arterial CVD (who have advanced atherosclerosis) have no longer procoagulant early atherosclerosis and are therefore not at increased venous thrombosis risk.

23.4   The role of statins in venous thrombosis

The underlying assumption beneath the use of a therapy for preventing both venous thrombosis and arterial CVD is that the biology of blood clot formation expresses some overlap. Early atherosclerosis, being procoagulant and being related

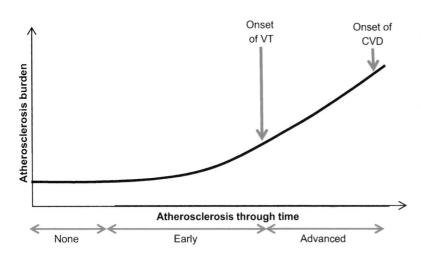

FIGURE 23.3 Hypothesis: Why venous thrombosis leads to arterial cardiovascular disease and not vice versa. The underlying atherosclerosis is procoagulant at time of venous thrombosis onset, while patients with arterial cardiovascular disease have advanced atherosclerosis, which is no longer procoagulant and therefore not able to lead to venous thrombosis.

to many conditions that also increase the risk of venous thrombosis (like obesity, inflammation, chronic kidney disease), could be the common denominator as statins can halt or even reverse the progression of atherosclerosis [31]. Recent advances in the Statins Reduce Thrombophilia (START) randomized trial have shown an effect of rosuvastatin decreasing the levels of coagulation factors, most notably factor VIII [32,33]. Interestingly, in START the decrease in coagulation factors appeared strongest in participants with unprovoked venous thrombosis [32]. This could make sense from a pathophysiological perspective as previous studies showed that patients with unprovoked venous thrombosis have endothelial dysfunction more often than patients with either provoked venous thrombosis or control individuals [34,35]. Statins can modify endothelial function as early as after 1 month of treatment, while endothelial dysfunction/early atherosclerosis is associated with a procoagulant state [18]. This result from START therefore suggests that statins may have the strongest potential to decrease venous thrombosis risk by halting/decreasing atherosclerosis, leading to a less procoagulant state in individuals with atherosclerosis. This also follows results from the JUPITER trial where a notable benefit was observed in the high-risk subgroups of participants with elevated waist circumference and metabolic syndrome [10]. Although START has provided stronger evidence that statins may affect coagulation in humans, the exact mechanism by which this happens has not been elucidated yet.

23.5 Potential

Venous thrombosis is a serious health problem with an immediate death rate of 4% and recurrence rates of 25% within 5 years [1,4]. In addition, other major illnesses such as liver disease, kidney disease, rheumatoid arthritis, multiple sclerosis, heart failure and hemorrhagic stroke have been reportedly associated with an increased risk of venous thrombosis. These risks were most pronounced at the time of immobilization or in the presence of thrombophilia [36]. Furthermore, more than 30% of patients with venous thrombosis are invalidated with postthrombotic syndrome [37]. Although anticoagulant treatment is effective to prevent venous thrombosis, it significantly increases long-term healthcare costs and has serious side effects such as major (organ) bleeding [2]. One estimate for the European Union arrived at a death toll of 500,000 venous thrombosis-related deaths per year. Notably, venous thrombosis related to hospital admissions is the leading cause of loss of disability-adjusted life-years [38].

Statins may form a suitable and safe alternative for long-term treatment in the prevention of (recurrent) venous thrombosis [11,39]. Although serious adverse drug reactions with statins are uncommon as they do not have the side effect of major (organ)bleeding that oral anticoagulants have, statins too carry significant potential side effects including muscle injury, pancreatic and liver impairment, neuropathy, cognitive loss, and sexual dysfunction [40]. Many inexpensive generic statins are now widely available and observational studies showed that currently statin usage is associated with a 20%−30% lower risk of recurrent venous thrombosis as compared to nonstatin users [41]. So when cheap and available drugs, like statins, might prevent venous thrombosis why not try it [42]? The answer is that we must have independent confirmation from other large trials that statins, like in START [32], can reduce coagulation. When confirmed, we also need to know *how* statins can decrease coagulation. For this we should have to set up a study that investigates whether early atherosclerosis as measured in vivo (e.g., with MRI) is indeed procoagulant. If so, participants with early atherosclerosis should be randomized to statin versus no statin treatment with follow-up on their coagulation and atherosclerosis status to confirm that treatment of atherosclerosis leads to lower levels of coagulation. Finally, a randomized trial with clinical endpoints is needed and where the potential benefits of statin treatmentcan be directly compared with the side effects.

23.6   Future perspectives

Since the question as to whether a therapy is effective is answered by the number-needed-to-treat per year , evaluation of the effectivity of statins in prevention of venous thrombosis after an unprovoked event is easily feasible given the high recurrent venous thrombosis event rates. It is in this aspect interesting to note that the current international guidelines suggest to discontinue anticoagulant treatment in patients with venous thrombosis who are considered to be at high risk of anticoagulation related bleeding [3]. Although statins are unlikely to be as effective as anticoagulant drugs, they do have the major advantage over anticoagulants that they do not induce bleeding. Therefore, a drug to prevent recurrent venous thrombosis in patients at risk of bleeding that does not induce bleeding and in which the number-needed-to-treat for the prevention of recurrence is sufficiently high, is a remedy that we should continue to look for, and for which statin therapy might be a suitable candidate.

Conflict of interest

None

References

[1]   Naess IA, Christiansen SC, Romundstad P, et al. Incidence and mortality of venous thrombosis: a population-based study. J Thromb Haemost 2007;5:692—9.

[2]   van Rein N, Lijfering WM, Bos MH, et al. Objectives and design of BLEEDS: a cohort study to identify new risk factors and predictors for major bleeding during treatment with vitamin K antagonists. PLoS One 2016;11:e0164485.

[3]   Kearon C, Akl EA, Ornelas J, Blaivas A, et al. Antithrombotic therapy for VTE disease: CHEST guideline and expert panel report. Chest. 2016;149:315—52.

[4]   Prandoni P, Noventa F, Ghirarduzzi A, et al. The risk of recurrent venous thromboembolism after discontinuing anticoagulation in patients with acute proximal deep vein thrombosis or pulmonary embolism. A prospective cohort study in 1,626 patients. Haematologica. 2007;92:199—205.

[5]   Heit JA, Silverstein MD, Mohr DN, et al. Risk factors for deep vein thrombosis and pulmonary embolism: a population-based case-control study. Arch Intern Med 2000;160:809—15.

[6]   Kearon C, Gent M, Hirsh J, et al. A comparison of three months of anticoagulation with extended anticoagulation for a first episode of idiopathic venous thromboembolism. N Engl J Med 1999;340:901—7.

[7]   Fauci AS, Braunwald E, Kasper DL, et al. Harrison's principles of internal medicine. 17th ed. New York, NY: McGraw-Hill; 2008.

[8]   Virchow R. Phlogose und Thrombose im Gefässystem. In: Gesammelte Abhandlungen zur zur Wissenschaftlichen Medicin. 1865.

[9]   Goff Jr DC, Lloyd-Jones DM, Bennett G, et al. 2013 ACC/AHA guideline on the assessment of cardiovascular risk: a report of the American College of Cardiology/American Heart Association Task Force on Practice Guidelines. Circulation 2014;129(25 Suppl 2):S49—73.

[10]   Glynn RJ, Danielson E, Fonseca FA, et al. A randomized trial of rosuvastatin in the prevention of venous thromboembolism. N Engl J Med 2009;360:1851—61.

[11]   Chaffey P, Thompson M, Pai AD, et al. Usefulness of statins for prevention of venous thromboembolism. Am J Cardiol 2018;121:1436—40.

[12]   Ross R, Glomset JA. Atherosclerosis and the arterial smooth muscle cell: proliferation of smooth muscle is a key event in the genesis of the lesions of atherosclerosis. Science. 1973;180:1332—9.

[13]   Ross R. Atherosclerosis — an inflammatory disease. N Engl J Med 1999;340:115—26.

[14]   DeWood MA, Spores J, Notske R, et al. Prevalence of total coronary occlusion during the early hours of transmural myocardial infarction. N Engl J Med 1980;303:897—902.

[15]   Christiansen SC, Lijfering WM, Naess IA, et al. The relationship between body mass index, activated protein C resistance and risk of venous thrombosis. J Thromb Haemost 2012;10:1761—7.

[16]   Ocak G, Vossen CY, Lijfering WM, et al. Role of hemostatic factors on the risk of venous thrombosis in people with impaired kidney function. Circulation. 2014;129:683—91.

[17]   Roach RE, Lijfering WM, Rosendaal FR, et al. Sex difference in risk of second but not of first venous thrombosis: paradox explained. Circulation. 2014;129:51—6.

[18]   Borissoff JI, Heeneman S, Kilinç E, et al. Early atherosclerosis exhibits an enhanced procoagulant state. Circulation. 2010;122:821—30.

[19]   Borissoff JI, Spronk HM, ten Cate H. The hemostatic system as a modulator of atherosclerosis. N Engl J Med 2011;364:1746—60.

[20]   Sørensen HT, Horvath-Puho E, Pedersen L, et al. Venous thromboembolism and subsequent hospitalisation due to acute arterial cardiovascular events: a 20-year cohort study. Lancet. 2007;370:1773—9.

[21]   Klok FA, Mos IC, Broek L, et al. Risk of arterial cardiovascular events in patients after pulmonary embolism. Blood. 2009;114:1484—8.

[22] Roach RE, Lijfering WM, Flinterman LE, et al. Increased risk of CVD after VT is determined by common etiologic factors. Blood. 2013;121:4948−54.

[23] Sørensen HT, Horvath-Puho E, Søgaard KK, Christensen S, Johnsen SP, Thomsen RW, et al. Arterial cardiovascular events, statins, low-dose aspirin and subsequent risk of venous thromboembolism: a population-based case-control study. J Thromb Haemost 2009;7:521−8.

[24] Rinde LB, Lind C, Småbrekke B, Njølstad I, Mathiesen EB, Wilsgaard T, et al. Impact of incident myocardial infarction on the risk of venous thromboembolism: the Tromsø Study. J Thromb Haemost 2016;14:1183−91.

[25] Kalz J, Ten Cate H, Spronk HM. Thrombin generation and atherosclerosis. J Thromb Thrombolysis 2014;37:45−55.

[26] Zhang Y, Guallar E, Qiao Y, Wasserman BA. Is carotid intima-media thickness as predictive as other noninvasive techniques for the detection of coronary artery disease? Arterioscler Thromb Vasc Biol 2014;34:1341−5.

[27] Småbrekke B, Rinde LB, Hald EM, et al. Repeated measurements of carotid atherosclerosis and future risk of venous thromboembolism: the Tromsø Study. J Thromb Haemost 2017;15:2344−51.

[28] Reich LM, Folsom AR, Key NS, et al. Prospective study of subclinical atherosclerosis as a risk factor for venous thromboembolism. J Thromb Haemost 2006;4:1909−13.

[29] Hald EM, Lijfering WM, Mathiesen EB, et al. Carotid atherosclerosis predicts future myocardial infarction but not venous thromboembolism: the Tromso study. Arterioscler Thromb Vasc Biol 2014;34:226−30.

[30] Cai JM, Hatsukami TS, Ferguson MS, et al. Classification of human carotid atherosclerotic lesions with in vivo multicontrast magnetic resonance imaging. Circulation. 2002;106:1368−73.

[31] Nissen SE, Nicholls SJ, Sipahi I, et al. Effect of very high-intensity statin therapy on regression of coronary atherosclerosis: the ASTEROID trial. JAMA. 2006;295:1556−65.

[32] Biedermann JS, Kruip MJHA, van der Meer FJ, et al. Rosuvastatin use improves measures of coagulation in patients with venous thrombosis. Eur Heart J 2018;39:1740−7.

[33] Orsi FA, Biedermann JS, Kruip MJHA, et al. Rosuvastatin use reduces thrombin generation potential in patients with venous thromboembolism: a randomized controlled trial. J Thromb Haemost 2019;17:319−28.

[34] Migliacci R, Becattini C, Pesavento R, et al. Endothelial dysfunction in patients with spontaneous venous thromboembolism. Haematologica. 2007;92:812−18.

[35] Prandoni P, Bilora F, Marchiori A, et al. An association between atherosclerosis and venous thrombosis. N Engl J Med 2003;348:1435−41.

[36] Ocak G, Vossen CY, Verduijn M, et al. Risk of venous thrombosis in patients with major illnesses: results from the MEGA study. J Thromb Haemost 2013;11(1):116−23.

[37] Brandjes DP, Buller HR, Heijboer H, et al. Randomised trial of effect of compression stockings in patients with symptomatic proximal-vein thrombosis. Lancet 1997;349:759−62.

[38] Cohen AT, Agnelli G, Anderson FA, et al. Venous thromboembolism (VTE) in Europe. The number of VTE events and associated morbidity and mortality. Thromb Haemost 2007;98(4):756−64.

[39] Lijfering WM, Biedermann JS, Kruip MJ, et al. Can we prevent venous thrombosis with statins: an epidemiologic review into mechanism and clinical utility. Expert Rev Hematol 2016;9:1023−30.

[40] Lippi G, Favaloro EJ. Statins for preventing venous thrombosis: for or against? Semin Thromb Hemost 2019;45:834−6. Available from: http://doi.org/10.1055/s-0039-1687912.

[41] Kunutsor SK, Seidu S, Khunti K. Statins and secondary prevention of venous thromboembolism: pooled analysis of published observational cohort studies. Eur Heart J 2017;38:1608−12.

[42] Rodger M. Is it time to try or to trial statins to prevent recurrent venous thromboembolism? J Thromb Haemost 2014;12:1204−6.

Index